GENERALIZATIONS OF THOMAE'S FORMULA FOR Z_n CURVES

Developments in Mathematics

VOLUME 21

Series Editors:

Krishnaswami Alladi, *University of Florida*
Hershel M. Farkas, *Hebrew University of Jerusalem*
Robert Guralnick, *University of Southern California*

For other titles published in this series, go to
www.springer.com/series/5834

GENERALIZATIONS OF THOMAE'S FORMULA FOR Z_n CURVES

HERSHEL M. FARKAS
The Hebrew University, Institute of Mathematics

SHAUL ZEMEL
The Hebrew University, Institute of Mathematics

 Springer

Hershel M. Farkas
Einstein Institute of Mathematics
The Hebrew University of Jerusalem
Edmond J. Safra Campus, Givat Ram
Jerusalem, 91904, Israel
farkas@math.huji.ac.il

Shaul Zemel
Einstein Institute of Mathematics
The Hebrew University of Jerusalem
Edmond J. Safra Campus, Givat Ram
Jerusalem, 91904, Israel
zemels@math.huji.ac.il

ISSN 1389-2177
ISBN 978-1-4614-2758-2   ISBN 978-1-4419-7847-9 (eBook)
DOI 10.1007/978-1-4419-7847-9
Springer New York Dordrecht Heidelberg London

Mathematics Subject Classification (2010): 20F10, 14H42

© Springer Science+Business Media, LLC 2010
Softcover reprint of the hardcover 1st edition 2010
All rights reserved. This work may not be translated or copied in whole or in part without the
written permission of the publisher (Springer Science+Business Media, LLC, 233 Spring Street,
New York, NY 10013, USA), except for brief excerpts in connection with reviews or scholarly analysis.
Use in connection with any form of information storage and retrieval, electronic adaptation, computer
software, or by similar or dissimilar methodology now known or hereafter developed is forbidden.
The use in this publication of trade names, trademarks, service marks, and similar terms, even if they
are not identified as such, is not to be taken as an expression of opinion as to whether or not they are
subject to proprietary rights.

Printed on acid-free paper

Springer is part of Springer Science+Business Media (www.springer.com)

To our wives, Sara and Limor

Introduction

In this book we shall present the background necessary to understand and then prove Thomae's formulae for Z_n curves. It is not our intention to actually give the proofs of all the necessary results concerning compact Riemann surfaces and their associated theta functions, but rather to give the statements which are necessary and refer the reader to the book [FK] for many of the proofs. Sometimes we shall sketch some proofs for the benefit of the reader.

We begin with the objects that we shall be studying. A *(fully ramified) Z_n curve* is the compact Riemann surface associated to an equation

$$w^n = \prod_{i=1}^{t} (z - \lambda_i)^{\alpha_i},$$

where $\lambda_i \neq \lambda_j$ if $i \neq j$ and the powers α_i, $1 \leq i \leq t$ satisfy $1 \leq \alpha_i \leq n - 1$, are relatively prime to n, and their sum is either divisible by n or relatively prime to n. Thus a concrete representation of the curve as a compact Riemann surface is that of an n-sheeted branched cover of the sphere with full branching over each point λ_i. We shall generally assume that there is no branching over the point at ∞, which is equivalent to the statement that $\sum_{i=1}^{t} \alpha_i$ is divisible by n. The Riemann–Hurwitz formula will now give that $2(g + n - 1) = (n - 1)t$, so that the genus of the Z_n curve is $g = (n - 1)(t - 2)/2$.

Viewed as a compact Riemann surface of genus g, when $g \geq 1$ we then have the possibility of constructing a *canonical homology basis* for the surface, i.e., a basis of $2g$ cycles, a_i, $1 \leq i \leq g$, and b_i, $1 \leq i \leq g$, on the Riemann surface which satisfy the conditions that the intersection number of a_i with a_j and of b_i with b_j is 0 and the intersection number of a_i with b_j is the Kronecker delta symbol δ_{ij}. Integration of a basis of the g-dimensional vector space of holomorphic differentials ω_i, $1 \leq i \leq g$, over the cycles of the canonical homology basis gives rise to a g by $2g$ matrix (A, B) where the ijth entry of A is $\int_{a_j} \omega_i$ and the ijth entry of B is $\int_{b_j} \omega_i$. One can then find a dual basis θ_i, $1 \leq i \leq g$, for the holomorphic differentials which satisfies $\int_{a_j} \theta_i = \delta_{ij}$ (i.e., with $A = I$), and the corresponding matrix B is traditionally denoted by Π and is called the *period matrix* corresponding to the chosen homology basis. The matrix Π is known to be symmetric, and its imaginary part is positive definite.

With the matrix Π in hand, we can define a holomorphic function of g complex variables, $\theta(\zeta, \Pi) \in \mathbb{C}$ for $\zeta \in \mathbb{C}^g$, and, more generally, for two vectors ε and ε' in $\mathbb{R}^g$, we define the corresponding function $\theta \begin{bmatrix} \varepsilon \\ \varepsilon' \end{bmatrix}$ again with arguments (ζ, Π) as above. These parameters $\begin{bmatrix} \varepsilon \\ \varepsilon' \end{bmatrix}$ are called *characteristics*.

We then look at two objects: one is the complex numbers λ_i from the algebraic equation of the Z_n curve, and the other is the values at $\zeta = 0$ of the holomorphic theta functions with characteristics $\theta \begin{bmatrix} \varepsilon \\ \varepsilon' \end{bmatrix} (\zeta, \Pi)$, called *theta constants*, which depend on the matrix Π. The Thomae formulae will be a bridge connecting these two different sets of parameters. We will indicate how to construct a finite set of characteristics $\begin{bmatrix} \varepsilon \\ \varepsilon' \end{bmatrix}$ with ε and ε' taken from $\mathbb{Q}^g$ and a polynomial $P_{\begin{bmatrix} \varepsilon \\ \varepsilon' \end{bmatrix}}$ for each such characteristic such that the value of the quotient

$$\frac{\theta^N \begin{bmatrix} \varepsilon \\ \varepsilon' \end{bmatrix} (0, \Pi)}{P_{\begin{bmatrix} \varepsilon \\ \varepsilon' \end{bmatrix}} (\lambda_1, \dots, \lambda_t)}$$

is independent of the characteristic $\begin{bmatrix} \varepsilon \\ \varepsilon' \end{bmatrix}$ for some power N.

Thomae did this in [T1] and [T2] for the Z_2 curves associated with the equation

$$w^2 = \prod_{i=0}^{2g+1} (z - \lambda_i),$$

where $\lambda_i \neq \lambda_j$ for $i \neq j$, i.e., gave the polynomials $P_{\begin{bmatrix} \varepsilon \\ \varepsilon' \end{bmatrix}}$ and proved that the value of

$$\frac{\theta^8 \begin{bmatrix} \varepsilon \\ \varepsilon' \end{bmatrix} (0, \Pi)}{P_{\begin{bmatrix} \varepsilon \\ \varepsilon' \end{bmatrix}} (\lambda_1, \dots, \lambda_{2g+2})}$$

is independent of the characteristic $\begin{bmatrix} \varepsilon \\ \varepsilon' \end{bmatrix}$ for all the characteristics $\begin{bmatrix} \varepsilon \\ \varepsilon' \end{bmatrix}$ with ε and ε' taken from $\mathbb{Z}^g$ such that $\theta \begin{bmatrix} \varepsilon \\ \varepsilon' \end{bmatrix} (0, \Pi) \neq 0$.

The most straightforward generalization of this case is to the Z_n curves associated to the equations

$$w^n = \prod_{i=0}^{nr-1} (z - \lambda_i),$$

where $\lambda_i \neq \lambda_j$ for $i \neq j$. Bershadsky and Radul treated this case (which has come to be called the *nonsingular* case) in [BR], and indeed generalized Thomae's work to this case. Nakayashiki reworked in [N] some of the ideas in [BR], replacing some of their ideas with the work of Fay [F] on the Szegő Kernel function.

It is at this point that the senior author of this book entered the picture. He was approached by his former student Y. Kopeliovich who asked whether one could derive the Thomae formulae from "first principles". This means just using Riemann's theory of theta functions on Riemann surfaces.

In a seminar held at the Hebrew University the senior author reworked Thomae's formulae for the case $n = 2$ and $g = 2$. One of the students succeeded in generalizing to $n = 2$ and $g = 3$, and then in [EiF] Eisenmann and Farkas re-derived the Thomae formulae for all Z_2 curves and obtained a generalization for nonsingular Z_3 curves as well. The case $n \geq 4$ required some new ideas.

In joint work with David Ebin [EbF] the senior author was able to complete the case of nonsingular Z_n curves for all n. The main tool used was the construction of some operators defined on the set of non-special integral divisors of degree g supported on the set of branch points. These operators gave rise to what we call in this book the *Poor Man's Thomae (PMT)* formulae and what was called *Limited Thomae Formulae (LTF)* in [EbF]. This work was then exposed in another seminar held at the Hebrew University and in addition some examples of singular Z_n curves were also treated. The examples of singular curves included those associated to the equations

$$w^3 = (z - \lambda_1)(z - \lambda_2)(z - \lambda_3)(z - \mu_1)^2 (z - \mu_2)^2 (z - \mu_3)^2$$

(of genus $g = 4$), and the family of Z_n curves associated to the equations

$$w^n = (z - \lambda_1)(z - \lambda_2)(z - \mu_1)^{n-1}(z - \mu_2)^{n-1}$$

(of genus $g = n - 1$).

It was at this point that the junior author entered the picture. He worked out different proofs for some of the statements along the lines of the proof of the nonsingular case, and showed that these proofs could be generalized to hold for the singular case

$$w^n = \prod_{i=0}^{m-1} (z - \lambda_i) \prod_{i=1}^{m} (z - \mu_i)^{n-1}.$$

These singular curves were introduced by Enolski and Grava in [EG], who also worked out Thomae type formulae for them. In addition, we note the work of Matsumoto [M] and the work of Matsumoto and Terasoma [MT] on singular curves with $n = 3$. The ideas presented here for this family of Z_n curves generalize with minor cosmetic changes to the more general family of the Z_n curves associated to the equations

$$w^n = \prod_{i=0}^{m+nr-1} (z - \lambda_i) \prod_{i=1}^{m} (z - \mu_i)^{n-1}$$

for m and r satisfying $m \geq 0$ and $r > -m/n$.

In all of the above cases it turns out, at least according to our current point of view, that a method of constructing non-special integral divisors supported on the set of branch points is crucial, and then one has to define an appropriate set of operators on this set. A particularly pleasant aspect of this approach is that it leads to a graphical representation and construction of the requisite polynomials in terms of a matrix whose rows and columns depend on the integral divisor. We hasten to admit that it is not quite clear how to do this for an arbitrary Z_n curve. This is especially true since we have examples of singular Z_n curves where we have *no* such divisors. In fact Gonzalez-Diez and Torres have exhibited explicitly in [GDT] an infinite family of Z_n curves with any number ≥ 3 of branch points such that no set of divisors of this type exists. Another point that needs mentioning is the difference of our theory from the original theorem of Thomae coming from the fact that for the Z_2 curves treated by Thomae this method captures all the characteristics taken from a certain subset of $\mathbb{Q}^g$ such that the corresponding theta constants do not vanish. The validity of this statement does not extend to any of the more general cases.

Drafts of Chapters 1, 2, 3 and 4 were already prepared by the senior author before the junior author entered the picture. Chapter 5 and also those parts of Chapter 2 which deal with the singular case are due to the junior author. The examples we present in Chapter 6 were done jointly. Of course the entire book, including the first four chapters, was rewritten by both authors.

The book has been written with at least two different kinds of readers in mind. The classical material concerning compact Riemann surfaces, algebraic curves and theta functions were written for graduate students of mathematics and young researchers whose interests are related to complex analysis or algebraic geometry. In addition we had in mind also the researchers in mathematical physics and those physicists whose interests include conformal field theory. In fact the renaissance of the subject is largely due to the interest of the mathematical physics community, who went back to the Thomae formulae which lay dormant for over a hundred years. We have attempted to carry out each calculation to the bitter end. No doubt some readers will be pleased but some may feel that we overdid it. We apologize to those and wish to repeat the words of Lars Ahlfors to the effect that "theta functions are not a spectator sport". We heartily recommend that every reader go through all the calculations on his own, for this is the way to appreciate the beauty of the subject.

The theory of theta functions themselves lay dormant for many years. In the 1960s Lewittes in his paper [L] and then Mumford, Rauch and several others including the current senior author started writing papers centered around theta functions. We hope that perhaps this book will be a catalyst once again for the study of the fascinating subject of theta functions.

Chapter 1 is a review of the theory of compact Riemann surfaces, including the theory of theta functions on the surface. Chapter 2 summarizes the material one needs to know about Z_n curves. Chapter 3 gives the examples that are meant to motivate and help in the understanding of the general case of nonsingular Z_n curves and the singular ones we deal with. Chapter 4 is the general case of the nonsingular curves, while Chapter 5 is the general case of the singular curves for which we have a theory. Chapter 6 provides some examples for which we do not yet have a theory but yet we do work out the relevant formulae. Appendix A contains the ideas behind the construction of some expressions in Chapters 4 and 5 and presents the process for the more general family presented above. Appendix B presents the corresponding construction for Chapter 6, and contains information necessary to understand the process of basepoint change there. These appendices will give the curious reader some more insight into how some of the technical theorems arose but are strictly speaking not necessary for the presentation. The two appendices are essentially due to the junior author, and are in fact part of his Ph.D. work. We believe that these methods were not yet used to their full extent, and that one can modify them to prove Thomae formulae for more general families.

Jerusalem, Israel   *Hershel M. Farkas*
August 2010   *Shaul Zemel*

Contents

Chapter 1

Riemann Surfaces

In this chapter we gather the information one needs to know about the general theory of Riemann surfaces and theta functions. A good reference for this material is [FK].

1.1 Basic Definitions

In this section we give the basic definitions and first properties of compact Riemann surfaces and points on them.

1.1.1 First Properties of Compact Riemann Surfaces

We begin with the basic definition. A *Riemann surface* is a connected second countable Hausdorff space X together with an open cover $\{U_\alpha\}_\alpha$ of X and homeomorphisms ϕ_α of U_α onto open subsets of the complex plane $\mathbb{C}$ satisfying the following condition: If U_α and U_β have nonempty intersection then, the homeomorphism $\phi_\alpha \circ \phi_\beta^{-1}$ of $\phi_\beta(U_\alpha \cap U_\beta)$ onto $\phi_\alpha(U_\alpha \cap U_\beta)$ is a complex analytic function between open subsets of $\mathbb{C}$. The sets U_α are called *coordinate discs* or *parametric discs*, the functions ϕ_α are called *coordinate charts*, and the complex variables $z_\alpha = \phi_\alpha(P)$, $P \in X$ are called *local coordinates* or *local parameters*. In this book we shall be interested only in compact Riemann surfaces, and it is well known that every nontrivial algebraic function of two variables z and w has a unique (up to isomorphism, which we shall not deal with) associated compact Riemann surface.

A *holomorphic differential* on X is a collection of holomorphic functions on open subsets of X, such that for every α the function μ_α is defined on the set U_α of the open cover, and we demand that if $T = \phi_\alpha \circ \phi_\beta^{-1}$ is a change of local coordinate, then

H.M. Farkas and S. Zemel, *Generalizations of Thomae's Formula for Z_n Curves*,
Developments in Mathematics 21, DOI 10.1007/978-1-4419-7847-9_1,
© Springer Science+Business Media, LLC 2011

on the intersection $U_\alpha \cap U_\beta$, we have $\mu_\beta = (\mu_\alpha \circ T)T'$ where T' is the derivative of T as a holomorphic function between open subsets of $\mathbb{C}$. In order to express this transition law, we write differentials in local coordinates as $\mu_\alpha(z_\alpha)dz_\alpha$ (z_α being the local coordinate).

If f is a continuous map between the Riemann surfaces X and Y, we say that f is a *holomorphic* map provided that f is complex analytic in terms of the local parameters at each point. This means explicitly that if ϕ_α is a local coordinate around a point $p \in X$ and ψ_β is a local coordinate around the point $f(p) \in Y$ then the map $\psi_\beta \circ f \circ \phi_\alpha^{-1}$ is local analytic as a function between open sets in $\mathbb{C}$. This is independent of the local coordinates chosen. It is an easy exercise using connectivity and the open mapping theorem to show that if $f : X \to Y$ is a holomorphic map of Riemann surfaces and X is compact, then either f is constant (i.e., maps everything to a point) or Y is also compact and f is surjective. We shall indicate why this is so at the end of this section.

The simplest example of a compact Riemann surface is the *Riemann sphere*, which can clearly be made into a Riemann surface by assigning the following open cover and local coordinates to it: We take U_1 to be the sphere without the "north pole" and U_2 to be the sphere without the "south pole", and the mapping $\phi_1 \circ \phi_2^{-1}$ is simply $z \mapsto 1/z$. By choosing the local parameter z of U_1 one can identify the Riemann sphere with $\mathbb{C} \cup \{\infty\}$. A holomorphic map of a Riemann surface into the sphere is called a *meromorphic* function. Using the identification of the Riemann sphere with $\mathbb{C} \cup \{\infty\}$ one can easily see that a map $f : X \to \mathbb{C} \cup \{\infty\}$ (X a compact Riemann surface) is meromorphic if and only if it is holomorphic on X minus a finite number of points (those mapping to ∞), and at each of these points, say P, the singularity is a *pole*, i.e., in local coordinates on a small neighborhood of P the map $f \circ \phi_\alpha^{-1}$ is of the form $g(z)/(z - z(P))^n$ where g is a holomorphic function in the neighborhood of $z(P)$ such that $g(P) \neq 0$ and n is an integer.

Having defined meromorphic functions, we can now generalize the notion of holomorphic differentials. If the functions μ_α defining the differential are not holomorphic but just meromorphic (as functions from open subsets of $\mathbb{C}$ to $\mathbb{C} \cup \{\infty\}$), we call the differential a *meromorphic differential*. Now, to each meromorphic function f on a Riemann surface X one can associate its differential df to be defined by the functions $\mu_\alpha = (f \circ \phi_\alpha^{-1})'$, i.e., in local coordinates $df = f'(z_\alpha)dz_\alpha$. It is clear that this gives a meromorphic differential on X.

Compact Riemann surfaces are classified topologically according to a nonnegative integer called the *genus*. The Riemann sphere has genus 0. The genus is most easily defined by the rank of the first homology group of the surface or by the dimension of the linear space of holomorphic differentials on the surface. Once again we stress that it is not our intention to prove explicitly all the well-known results concerning Riemann surfaces, since the reader can refer to the book [FK] for these facts and ideas. We shall however give in many cases either hints or try to explain the results without actually giving a proof. We shall always give a reference, usually to the book [FK].

It will be convenient to introduce the following notation. A *divisor* on the Riemann surface is a formal symbol $\prod_{P \in X} P^{l_P}$, where l_P is an integer called the order of the divisor at P and this product is understood to be finite, i.e., $l_P = 0$ except for finitely many points $P \in X$. The set of divisors has the structure of an abelian group. If Δ and $\varXi$ are two divisors, $\Delta = \prod_{P \in X} P^{l_P}$ and $\varXi = \prod_{P \in X} P^{k_P}$, then

$$\Delta \varXi = \prod_{P \in X} P^{l_P + k_P}.$$

The identity element of the group is the divisor with $l_P = 0$ for all P and the element inverse to Δ is just $\prod_{P \in X} P^{-l_P}$. The *degree* of the divisor Δ is defined to be just the sum $\sum_P l_P$, which is well defined since $l_P \neq 0$ only for finitely many points P. The divisor Δ is called *integral* if $l_P \geq 0$ for every $P \in X$, and the degree of an integral divisor is clearly nonnegative (and positive if the divisor is not with $l_P = 0$ for all P). The divisor $\varXi$ is called a *multiple* of Δ if $\varXi/\Delta$ is integral, i.e., if $\varXi$ is a multiple of Δ by some integral divisor, and then the degree of $\varXi$ must be at least the one of Δ (with equality only if $\varXi = \Delta$).

If f is a meromorphic function on the Riemann surface, not identically vanishing, one defines the *divisor of f* to be

$$\text{div}(f) = \prod_{P \in X} P^{\text{Ord}_P f},$$

where $\text{Ord}_P f$ is the order of the meromorphic function f at the point P defined to be m if f has a zero of order $m \geq 1$ at P, $-m$ if f has a pole of order $m \geq 1$ at P, and 0 if P is neither a zero nor a pole of f. In other words the divisor of a meromorphic function is a concise way of writing down the zeros and poles of the function. There are no questions of convergence here since all sums are finite. If there were an infinite number of zeros or poles, then by compactness they would have to have a limit point, and thus the function would have to be constant, which is not interesting. Hence the divisor of a meromorphic function is indeed a divisor. These divisors are called *principal* divisors. One defines similarly for every meromorphic differential ω the *divisor of ω*, which is similarly denoted by $\text{div}(\omega)$. This is also well defined since the condition satisfies by the functions $\{\mu_\alpha\}_\alpha$ there implies that for $P \in U_\alpha \cap U_\beta$ we have $\text{Ord}_P \mu_\alpha = \text{Ord}_P \mu_\beta$ since the transition function T there and its inverse are both holomorphic. The divisor of a differential is also well defined (i.e., the corresponding sum is finite) by exactly the same argument.

Since it is clear that for two meromorphic functions f and g on X we have $\text{div}(fg) = \text{div}(f)\text{div}(g)$ we find in particular that the set of principal divisors is a subgroup of the group of divisors of X. We shall say that two divisors Δ and $\varXi$ are *equivalent* if their quotient is principal, which is the same as saying that they lie in the same coset of the quotient group. We remark that in the literature such divisors are called *linearly* equivalent (to distinguish from other possible equivalence relations), but we shall stick here to the shorter term with no possible confusion.

1.1.2 Some Examples

Let us now pause for an example which hopefully demonstrates some of the previous remarks. Let z be the natural parameter for the complex plane $\mathbb{C}$ and the Riemann sphere, let $g \geq 1$ be an integer, and let us consider a function w defined by

$$w^2 = z(z-1) \prod_{i=1}^{2g-1} (z - \lambda_i),$$

with the numbers λ_i, $1 \leq i \leq 2g - 1$ distinct from 0, from 1, and from one another. w is not a well-defined function on the complex plane since if one continues it analytically along little circles surrounding 0, 1, λ_i for $1 \leq i \leq 2g - 1$ or ∞ one does not return to the original value but rather to the negative of the original value. As the reader recalls from elementary complex variables the way to solve this problem of w being multi-valued is to construct a 2-sheeted cover of the sphere branched over the points 0, 1, λ_i for $1 \leq i \leq 2g - 1$ and ∞, so that when one traverses the boundary of a small disc containing any of these points, one does not return to the original point but rather to the other sheet. This is indeed a Riemann surface, whose genus turns out to be g, and the branched covering map is just z. If P is any point on this surface other than the ones lying over 0, 1, λ_i for $1 \leq i \leq 2g - 1$ or ∞, the projection map z to the complex plane is a good local coordinate. If the point P satisfies $z(P) = 0$, then $t = \sqrt{z}$ is a good local coordinate. If the point P satisfies $z(P) = 1$, then $t = \sqrt{z-1}$ is a good local coordinate. If the point P satisfies $z(P) = \lambda_i$ for $1 \leq i \leq 2g - 1$, then $t = \sqrt{z - \lambda_i}$ is a good local coordinate, and finally if $z(P) = \infty$, then $1/\sqrt{z}$ is a good local coordinate.

There are two functions naturally associated with this surface. One of course is the function z already discussed. It is a 2-to-1 map of the surface onto the sphere and is clearly a meromorphic function on the surface. It has a zero at the point above 0 and a pole at the point above ∞, and it is clear from the local coordinates that the order of zero and pole is 2. The second function is w, the object for which the surface was constructed. In terms of the local coordinates described above we see that the zeros of w are the points P lying over 0, 1, and λ_i for $1 \leq i \leq 2g - 1$ and that w has a pole of order $2g + 1$ at the point over ∞. Let us agree to denote the points lying over 0 and 1 by P_0 and P_1 respectively, the point lying over λ_i for $1 \leq i \leq 2g - 1$ by P_{λ_i}, and the point over ∞ by P_∞. In this case the divisors of the functions z and w are just

$$\mathrm{div}(z) = \frac{P_0^2}{P_\infty^2}, \qquad \mathrm{div}(w) = \frac{P_0 P_1 \prod_{i=1}^{2g-1} P_{\lambda_i}}{P_\infty^{2g+1}}.$$

We now observe that we can use the functions z and w in order to construct holomorphic differentials on the surface. The first observation is that for the differential dz of the meromorphic function z we have

$$\mathrm{div}(dz) = \frac{P_0 P_1 \prod_{i=1}^{2g-1} P_{\lambda_i}}{P_\infty^3}.$$

We have already computed the divisor of the function w so it is clear that the divisor of the meromorphic differential dz/w has divisor given by

$$\mathrm{div}\left(\frac{dz}{w}\right) = P_\infty^{2g-2}$$

showing that this meromorphic differential is in fact a holomorphic differential. Clearly it follows that the differential $z^l dz/w$ is also holomorphic with divisor $P_0^{2l}P_\infty^{2g-2-2l}$ for every $0 \le l \le g - 1$. Hence we have produced in this way g linearly independent holomorphic differentials on the surface and have observed that each of these holomorphic differentials has precisely $2g - 2$ zeros.

A compact Riemann surface of genus 0 (the simplest possibility) is the Riemann sphere. A meromorphic function on the sphere is a map from the sphere into the sphere, but in complex variables one generally shows that the map is in fact onto the sphere (if not constant). Let us recall that a meromorphic function on the sphere is a rational function that is a quotient of polynomials. If the quotient is given by f/g where f is a polynomial of degree m and g is a polynomial of degree n with no common root (so that the quotient is reduced) then clearly our rational function has n poles in the complex plane (the zeros of the denominator g) and m zeros (the zeros of the numerator f). This is all true in the plane, but on the sphere one needs check what is happening at ∞. What happens at ∞ depends on the relation between m and n. If $m = n$, then the function has a finite value at ∞, this value being equal to the quotient of the leading coefficients of the polynomials f and g. If $m > n$, then the function has an additional pole of order $m - n$ at ∞, and if $m < n$, the function has an additional zero of order $n - m$ at ∞. The reader who is not familiar with these statements should verify them in local coordinates. In particular the number of zeros is equal to the number of poles, and applying this argument to the rational function $f - c$ where c is any complex number gives that every value is assumed the same number of times. Thus we see that not only is the map onto, but it also covers each point on the sphere the same number of times (counted with multiplicities).

In the above example we have seen that the meromorphic function w has the same number of zeros as poles and of course the meromorphic function z has the same property. This is true in full generality. Let X be a compact Riemann surface of genus g and let f be a nonconstant holomorphic map of X into a Riemann surface Y. Then Y is compact and the map is onto. In fact every value is assumed the same number of times (again counting with multiplicities), a number which is called the *degree* of the map f.

Let us explain why the first statement is true. A holomorphic map is continuous so the image of X needs be compact. It also needs to be closed since compact subsets of Hausdorff spaces are closed. It needs to be open as well if the map is not constant since a holomorphic map is an open map. It is thus open and closed and therefore the entire space. The second statement is really not much harder to prove, but we will settle for a proof in which the holomorphic mapping is to the Riemann sphere and is therefore given by a meromorphic function. The crucial point is that we can triangulate the surface in such a way that if P_i, $1 \le i \le l$, are the poles of the function and Q_i, $1 \leq i \leq m$, are the zeros of the function, then each P_i and each Q_i lies in the interior of a distinct triangle of the triangulation. We now look at the expression

$$\frac{1}{2\pi i} \sum_j \int_{\Delta_j} \frac{df}{f}$$

where Δ_j runs over the triangles of the triangulation. On the one hand, this gives us the number of zeros of the function f minus the number of poles of the function f, counted with multiplicities, as the sum of the residues of this meromorphic differential. On the other hand, this integral is clearly zero since we are integrating twice over each side of each triangle in opposite directions. It follows therefore that the number of zeros equals the number of poles, and therefore the divisor of a meromorphic function on the surface defined above is a divisor of degree 0. Applying the same considerations for the function $f - c$, $c \in \mathbb{C}$, whose poles are the same as those of f and whose zeros are the points where f attains the value c, we obtain the result that every element of the Riemann sphere is attained by f the same number of times.

1.2 The Abel Theorem, the Riemann–Roch Theorem and Weierstrass Points

In this section we describe the statements of some basic important theorems about Riemann surfaces, and some of their proofs.

1.2.1 The Abel Theorem and the Jacobi Inversion Theorem

The first theorem we present is the Abel Theorem, which describes the divisor of a meromorphic function in the sense that the zeros and poles are not arbitrary. We have already seen that a meromorphic function on a compact Riemann surface has the property that it has the same number of zeros as poles (counting multiplicities). We have already defined the divisor of a meromorphic function f to be $\prod_{P \in X} P^{\text{Ord}_P f}$ and thus we have $\sum_{P \in X} l_P = 0$. Therefore the degree of every principal divisor is zero, and consequently two equivalent divisors have the same degree. As we have already observed, this property is a general property on all compact Riemann surfaces and is well known for the sphere. On the sphere that is the only condition. In other words, given any set of n points P_i, $1 \leq i \leq n$, on the sphere (not necessarily distinct) and any other set of n (again not necessarily distinct) points Q_i, $1 \leq i \leq n$, such that $P_i \neq Q_j$ for any i and j, there is a meromorphic function on the sphere with those points as zeros and poles (i.e., every degree 0 divisor on the sphere is principal). This is no longer true when the genus is positive. It is the *Abel Theorem* which describes the conditions under which the statement is true. We shall rephrase the question to ask what is the condition on a divisor which is equivalent to the divisor being principal. We already know that one necessary condition is that the divisor is of degree 0.

The condition will be in terms of the Abel–Jacobi map from Definition 1.1 below. We recall that a basis a_i, $1 \le i \le g$, and b_i, $1 \le i \le g$, for the homology for the Riemann surface X is called *canonical* if the intersection number of a_i with a_j and of b_i with b_j is 0 for every i and j, and the intersection number of a_i with b_j is the Kronecker delta symbol δ_{ij}. A basis θ_i, $1 \le i \le g$, for the holomorphic differentials on X is called *the dual basis* corresponding to the given canonical homology basis if $\int_{a_i} \theta_j = \delta_{ij}$ for every i and j. We also define Π to be the matrix whose ith entry is $\int_{b_i} \theta_j$. The matrix Π is called the *period matrix* of X with respect to this homology basis, and one can show that Π is symmetric and that its imaginary part is positive definite as a real bilinear form. We also define the *jacobian variety* (or simply *jacobian*) $J(X)$ of X to be the quotient of $\mathbb{C}^g$ modulo the group of translations generated by the columns of the $g \times g$ identity matrix I and by the columns of the matrix Π (or equivalently by the subgroup of $\mathbb{C}^g$ generated by these vectors—it is a full lattice in $\mathbb{C}^g$). This is a complex torus of dimension g. Now let $P_0 \in X$ be arbitrary, and we define

Definition 1.1. The *Abel–Jacobi map* $\varphi_{P_0} : X \to J(X)$ with basepoint P_0 is the function defined by

$$
\varphi_{P_0}(P) = \int_{P_0}^{P} \begin{pmatrix} \theta_1 \\ \cdot \\ \cdot \\ \cdot \\ \theta_g \end{pmatrix}.
$$

Note that in Definition 1.1 there is an integral between P_0 and P in which one does not specify the explicit path on the Riemann surface taken from P_0 to P. The value in $\mathbb{C}^g$ obtained by φ_{P_0} on a given point $P \in X$ is indeed not well defined and depends on the chosen path from P_0 to P, but the important point is that changing the path can change the value only by some integral linear combination of columns of I and Π. Hence in $J(X)$ this value is indeed well defined. For more details see [FK, Chap. 3]. In particular it is clear that $\varphi_{P_0}(P_0) = 0$ (take a constant path), a fact which will be very useful throughout the book. The reader should note that the choice of a canonical basis for the homology of X is not unique, and changing the canonical basis changes both the matrix Π and hence our representation of $J(X)$; we shall see later that $J(X)$ itself has a more intrinsic presentation, independent of homology bases. There is a very interesting theory for the behavior of jacobians and Abel–Jacobi maps under changes of the homology basis, but we shall not deal with this subject. Moreover, throughout this book we shall usually assume that the Riemann surface X comes with a specific choice of a canonical basis for its homology.

We can clearly extend the definition of the Abel–Jacobi map from points to divisors. For a divisor $\Delta = \prod_{P \in X} P^{l_P}$ we define $\varphi_{P_0}(\Delta) = \sum_{P \in X} l_P \varphi_{P_0}(P)$ (a finite sum since Δ is a divisor). The Abel Theorem now can be formulated:

Theorem 1.2. *A necessary and sufficient condition for a divisor Δ of degree 0 to be a principal divisor, that is the divisor of a meromorphic function on the Riemann surface is that $\varphi_{P_0}(\Delta) = 0$.*

The proof of this theorem given in [FK, Chap. 3] is based on the fact that on any compact Riemann surface it is true that given any two distinct points P and Q on the Riemann surface, one can find a meromorphic differential on the surface with simple poles at the points P and Q with residues (we did not define residues since we shall not use them in this book—for their definition and properties see, for example, [FK]) 1 at P and -1 at Q and holomorphic everywhere else. One then uses these differentials to construct a meromorphic function on the surface corresponding to a divisor of degree 0 with the Abel property. Conversely, if there is such a function on the surface one can easily show (using simple integration) that it satisfies the Abel property.

When examining the Abel Theorem, one encounters a natural question: The question whether a divisor of degree 0 is principal or not depends only on the divisor, while its image by the map φ_{P_0} depends also on the basepoint P_0. For this (and for following purposes) we wish to see what happens when we change the basepoint to some other point $Q \in X$. Since the definition of the Abel–Jacobi map is based on integration, it is easy to see that for every $P \in X$ one has the equality $\varphi_Q(P) = \varphi_Q(P_0) + \varphi_{P_0}(P)$, meaning that the two different Abel–Jacobi maps differ by an additive constant. Applying this to a general divisor Δ of degree k, we find that

$$\varphi_Q(\Delta) = k\varphi_Q(P_0) + \varphi_{P_0}(\Delta). \tag{1.1}$$

In particular we see that for divisors of degree 0 the image of the Abel–Jacobi map does not depend on the basepoint.

Now, the divisors of degree 0 form a subgroup of the group of all divisors on the surface. The map φ_{P_0} is a homomorphism of the group of divisors into the group $J(X)$ which is a complex torus of dimension g, and its restriction to the subgroup of divisors of degree 0 was seen to be independent of the basepoint P_0. The Abel Theorem describes the kernel of this restricted map as simply the group of principal divisors (which we already know that is contained in the group of divisors of degree 0). The point that is missing thus far is the fact that this restricted map is onto. This is a consequence of the *Jacobi Inversion Theorem* which states

Theorem 1.3. *The map φ_{P_0} defined on the set of integral divisors of degree g is a surjection to $J(X)$.*

This result of course implies the statement that the map is surjective when the domain is divisors of degree 0, hence also independent of the basepoint. This follows because, given any integral divisor $\prod_{i=1}^g P_i$ of degree g, the divisor $\left(\prod_{i=1}^g P_i\right)/P_0^g$ is of degree 0 and has the same image under φ_{P_0}. Now, as a consequence of the fundamental theorem on group homomorphisms, we conclude that the group $J(X)$ is isomorphic to the quotient group of divisors of degree 0 modulo principal divisors. This is the more intrinsic way of looking at the jacobian of X without depending on the choice of the canonical homology basis. This way of looking at $J(X)$ also assures us that we get the same jacobian (up to isomorphism, which again we shall not define exactly) regardless of the choice of the canonical homology basis, even though the group of translations by which we divide $\mathbb{C}^g$ in order to obtain $J(X)$ depends on Π and thus on the choice of the basis.

1.2.2 The Riemann–Roch Theorem and the Riemann–Hurwitz Formula

The Abel Theorem is descriptive in the sense that it describes the divisor of a mero-morphic function. A related theorem is the *Riemann–Roch Theorem*, which for integral divisors $\Delta = \prod_{i=1}^{r} P_i^{l_i}$ (with $l_i \geq 0$) tells us the dimension of the vector space of meromorphic functions on the surface that has poles at most at the set of points P_i, $1 \leq i \leq r$, on the surface and such that the order of the pole at P_i is bounded by the positive integer l_i. The set of these meromorphic functions form a finite-dimensional complex vector space which is denoted by $L(1/\Delta)$, and its dimension is denoted by $r(1/\Delta)$. Similarly, we define $\Omega(\Delta)$ to be the space of holomorphic differentials which vanishes at each point P_i to order at least l_i, and we denote its dimension by $i(\Delta)$.

The Riemann–Roch Theorem asserts:

Theorem 1.4. *If* $\Delta = \prod_{i=1}^{r} P_i^{l_i}$ *is a divisor on the Riemann surface, then*

$$r(1/\Delta) = \deg(\Delta) - g + 1 + i(\Delta).$$

We remark that we have stated the Riemann–Roch Theorem for every divisor on the Riemann surface, not only integral. It holds in general when the definition is that if l_i is negative for some $1 \leq i \leq r$, one demands that the functions in $L(1/\Delta)$ must vanish at P_i to order at least $|l_i|$, and the differentials in $\Omega(\Delta)$ can be meromorphic and have a pole at P_i, but to order not exceeding $|l_i|$. With these generalizations the Riemann–Roch Theorem holds as stated, without the integrality assumption on Δ.

One important consequence of the Riemann–Roch Theorem is that the degree of the divisor of a meromorphic differential (these divisors are called *canonical divisors*) on X is $2g - 2$, as can be checked in all the previous examples. As two meromorphic differentials are obtained from one another by multiplication by a meromorphic function (just divide two differentials and see), whose divisor has degree 0, the divisors of all the meromorphic differentials on X are equivalent and have the same degree $2g - 2$. It is also worth mentioning that because all the meromor-phic differentials are multiples of a given meromorphic differential by meromorphic functions and the image of the principal divisors vanish under the Abel–Jacobi map φ_{P_0} (for every basepoint P_0), one sees that φ_{P_0} attains the same value on every canon-ical divisor. This value is called *the canonical point on $J(X)$ with respect to P_0*; we shall have more to say about the canonical point later.

At this point we also observe that the fact that the divisor of a meromorphic differential has degree $2g - 2$ gives us the *Riemann–Hurwitz formula*. This formula states that given a holomorphic function f of degree n between compact Riemann surfaces X and Y of genera g_X and g_Y respectively, we have the equality

$$2g_X - 2 = n(2g_Y - 2) + B, \quad B = \sum_{x \in X} b_x,$$

where b_x is the *order of branching* of f at x (meaning that on some punctured neighborhoods of x in X and of $f(x)$ in Y, f is $(b_x + 1)$-to-1, in particular f is one-to-one on such a neighborhood if and only if x is not a branch point of f), and B (which is an essentially finite sum) is called the *total branching number* or *total ramification* of f. We shall be using the Riemann–Hurwitz formula only for meromorphic functions on compact Riemann surfaces, which is the same as holomorphic functions from the Riemann surface X (whose genus will now be simply g) to the Riemann sphere (whose genus is $g_Y = 0$). For such a function f (again of degree n) the Riemann–Hurwitz formula becomes simply $2g - 2 = B - 2n$, or equivalently $B = 2(g + n - 1)$. In this way we shall find the genera of all the Riemann surfaces we encounter in what follows. We remark that a Riemann surface admitting a meromorphic function of degree 2 is called *hyperelliptic*, and for these surfaces there is a vast theory (with which we shall not deal here).

1.2.3 Weierstrass Points

Another interesting consequence of this theorem is the notion of *Weierstrass points* on a compact Riemann surface.

Definition 1.5. A *Weierstrass point* on a compact Riemann surface X of genus g is a point P on X such that there exists a nonconstant meromorphic function on X whose only singularity is a pole at P of order at most g.

This is equivalent, via Riemann–Roch, to the assertion that $i(P^g) \geq 1$, i.e., to the existence of a holomorphic differential vanishing at P to order at least g. This means in particular that Weierstrass points do not exist on Riemann surfaces of genus 0 (since there are no holomorphic differentials on the Riemann sphere) or 1 (on a complex torus all the holomorphic differentials are scalar multiples of one another, and none of them have any zeros); hence we assume now that $g \geq 2$.

Now fix a point $P \in X$. We know that for every k, $r(1/P^k)$ can be either $r(1/P^{k-1})$ or $r(1/P^{k-1}) + 1$, the latter expressing the existence of a meromorphic function on X with pole only at P and to order exactly k. This is so since $L(1/P^{k-1}) \subseteq L(1/P^k)$ and since the quotient space $L(1/P^k)/L(1/P^{k-1})$ is clearly at most 1-dimensional. Since we know that $r(1/P^0) = r(1) = 1$ (only constant functions) and by Riemann–Roch $r(1/P^{2g}) = g + 1$ as $i(P^{2g}) = 0$ (the degree of divisors of differentials is only $2g - 2$, so they cannot be multiples of a divisor of degree $2g$), it follows that there are precisely g orders between 1 and $2g$ which are assumed as the exact orders of poles of meromorphic functions whose only pole is at P and g orders which are not assumed. The assumed numbers are called *non-gaps*, while the unassumed numbers are called *gaps*, and the list of the g gaps at a point $P \in X$ in increasing order is called the *gap sequence* at P. Thus we see that $P \in X$ is a Weierstrass point if and only if its gap sequence is not simply the numbers k, $1 \le k \le g$. The theory of Weierstrass points is explained in detail in [FK, Chap. 3], and there one can learn that there are always Weierstrass points when $g \ge 2$, but there are only a finite number of such points. This number is between $2g + 2$ and $g^3 - g$, and counting with multiplicities (for the definition of the multiplicity of a Weierstrass point see [FK, Chap. 3], as we shall not deal with this notion here) it always equals $g^3 - g$. On the other hand, for hyperelliptic Riemann surfaces (with $g \ge 2$) this number is always $2g + 2$.

There are two statements made above concerning Weierstrass points. The first is the statement that there is a meromorphic function on the surface with a pole at one point P whose order is at most g, the genus. The second statement which is equivalent is about the existence of a holomorphic differential with a zero of order at least g, which is the dimension of the space of holomorphic differentials, at the point P. This allows us to generalize the notion of Weierstrass point to other linear spaces. The interested reader can pursue this subject in [FK]. Here we simply remark that the gaps at P will be denoted by γ_i, $1 \le i \le g$, with $\gamma_{i+1} > \gamma_i$ for $1 \le i \le g - 1$, $\gamma_1 = 1$ and $\gamma_g \le 2g - 1$. The complement of this sequence with respect to the set of integers between 1 and $2g$ (both sides included) form the sequence α_i, $1 \le i \le g$, of non-gaps with $\alpha_{i+1} > \alpha_i$ for $1 \le i \le g - 1$, $\alpha_1 > 1$ and $\alpha_g = 2g$. In terms of these quantities P is a Weierstrass point provided that it has a non-gap less than or equal to g, or equivalently, has a gap which is greater than g. In other words a point P is a Weierstrass point provided that the gap sequence is not simply $\gamma_i = i$, $1 \le i \le g$. We shall say that a basis ω_i, $1 \le i \le g$, for the space of holomorphic differentials is *adapted to the point* P if $\mathrm{Ord}_P \omega_1 = 0$ and $\mathrm{Ord}_P \omega_i < \mathrm{Ord}_P \omega_{i+1}$ for every $1 \le i \le g - 1$.

At this point we mention the useful observation that if one is given g holomorphic differentials on a Riemann surface X whose orders at a given point P are all different, then these differentials must be linearly independent and hence form a basis for the holomorphic differentials on X (and thus are a basis adapted to P). Indeed, the order of a nonzero linear combination of them at P must be the minimal order of the differentials which appear in the combination with nonzero coefficients. This is because the other differentials have higher order at P, and hence cannot cancel the coefficient preceding the minimal power at the expansion of the function describing the differential at P.

The property that we shall find useful in the sequel is the following:

Proposition 1.6. *Let P be a point on the Riemann surface X and let ω_i, $1 \le i \le g$, be a basis adapted to P. Then $\mathrm{Ord}_P \omega_i = \gamma_i - 1$ where γ_i, $1 \le i \le g$, is the gap sequence at P. In particular if α is a non-gap at P, there is no holomorphic differential which vanishes to order $\alpha - 1$ at P.*

Proof. Suppose γ is a gap at P. Then one has $r(1/P^\gamma) = r(1/P^{\gamma-1})$ by the definition of gaps, and thus by the Riemann–Roch theorem $i(P^\gamma) = i(P^{\gamma-1}) - 1$. If no differential vanishes to order exactly $\gamma - 1$ at P, then every element of $\Omega(P^{\gamma-1})$ would actually be in $\Omega(P^\gamma)$ (as the order of the zero it has at P would be at least γ) which contradicts the above equality. It thus follows that each gap γ gives rise in this way to a differential which vanishes to order $\gamma - 1$ at P, and thus we obtain a basis for the holomorphic differentials on X (which is of course adapted to the point P). Finally, it is clear that these are the only orders of vanishing at the point P, so that if α is a non-gap, one cannot have order of vanishing $\alpha - 1$ at P. $\square$

Let us now return to our example of a compact Riemann surface, the Riemann surface of the algebraic equation

$$w^2 = z(z-1) \prod_{i=1}^{2g-1} (z - \lambda_i),$$

where the numbers λ_i, $1 \le i \le 2g - 1$, are distinct from 0, 1, and from one another. The first important remark we wish to make is that if we take the basepoint P_0 of the Abel–Jacobi map to be any one of the branch points P_0, P_1, P_{λ_i} for $1 \le i \le 2g - 1$ and P_∞, then the image of any of the branch points is either 0 or a point of order 2 in $J(X)$. This is an immediate consequence of the Abel Theorem, because as already explained, the function z is a meromorphic function on this surface which represents the surface as a two sheeted branched cover of the sphere branched over the points 0, 1, λ_i for $1 \le i \le 2g - 1$ and ∞. Hence for any choice of two of these points, say λ_i and λ_j with $i \ne j$, we have that $(z - \lambda_i)/(z - \lambda_j)$ is a meromorphic function with a double zero at P_i and a double pole at P_j (and in particular we see that this Riemann surface is hyperelliptic). Hence $2\varphi_{P_0}(P_i) - 2\varphi_{P_0}(P_j) = 0$ by the Abel Theorem. On the other hand, choosing either P_i or P_j as P_0 gives that each point P_i over λ_i has the same property, i.e., we see that $2\varphi_{P_i}(P_j) = 0$ for any pair i, j. Similar results are obtained with the points above 0, 1 and ∞, where if P_∞ is involved we shall have to use the functions z and $1/z$ (or more generally $z - \lambda_i$ and $1/(z - \lambda_i)$).

The same sort of argument shows that in the case of the compact Riemann surface associated to

$$w^n = \prod_{i=0}^{nr-1} (z - \lambda_i),$$

with $\lambda_i \ne \lambda_j$ when $i \ne j$, we would find that if P_i and P_j are branch points, then $\varphi_{P_i}(P_j)$ is necessarily a point of order n in $J(X)$. For the careful reader we now remark that for us throughout the book a *point of order N* in $J(X)$ is an element y of $J(X)$ such that $Ny = 0$, without the assumption usually made that N is the minimal number such that $Ny = 0$. Those who reject this abuse of notation can replace "of order N" by "of order *dividing N*" anywhere in the book, although we consider this abuse of notation very convenient for our purposes.

We now show that we can explicitly write a basis for the holomorphic differentials and explicitly write down their divisors. As an example we do this for the case $n = 3$ and $r = 2$. Consider therefore the compact Riemann surface associated to the equation

$$w^3 = z(z-1)(z-\lambda_1)(z-\lambda_2)(z-\lambda_3).$$

This is a 3-sheeted cover of the sphere branched over the six points 0, 1, λ_1, λ_2, λ_3 and ∞. Since z, w and dz have divisors P_0^3/P_∞^3, $P_0 P_1 P_{\lambda_1} P_{\lambda_2} P_{\lambda_3}/P_\infty^5$ and $P_0^2 P_1^2 P_{\lambda_1}^2 P_{\lambda_2}^2 P_{\lambda_3}^2/P_\infty^4$ respectively, it is clear that the four differentials

$$\frac{dz}{w}, \quad \frac{dz}{w^2}, \quad z\frac{dz}{w^2}, \quad z^2\frac{dz}{w^2}$$

are holomorphic on this surface with divisors

$$P_0 P_1 P_{\lambda_1} P_{\lambda_2} P_{\lambda_3} P_\infty, \quad P_\infty^6, \quad P_0^3 P_\infty^3, \quad P_0^6$$

respectively. The genus of the surface is 4, either by the Riemann–Hurwitz formula (with $n = 3$ and $B = 12$ for 6 points of branching number 2) or since the degree of these divisors are 6. Furthermore one can construct from these expressions mero-morphic functions on the surface all of whose poles are at P_0 such that the orders of the pole are 3 ($1/z$ has divisor P_∞^3/P_0^3), 5 (w/z^2 has divisor $P_1 P_{\lambda_1} P_{\lambda_2} P_{\lambda_3} P_\infty/P_0^5$), 6 ($1/z^2$ has divisor P_∞^6/P_0^6) and 8 (w/z^3 has divisor $P_1 P_{\lambda_1} P_{\lambda_2} P_{\lambda_3} P_\infty^4/P_0^8$), and one can-not construct functions with all poles at P_0 of orders 1,2,4,7 (the reader can either try and see why, or just note that there are only $g = 4$ orders which can be non-gaps and we already found all of them). In other words the numbers 1,2,4,7 are all gaps at P_0 and the complementary set 3,5,6,8 are the non-gaps at P_0. For more information on this, see [FK] on Weierstrass points. In fact this is a basis adapted either to the point P_0 or to P_∞, as one sees that the orders of these differentials at P_0 or at P_∞ are a strictly increasing sequence (when ordered correctly). In fact in this case we have shown in Proposition 1.6 that these orders are in fact the sequence $\gamma_i - 1$, $1 \leq i \leq g$, where γ_i is the ith gap at P_0. It is easy to see that this is the sequence of gaps at any one of the branch points.

1.3 Theta Functions

In this section we present the main object investigated in this book in the setting of general compact Riemann surfaces. A good reference is [FK, Chap. 6].

1.3.1 First Properties of Theta Functions

In order to define the theta function one does not have to begin with Riemann sur-faces but rather with the Siegel upper halfplane. For a natural number $g \geq 1$, the

Siegel upper halfplane of degree g is the set consisting of all the complex symmetric $g \times g$ matrices Π such that their imaginary part (viewed as a real symmetric $g \times g$ bilinear form) is positive definite. This is a complex manifold of dimension $g(g+1)/2$ which is denoted by $\mathcal{H}_g$. For such a matrix Π we define the following function, where we introduce the useful notation $\mathbf{e}(z) = e^{2\pi i z}$ for any complex number z.

Definition 1.7. Given two vectors ε and ε' in $\mathbb{R}^g$, the *theta function with characteristic* $\begin{bmatrix} \varepsilon \\ \varepsilon' \end{bmatrix}$ is the function $\theta \begin{bmatrix} \varepsilon \\ \varepsilon' \end{bmatrix} : \mathbb{C}^g \times \mathcal{H}_g \to \mathbb{C}$ defined by

$$\theta \begin{bmatrix} \varepsilon \\ \varepsilon' \end{bmatrix} (\zeta, \Pi) = \sum_{N \in \mathbb{Z}^g} \mathbf{e} \left[\frac{1}{2} \left(N + \frac{\varepsilon}{2} \right)^t \Pi \left(N + \frac{\varepsilon}{2} \right) + \left(N + \frac{\varepsilon}{2} \right)^t \left(\zeta + \frac{\varepsilon'}{2} \right) \right].$$

The above definition of the theta function is a generalization of the classical *Riemann theta function,*

$$\theta(\zeta, \Pi) = \sum_{N \in \mathbb{Z}^g} \mathbf{e} \left(\frac{N^t \Pi N}{2} + N^t \zeta \right),$$

which is simply $\theta \begin{bmatrix} 0 \\ 0 \end{bmatrix}$ in Definition 1.7. It is not hard to see that the theta functions with characteristics can be obtained from the classical theta function by the relation

$$\theta \begin{bmatrix} \varepsilon \\ \varepsilon' \end{bmatrix} (\zeta, \Pi) = \mathbf{e} \left(\frac{\varepsilon^t \Pi \varepsilon}{8} + \frac{\varepsilon^t \zeta}{2} + \frac{\varepsilon^t \varepsilon'}{4} \right) \theta \left(\zeta + \Pi \frac{\varepsilon}{2} + I \frac{\varepsilon'}{2}, \Pi \right)$$

(see Eq. (1.3) below). For a fixed matrix $\Pi \in \mathcal{H}_g$, the value of the theta function $\theta \begin{bmatrix} \varepsilon \\ \varepsilon' \end{bmatrix} (\zeta, \Pi)$ at $\zeta = 0$ is called the *theta constant* with characteristic $\begin{bmatrix} \varepsilon \\ \varepsilon' \end{bmatrix}$ with respect to the matrix Π.

The following properties of the theta functions with characteristics are easily derivable from the definition. For every two vectors M and M' in $\mathbb{Z}^g$,

$$\theta \begin{bmatrix} \varepsilon \\ \varepsilon' \end{bmatrix} (\zeta + \Pi M + IM', \Pi) = \mathbf{e} \left(\frac{\varepsilon^t M' - M^t \varepsilon'}{2} - \frac{M^t \Pi M}{2} - M^t \zeta \right) \theta \begin{bmatrix} \varepsilon \\ \varepsilon' \end{bmatrix} (\zeta). \tag{1.2}$$

Also, one has for every two vectors μ and μ' in $\mathbb{R}^g$

$$\theta \begin{bmatrix} \varepsilon + \mu \\ \varepsilon' + \mu' \end{bmatrix} (\zeta, \Pi) = \mathbf{e} \left(\frac{\mu^t \Pi \mu}{8} + \frac{\mu^t (\varepsilon' + \mu')}{4} + \frac{\mu^t \zeta}{2} \right) \theta \begin{bmatrix} \varepsilon \\ \varepsilon' \end{bmatrix} \left(\zeta + \Pi \frac{\mu}{2} + I \frac{\mu'}{2}, \Pi \right), \tag{1.3}$$

in which taking $\varepsilon = \varepsilon' = 0$ gives the relation connecting the theta functions with characteristics to the classical theta function described above. When combining equations (1.2) and (1.3) with the choice of $\mu = 2M$ and $\mu' = 2M'$ (with M and M' in $\mathbb{Z}^g$) one obtains

$$\theta \begin{bmatrix} \varepsilon + 2M \\ \varepsilon' + 2M' \end{bmatrix} (\zeta, \Pi) = \mathbf{e}\left(\frac{\varepsilon' M'}{2}\right) \theta \begin{bmatrix} \varepsilon \\ \varepsilon' \end{bmatrix} (\zeta, \Pi). \tag{1.4}$$

It is also easy to see from Definition 1.7 of the theta functions with characteristics that

$$\theta \begin{bmatrix} -\varepsilon \\ -\varepsilon' \end{bmatrix} (\zeta, \Pi) = \theta \begin{bmatrix} \varepsilon \\ \varepsilon' \end{bmatrix} (-\zeta, \Pi),$$

a fact which in the classical case $\varepsilon = \varepsilon' = 0$ reduces to the assertion that the classical Riemann theta function is even as a function of the variable ζ. Substituting $\zeta = 0$ in the last equation yields

$$\theta \begin{bmatrix} -\varepsilon \\ -\varepsilon' \end{bmatrix} (0, \Pi) = \theta \begin{bmatrix} \varepsilon \\ \varepsilon' \end{bmatrix} (0, \Pi), \tag{1.5}$$

an identity between theta constants which will be very useful in the sequel.

From the above properties it is clear that the theta functions with characteristics are well-defined holomorphic functions on $\mathbb{C}^g \times \mathscr{H}_g$, hence for fixed Π they are well-defined holomorphic functions on $\mathbb{C}^g$. However, for fixed Π they are not well defined on the quotient J of $\mathbb{C}^g$ modulo the group of translations generated by

$$\zeta \mapsto \zeta + e^i, \qquad \zeta \mapsto \zeta + \Pi^i$$

(which in Riemann surface theory is the quotient $J(X)$). While the function is not well defined it is self-evident that the zeros of the function are indeed well defined, since whether a point in the quotient space is or is not a zero is independent of the representation. This is so because the values at two points in $\mathbb{C}^g$, which are the same in J, differ only by multiplication by the exponent of something holomorphic, and the exponential of a holomorphic function does not vanish and is holomorphic. Therefore the set of zeros of the (classical) theta function on J is a well-defined subset of J, and since J is a complex manifold of dimension g and the theta function is locally holomorphic, this set of zeros forms a complex submanifold of dimension $g - 1$, called the *theta divisor*.

This point led Riemann to consider the following expression. For e, an arbitrary point in $\mathbb{C}^g$, and for P_0, an arbitrary fixed point on the Riemann surface, consider

$$f(P) = \theta(\varphi_{P_0}(P) - e, \Pi).$$

This is not a well-defined function on the surface since $\varphi_{P_0}(P)$ depends on the path of integration one takes from P_0 to P. Its zeros however are well defined by what we said above.

1.3.2 Quotients of Theta Functions

At this point, however, it is important to observe that one can indeed use the theta functions to construct meromorphic single-valued functions on the quotient space, the complex torus of dimension g, or ultimately on the Riemann surface. We indicate how this can be done in a very important example, which will serve us throughout the book.

Let $f(\zeta) = \dfrac{\theta\begin{bmatrix} \varepsilon \\ \varepsilon' \end{bmatrix}(\zeta, \Pi)}{\theta\begin{bmatrix} \delta \\ \delta' \end{bmatrix}(\zeta, \Pi)}$. By Eq. (1.2) we see that for M and M' in $\mathbb{Z}^g$ we have

$$\frac{\theta\begin{bmatrix} \varepsilon \\ \varepsilon' \end{bmatrix}(\zeta + \Pi M + IM', \Pi)}{\theta\begin{bmatrix} \delta \\ \delta' \end{bmatrix}(\zeta + \Pi M + IM', \Pi)} = \mathbf{e}\left[-\frac{M^t(\varepsilon' - \delta') - M'^t(\varepsilon - \delta)}{2} \right] f(\zeta).$$

If we now assume that $\varepsilon = \alpha/n$, $\varepsilon' = \alpha'/n$, $\delta = \beta/n$, and $\delta' = \beta'/n$ where α, α', β and β' are in $\mathbb{Z}^g$, then the multiplier of $f(\zeta)$ in the above formula is at worse a $2n$th root of unity. Hence if we take the quotient to the nth power and assume further that $\alpha - \beta$ and $\alpha' - \beta'$ are in $2\mathbb{Z}^g$, then this root of unity will become 1 and the function will be invariant under these translations, a fact that will be written as

$$\frac{\theta^n\begin{bmatrix} \alpha/n \\ \alpha'/n \end{bmatrix}(\zeta + \Pi M + IM', \Pi)}{\theta^n\begin{bmatrix} \beta/n \\ \beta'/n \end{bmatrix}(\zeta + \Pi M + IM', \Pi)} = \frac{\theta^n\begin{bmatrix} \alpha/n \\ \alpha'/n \end{bmatrix}(\zeta, \Pi)}{\theta^n\begin{bmatrix} \beta/n \\ \beta'/n \end{bmatrix}(\zeta, \Pi)}. \tag{1.6}$$

This means that the nth power of our function $f(\zeta)$ is well defined not only on $\mathbb{C}^g$ but also on the quotient $J(X)$.

We now wish to make a further observation which will be very useful in the sequel. If we consider now $f(\zeta) = \dfrac{\theta\begin{bmatrix} \alpha/n \\ \alpha'/n \end{bmatrix}(\zeta, \Pi)}{\theta\begin{bmatrix} \beta/n \\ \beta'/n \end{bmatrix}(\zeta, \Pi)}$ where still α, α', β and β' are

in $\mathbb{Z}^g$, then Eq. (1.3) shows that for arbitrary μ and μ' in $\mathbb{R}^g$ we have

$$f\left(\zeta + \Pi\frac{\mu}{2} + I\frac{\mu'}{2}\right) = \mathbf{e}\left[-\frac{\mu^t}{4}\left(\frac{\alpha' - \beta'}{n}\right) \right] \frac{\theta\begin{bmatrix} \mu + \alpha/n \\ \mu' + \alpha'/n \end{bmatrix}(\zeta, \Pi)}{\theta\begin{bmatrix} \mu + \beta/n \\ \mu' + \beta'/n \end{bmatrix}(\zeta, \Pi)}.$$

If in addition we also assume that $\mu = \gamma/n$ and $\mu' = \gamma'/n$ with γ and γ' in $\mathbb{Z}^g$, then the previous expression becomes

$$
\mathbf{e}\left[-\frac{\gamma'}{4n}\left(\frac{\alpha'-\beta'}{n}\right)\right]\frac{\theta\left[\begin{matrix}(\alpha+\gamma)/n\\(\alpha'+\gamma')/n\end{matrix}\right](\zeta,\Pi)}{\theta\left[\begin{matrix}(\beta+\gamma)/n\\(\beta'+\gamma')/n\end{matrix}\right](\zeta,\Pi)},
$$

which is a root of unity of order $4n^2$. If we take the quotient to the power n^2, we get the previous expression equal to

$$
\mathbf{e}\left[-\frac{\gamma'}{4}(\alpha'-\beta')\right]\frac{\theta^{n^2}\left[\begin{matrix}(\alpha+\gamma)/n\\(\alpha'+\gamma')/n\end{matrix}\right](\zeta,\Pi)}{\theta^{n^2}\left[\begin{matrix}(\beta+\gamma)/n\\(\beta'+\gamma')/n\end{matrix}\right](\zeta,\Pi)}.
$$

The final observation for now is that if γ and the difference $\alpha'-\beta'$ are in $2\mathbb{Z}^g$, then the multiplier disappears, a fact which will be written as

$$
\frac{\theta^{n^2}\left[\begin{matrix}\alpha/n\\\alpha'/n\end{matrix}\right]\left(\zeta+\Pi\frac{\gamma}{2n}+I\frac{\gamma'}{2n},\Pi\right)}{\theta^{n^2}\left[\begin{matrix}\beta/n\\\beta'/n\end{matrix}\right]\left(\zeta+\Pi\frac{\gamma}{2n}+I\frac{\gamma'}{2n},\Pi\right)}=\frac{\theta^{n^2}\left[\begin{matrix}(\alpha+\gamma)/n\\(\alpha'+\gamma')/n\end{matrix}\right](\zeta,\Pi)}{\theta^{n^2}\left[\begin{matrix}(\beta+\gamma)/n\\(\beta'+\gamma')/n\end{matrix}\right](\zeta,\Pi)}. \tag{1.7}
$$

1.3.3 Theta Functions on Riemann Surfaces

Riemann's seminal idea was to study the theta function as a "function" on the Riemann surface. His idea was to use the Abel–Jacobi map φ_{P_0} already defined and to consider $f(P)=\theta(\varphi_{P_0}(P)-e,\Pi)$ as a locally defined holomorphic function on the surface. It is not globally defined since φ_{P_0} depends on the path of integration and θ is not quite invariant under $\zeta\mapsto\zeta+\Pi M+IN$. However, as mentioned above its zeros are well defined, and regarding them Riemann proved by contour integration (see [FK, Chap. 6]) the following theorem (called the *Riemann Vanishing Theorem*), stating that there are two possibilities.

Theorem 1.8. *For $e\in\mathbb{C}^g$ and $f(P)$ as above, either $f(P)$ vanishes identically on the Riemann surface X, or if not, $f(P)$ has precisely g zeros Q_i, $1\leq i\leq g$, on the surface (not necessarily distinct but counted with multiplicities). In the latter case the zeros satisfy the relation*

$$
e=\varphi_{P_0}\left(\prod_{i=1}^{g}Q_i\right)+K_{P_0},
$$

where K_{P_0} is a vector in $\mathbb{C}^g$ called the vector of Riemann constants. *It depends on the Riemann surface, the period matrix Π and the basepoint P_0 but is indepen*

dent of the point e. Furthermore the integral divisor $\prod_{i=1}^{g} Q_i$ satisfies the condition
$i\left(\prod_{i=1}^{g} Q_i\right) = 0.$

We refer the reader to [FK] for the complete proof and only remark here that the first case of identical vanishing simply means that the entire Riemann surface is mapped by the Abel–Jacobi map into the theta divisor, while the alternative case is that the image of the surface under the Abel–Jacobi map intersects the theta divisor in precisely g points. Note that this image depends on the basepoint P_0, and therefore both the option chosen in the theorem (identical vanishing or not) and the points Q_i, $1 \leq i \leq g$, in the second option depend on the choice of P_0 as well. We note that even though it seems that we found a function on the Riemann surface whose divisor is of degree g (g zeros and no poles), this is not a contradiction to the fact that a meromorphic functions on X must have divisors of degree 0 since this function is not a well-defined meromorphic function on X.

The following material is quite classical and can be found in [FK] but for the sake of completeness and getting the reader into the proper spirit we shall include some proofs and outlines of proofs. We are always in the situation where we are given a Riemann surface X of genus $g \geq 1$ together with a canonical homology basis a_i, $1 \leq i \leq g$, and b_i, $1 \leq i \leq g$, which gives the matrix Π with respect to which we are taking the theta function of X as a function of $\zeta \in \mathbb{C}^g$ and the representation of $J(X)$. Again we fix a basepoint P_0 for the Abel–Jacobi map, and the first result, which is due to Riemann, asserts:

Theorem 1.9. *For $e \in J(X)$ we have $\theta(e,\Pi) = 0$ if and only if there is an integral divisor $\prod_{i=1}^{g-1} P_i$ such that $e = \varphi_{P_0}\left(\prod_{i=1}^{g-1} P_i\right) + K_{P_0}$.*

Proof. We shall first show that $\theta\left(\varphi_{P_0}\left(\prod_{i=1}^{g-1} P_i\right) + K_{P_0}, \Pi\right)$ vanishes on an open set of the symmetric product of the surface with itself $g - 1$ times. This will imply that the function vanishes identically on the space since the function is holomorphic. In order to demonstrate the above we begin by choosing g points R_i, $1 \leq i \leq g$, on the Riemann surface X such that $R_i \neq R_j$ for $i \neq j$ and also satisfying the condition that $i\left(\prod_{i=1}^{g} R_i\right) = 0$. We leave for the reader the problem of showing that

$$i\left(\prod_{i=1}^{g} R_i\right) = 0 \iff \det(\omega_i(R_j))_{ij} \neq 0$$

for any basis ω_i, $1 \leq i \leq g$, for the holomorphic differentials on X. From this it will follow that the set of points R_i, $1 \leq i \leq g$, satisfying the condition is an open set U on the symmetric product of the surface with itself g times.

With this as background consider now $e = \varphi_{P_0}\left(\prod_{i=1}^{g} R_i\right) + K_{P_0}$ and the function $f(P) = \theta(\varphi_{P_0}(P) - e, \Pi)$. By the Riemann Vanishing Theorem (Theorem 1.8) this function either vanishes identically, and if it does not, then its zeros are some g points Q_i, $1 \leq i \leq g$, and $e = \varphi_{P_0}\left(\prod_{i=1}^{g} Q_i\right) + K_{P_0}$. Assume first that the function does not vanish identically. By comparing the two values of e we obtain from the Abel Theorem that $\prod_{i=1}^{g} Q_i / \prod_{i=1}^{g} R_i$ is the divisor of a meromorphic function h on

X, which clearly belongs to $L\left(1/\prod_{i=1}^{g} R_i\right)$. But the condition $i\left(\prod_{i=1}^{g} R_i\right) = 0$ implies from the Riemann–Roch Theorem that this space is 1-dimensional, hence consists only of the constant functions, so that h is constant and the zeros Q_i, $1 \le i \le g$, of $f(P)$ are R_i, $1 \le i \le g$. It follows that for each $1 \le j \le g$,

$$0 = \theta(\varphi_{P_0}(R_j) - e, \Pi) = \theta\left(-\varphi_{P_0}\left(\prod_{i \ne j} R_i\right) - K_{P_0}, \Pi\right)$$

(recall the value of e!), and we clearly get to this conclusion also if we assume that $f(P)$ does vanish identically. Therefore $\theta\left(\varphi_{P_0}\left(\prod_{i=1}^{g-1} P_i\right) + K_{P_0}, \Pi\right)$ vanishes on an open set of the $g-1$ symmetric product of the surface with itself (clearly the set of elements in this symmetric product to which we can add a point and get an element of U from the previous paragraph is open) and therefore vanishes identically there.

For the converse we need to be a bit more clever. Assume therefore that r is a number such that

$$\theta\left(\varphi_{P_0}\left(\prod_{i=1}^{s} P_i\right) - \varphi_{P_0}\left(\prod_{i=1}^{s} Q_i\right) - e, \Pi\right) = 0$$

for all $0 \le s \le r-1$ and points P_i, $1 \le i \le s$, and Q_i, $1 \le i \le s$, and that there are points T_i, $1 \le i \le r$, and S_i, $1 \le i \le r$, such that

$$\theta\left(\varphi_{P_0}\left(\prod_{i=1}^{r} T_i\right) - \varphi_{P_0}\left(\prod_{i=1}^{r} S_i\right) - e, \Pi\right) \ne 0.$$

The assumption we are making is identical vanishing for all divisors of degree $r-1$ or less, but not for all divisors of degree r. Such r always exists and satisfies $r \le g$ by the Jacobi Inversion Theorem. Since $\theta(e, \Pi) = 0$, we have $r \ge 1$. By the continuity of θ and φ_{P_0} we can assume that the points T_i, $1 \le i \le r$, and S_i, $1 \le i \le r$, are all distinct. From the choice of the points S_i, $1 \le i \le r$, and T_i, $1 \le i \le r$, it follows that the function

$$\theta\left(\varphi_{P_0}(P) + \varphi_{P_0}\left(\prod_{i=1}^{r-1} T_i\right) - \varphi_{P_0}\left(\prod_{i=1}^{r} S_i\right) - e, \Pi\right)$$

does not vanish identically (set $P = T_r$) and therefore from the Riemann Vanishing Theorem it has g zeros Q_i, $1 \le i \le g$, and

$$e + \varphi_{P_0}\left(\prod_{i=1}^{r} S_i\right) - \varphi_{P_0}\left(\prod_{i=1}^{r-1} T_i\right) = \varphi_{P_0}\left(\prod_{i=1}^{g} Q_i\right) + K_{P_0}.$$

But it follows from the hypothesis on r that S_i, $1 \le i \le r$, are r distinct zeros of the above function. Hence assume, without loss of generality, that $S_j = Q_{g-r+j}$ for all $1 \le j \le r$ and obtain immediately that

$$e = \varphi_{P_0}\left(\prod_{i=1}^{r-1} T_i \prod_{i=1}^{g-r} Q_i\right) + K_{P_0},$$

hence e is of the desired form.   □

One of the reasons we have given the proof of Theorem 1.9 is that we can now justify the statement $i\left(\prod_{i=1}^{g} Q_i\right) = 0$ in the Riemann Vanishing Theorem where $\prod_{i=1}^{g} Q_i$ is the divisor of zeros there. This is justified by the following

Proposition 1.10. *Suppose* $e = \varphi_{P_0}\left(\prod_{i=1}^{g} Q_i\right) + K_{P_0}$. *Then* $\theta(\varphi_{P_0}(P) - e, \Pi)$ *does not vanish identically on the surface if and only if* $i\left(\prod_{i=1}^{g} Q_i\right) = 0$. *In this case of* $i\left(\prod_{i=1}^{g} Q_i\right) = 0$, *the zero divisor of this function is the integral divisor* $\Delta = \prod_{i=1}^{g} Q_i$ *of degree g.*

Proof. In any case it is clear that if $P = Q_i$, then $\theta(\varphi_{P_0}(P) - e, \Pi) = 0$, since we know that $e - \varphi_{P_0}(Q_i)$ is $\varphi_{P_0}(\Delta) + K_{P_0}$ where Δ is integral of degree $g - 1$—the remaining Q_js—and use Theorem 1.9. Now, if $i\left(\prod_{i=1}^{g} Q_i\right) \geq 1$, then by the Riemann–Roch Theorem we see that the dimension of $L\left(1/\prod_{i=1}^{g} Q_i\right)$ is ≥ 2, so that there is a nonconstant function h in this space. If P is not any of the points Q_i, $1 \leq i \leq g$, then choosing a number $c \in \mathbb{C}$ such that P is a zero of $h - c$ yields an integral divisor Δ of degree g containing P such that $\varphi_{P_0}(\Delta) = \varphi_{P_0}\left(\prod_{i=1}^{g} Q_i\right)$ (by the Abel Theorem), and hence $e - \varphi_{P_0}(P)$ has the presentation $\varphi_{P_0}(\Delta/P) + K_{P_0}$ with Δ/P integral of degree $g - 1$. Thus Theorem 1.9 yields (together with the fact that the theta function is even) the identical vanishing of $\theta(\varphi_{P_0}(P) - e, \Pi) = 0$.

On the other hand, suppose $\theta(\varphi_{P_0}(P) - e, \Pi)$ was identically vanishing on the surface. By choosing a point $P \neq Q_i$ for all $1 \leq i \leq g$, we would clearly have $\theta(\varphi_{P_0}(P) - e, \Pi) = 0$, and since the theta function is even we would also have $\theta(e - \varphi_{P_0}(P), \Pi) = 0$. It would follow from Theorem 1.9 above that

$$e - \varphi_{P_0}(P) = \varphi_{P_0}\left(\prod_{i=1}^{g-1} P_i\right) + K_{P_0}$$

for some choice of P_i, $1 \leq i \leq g - 1$; comparing the two values of e would yield $\varphi_{P_0}\left(\prod_{i=1}^{g} Q_i\right) = \varphi_{P_0}\left(P\prod_{i=1}^{g-1} P_i\right)$. The Abel Theorem would then imply the existence of a meromorphic function h on the surface with poles at $\prod_{i=1}^{g} Q_i$ and zeros at $P\prod_{i=1}^{g-1} P_i$, and since h cannot be constant (since $P \neq Q_i$ for all $1 \leq i \leq g$ we know that its divisor of zeros is different from its divisor of poles) we would obtain $r\left(1/\prod_{i=1}^{g} Q_i\right) \geq 2$. Thus by the Riemann–Roch Theorem $i\left(\prod_{i=1}^{g} Q_i\right) \geq 1$, as we wanted. Note that it suffices to have $f(P) = 0$ for P distinct from all the points Q_i, $1 \leq i \leq g$, to obtain this. Hence if the function is not identically vanishing, then $i\left(\prod_{i=1}^{g} Q_i\right) = 0$ and the function has g zeros, and each of these zeros has to be one of the points Q_i with $1 \leq i \leq g$. Conversely, we have seen that every such point Q_i is a zero of this function.

We now use this information to show that when $i\left(\prod_{i=1}^{g} Q_i\right) = 0$ the divisor of zeros of the function in question is exactly $\prod_{i=1}^{g} Q_i$. First, assume that all the points Q_i,

$1 \leq i \leq g$, are distinct. Then the zeros of this function are all these points, and each with multiplicity one. Since $L(1/\prod_{i=1}^{g} Q_i)$ is 1-dimensional (Riemann–Roch again) and consisting of only the constant functions, the presentation of e as $\varphi_{P_0}(\Delta) + K_{P_0}$ is unique with $\Delta = \prod_{i=1}^{g} Q_i$—the divisor of zeros of the function $\theta(\varphi_{P_0}(P) - e, \Pi)$. For the general case we use an argument of continuity. The condition $i(\prod_{i=1}^{g} Q_i) = 0$ is an open condition on the symmetric product of X with itself g times (an indication for that was seen in the proof of Theorem 1.9), so that we can find a sequence of elements Δ_n, $n \in \mathbb{N}$ of this symmetric product which converges to $\prod_{i=1}^{g} Q_i$ and such that each Δ_n is composed of g distinct points and satisfies $i(\Delta_n) = 0$. Letting $e_n = \varphi_{P_0}(\Delta_n)$ we find that $\theta(\varphi_{P_0}(P) - e_n, \Pi)$ does not vanish identically on X and its divisor of zeros is Δ_n. Taking the limit $n \to \infty$, we obtain the desired result for our point e and divisor $\prod_{i=1}^{g} Q_i$. This completes the proof of the proposition. $\square$

We therefore have reached the following conclusion:

Corollary 1.11. *Suppose* $e = \varphi_{P_0}\left(\prod_{i=1}^{g} Q_i\right) + K_{P_0}$ *with* $i\left(\prod_{i=1}^{g} Q_i\right) = 0$, *and suppose that* $-e = \varphi_{P_0}\left(\prod_{i=1}^{g} R_i\right) + K_{P_0} = \Pi \varepsilon/2 + I\varepsilon'/2$ *for some* g *points* R_i, $1 \leq i \leq g$, *on* X *and vectors* ε *and* ε' *in* $\mathbb{R}^g$. *Then* $\theta\begin{bmatrix} \varepsilon \\ \varepsilon' \end{bmatrix}(\varphi_{P_0}(P), \Pi)$ *does not vanish identically on the surface and the zeros of this function are precisely the points* Q_i, $1 \leq i \leq g$.

Proof. Since by Eq. (1.3) we have $\theta\begin{bmatrix} \varepsilon \\ \varepsilon' \end{bmatrix}(\varphi_{P_0}(P), \Pi) = \mathbf{e}(\cdot)\theta(\varphi_{P_0}(P) - e, \Pi)$, and since $e = \varphi_{P_0}\left(\prod_{i=1}^{g} Q_i\right) + K_{P_0}$ and $i\left(\prod_{i=1}^{g} Q_i\right) = 0$, the result is immediate from Proposition 1.10. $\square$

Theorem 1.9 has another useful corollary, which is that the canonical point with respect to P_0 (which we recall to be the image of a canonical divisor under the map φ_{P_0} and is the same for every canonical divisor) equals $-2K_{P_0}$. For this let us construct an integral canonical divisor (we remark that this can be done only for genus $g \geq 1$, but for genus 0 the jacobian $J(X)$ reduces to one point and the claim is trivial) such that its image will be $-2K_{P_0}$. We take a point e such that $\theta(e, \Pi) = 0$, and then Theorem 1.9 gives that we can write $e = \varphi_{P_0}\left(\prod_{i=1}^{g-1} P_i\right) + K_{P_0}$ for some points P_i, $1 \leq i \leq g - 1$. Since the theta function is even we see that $\theta(-e, \Pi) = 0$ as well, so that Theorem 1.9 gives that $-e = \varphi_{P_0}\left(\prod_{i=1}^{g-1} Q_i\right) + K_{P_0}$ for some points Q_i, $1 \leq i \leq g - 1$. Addition gives $\varphi_{P_0}\left(\prod_{i=1}^{g-1} P_i \prod_{i=1}^{g-1} Q_i\right) = -2K_{P_0}$, and we want to show that this divisor $\Delta = \prod_{i=1}^{g-1} P_i \prod_{i=1}^{g-1} Q_i$ is canonical. Since its degree is $2g - 2$, it suffices to show that $i(\Delta) \geq 1$ (and then equality holds, but this does not matter to us), or equivalently by the Riemann–Roch Theorem, that $r(1/\Delta) \geq g$. Now, the points P_i, $1 \leq i \leq g - 1$, were arbitrary, meaning that moving them a bit to points R_i, $1 \leq i \leq g - 1$, distinct from all the points in the support of Δ, we obtain in this way another divisor $\Xi = \prod_{i=1}^{g-1} R_i \prod_{i=1}^{g-1} S_i$ whose image by φ_{P_0} is again $-2K_{P_0}$. Thus Ξ/Δ is a divisor of degree 0 with φ_{P_0}-image 0, and hence by the Abel Theorem there exists a function whose divisor is Ξ/Δ. In particular we can find in $L(1/\Delta)$ functions whose zeros contain the $g - 1$ arbitrary points R_i, $1 \leq i \leq g - 1$, of Ξ, which gives $g-1$ degrees of freedom in $L(1/\Delta)$. This increases $r(1/\Delta)$ from 1 to at least g, as needed by Riemann–Roch to show that Δ is canonical. Therefore for every canonical divisor Δ (integral or not) we have $\varphi_{P_0}(\Delta) = -2K_{P_0}$, a fact which will be very useful in this book.

We have seen that for an integral divisor Δ of degree g on a Riemann surface there is a significant distinction between the cases $i(\Delta) = 0$ and $i(\Delta) \geq 1$. Since we have seen in the proof of Theorem 1.9 that the condition $i(\Delta) = 0$ is an open condition on the set of integral divisors of degree g, we say that such a divisor is *special* if $i(\Delta) \geq 1$, and *non-special* if $i(\Delta) = 0$. Since we will be interested in theta functions which do not vanish identically on the surface, we understand from Proposition 1.10 that we shall be interested in non-special divisors.

1.3.4 Changing the Basepoint

As in the Abel Theorem, we encounter again a situation in which a notion, on the one hand, depends on some choice of basepoint, and on the other hand, it does not. Explicitly, the function $\theta(\zeta, \Pi)$, with Π being a fixed element of the Siegel upper halfplane of degree g, should not depend on Riemann surface theory considerations, and surely its zeros should not depend on the choice of a basepoint P_0. We have however proved that in the case in which Π comes from a Riemann surface X the divisor of zeros of this function is the set of points $\varphi_{P_0}(\Delta) + K_{P_0}$, where Δ runs over the set of integral divisors of degree $g-1$ on the Riemann surface, and where P_0 is an arbitrarily chosen basepoint for the map of the set of integral divisors of degree $g-1$ into $J(X)$. Clearly if we choose a different basepoint Q we would find that the zeros of the function $\theta(\zeta, \Pi)$ would all be of the form $\varphi_Q(\Delta) + K_Q$. It is therefore clear that the set of points $\varphi_Q(\Delta) + K_Q$ where Δ runs over the set of integral divisors of degree $g-1$, as a subset of $J(X)$, must be independent of the basepoint Q. In the sequel though we shall need to know more.

Theorem 1.12. *Let Δ be a divisor of degree $g-1$ on X. Then the point $\varphi_Q(\Delta) + K_Q$ in $J(X)$ is independent of the choice of the point Q.*

Proof. Begin by choosing a point $e \in \mathbb{C}^g$ such that $\theta(-e, \Pi) \neq 0$. Then the function $\theta(\varphi_{P_0}(P) - e, \Pi)$ does not vanish identically on X, and thus by the Riemann Vanishing Theorem its zeros are at g (not necessarily distinct) points P_i, $1 \leq i \leq g$. In addition, the point e equals $\varphi_{P_0}\left(\prod_{i=1}^g P_i\right) + K_{P_0}$. Let Q be some other basepoint on the surface. Then by Eq. (1.1) (with $\Delta = P$ and of degree 1) we clearly have

$$\theta(\varphi_{P_0}(P) - e, \Pi) \equiv \theta(\varphi_Q(P) - \varphi_Q(P_0) - e, \Pi)$$

for every point $P \in X$, and since $\theta(-e, \Pi) \neq 0$ we can apply the Riemann Vanishing Theorem again to the right-hand side description of this function and obtain

$$e + \varphi_Q(P_0) = \varphi_Q\left(\prod_{i=1}^{g} P_i\right) + K_Q.$$

Combining this with the known value $\varphi_{P_0}\left(\prod_{i=1}^{g} P_i\right) + K_{P_0}$ of e we obtain

$$\varphi_{P_0}\left(\prod_{i=1}^{g} P_i\right) + K_{P_0} = \varphi_Q\left(\prod_{i=1}^{g} P_i\right) - \varphi_Q(P_0) + K_Q,$$

which by Eq. (1.1) equals

$$g\varphi_Q(P_0) + \varphi_{P_0}\left(\prod_{i=1}^{g} P_i\right) - \varphi_Q(P_0) + K_Q = \varphi_{P_0}\left(\prod_{i=1}^{g} P_i\right) + (g-1)\varphi_Q(P_0) + K_Q.$$

This allows us to deduce the equality

$$K_{P_0} = (g-1)\varphi_Q(P_0) + K_Q$$

for any two basepoints P_0 and Q on the Riemann surface X. Now, applying Eq. (1.1) again to our divisor Δ of degree $g-1$ (integral or not, this does not matter) we can easily obtain

$$\varphi_Q(\Delta) + K_Q = (g-1)\varphi_Q(P_0) + \varphi_{P_0}(\Delta) + K_Q = \varphi_{P_0}(\Delta) + K_{P_0},$$

which completes the proof. □

We remark that the main assertion of Theorem 1.12 is the relation between the two vectors of Riemann constants K_{P_0} and K_Q which appears in the proof, regardless of the divisor Δ. However, the applications of Theorem 1.12 will involve divisors Δ of degree $g-1$, and this is why its assertion is formulated in this way. In the next paragraph we apply Theorem 1.12 for integral divisors, but later we shall make use of Theorem 1.12 for non-integral divisors as well, and it is good to know that it holds equally there (and without any additional complication in the proof).

Now, we already knew that the presentation of the set of zeros of the theta function as $\varphi_{P_0}(\Delta) + K_{P_0}$ where Δ runs over the set of integral divisors of degree $g-1$ on X should be independent (as a set) of the basepoint P_0 (as the theta function does not know about it). However, applying Theorem 1.12 to integral divisors (of degree $g-1$) shows us that the presentation of any point $e \in J(X)$ which is a zero of the theta function as $\varphi_{P_0}(\Delta) + K_{P_0}$ is independent of P_0. This means that now only this presentation of the set of zeros of the theta function does not depend on the basepoint chosen, but the presentation of every element in this zero set is independent of it.

Most of the expressions we later obtain will be of the form $\varphi_{P_0}(\Delta) + K_{P_0}$ with Δ an integral divisor of degree g (non-special, but this is of no importance at this point) whose support does not include P_0. In order to be able to change the basepoint, we must replace Δ with some divisor of degree $g-1$. Now, since $\varphi_{P_0}(P_0) = 0$, we find that our expression equals $\psi_{P_0}(\Delta/P_0) + K_{P_0}$, and since Δ/P_0 is a divisor of degree $g - 1$ (non-integral, but as we said before this does not matter), we can apply Theorem 1.12 and obtain that this has the same value in $J(X)$ as $\varphi_Q(\Delta/P_0) + K_Q$ for any point Q in X.

In the sequel however we will be in a situation where P_0 is one point of a finite set of points, and the others will be denoted by P_i with $1 \leq i \leq t$. We will also have the property that for some natural number n, P_i^n/P_j^n is a principal divisor on the Riemann surface X for any i and j. This of course implies that $\varphi_{P_i}(P_j^n) = 0$ for any i and j. We further assume that the divisor Δ (which is still assumed to be integral and of degree g) is supported only on the points P_i, $1 \leq i \leq t$, and that no point appears in Δ to a power n or higher. Let us denote the set of points appearing in Δ to the power $n - 1$ by A, so that $\Delta = A^{n-1} \Xi$ with Ξ being a divisor supported only on the points P_i, $1 \leq i \leq t$, and such that no point appears in it to a power exceeding $n - 2$ (the set A may, in general, be empty—we now assume that it is not). In this expression we have introduced the notation Y^k with Y a set of points on the Riemann surface X and k is an integer to mean the product of the kth powers of the points in Y. We shall usually write divisors as products of such expressions with distinct sets (and then Y^k means that every point in Y appears in the divisor exactly to the kth power), but in general in an expression of the sort $Y^k Z^l$ with Y and Z intersecting the points in the intersection are understood to appear to the power $k + l$. Let us define, for any point $S \in A$, the divisor Γ_S to be $(A \setminus S \cup P_0)^{n-1} \Xi$, which is integral and of the same degree g as Δ (as $\Gamma_S = (P_0^{n-1}/S^{n-1})\Delta$ and S appears in Δ to the $(n - 1)$th power). Note that here we have introduced the shortened notation $Y \cup y$ for $Y \cup \{y\}$ and $Y \setminus y$ for $Y \setminus \{y\}$. We shall be using this shortened notation a lot throughout the book. Now, Theorem 1.12 has the following

Corollary 1.13. *In this notation we have, under these assumptions, the equality* $\varphi_{P_0}(\Delta) + K_{P_0} = \varphi_S(\Gamma_S) + K_S$. *Also, for a divisor Ξ of degree $g + n - 1$ we find the* $\varphi_{P_i}(\Xi) + K_{P_i}$ *gives the same value for every basepoint P_i.*

Proof. As explained above, Theorem 1.12 gives $\varphi_{P_0}(\Delta) + K_{P_0} = \varphi_Q(\Delta/P_0) + K_Q$ for any basepoint Q, and in particular for $Q = S$. Multiplying the numerator and denominator by P_0^{n-1} we find that this is the same as $\varphi_S(P_0^{n-1}\Delta) + K_S$, since $\varphi_S(P_0^n) = 0$. Now, since we have $P_0^{n-1}\Delta = S^{n-1}\Gamma_S$ (clear from the definition) and since $\varphi_S(S) = 0$, we find that our expression is $\varphi_S(\Gamma_S) + K_S$, as desired.

For the second statement we see that $\varphi_{P_i}(\Xi) + K_{P_i} = \varphi_{P_i}(\Xi/P_j^n) + K_{P_i}$ for any fixed branch point P_j (as $\varphi_{P_i}(P_j^n) = 0$), and since the divisor Ξ/P_j^n is of degree $g - 1$, the assertion follows immediately from Theorem 1.12. $\square$

Note that a good way to understand Corollary 1.13 is that when applying the map $\varphi_{P_i}(\cdot) + K_{P_i}$ to the divisor $P_0^{n-1}\Delta = S^{n-1}\Gamma_S$ then the value is the same for every point P_i. This is actually the proof of Corollary 1.13: We divide this divisor by P_j^n in order to obtain a divisor of degree $g - 1$, apply Theorem 1.12, and use the fact that $\varphi_{P_i}(P_j^n) = 0$ for every i and j (which removes the dependence on j).

1.3.5 Matching Characteristics

When proving the Thomae formulae, which is the main result of this book, there will be an emphasized basepoint throughout most of the proof. Corollary 1.13 will be the tool to remove this dependence on the basepoint. However, the role of these divisors in the Thomae formulae will be to represent characteristics of theta functions and constants via the maps $\varphi_{P_i}(\cdot) + K_{P_i}$ for the various points P_i, and the assertion $\varphi_{P_0}(\Delta) + K_{P_0} = \varphi_S(\Gamma_S) + K_S$ of Corollary 1.13 can be interpreted to mean that the divisor Δ with basepoint P_0 represents the same characteristic as the divisor Γ_S with basepoint S. This was proved in Corollary 1.13 only in the case where S appeared in Δ to the power $n - 1$. We now want to show that in general (without any assumption on the power to which P_i appears in Δ), for any integral divisor Δ of degree g supported on the points P_i, $1 \leq i \leq t$, and of some specific form with basepoint P_0, and any other choice of basepoint S, there is an integral divisor Γ_S of degree g supported on the points P_i, $0 \leq i \leq t$, distinct from S and of the same form as Δ such that the equality $\varphi_{P_0}(\Delta) + K_{P_0} = \varphi_S(\Gamma_S) + K_S$ holds. Our assumptions do not suffice to prove such a claim in the general setting, but in the two situations in which we shall make use of this claim, we will also have extra assumptions. We now prove this assertion in two different situations, and we keep the assumptions $\varphi_{P_i}(P_j^n) = 0$ for every i and j in the following two propositions without writing them. We shall use throughout the book the notation $|Y|$ for the cardinality of the (finite) set Y.

The first situation is presented in

Proposition 1.14. *Assume that the number $t + 1$ of points (recall that we start from 0) is a multiple nr of n with $r \geq 1$, that the genus is $g = (n - 1)(nr - 2)/2$, and that $\varphi_{P_i}\left(\prod_{t=0}^{nr-1} P_t\right) = 0$ for every $0 \leq i \leq nr - 1$. Assume that the divisor Δ is $A^{n-1} \prod_{j=0}^{n-2} C_j^{n-2-j}$ where the sets A and C_j, $0 \leq j \leq n-2$, form a partition of the set of points P_i, $1 \leq i \leq nr - 1$, into n sets such that $|A| = r - 1$ and $|C_j| = r$ for every $0 \leq j \leq n - 2$ (note that the degree of Δ is indeed g). Then for every point P_i there is a divisor Γ_{P_i}, of the same form as Δ but with P_i removed instead of P_0, such that the equality $\varphi_{P_0}(\Delta) + K_{P_0} = \varphi_{P_i}(\Gamma_{P_i}) + K_{P_i}$ holds.*

Proof. Corollary 1.13 finishes the proof if $P_i \in A$, and without using the extra assumption $\varphi_{P_i}\left(\prod_{t=0}^{nr-1} P_t\right) = 0$. This is because the divisor Γ_{P_i} there has the same sets C_j with $0 \leq j \leq n - 2$ (hence still with cardinality r and do not contain P_i) and instead of the set A now appears $A \cup P_0 \setminus P_i$, which is of cardinality $r - 1$ and does not contain P_i. Therefore we assume that P_i is in C_k for some $0 \leq k \leq n - 2$.

As we saw in Corollary 1.13, it is more convenient to work with the divisor $P_0^{n-1}\Delta$, which yields the same value under φ_{P_0} and has degree $g + n - 1$. This means that

$$\varphi_{P_0}(\Delta) + K_{P_0} = \varphi_{P_0}(P_0^{n-1}\Delta) + K_{P_0} = \varphi_{P_i}(P_0^{n-1}\Delta) + K_{P_i}$$

by the second assertion of Corollary 1.13. We now recall that having P_i to the power $n - 1$ helps us in finding the divisor Γ_{P_i}, and in Δ (or in $P_0^{n-1}\Delta$) it appears only to the power $n - 2 - k$ (being in C_k). This suggests adding $k + 1$ times the expression

$\varphi_{P_i}\left(\prod_{t=0}^{nr-1} P_t\right)$ (which is 0 and does not change the value), but the divisor thus obtained is of degree $g + n - 1 + (k+1)rn$ and contains some points (P_0, the points in A, and the points in every set C_j with $0 \le j \le k-1$) to the power n or higher. Therefore we shall subtract $\varphi_{P_i}(P_l^n)$ for every such point P_l which now appears to the power n or higher (which is again 0 and does not change the value), and obtain

$$\varphi_{P_0}(\Delta) + K_{P_0} = \varphi_{P_i}\left(P_0^{n-1} \Delta \prod_{t=0}^{nr-1} P_t^{k+1}\right) + K_{P_i} = \varphi_{P_i}\left(P_0^{n-1}\Delta \frac{\prod_{t=0}^{nr-1} P_t^{k+1}}{P_0^n A^n \prod_{j=0}^{k-1} C_j^n}\right) + K_{P_i}.$$

Note that we have subtracted $\varphi_{P_i}(P_l^n)$ for $(k+1)r$ points ($|A| = r-1$, $|C_j| = r$ for $0 \le j \le k-1$ and are k sets, and P_0 is one extra point), which means that the divisor on the right is again of degree $g + n - 1$. This divisor is clearly integral, and since P_i appears in it to the power $n-1$ it is of the form $P_i^{n-1} \Gamma_{P_i}$ with Γ_{P_i} integral, of degree g, and supported on the points P_l, $l \ne i$.

It remains to show that Γ_{P_i} has the same form as Δ, for which we would like to write it explicitly. Following our calculations we see that the points in C_j with $j > k$ and the points in C_k distinct from P_i appear in Γ_{P_i} to a power higher by $k+1$ from the power to which they appeared in Δ, while the points in A and in C_j with $j < k$ appear in Γ_{P_i} to a power lower by $n-1-k$ from the power to which they appeared in Δ. Since P_0 now appears to the power k we can altogether write

$$\Gamma_{P_i} = (C_k \setminus P_i)^{n-1} \left(\prod_{j=k+1}^{n-2} C_j^{n+k-1-j}\right)(A \cup P_0)^k \prod_{j=0}^{k-1} C_j^{k-1-j},$$

(where the parentheses are entered to state on which expressions one should take the products) with no point appearing to the power n or higher. In order to make the comparison of forms easier, we replace the index j by $j - k - 1$ in the first product and by $j + n - k - 1$ in the second product, giving us

$$\Gamma_{P_i} = (C_k \setminus P_i)^{n-1} \left(\prod_{j=0}^{n-3-k} C_{j+k+1}^{n-2-j}\right)(A \cup P_0)^{n-2-(n-2-k)} \prod_{j=n-1-k}^{n-2} C_{j+k+1-n}^{n-2-j}.$$

Now, since $|C_k \setminus P_i| = r - 1$, $|C_j| = r$ for all $j \ne k$ and $|A \cup P_0| = r$, we see that Γ_{P_i} is indeed of the same form as Δ. The fact that $\varphi_{P_i}(P_i) = 0$ shows that we have proved that $\varphi_{P_0}(\Delta) + K_{P_0} = \varphi_{P_i}(\Gamma_{P_i}) + K_{P_i}$ with Γ_{P_i} a divisor with all the desired properties, which proves the proposition.   □

We note that by working with $P_0^{n-1} \Delta$ instead of Δ and $P_i^{n-1} \Gamma_{P_i}$ instead of Γ_{P_i} we simply get a "rotation" in the roles of the sets, and denoting $A \cup P_0$ by C_{-1} gives a more "symmetric" description of the sets participating in the process and of the process itself. We shall indeed adopt this notation C_{-1} later. If we look at the special case where $k = 0$ (i.e., P_i is in C_0), then the action we did in order to move from $P_0^{n-1} \Delta$ to $P_i^{n-1} \Gamma_{P_i}$ is multiplication by $\left(\prod_{i=0}^{nr-1} P_i\right)/C_{-1}^n$, an action which we denote by M. The reader is advised to verify that in the general case what we did was simply to apply this operation M successively $k+1$ times (note the changing of the set C_{-1}

after every application!). It is important to note here that applying M to a divisor Ξ (which we will think of as $P_0^{n-1}\Delta$) satisfying the cardinality conditions $|C_j| = r$ for all $-1 \leq j \leq n-2$ imposed on it by those satisfied by Δ takes Ξ to another divisor which also satisfies these cardinality conditions. Note again that in the special case where P_i is in A, the divisors $P_0^{n-1}\Delta$ and $P_i^{n-1}\Gamma_{P_i}$ are equal, as we saw in the proof of Corollary 1.13. Using $C_{-1} = A \cup P_0$ we see that in the proof of Proposition 1.14 we do not have to separate the case $P_i \in A \subseteq C_{-1}$ from the others—simply substitute $k = -1$ throughout and it all works. We also note that n successive applications of M brings us back to the original divisor (as we multiplied it by $\prod_{t=0}^{nr-1} P_t^n$ and divided it by C_j^n for all the sets C_j, $-1 \leq j \leq n-2$), so that as an operator M is of order n.

We now present the second situation.

Proposition 1.15. *Assume that the number $t+1$ is even and write it as $2m$. Assume further that the genus is $g = (n-1)(m-1)$. Denote the point P_{m-1+i}, $1 \leq i \leq m$ by Q_i, and assume that $\prod_{i=0}^{m-1} P_i / \prod_{i=1}^{m} Q_i$ is a principal divisor. Assume that the divisor Δ is $A^{n-1}D_{-1}^0 \prod_{j=0}^{n-2} C_j^{n-2-j} D_j^{j+1}$ where the sets A and C_j, $0 \leq j \leq n-2$, form a partition of the set of points P_i, $1 \leq i \leq m-1$, into n sets and the sets D_j, $-1 \leq j \leq n-2$, form a partition of the set of points Q_i, $1 \leq i \leq m$, into n sets such that $|D_{-1}| = |A|+1$ and $|C_j| = |D_j|$ for every $0 \leq j \leq n-2$. Then for every point P_i there is a divisor Γ_{P_i}, of the same form as Δ but with P_i removed instead of P_0, such that the equality $\varphi_{P_0}(\Delta) + K_{P_0} = \varphi_{P_i}(\Gamma_{P_i}) + K_{P_i}$ holds, and for every point Q_i there is a divisor Γ_{Q_i}, of the same form as Δ but with Q_i removed instead of P_0, such that the equality $\varphi_{P_0}(\Delta) + K_{P_0} = \varphi_{Q_i}(\Gamma_{Q_i}) + K_{Q_i}$ holds.*

Proof. Here also Corollary 1.13 finishes the proof if $P_i \in A$, and without using the extra assumption that $\prod_{i=0}^{m-1} P_i / \prod_{i=1}^{m} Q_i$ is principal (note that in this case the set which was denoted by A before proving Corollary 1.13 is now $A \cup D_{n-2}$). We leave this for the reader to verify, but it should be noticed in the process that when Q_i is in D_{n-2} the roles of the sets change since the basepoint is now of the other "type". As the reader can indeed check, after doing this basepoint change the set C_j with $0 \leq j \leq n-2$ plays the role of D_{n-3-j} and the set D_j with $-1 \leq j \leq n-3$ plays the role of C_{n-3-j}, and little extra work deals also with A and D_{n-2}. We thus now assume that P_i is in C_k or that Q_i is in D_{k-1} for some $0 \leq k \leq n-2$.

Once again it is more convenient to work with the divisor $P_0^{n-1}\Delta$, which yields the same value under φ_{P_0} and has degree $g+n-1$. This means that

$$\varphi_{P_0}(\Delta) + K_{P_0} = \varphi_{P_0}(P_0^{n-1}\Delta) + K_{P_0} = \varphi_{P_i}(P_0^{n-1}\Delta) + K_{P_i} = \varphi_{Q_i}(P_0^{n-1}\Delta) + K_{Q_i}$$

by the second assertion of Corollary 1.13. We again would like the point P_i to appear to the power $n-1$ or the point Q_i to appear to the power $n-1$ (or simply -1 and then corrected), and in Δ (or in $P_0^{n-1}\Delta$) they appear to the powers $n-2-k$ and k (being in C_k and D_{k-1} respectively). This again suggests adding $k+1$ times the expression $\varphi_S\left(\prod_{i=0}^{m-1} P_i / \prod_{i=1}^{m} Q_i\right)$ for S being P_i or Q_i (which is just 0), but the divisor thus obtained, however still of degree $g+n-1$, is no longer integral (the points in the sets D_j with $-1 \leq j \leq k-1$ appear now to negative powers) and contains some points (P_0, the points in A, and the points in every set C_j with $0 \leq j \leq k-1$) to the power n or higher. Therefore we shall again subtract $\varphi_S(P_l^n)$ for every such point P_l which now appears to the power n or higher (another 0) and add $\varphi_S(Q_l^n)$ for every such point Q_l which now appears to negative power (yet another 0), and obtain

$$\varphi_{P_0}(\Delta) + K_{P_0} = \varphi_S\left(P_0^{n-1}\Delta \frac{\prod_{i=0}^{m-1}P_i^{k+1}}{\prod_{i=1}^{m}Q_i^{k+1}} \frac{\prod_{j=-1}^{k-1}D_j^n}{P_0^n A^n \prod_{j=0}^{k-1}C_j^n}\right) + K_S.$$

Note that we have subtracted $\varphi_S(P_l^n)$ for the same number of points P_l as the number of points Q_l such that we have added $\varphi_S(Q_l^n)$ (this is clear since $|C_j| = |D_j|$ for every $0 \leq j \leq k-1$, $|D_{-1}| = |A| + 1$, and P_0 covers this difference of 1), which means that the divisor on the right is still of degree $g + n - 1$. Note that in both cases $S = P_i$ and $S = Q_i$ this divisor is integral and we have that S appears in the divisor on the right to the power $n - 1$, and thus we can write this divisor as $S^{n-1}\Gamma_S$ with Γ_S integral, of degree g, and supported on the points P_i, $0 \leq i \leq m-1$ and Q_i, $1 \leq i \leq m$ distinct from S.

Again we would like to write the divisor Γ_S explicitly in order to show that it has the same form as Δ. The calculations now show that the points in C_j with $j \geq k$ (from C_k we take only the points distinct from P_i if $S = P_i$) appear in Γ_S to a power higher by $k+1$ from the power to which they appeared in Δ, while the points in A and in C_j with $j < k$ appear in Γ_S to a power lower by $n - 1 - k$ from the power to which they appeared in Δ. Moreover, they show that the points in D_j with $j < k$ (from D_{k-1} we take only the points distinct from Q_i if $S = Q_i$) appear in Γ_S to a power higher by $n - 1 - k$ from the power to which they appeared in Δ, while the points in D_j with $j \geq k$ appear in Γ_S to a power lower by $k+1$ from the power to which they appeared in Δ. Since P_0 now appears to the power k we can altogether write that

$$\Gamma_{P_i} = (C_k \setminus P_i)^{n-1} D_k^0 \left(\prod_{j=k+1}^{n-2} C_j^{n+k-1-j} D_j^{j-k}\right)(A \cup P_0)^k D_{-1}^{n-k-1} \prod_{j=0}^{k-1} C_j^{k-1-j} D_j^{n-k+j}$$

and that

$$\Gamma_{Q_i} = (D_{k-1} \setminus Q_i)^{n-1} C_{k-1}^0 \left(\prod_{j=0}^{k-2} D_j^{j+n-k} C_j^{k-1-j}\right) D_{-1}^{n-k-1}(A \cup P_0)^k \prod_{j=k}^{n-2} D_j^{j-k} C_j^{n+k-1-j}$$

(again with the parentheses entered to state on which expressions one should take the products), both with no point appearing to the power n or higher. Once again we change the index j to $j - k - 1$ in the first product and to $j + n - k - 1$ in the second product of the expression for Γ_{P_i}, while in the expression for Γ_{Q_i} we change j to $k - 2 - j$ in the first product and to $n - 2 - k - j$ in the second product. This allows us to write

$$\Gamma_{P_i} = (C_k \setminus P_i)^{n-1} D_k^0 \left(\prod_{j=0}^{n-3-k} C_{j+k+1}^{n-2-j} D_{j+k+1}^{j+1} \right) (A \cup P_0)^{n-2-(n-2-k)} D_{-1}^{(n-2-k)+1}$$

$$\times \prod_{j=n-1-k}^{n-2} C_{j+k+1-n}^{n-2-j} D_{j+k+1-n}^{j+1}$$

and

$$\Gamma_{Q_i} = (D_{k-1} \setminus Q_i)^{n-1} C_{k-1}^0 \left(\prod_{j=0}^{k-2} D_{k-2-j}^{n-2-j} C_{k-2-j}^{j+1} \right) D_{-1}^{n-2-(k-1)} (A \cup P_0)^{(k-1)+1}$$

$$\times \prod_{j=k}^{n-2} D_{n+k-2-j}^{n-2-j} C_{n+k-2-j}^{j+1}$$

(note that with the basepoint Q_i the sets C_j and D_j change their roles). Now, since $|C_k \setminus P_i| = |D_k| - 1$, $|C_j| = |D_j|$ for $j \neq k$ and $|A \cup P_0| = |D_{-1}|$, we see that Γ_{P_i} is indeed of the same form as Δ. Similarly, since $|D_{k-1} \setminus Q_i| = |C_{k-1}| - 1$, $|D_j| = |C_j|$ for $j \neq k-1$ and $|A \cup P_0| = |D_{-1}|$, we see that Γ_{Q_i} is also of the same form as Δ. Since $\varphi_{P_i}(P_i) = 0$ and $\varphi_{Q_i}(Q_i) = 0$ we have proved that $\varphi_{P_0}(\Delta) + K_{P_0} = \varphi_{P_i}(\Gamma_{P_i}) + K_{P_i}$ with Γ_{P_i} a divisor with all the desired properties and $\varphi_{P_0}(\Delta) + K_{P_0} = \varphi_{Q_i}(\Gamma_{Q_i}) + K_{Q_i}$ with Γ_{Q_i} a divisor with all the desired properties, which proves the proposition. $\qquad\square$

Again let us see how this process looks if we work with $P_0^{n-1}\Delta$ instead of Δ and $P_i^{n-1}\Gamma_{P_i}$ and $Q_i^{n-1}\Gamma_{Q_i}$ instead of Γ_{P_i} and Γ_{Q_i}. In changing the basepoint to P_i we again get a similar "rotation" in the roles of the sets, while in changing the base-point to Q_i we get an interchange in the roles of the points P_l and Q_l and the sets C_j and D_j. Again denoting $A \cup P_0$ by C_{-1} makes this description more "symmetric", and it settles well also with the description of the interchange of roles between the sets when the basepoint is changed to some Q_i. As mentioned above, this notation C_{-1} will indeed be used later. Again if we denote the action of the special case where $k = 0$ (i.e., P_i is in C_0 or Q_i is in D_{-1}) by M, then we see that M multiplies the divisor $P_0^{n-1}\Delta$ by $\left(\prod_{i=0}^{m-1} P_i / \prod_{i=1}^m Q_i \right) \left(D_{-1}^n / C_{-1}^n \right)$ and that in the general case we again apply M successively $k+1$ times (the sets C_{-1} and D_{-1} again change with every application!). In the language of the divisors Ξ (which again should be thought of as $P_0^{n-1}\Delta$), where the cardinality conditions which they satisfy (imposed by those satisfied by Δ) are now $|C_j| = |D_j|$ for all $-1 \leq j \leq n-2$, the M-image of such a divisor also satisfies these cardinality conditions. If P_i is in A or Q_i is in D_{n-2}, we again have equality between the divisor $P_0^{n-1}\Delta$ and the divisor $P_i^{n-1}\Gamma_{P_i}$ or $Q_i^{n-1}\Gamma_{Q_i}$ (see the proof of Corollary 1.13 again). We again write $C_{-1} = A \cup P_0$ and see that the case $P_i \in A \subseteq C_{-1}$ in the proof of Proposition 1.15 works just as well with the substitution $k = -1$. As for changing the basepoint to some Q_i which is in D_{n-2}, we actually have $k = n-1$ rather than $k = -1$, and direct substitution does not give directly multiplication by one but multiplication by $\left(\prod_{i=0}^{m-1} P_i^n / \prod_{i=1}^m Q_i^n \right) \left(\prod_{j=-1}^{n-2} D_j^n / P_0^n A^n \prod_{j=0}^{n-2} C_j^n \right)$. Since, however, we have all the sets and all the points, this indeed brings us back to $P_0^{n-1}\Delta = Q_i^{n-1}\Gamma_{Q_i}$ in this case.

The last calculation shows that also here we have that n successive applications of M brings us back to the original divisor, so that M is again an operator of order n.

We conclude this discussion with a uniqueness assertion. We shall later see that the conditions imposed on the divisors in Propositions 1.14 and 1.15 force these divisors to be non-special, i.e., to satisfy the condition that no integral multiple of them can be canonical. This means that the presentation of the characteristic $\varphi_S(\Gamma) + K_S$ with any basepoint S determines the divisor Γ. This is so because Γ is of degree g, and $i(\Gamma) = 0$ (as we assume) yields $r(1/\Gamma) = 1$. Since Γ is integral we find that $L(1/\Gamma)$ consists only of the constant functions. Therefore if another divisor Λ (again integral of degree g with support not including P_0) gives the same value $\varphi_S(\Lambda) + K_S = \varphi_S(\Gamma) + K_S$ then $\varphi_S(\Lambda/\Gamma) = 0$ and there is a function h whose divisor is Λ/Γ. Therefore this h is in $L(1/\Gamma)$ and hence constant, which means that $\Lambda = \Gamma$ as asserted. Therefore we could have added "unique" to the assertions of Propositions 1.14 and 1.15, and we also know that when the divisor Δ varies over the set of non-special integral divisors of degree g supported on some set of points with one basepoint (say P_0) removed (as it later will), all the characteristics $\varphi_{P_0}(\Delta) + K_{P_0}$ thus obtained are distinct.

Chapter 2

Z_n Curves

Definition 2.1. A Z_n *curve* is a compact Riemann surface which is associated with the algebraic equation

$$w^n = \prod_{i=0}^{rn-1} (z - \lambda_i).$$

If we also assume that $\lambda_i \neq \lambda_j$ whenever $i \neq j$, we shall call the curve a *nonsingular* Z_n *curve*; otherwise we shall call the curve a *singular* Z_n *curve*.

Definition 2.1 assumes that none of the branch points of the projection map z lies over ∞. If we want some of the branch points to lie over ∞, we write the algebraic equation as

$$w^n = \prod_{i=0}^{rn-1-s} (z - \lambda_i),$$

for some number $1 \leq s \leq n-1$. All of this will be written more explicitly and with more details and explanations when needed.

2.1 Nonsingular Z_n Curves

From Definition 2.1 we see that on a nonsingular Z_n curve there are precisely rn branch points, each of order $n-1$. The case where we call a Z_n curve with a branch point over ∞ nonsingular is only when the algebraic equation is

$$w^n = \prod_{i=0}^{rn-2} (z - \lambda_i)$$

with the λ_i distinct, and then we still have precisely rn branch points each of order $n-1$. It is also sometimes convenient to normalize and assume that the points λ_0

H.M. Farkas and S. Zemel, *Generalizations of Thomae's Formula for Z_n Curves*,
Developments in Mathematics 21, DOI 10.1007/978-1-4419-7847-9_2,
© Springer Science+Business Media, LLC 2011

and λ_1 are the points 0 and 1 respectively, but we shall hardly use this normalization in the book.

2.1.1 Functions, Differentials and Weierstrass Points

Since in any event we are dealing with an n-sheeted branched cover of the sphere and the total branching order of this cover equals $rn(n-1)$, the genus g of this curve satisfies $rn(n-1) = 2(n+g-1)$ by the Riemann–Hurwitz formula and hence $g = (n-1)(rn-2)/2$. We shall be assuming that $n \geq 2$. The case $n = 2$ is a well-studied case, the case of hyperelliptic surfaces, with which Thomae dealt and for which he proved the formulae bearing his name. In the end we shall see that if $r = 1$, then certain things do not work, but we can still prove Thomae type formulae for this case as well using the tool of basepoint change.

We now show that we can write down explicitly a basis for the holomorphic differentials on our Z_n curve X. For this we recall that the function z on this surface is just the projection map of the Riemann surface onto the sphere and is the function which gives rise to the representation of this surface as a branched n-sheeted cover of the sphere with rn branch points. We shall denote the unique point over the complex number λ_i by P_i, and if ∞ happens to be the image of a branch point, the point over it will be denoted by P_∞. In order to slightly simplify what follows we shall assume that $\lambda_0 = 0$ and that the point over ∞ is also a branch point. Later we describe what changes when we do not make these assumptions.

These assumptions allow us to calculate the divisors

$$\operatorname{div}(z) = \frac{P_0^n}{P_\infty^n}, \quad \operatorname{div}(w) = \frac{\prod_{i=0}^{rn-2} P_i}{P_\infty^{rn-1}}, \quad \operatorname{div}(dz) = \frac{\prod_{i=0}^{rn-2} P_i^{n-1}}{P_\infty^{n+1}}$$

from which we deduce

$$\operatorname{div}\left(z^l \frac{dz}{w^k}\right) = P_0^{ln+n-1-k} \left(\prod_{i=1}^{rn-2} P_i^{n-1-k}\right) P_\infty^{(kr-l-1)n-1-k}.$$

We observe that if we take k and l which satisfy the inequalities $1 \leq k \leq n-1$ and $0 \leq l \leq rk-2$, then all the powers which appear here are nonnegative, and hence the corresponding differential $z^l dz/w^k$ is holomorphic. We note that if $r = 1$ and $k = 1$, then the inequality for l is $0 \leq l \leq -1$ and there is no such l, which means that no holomorphic differential exists of that form with $k = 1$. This may already signal that the case $r = 1$ is more problematic, although the real problems we encounter in this case are not related to this fact.

We now prove the following

Lemma 2.2. *The set of differentials $z^l dz/w^k$ where k and l run through all the pairs of integers satisfying $1 \leq k \leq n-1$ and $0 \leq l \leq rk-2$ form a basis for the holomor-*

phic differentials on the Z_n curve X. Furthermore, this basis is a basis adapted to the point P_0.

Proof. We first show that we indeed have g differentials in our set. This is so because for each $1 \leq k \leq n-1$ we have $rk-1$ differentials (note that this holds also when $r=1$ and $k=1$). Hence summing over k we have

$$\sum_{k=1}^{n-1}(rk-1) = r\frac{n(n-1)}{2} - (n-1) = \frac{n-1}{2}(rn-2) = g.$$

This shows that the number of differentials is indeed correct, and it remains to show that these differentials are linearly independent.

Perhaps the simplest way of showing that they are independent is to compute the orders of the differentials at the point P_0. Clearly we see from our previous divisor calculations that $\mathrm{Ord}_{P_0}(z^l dz/w^k)$ is just $nl+n-1-k$. We show that from this number we can find the pair k and l which gave rise to it. We denote for a real number x its integral part (the largest integer t such that $t \leq x$) by $\lfloor x \rfloor$ and its fractional part (which is simply $x - \lfloor x \rfloor$) by $\{x\}$, and we consider this number $nl+n-1-k$ again. If we divide it by n and take into consideration the fact that $1 \leq k \leq n-1$, then we find that

$$\left\lfloor \frac{nl+n-1-k}{n} \right\rfloor = l, \qquad \left\{ \frac{nl+n-1-k}{n} \right\} = 1 - \frac{1+k}{n},$$

which gives us back the values of k and l. This means that these orders, the numbers $nl+n-1-k$ for the described pairs of k and l, are indeed all different, which suffices to show linear independence of the holomorphic differentials as we saw in the paragraph preceding Proposition 1.6. Since we found a basis composed of g differentials whose orders at P_0 are all distinct, this basis is adapted to the point P_0, which concludes the proof of the lemma. □

In fact we have actually shown a lot more. In place of the differentials we chose, $z^l dz/w^k$, we could just as well have chosen the differentials $(z-\lambda_i)^l dz/w^k$ for some $1 \leq i \leq nr-2$ and the conclusion would be a basis adapted to the point P_i. When writing the differentials this way we need not assume that $\lambda_0 = 0$ and the conclusion holds just as well. We also note that for any such i, k and l, the order at P_∞ of $(z-\lambda_i)^l dz/w^k$ is $(kr-l-1)n-1-k$, and dividing it by n and again checking the integral and fractional values of the result gives

$$\left\lfloor \frac{(kr-l-1)n-1-k}{n} \right\rfloor = kr-2-l, \qquad \left\{ \frac{(kr-l-1)n-1-k}{n} \right\} = 1 - \frac{1+k}{n}.$$

Thus by similar consideration any such basis (for any i) is also adapted to P_∞. Furthermore we see that no holomorphic differential vanishes to order congruent to $n-1$ modulo n at any of the branch points P_i, since this fractional number $1 - (1+k)/n$ never equals $(n-1)/n$ if $1 \leq k \leq n-1$. This will be important later.

We recall, however, that by Proposition 1.6 the orders of the zeros at the point P_i in a basis adapted to P_i are in fact the numbers $\gamma_j - 1$ where γ_j is the jth gap at P_i. This gives us the following important information.

Lemma 2.3. *The gaps at any branch point P_i are the numbers $nl + n - k$ for k and l satisfying $1 \leq k \leq n - 1$ and $0 \leq l \leq rk - 2$.*

Proof. The truth of the lemma for P_0 follows immediately from the calculations we did, and the truth for any other branch point which does not lie over ∞ follows from the remarks above about how things work exactly the same for all these points. For P_∞ we saw that the orders were $(kr - l - 1)n - 1 - k$, which can be written as $(kr - l - 2)n + n - 1 - k$, for such k and l. When l goes from 0 to $kr - 2$ so does $kr - 2 - l$, and this proves it also for P_∞. Another way to see this is from the fact that the gap sequence is an intrinsic property of the point, and there is no way to distinguish between P_∞ and the other branch points (just change the function z). $\square$

Let us show explicitly, for every $0 \leq l \leq r(n-1) - 2$, which numbers are between $nl + 1$ and $nl + n - 1$ are gaps at every point P_i, $0 \leq i \leq rn - 1$. We claim that these are the numbers between $nl + 1$ and $nl + n - 1 - \lfloor (l+1)/r \rfloor$. Indeed, the inequality $l \leq rk - 2$ is equivalent to $k \geq (l+2)/r$, and hence k is at least the smallest integer that is no less than $(l+2)/r$. However, for every real number x we have that the smallest integer which is no less than x is $\lfloor x \rfloor = x$ is $x \in \mathbb{Z}$ and $\lfloor x \rfloor + 1$ otherwise. Since here we deal with integral multiples of $1/r$ we find that this integral lower bound of k is $\lfloor (l+1)/r \rfloor + 1$. As the gaps are $nl + n - k$ with these k and l (and $k \leq n - 1$) we see that the gap sequence is indeed as stated).

2.1.2 Abel–Jacobi Images of Certain Divisors

There is additional information which we can extract from the above. We have already commented in Chapter 1 that the image of the divisor of a meromorphic differential under the Abel–Jacobi map with basepoint Q is the point $-2K_Q$ in the jacobian variety of the surface. Recall that the vector K_Q arises in the study of theta functions on Riemann surfaces and is the constant by which the image of integral divisors of degree $g - 1$ by φ_Q has to be translated to get the theta divisor, i.e., the zero divisor of $\theta(\zeta, \Pi)$.

Lemma 2.4. *Let P_i be any of the rn branch points of the projection map z. Then K_{P_i} is a point of order 2 in the jacobian variety. Furthermore, the image of any other branch point P_j by the Abel–Jacobi map φ_{P_i} with basepoint P_i is a point of order n in the jacobian variety.*

Proof. The divisor of the holomorphic differential dz/w^{n-1} is $P_\infty^{(rn-1)(n-1)-n-1}$. The power is just $rn^2 - rn - 2n$ which equals $2g - 2$. More important for us though is that the power is a multiple of n. Hence we have $\varphi_{P_i}(P_\infty^{n(rn-r-2)})$ for the image of this divisor in the jacobian variety, which means

$$\varphi_{P_i}(P_\infty^{n(rn-r-2)}) = -2K_{P_i}.$$

It follows from the Abel Theorem that since $z - \lambda_i$ is a meromorphic function on the surface with divisor P_i^n/P_∞^n we have for any basepoint, in particular for the basepoint P_i, that

$$\varphi_{P_i}(P_\infty^n) = \varphi_{P_i}(P_i^n) = 0.$$

Thus also $\varphi_{P_i}(P_\infty^{2g-2}) = \varphi_{P_i}(P_\infty^{n(rn-r-2)}) = 0$. Hence we conclude that for any branch point P_i we have $-2K_{P_i} = 0$ and thus K_{P_i} is a point of order 2. On the way we have also observed that $\varphi_{P_i}(P_\infty^n) = 0$ and since $z - \lambda_j$ is a meromorphic function on the surface with divisor P_j^n/P_∞^n we also find that $\varphi_{P_i}(P_j^n) = \varphi_{P_i}(P_\infty^n) = 0$, so that the image of any branch point P_j under the Abel–Jacobi map with basepoint any other branch point P_i is a point of order n in the jacobian variety. $\qquad\square$

An additional bit of information, which will be quite useful to us, is the following

Lemma 2.5. *If P_i, $0 \le i \le rn - 1$, are the $rn - 1$ branch points with P_{rn-1} being P_∞, then $\varphi_{P_0}\left(\prod_{i=1}^{rn-1} P_i\right) = \varphi_{P_0}\left[\left(\prod_{i=1}^{rn-2} P_i\right)P_\infty\right] = 0$.*

Proof. The divisor of the function w is $\left(\prod_{i=0}^{rn-2} P_i\right)/P_\infty^{rn-1}$, and hence the image of this divisor by φ_{P_0} is 0. By the previous lemma we also have $\varphi_{P_0}(P_\infty^n) = 0$ and thus also $\varphi_{P_0}(P_\infty^{rn}) = 0$, and it is also clear that $\varphi_{P_0}(P_0) = 0$. Combining all this we obtain the assertion of the lemma. $\qquad\square$

We note here that what we shall really need is the vanishing of the expression $\varphi_{P_0}\left[\left(\prod_{i=1}^{rn-2} P_i^{n-2}\right)P_\infty^{n-2}\right]$. This fact can be also proved in another way, using the differential $(z - \lambda_0)^{r-2}dz/w$, whose (canonical) divisor is simply the divisor $P_0^{n(r-1)-2}\left(\prod_{i=1}^{rn-2} P_i^{n-2}\right)P_\infty^{n-2}$ (this is not an integral divisor if $r = 1$, but it makes no difference). Then the fact that $\varphi_{P_0}(P_0) = 0$, together with Lemma 2.4, which states that K_{P_0} is of order 2 in the jacobian variety, finishes the proof.

In the proofs above we have assumed that one of the branch points (chosen to be P_{rn-1}) lies over ∞. We did this to simplify certain expressions. If this is not the case and there is no branching over ∞, then everything still holds except for the explicit formulas of several divisors we used here. This is of course true since different choice of the function z takes one form to another, but since we prefer to prove things more directly, we list the changes that should be done for this case. The genus is of course the same. We denote the n distinct points on X which lie over ∞ by ∞_h with $1 \le h \le n$.

Now, let us as before assume that $\lambda_0 = 0$, and then the divisors of the functions z and w, and of the differential dz, are

$$\frac{P_0^n}{\prod_{h=1}^n \infty_h}, \qquad \frac{\prod_{i=0}^{rn-1} P_i}{\prod_{h=1}^n \infty_h^r}, \qquad \frac{\prod_{i=0}^{rn-1} P_i^{n-1}}{\prod_{h=1}^n \infty_h^2}$$

respectively (and the degree of div(dz) is still $2g - 2$). Furthermore, for any k and l the divisor of the differential $z^l dz / w^k$ is

$$\text{div}\left(z^l \frac{dz}{w^k} \right) = P_0^{ln+n-1-k} \prod_{i=1}^{rn-1} P_i^{n-1-k} \prod_{h=1}^{n} \infty_h^{kr-2-l}.$$

Again we see that this differential is holomorphic if we take $1 \le k \le n - 1$ and $0 \le l \le rk - 2$ (with the same remark about the case $r = 1$ and $k = 1$), and now Lemma 2.2, its proof and the remarks following it go through word-for-word (except for the part dealing with P_∞, which now becomes unnecessary). The remark about looking at another branch point P_i, replacing every z by $z - \lambda_i$, and then removing the assumption of $\lambda_0 = 0$ is valid also here, and Lemma 2.3 and its proof remain true untouched (except for omitting the part concerning P_∞). Lemma 2.4 is again true, and the proof is similar except that the divisor of dz/w^{n-1} is now $\prod_{h=1}^{n} \infty_h^{(n-1)r-2}$. Examining the divisor of the function $z - \lambda_i$ shows that $\varphi_{P_i}(\prod_{h=1}^{n} \infty_h) = 0$, and the last assertion of the lemma is most easily seen using the meromorphic function $(z - \lambda_j)/(z - \lambda_i)$. Lemma 2.5 remains true without the assumption $P_{rn-1} = P_\infty$ (and thus without the expression including P_∞), with a similar proof based on the divisor of w being $\left(\prod_{i=0}^{rn-1} P_i \right) / \left(\prod_{h=1}^{n} \infty_h^r \right)$ and the fact that $\varphi_{P_i}(\prod_{h=1}^{n} \infty_h) = 0$. The remark made after Lemma 2.5 also holds without the assumption $P_{rn-1} = P_\infty$ once we write P_{rn-1} instead of P_∞.

2.2 Non-Special Divisors of Degree g on Nonsingular Z_n Curves

Our objective now is to construct integral divisors of degree g on our Z_n curve. The supports of the divisors are to lie in the set P_i, $1 \le i \le rn - 1$, of branch points with P_0 omitted. We have purposely deleted the point P_0 from our set. The reason is that we will be looking at φ_{P_0} of these divisors and thus we find it more convenient to remove this point from the discussion. We just note that until the Abel–Jacobi map φ_{P_0} enters the picture, all the results equally hold when P_0 appears in the divisors. Indeed, ultimately the point P_0 will return. We do not assume at this point either that the point lying over ∞ is a branch point or that it is not. It will make no difference. We are interested, among these divisors, in those which are *non-special*, i.e., those divisors Δ which satisfy $i(\Delta) = 0$. The reason will be clearer later, but the main reason for this lies in Proposition 1.10 and its corollary. We now present an algorithm for their construction.

We partition the set of points P_i, $1 \le i \le rn - 1$, into n sets, where we call one set A and the others C_j, $0 \le j \le n - 2$. In all our partitions the set A will satisfy $|A| = r - 1$, while the other sets C_j with $0 \le j \le n - 2$ all satisfy $|C_j| = r$. We thus have altogether $r - 1 + (n - 1)r$ points which are indeed all the $rn - 1$ points in the set. We now use this partition to construct an integral divisor of degree g on the surface.

Given such a partition into A and C_j, $0 \le j \le n-2$, the divisor we construct from it is $\Delta = A^{n-1} \prod_{j=0}^{n-2} C_j^{n-2-j}$, where we recall that Y^k means that every element of the set Y appears in the divisor Δ to the kth power. This means that, in Δ, the elements of A appear to the power $n-1$, and the elements of the set C_j with $0 \le j \le n-2$ appear to the power $n-2-j$. In particular this means that the elements of C_{n-2} do not appear at all, and thus can be omitted when writing Δ explicitly. When we calculate the degree of the divisor Δ we find that it is

$$(n-1)(r-1)+r\sum_{j=0}^{n-2} j = (n-1)(r-1)+r\frac{(n-2)(n-1)}{2} = \frac{(n-1)(rn-2)}{2} = g.$$

Our next goal is to prove that the divisors Δ of the type constructed above are non-special, i.e., that $i(\Delta) = 0$, which means explicitly that no holomorphic differential has a divisor which is a multiple of Δ. Before proving this result we shall give an example which explains the main idea of the proof.

2.2.1 An Example with $n = 3$ and $r = 2$

We now explain in detail the simple case with $n = 3$ and $r = 2$. Here the surface is given by the equation

$$w^3 = \prod_{i=0}^{5}(z - \lambda_i).$$

We shall, for the example, normalize so that $\lambda_0 = 0$, $\lambda_1 = 1$ and $\lambda_5 = \infty$ and the other three branch points remain general and will still be denoted λ_2, λ_3, and λ_4. The genus of this surface is $(3-1)(6-2)/2 = 4$, and Lemma 2.2 yields that the four differentials

$$\frac{dz}{w}, \quad \frac{dz}{w^2}, \quad z\frac{dz}{w^2}, \quad z^2\frac{dz}{w^2}$$

form a basis for the space of holomorphic differentials on the Z_3 curve X (the limits on k and l are $1 \le k \le 2$ and $0 \le l \le 2k-2$), and that this basis is adapted to the branch points P_0 and P_∞. Lemma 2.3 yields (examine the number $nl + n - k$ with these limitations on k and l) that the gap sequence at each of the branch points of the curve is the sequence $1,2,4,7$. This indeed fits the results we obtained for this surface by direct calculations in Chapter 1. Of course, as we explained after the proof of Lemma 2.2, this remains true without the assumptions on the specific values of λ_0, λ_1, and λ_5 (with the minor changes needed for the bases to be adapted to the branch points), so we remove these assumptions now.

As for the divisors we construct, we partition the 5 points P_i, $1 \le i \le 5$, into 3 sets A, C_0, and C_1 where $|A| = 1$ and $|C_0| = |C_1| = 2$. We let i, j, k, l and m run through the 5 indices $1,2,3,4,5$ and we assume that they are all distinct. Therefore if A contains the point P_i and C_0 contains the points P_j and P_k, then C_1 contains the two points in the complement of these two sets, i.e., P_l and P_m. The general construction described above yields here the divisor $\Delta = A^2 C_0^1 C_1^0 = P_i^2 P_j P_k$. We now want to show that there is no holomorphic differential whose divisor is a multiple of Δ, and the ideas of the proof here will be used in the general case. We hope that after seeing this example and the one following it the general case will be more understandable.

The genus of X is 4, and hence the divisor of any meromorphic differential on X is $2g - 2 = 6$. This means that if there were a holomorphic differential whose divisor is a multiple of Δ, its divisor would necessarily be $P_i^2 P_j P_k QR$ for some two points Q and R on the surface. Since no holomorphic differential vanishes to order 2 at any of the branch points (since 3 is a non-gap at any one of them) it follows that either Q or R must be P_i. Hence this differential would have to vanish to order 3 at the point P_i. It would thus follow that $P_i^3 P_j P_k Q$ is the divisor of a holomorphic differential. Since we have seen in Section 1.3 that the image of any canonical divisor by φ_{P_t} is $-2K_{P_t}$ for any basepoint (and in particular for a branch point P_t with $0 \leq t \leq 5$) and by Lemma 2.4, K_{P_t} is of order 2 in $J(X)$, we conclude that $\varphi_{P_t}(P_i^3 P_j P_k Q) = 0$. Since $\varphi_{P_t}(P_i^3) = 0$ we also have $\varphi_{P_t}(P_j P_k Q) = 0$. This is true for any basepoint which is one of the branch points of the curve, and in particular taking the basepoint to be P_j, we would find that $\varphi_{P_j}(P_k Q) = 0$. This implies that

$$\varphi_{P_j}\left(\frac{P_k Q}{P_j^2}\right) = 0$$

and then the Abel Theorem yields that $P_k Q / P_j^2$ is the divisor of a meromorphic function on X. But since we saw that 1 and 2 are gaps at the point P_j we find that $r(1/P_j^2) = 1$ and thus $L(1/P_j^2)$ consists only of the constant functions. Therefore no meromorphic function on X can have $P_k Q / P_j^2$ as its divisor (being in $L(1/P_j^2)$ it must be constant and with the trivial divisor 1, which does not equal $P_k Q / P_j^2$ since by assumption $P_k \neq P_j$), so that we obtain the contradiction required to show that our divisor $\Delta = P_i^2 P_j P_k$ is non-special.

2.2.2 An Example with $n = 5$ and $r = 3$

The above example is perhaps too simple so we give another example, which will explain the whole process of the general proof with more details. We take the example of $n = 5$ and $r = 3$, which gives a Z_5 curve of genus $(5-1)(15-2)/2 = 26$. Lemma 2.3 gives the gaps at any of the branch points P_i, which are (the limitations on k and l here are $1 \leq k \leq 4$ and $0 \leq l \leq 3k - 2$) the numbers in the following list: 1,6,11,16,21,26,31,36,41,46,51 (which are congruent to 1 modulo 5), 2,7,12,17,22,27,32,37 (which are congruent to 2 modulo 5), 3,8,13,18,23 (which are congruent to 3 modulo 5) and 4 and 9 (which are congruent to 4 modulo 5).

We now want to examine the divisors of degree $g = 26$ constructed above. Let us take the first partition we can think of (the first 2 branch points P_1 and P_2 will be taken to be the set A, the next 3 branch points P_3, P_4 and P_5 will be to the set C_0, etc.), and we obtain the divisor $\Delta = P_1^4 P_2^4 P_3^3 P_4^3 P_5^3 P_6^2 P_7^2 P_8^2 P_9 P_{10} P_{11}$, which is indeed of degree 26, and we wish to show that it is non-special. We start as before—the assumption $i(\Delta) \geq 1$ implies the existence of an integral divisor Ξ_1 such that $\Delta \Xi_1$ is canonical, and as any canonical divisor is of degree $2g - 2 = 50$ and Δ is of degree 26, Ξ_1 must be of degree 24. Hence we begin with the assumption that

$$P_1^4 P_2^4 P_3^3 P_4^3 P_5^3 P_6^2 P_7^2 P_8^2 P_9 P_{10} P_{11} \Xi_1$$

is the divisor of a holomorphic differential. Here Ξ_1 plays the role of the degree 2 divisor QR in the previous example. Since 5 is a non-gap at every branch point, no holomorphic differential can vanish to order 4 at a branch point. This implies that Ξ_1 must contain the points P_1 and P_2, so that the above divisor must actually be of the form

$$P_1^5 P_2^5 P_3^3 P_4^3 P_5^3 P_6^2 P_7^2 P_8^2 P_9 P_{10} P_{11} \Xi_2$$

with Ξ_2 being of degree 22.

We have also already seen (Section 1.3 and Lemma 2.4) that the map φ_{P_i} with any branch point P_i as the basepoint vanishes on the divisor of any meromorphic differential. Since by Lemma 2.4 it is also true that $\varphi_{P_j}(P_k^5) = 0$ for every j and k, we can conclude that

$$\varphi_{P_i}\left(P_3^3 P_4^3 P_5^3 P_6^2 P_7^2 P_8^2 P_9 P_{10} P_{11} \Xi_2 \right) = 0,$$

and thus it follows that the divisor of degree 0

$$\frac{P_3^3 P_4^3 P_5^3 P_6^2 P_7^2 P_8^2 P_9 P_{10} P_{11} \Xi_2}{P_i^{40}}$$

(since the degree of Ξ_2 is 22, the numerator has indeed degree 40) has image 0 under φ_{P_i} whenever P_i is one of the branch points. In particular choosing P_i as one of the branch points in the set $C_0 = \{P_3, P_4, P_5\}$ yields that we have a meromorphic function whose only pole is at P_i and the order is 37. This is a gap, and this means that there must be a copy of this point P_i in Ξ_2. This would give a pole of order 36 at P_i which is still impossible since 36 is also a gap. The conclusion therefore is that each of the points P_3, P_4, and P_5 must appear in the divisor Ξ_2 and must appear there at least twice. Therefore the original divisor $\Delta \Xi_1$ must be

$$P_1^5 P_2^5 P_3^5 P_4^5 P_5^5 P_6^2 P_7^2 P_8^2 P_9 P_{10} P_{11} \Xi_3$$

with Ξ_3 being of degree 16.

We now repeat the above with basepoint P_i from the set $C_1 = \{P_6, P_7, P_8\}$. The conclusion will be

$$\varphi_{P_i}\left(\frac{P_6^2 P_7^2 P_8^2 P_9 P_{10} P_{11} \Xi_3}{P_i^{25}} \right) = 0$$

(the degree of Ξ_3 is 16), and since 23,22,21 are all gaps at any branch point the conclusion will be that P_6, P_7 and P_8 must all appear in Ξ_3 and must appear there at least 3 times. This implies that the original divisor must be

$$P_1^5 P_2^5 P_3^5 P_4^5 P_5^5 P_6^5 P_7^5 P_8^5 P_9 P_{10} P_{11} \Xi_4$$

with Ξ_4 being of degree 7. The arguments given above now show that

$$\varphi_{P_i}\left(\frac{P_9 P_{10} P_{11} \Xi_4}{P_i^{10}}\right) = 0,$$

and choosing P_i from the set $C_2 = \{P_9, P_{10}, P_{11}\}$ gives rise to functions whose only singularity is at P_i of orders 9,8,7,6 at P_i all of which are gaps. This means that each of the points P_9, P_{10} and P_{11} must appear in Ξ_4 and must appear there at least 4 times, which is clearly impossible since the degree of Ξ_4 is only 7 and cannot contain 12 points (counted with multiplicities). Hence the conclusion is that $i(\Delta) = 0$.

The above examples hopefully will make the following proof of the general statement more transparent and palatable.

2.2.3 Non-Special Divisors

We now present the general statement for nonsingular Z_n curves.

Theorem 2.6. *Let* $\Delta = A^{n-1} \prod_{j=0}^{n-2} C_j^{n-2-j}$ *be an integral divisor on the Riemann surface with the sets A and C_j, $0 \leq j \leq n-2$ formed from a partition of the set of branch points as described above. Then* $i(\Delta) = 0$.

Proof. Assume conversely that $i(\Delta) = i\left(A^{n-1} \prod_{j=0}^{n-2} C_j^{n-2-j}\right) \geq 1$. This implies the existence of a canonical divisor $A^{n-1}\left(\prod_{j=0}^{n-2} C_j^{n-2-j}\right) \Xi_1$ with Ξ_1 an integral divisor of degree $g - 2$. Since no holomorphic differential can vanish to order $n - 1$ at a branch point (since n is not a gap at any branch point by Lemma 2.3), the canonical divisor must actually be of the form $A^n\left(\prod_{j=0}^{n-2} C_j^{n-2-j}\right) \Xi_2$ with Ξ_2 an integral divisor of degree $g - r - 1$. Note that for $n = 2$ this concludes the proof, since $g = r - 1$ in this case and hence the degree of Ξ_2 is -2, which is impossible. We thus assume for the rest of the proof that $n \geq 3$. We can also exclude in what follows the case $n = 3$ and $r = 1$, since then $g = 1$ and $g - r - 1 = -1$ is again negative, so this case is also complete. Now, recall that by Lemma 2.4 K_{P_i} for any branch point is a point of order 2 in the jacobian variety in our case, which implies that for any branch point P_i the image of a canonical divisor under the Abel–Jacobi map with basepoint P_i is 0. Since by Lemma 2.4 we also have that $\varphi_{P_i}(P_j^n) = 0$ for any two branch points P_i and P_j, we have

$$\varphi_{P_i}\left[\left(\prod_{j=0}^{n-2} C_j^{n-2-j}\right) \Xi_2\right] = 0$$

with the degree of $\left(\prod_{j=0}^{n-2} C_j^{n-2-j}\right) \Xi_2$ equal to

$$2g - 2 - n(r-1) = n[r(n-2) - 1]$$

(recall the value of g). It thus follows from the Abel Theorem that the divisor $\left(\prod_{j=0}^{n-2} C_j^{n-2-j}\right) \Xi_2 / P_i^{2g-2-n(r-1)}$, of degree 0, is principal. If we now choose the point P_i to lie in the set C_0, then we find that

$$\frac{(C_0 \setminus P_i)^{n-2} \left(\prod_{j=1}^{n-2} C_j^{n-2-j}\right) \Xi_2}{P_i^{n[(n-2)r-2]+2}}$$

is principal (note that we removed P_i^{n-2} from both the numerator and denominator).

Now, the statement that the above divisor is principal implies that we have a meromorphic function on the Riemann surface whose only singularity is a pole of order $n[(n-2)r-2] + 2$ at P_i, which is impossible since Lemma 2.3 shows that this order is a gap at each branch point P_i (it comes from the pair $k = n-2$ and $l = (n-2)r - 2$, which satisfy $l \leq rk - 2$, and since $n \geq 3$ and the case $n = 3$ and $r = 1$ is excluded, also $l \geq 0$). This means that the point P_i must be contained in the support of Ξ_2 for each P_i in C_0, and since $n[(n-2)r-2] + 1$ is also a gap at each branch point P_i (with $k = n-1$ and $l = (n-2)r - 2$, which satisfy $0 \leq l \leq rk-2$), we find that each P_i in C_0 must appear in Ξ_2 at least twice. Thus the original canonical divisor must be of the form $A^n C_0^n \left(\prod_{j=1}^{n-2} C_j^{n-2-j}\right) \Xi_3$ with the degree of Ξ_3 being $g - r - 1 - 2r = g - 3r - 1$ and the degree of $\left(\prod_{j=1}^{n-2} C_j^{n-2-j}\right) \Xi_3$ being

$$2g - 2 - n(r-1) - rn = n[r(n-3) - 1].$$

Note that this concludes the case $n = 3$, since then $g = 3r - 2$ and hence Ξ_3 is of degree -3, which is impossible. Therefore we now assume $n \geq 4$. We also note that for $n = 4$ and $r = 1$ the genus is $g = 3$ and $g - 3r - 1 = -1$ is again negative, completing this case as well and allowing us to exclude it from the rest of the proof.

A repeat of the previous arguments will now give that

$$\frac{\left(\prod_{j=1}^{n-2} C_j^{n-2-j}\right) \Xi_3}{P_i^{n[r(n-3)-1]}}$$

is principal for any branch point P_i (using again $\varphi_{P_i}(P_j^n) = 0$ and calculating the degrees), and then choosing P_i to lie in the set C_1 would give that

$$\frac{(C_1 \setminus P_i)^{n-3} \left(\prod_{j=2}^{n-2} C_j^{n-2-j}\right) \Xi_3}{P_i^{n[(n-3)r-2]+3}}$$

is principal (canceling P_i^{n-3}). The fact that Lemma 2.3 shows that the numbers $n[(n-3)r-2] + 3$, $n[(n-3)r-2] + 2$ and $n[(n-3)r-2] + 1$ are all gaps (with $l = (n-3)r - 2$ and k being $n-3$, $n-2$ and $n-1$ respectively—the inequality

$l \le rk - 2$ holds for each of them, and since $n \ge 4$ and we exclude the case $n = 4$ and $r = 1$ we also have $l \ge 0$) implies that each branch point P_i from C_1 must appear in Ξ_3, and must appear there at least 3 times. Therefore our original canonical divisor must be of the form $A^n C_0^n C_1^n \left(\prod_{j=2}^{n-2} C_j^{n-2-j} \right) \Xi_4$ with the degree of the divisor Ξ_4 being $g - 3r - 1 - 3r = g - 6r - 1$ and the degree of $\left(\prod_{j=2}^{n-2} C_j^{n-2-j} \right) \Xi_4$ being

$$2g - 2 - n(r - 1) - 2rn = n[r(n - 4) - 1].$$

Since for $n = 4$ the genus is $6r - 3$ and the degree of Ξ_4 is -4 and negative, this case is now also complete and will be excluded in what follows. Similarly for $n = 5$ and $r = 1$, the genus is 6 and the degree of Ξ_4 is $g - 6r - 1 = -1$ and negative, so that this case is done as well.

We now proceed by induction. If $n = 5$, then skip this paragraph since we are standing now at the point described in the beginning of the next paragraph. Alternatively, one can treat the last two paragraphs as two induction steps (as we see in a second) and suppose only that $n \ge 4$ and we are with the divisor denoted Ξ_2 (and for $n = 3$ one jumps to the following paragraph). Now, suppose that we have shown that our canonical divisor is of the form

$$A^n \prod_{j=0}^{s-1} C_j^n \left(\prod_{j=s}^{n-2} C_j^{n-2-j} \right) \Xi_{s+2}$$

for some $s \le n - 4$, with the degree of Ξ_{s+2} being $g - 1 - (s+1)(s+2)r/2$ and the degree of the divisor $\left(\prod_{j=s}^{n-2} C_j^{n-2-j} \right) \Xi_{s+2}$, the "remainder", being equal to

$$2g - 2 - n(r - 1) - srn = n[r(n - (s+2)) - 1]$$

(note that the last paragraph ended with exactly this conclusion for $s = 2$ and the previous one ended with exactly this conclusion for $s = 1$). Then for any branch point P_i in C_s we find that

$$\frac{(C_s \setminus P_i)^{n-(s+2)} \left(\prod_{j=s+1}^{n-2} C_j^{n-2-j} \right) \Xi_{s+2}}{P_i^{n[r(n-(s+2))-2]+s+2}}$$

is principal (cancel P_i^{n-2-s}). Now, the numbers between $n[r(n - (s+2)) - 2] + 1$ and $n[r(n - (s+2)) - 2] + s + 2$ are all gaps by Lemma 2.3, with $l = r(n - (s+2)) - 2$ and k taking all the values from $n - (s+2)$ to $n - 1$; we have $l \le rk - 2$ for any of these pairs, and since $n \ge s + 4$, we also have $l \ge 0$. Hence we find that each of these branch points P_i from C_s must appear in the divisor Ξ_{s+2}, and must appear at least $s + 2$ times in order to avoid all these gaps. This shows us that our canonical divisor is of the form

$$A^n \prod_{j=0}^{s} C_j^n \left(\prod_{j=s+1}^{n-2} C_j^{n-2-j} \right) \Xi_{s+3}$$

with the degree of Ξ_{s+3} being

$$g - \frac{(s+1)(s+2)}{2}r - 1 - (s+2)r = g - \frac{(s+2)(s+3)}{2}r - 1$$

and the degree of $\left(\prod_{j=s+1}^{n-2} C_j^{n-2-j}\right) \Xi_{s+3}$ being $2g - 2 - n(r-1) - (s+1)rn$. This sets us exactly at the next step of the induction, as we wanted.

Let us continue the induction until we reach $s = n - 3$, i.e., after doing induction step of $s = n - 4$. At this stage our canonical divisor is

$$A^n \prod_{j=0}^{n-4} C_j^n \left(\prod_{j=n-3}^{n-2} C_j^{n-2-j}\right) \Xi_{n-1} = A^n \left(\prod_{j=0}^{n-4} C_j^n\right) C_{n-3} \Xi_{n-1}$$

with the degree of Ξ_{n-1} being

$$g - \frac{(n-2)(n-1)}{2}r - 1 = (n-1)(r-1) - 1$$

and the degree of $C_{n-3} \Xi_{n-1}$ being

$$2g - 2 - n(r-1) - (n-3)rn = n(r-1)$$

(note that the difference between the degrees is indeed r). Now, for $r = 1$ the degree of Ξ_{n-1} is -1, which concludes the proof. Assume now $r \geq 2$, and we show how the induction step from $s = n - 3$ to $s = n - 2$ yields the desired contradiction. Again we find that for every branch point P_i the divisor $C_{n-3} \Xi_{n-1}/P_i^{n(r-1)}$ is principal, and if P_i is in C_{n-3}, then the divisor $(C_{n-3} \setminus P_i) \Xi_{n-1}/P_i^{n(r-1)-1}$ is principal. However, since by Lemma 2.3 the numbers between $n(r-2)+1$ and $n(r-2)+n-1 = n(r-1)-1$ are all gaps (with $l = r - 2 \geq 0$ and k taking all the values from 1 to $n - 1$, which means in particular $l \leq rk - 2$), we find that each of these branch points P_i from C_{n-3}, must appear in the divisor Ξ_{n-1} and must appear at least $n - 1$ times in order to avoid all these gaps. But this means that the degree of Ξ_{n-1} has to be at least $r(n-1)$, which contradicts the fact that this degree is only $(n-1)(r-1) - 1$. This contradiction shows that Δ cannot be special, and hence $i(\Delta) = 0$ as desired. ☐

There is perhaps a simpler proof that we can give. We recall that Lemma 2.2 shows that for a given branch point P_i the differentials $(z - \lambda_i)^l dz/w^k$ for which $1 \leq k \leq n - 1$ and $0 \leq l \leq rk - 2$ form a basis for the holomorphic differentials on the Z_n curve X which is adapted to the point P_i and that the order of such a differential at P_i is $nl + n - 1 - k$. Actually we can say that any differential of the form $p_k(z)dz/w^k$ with $p_k(z)$ a polynomial in z is holomorphic if the degree of p_k does not exceed $rk - 2$; it is easily seen from these considerations that for any branch point P_i we have

$$\mathrm{Ord}_{P_i}\left(p_k(z)\frac{dz}{w^k}\right) = n\mathrm{Ord}_{\lambda_i} p_k + n - 1 - k.$$

In particular we see that this order is congruent to $n - 1 - k$ modulo n. We also notice that if we have a branch point lying over ω, then the calculation of orders at P_∞ yields that

$$\mathrm{Ord}_{P_\infty}\left(p_k(z)\frac{dz}{w^k}\right) = n(kr - 2 - \deg p_k) + n - 1 - k.$$

In particular this order is also congruent to $n - 1 - k$ modulo n.

It will be easier to describe the second proof of Theorem 2.6 if we introduce some notation. For a given integral divisor $\Xi = \prod_P P^{l_P}$, we recall that $\Omega(\Xi)$ denotes the (finite-dimensional) vector space of holomorphic differentials whose divisors are multiples of Ξ. For $1 \le k \le n - 1$ denote by $\Omega_k(\Xi)$ the subspace of $\Omega(\Xi)$ composed of differentials of the form $p_k(z)dz/w^k$ with $p_k(z)$ a polynomial in z. In order for $p_k(z)dz/w^k$ to be holomorphic, it is necessary for the degree of the polynomial p_k to be bounded by $rk - 2$, otherwise the differential will have poles over the point $z = \infty$. It is clear that the subspaces $\Omega_k(\Xi)$ are disjoint, and thus we can write for any Ξ that $\bigoplus_{k=0}^{n-1} \Omega_k(\Xi) \subseteq \Omega(\Xi)$. In general this inclusion can be strict, but we will soon see that in the cases interesting us this inclusion will actually be an equality. Note for example that the space $\Omega(1)$ of holomorphic differentials can be decomposed as $\bigoplus_{k=1}^{n-1} \Omega_k(1)$ (so that for the divisor 1 we do have equality) and that $\Omega_k(1)$ consists of all the differentials $p_k(z)dz/w^k$ with p_k a polynomial of degree not exceeding $rk - 2$ in z.

We now prove

Lemma 2.7. *For any integral divisor Ξ with support in the branch points we have* $\Omega(\Xi) = \bigoplus_{k=1}^{n-1} \Omega_k(\Xi)$.

Proof. As noted above, the inclusion $\supseteq$ is clear. In order to prove $\subseteq$, we note that the elements of $\Omega(\Xi)$ are holomorphic differentials, and hence we can decompose any differential $\omega \in \Omega(\Xi)$ as $\omega = \sum_{k=1}^{n-1} \omega_k$ with $\omega_k \in \Omega_k(1)$. We would like to show that ω_k actually lies in $\Omega_k(\Xi)$, which would show that $\omega \in \bigoplus_{k=1}^{n-1} \Omega_k(\Xi)$, as we wish. For this we note that a differential in $\Omega(1)$ lies in $\Omega(\Xi)$ if and only if $\mathrm{Ord}_{P_i}\omega \ge l_i$ for every $0 \le i \le rn - 1$ where $\Xi = \prod_{i=0}^{rn-1} P_i^{l_i}$ (recall that the support of Ξ consists only of branch points), and clearly $\Omega_k(\Xi) = \Omega(\Xi) \cap \Omega_k(1)$. Therefore it suffices to prove that $\mathrm{Ord}_{P_i}\omega_k \ge l_i$ for every branch point P_i. We already saw that $\mathrm{Ord}_{P_i}\omega_k$ is congruent to $n - 1 - k$ modulo n, which means that at a given point P_i, the numbers $\mathrm{Ord}_{P_i}\omega_k$ with $1 \le k \le n - 1$ are all distinct (as their residues modulo n are all distinct). This means that $\mathrm{Ord}_{P_i}\omega$ must be exactly the minimal among the numbers $\mathrm{Ord}_{P_i}\omega_k$. Obviously $\mathrm{Ord}_{P_i}\omega$ is not smaller than this minimum, but since only differentials with the same order at P_i can "cancel each other" there and allow their sum to have a larger order there, this does not happen here and we have equality. Since we know that $\mathrm{Ord}_{P_i}\omega \ge l_i$ for every branch point P_i (as we started from $\omega \in \Omega(\Xi)$), it thus follows that $\min_{1 \le k \le n-1} \mathrm{Ord}_{P_i}\omega_k \ge l_i$ for every branch point P_i. Therefore we obtain that $\mathrm{Ord}_{P_i}\omega_k \ge l_i$ for every $1 \le k \le n - 1$ and branch point P_i. Hence $\omega_k \in \Omega(\Xi)$ for every $1 \le k \le n - 1$, and since clearly $\omega_k \in \Omega_k(1)$, we have $\omega_k \in \Omega_k(\Xi)$, as desired. $\square$

We now turn to the second proof of Theorem 2.6. Assume that Δ is the integral divisor $A^{n-1} \prod_{j=0}^{n-2} C_j^{n-2-j}$ with the cardinality conditions $|A| = r - 1$ and $|C_j| = r$ for all $0 \le j \le n - 2$. Lemma 2.7 shows that in order to prove that $i(\Delta) = 0$ it suffices to show that $\Omega_k(\Delta)$ is $\{0\}$ for every $1 \le k \le n-1$. First, we assume that there is no branch point lying over ∞. We choose any differential $\omega_k = p_k(z)\mathrm{d}z/w^k$ in $\Omega_k(\Delta)$ and we want to show that $\omega_k = 0$, i.e., that $p_k(z) = 0$. We know that $\mathrm{Ord}_{P_i}\omega_k$ is $n-1-k$ if $p_k(\lambda_i) \ne 0$ and is at least n if $p_k(\lambda_i) = 0$. The assumption that ω_k is in $\Omega_k(\Delta)$ implies that $\mathrm{Ord}_{P_i}\omega_k \ge n-1$ for every branch point P_i which lies in Λ and $\mathrm{Ord}_{P_i}\omega_k \ge n-2-j$ for every $0 \le j \le n-2$ and branch point P_i in C_j. Now, if P_i is in C_j with $k-1 \le j \le n-2$, then the order $n-1-k$ suffices for the point P_i in any case, but if P_i is in A or in C_j with $0 \le j \le k-2$ (this means only P_i from A when $k=1$) then $n-1-k$ cannot be the order of ω_k at P_i and thus every corresponding λ_i must be a root of the polynomial $p_k(z)$. But the set $A \cup \bigcup_{j=0}^{k-2} C_j$ contains $kr-1$ points (recall that $|A| = r-1$ and $|C_j| = r$ for every $0 \le j \le n-2$) [note that this holds for $k=1$ as well] and the degree of p_k cannot exceed $rk-2$. Since a polynomial of degree not exceeding $rk-2$ which vanishes at $rk-1$ distinct points must be the zero polynomial, we are done.

It remains to see what happens if there is a branch point which lies over ∞. Recall that the order of the differential ω_k at P_∞ is $n-1-k$ plus n times the "unused degrees" of p_k (this is a useful interpretation for the number $rk-2-\deg p_k$ which appears there). Now, if $P_\infty \in C_j$ with $k-1 \le j \le n-2$, then everything goes as before, since the order $n-1-k$ suffices for P_∞; but if P_∞ lies in A or in C_j with $0 \le j \le k-2$, then the polynomial p_k must vanish only at $rk-2$ points (those in the union $A \cup \bigcup_{j=0}^{k-2} C_j$ aside from P_∞). On the other hand, if this is the case with P_∞, then the order $n-1-k$ does not suffice for P_∞, and we must increase it by forcing $p_k(z)$ to have at least one "unused degree". Hence we remain with a polynomial p_k of degree not exceeding $rk-3$ which must vanish at $rk-2$ points, so that p_k is the zero polynomial and $\omega_k = 0$ as needed.

2.2.4 Characterizing All Non-Special Divisors

In fact the converse of the above theorem is also true. We begin with a lemma concerning partitions.

Lemma 2.8. *Assume that we have a partition of the $rn-1$ branch points P_i with $1 \le i \le rn-1$ into n sets, one denoted by A and the others by C_j, $0 \le j \le n-2$. Assume that*

$$(n-1)|A| + \sum_{j=0}^{n-3}(n-2-j)|C_j| = \frac{(n-1)(rn-2)}{2}.$$

Then either we have $|A| = r-1$ and $|C_j| = r$ for every $0 \le j \le n-2$, or the inequality

$$|A| + \sum_{j=0}^{k-2}|C_j| \le kr-2$$

holds for some $1 \leq k \leq n-1$ *(with $k=1$ giving the inequality $|A| \leq r-2$).*

Proof. Suppose none of the inequalities hold. Then for every $1 \leq k \leq n-1$ we have the reverse inequality (recall that we deal with integers)

$$|A| + \sum_{j=0}^{k-2} |C_j| \geq rk - 1$$

(i.e., $|A| \geq r-1$, $|A| + |C_0| \geq 2r-1$, etc.). When adding these inequalities we find that on the left-hand side, with the cardinalities, $|A|$ appears in all the $n-1$ inequalities and the set C_j with $0 \leq j \leq n-3$ appears in exactly $n-2-j$ inequalities, so that this side sums up to the left-hand side in the assumption of the lemma. On the right-hand side we have, on the other hand, the sum $\sum_{k=1}^{n-3}(rk-1)$, which sums up to $g = (n-1)(rn-2)/2$ as we saw when counting the differentials before proving Lemma 2.2. This yields the inequality

$$(n-1)|A| + \sum_{j=0}^{n-3} (n-2-j)|C_j| \geq \frac{(n-1)(rn-2)}{2},$$

and the only way for this to agree with the assumption of the lemma is if all the inequalities are in fact equalities. This means that $|A| = r-1$, and then by induction on l we can easily deduce that $|C_l| = r$ for all $0 \leq l \leq n-3$: if this is true for all $j < l$, then the $(l+2)$th equality yields $|A| + \sum_{j=0}^{l}|C_j| = rk - 1$, from which subtracting $|A| = r-1$ and $|C_j| = r$ for all $0 \leq j \leq l-1$ easily gives $|C_l| = r$. The remaining set C_{n-2} is the complement of the set $A \cup \bigcup_{j=0}^{n-3} C_j$ of cardinality $(n-1)r-1$ in the set of branch points P_i, $1 \leq i \leq rn-1$ of cardinality $rn-1$; hence $|C_{n-2}| = r$ as well. This proves the lemma. $\qquad\square$

We now prove the converse of Theorem 2.6.

Theorem 2.9. *Every non-special divisor of degree g with support in the branch points distinct from P_0 is of the form $\Delta = A^{n-1} \prod_{j=0}^{n-2} C_j^{n-2-j}$ where A and C_j, $0 \leq j \leq n-2$, is a partition of the set of branch points P_i, $1 \leq i \leq rn-1$, with $|A| = r-1$ and $|C_j| = r$ for all $0 \leq j \leq n-2$.*

Proof. We first have to prove that $\Delta = A^{n-1} \prod_{j=0}^{n-2} C_j^{n-2-j}$ for some sets of branch points A and C_j, $0 \leq j \leq n-2$. This means that no branch point can appear to power n or larger at a non-special divisor. Indeed, if P_i appears to the power n or larger in Δ, then we have $L(1/P_i^n) \subseteq L(1/\Delta)$ (clearly) and since we know that the former contains a nonconstant function $(1/(z-\lambda_i)$, for example, if $\lambda_i \neq \infty$, and z for example if $\lambda_i = \infty$), so does the latter, and thus $r(1/\Delta) \geq 2$. Since the degree of Δ is g, the Riemann–Roch Theorem yields that $i(\Delta) \geq 1$ and Δ is special. Hence if Δ is non-special, then no point appears to power n or larger in Δ, so we can denote the set of branch points appearing in Δ to the power $n-1$ by A and the set of branch points appearing in Δ to the power $n-2-j$ by C_j for every $0 \leq j \leq n-3$ and write $\Delta = A^{n-1} \prod_{j=0}^{n-3} C_j^{n-2-j}$. By calling the complement C_{n-2}, we obtain a partition of the set of branch points P_i, $1 \leq i \leq rn - 1$ into n sets, and since the points in C_{n-2} do not appear in Δ we can equally take the product describing it until $n - 2$ and not $n - 3$. It remains to prove that the cardinality conditions hold.

We now invoke Lemma 2.8. The equality one assumes there is satisfied since the degree of Δ is $g = (n - 1)(rn - 2)/2$, and if we contradict all the inequalities in it the lemma will finish the proof. First assume that no branch point lies over ∞. Now, if we assume that the kth inequality holds, then take $p_k(z)$ to be the product of the linear expressions $z - \lambda_i$ for every P_i which lies in A or in C_j with $0 \leq j \leq k - 2$ (again this is only A if $k = 1$) and define ω_k to be, as usual, $p_k(z)\mathrm{d}z/w^k$. Since by the kth inequality the degree of p_k does not exceed $rk - 2$, this is a holomorphic differential. As we have seen in the second proof of Theorem 2.6, for every point P_i which lies in C_j with $k - 1 \leq j \leq n - 2$, the order $n - 1 - k$ which ω_k has at P_i suffices, and for a point P_i in A or in C_j with $0 \leq j \leq k - 2$, the fact that $p_k(\lambda_i) = 0$ gives ω_k a high enough order as well. Since ω_k vanishes at every branch point to a high enough order, we find that ω_k is a nonzero differential in $\Omega(\Delta)$ (and actually in $\Omega_k(\Delta)$), and in particular $i(\Delta) \geq 1$ and Δ is special.

If we do have a branch point over ∞ and the kth inequality holds, then if P_∞ is in C_j with $k - 1 \leq j \leq n - 2$, then everything is as before (the order $n - 1 - k$ which ω_k has at P_∞ suffices), and if P_∞ is in A or in C_j with $0 \leq j \leq k - 2$, then we take $p_k(z)$ again to be the same polynomial but with the meaningless expression $z - \infty$ omitted (and ω_k as before). Now, the kth inequality assures that the degree of p_k is now at most $rk - 3$, the same considerations for the branch points distinct from P_∞ hold, and at P_∞ the order is at least n since $p_k(z)$ has an "unused degree". Therefore also here this ω_k is in $\Omega(\Delta)$ (and actually in $\Omega_k(\Delta)$) and Δ is special.

We found that any of the inequalities in Lemma 2.8 gives that Δ is special, and hence if Δ is non-special, then none of these inequalities can hold. Since the fact that the degree of Δ is g gives the equation which the lemma assumes, the lemma assures us that we have $|A| = r - 1$ and $|C_j| = r$ for every $0 \leq j \leq n - 2$, which finishes the proof. $\square$

We note that all the explicit differentials we found in $\Omega(\Delta)$ in the proof of Theorem 2.9 were in fact in $\Omega_k(\Delta)$ for some $1 \leq k \leq n - 1$, which emphasizes the assertion of Lemma 2.7.

We emphasize that for what we have done here the fact that the specific point P_0 was excluded from the support plays absolutely no role, and therefore all of our statements in this section still hold also when P_0 appears in the support. This is so since all the branch points are clearly symmetric. We also remark that in every non-special divisor supported on the branch points at least one has to be omitted. This is so, since $r(1/\prod_{i=0}^{rn-1} P_i) \geq 2$ (it contains the constant function and the function $1/w$), and therefore for any multiple Ξ of it we have $r(1/\Xi) \geq 2$. Hence if Δ is of degree g and contains every branch point in its support, then $r(1/\Delta) \geq 2$, and thus by Riemann–Roch $i(\Delta) \geq 1$ and Δ is special. Note that the condition that Δ contains every branch point in its support is impossible for $n \leq 3$, and also for $n = 4$ and $r = 1$, since $g < rn$ in these cases; hence then this claim is trivial, and the proof we gave deals with the remaining cases. Therefore we know that in the non-special divisors of degree g, at least one branch point must not appear, and until now it did not matter whether this missing branch point was P_0 or not. However, for what follows, the property of P_0 not being in the support of any divisor Δ will be needed, and therefore we shall keep assuming that P_0 does not appear in the support of any of our divisors. In the sequel, however, we will wish to have divisors Δ which in fact do contain the point P_0, and the reader should be aware that this is not a problem. As we have indicated, our removal of P_0 at this point is due to the fact that the basepoint of the Abel–Jacobi map will be taken as P_0, and this motivates removing this point from our considerations at this point.

2.3 Singular Z_n Curves

We now turn our attention to the case of singular Z_n curves. Definition 2.1 gives a very wide definition for singular Z_n curves, but we restrict our attention in this chapter to those that are associated with algebraic equations of the kind

$$w^n = \prod_{i=0}^{m-1}(z-\lambda_i) \prod_{i=1}^{m}(z-\mu_i)^{n-1},$$

where $\lambda_i \neq \lambda_j$ for $i \neq j$, $\mu_i \neq \mu_j$ for $i \neq j$, $\lambda_i \neq \mu_j$ for every i and j and $m \geq 2$ (the last assumption being made to make sure the genus is positive and not 0). One may want to assume that $n \geq 3$ here, for otherwise this is just a nonsingular Z_2 curve with $r = m$, but this will not be important for anything in the proof, so we do not assume it. Clearly all the points P_i, $0 \leq i \leq m-1$, are symmetric and all the points Q_i, $1 \leq i \leq m$, are symmetric, but we also claim that all these points are symmetric to one another. This is so since we can replace the function w by the function $\prod_{t=0}^{m-1}(z-\lambda_t)\big(\prod_{t=1}^{m}(z-\mu_t)\big)/w$, which satisfies

$$\left(\frac{\prod_{t=0}^{m-1}(z-\lambda_t)\prod_{t=1}^{m}(z-\mu_t)}{w}\right)^n = \prod_{i=0}^{m-1}(z-\lambda_i)^{n-1} \prod_{i=1}^{m}(z-\mu_i),$$

and interchanges the role of the points P_i, $0 \leq i \leq m-1$, with the role of the points Q_i, $1 \leq i \leq m$. We chose the indices in this way since we shall later take P_0 to be our basepoint, but it will be useful later to know that everything we do holds equally with any other choice of basepoint, either a P_i or a Q_i.

2.3.1 Functions, Differentials and Weierstrass Points

This Riemann surface is an n-sheeted cover of the sphere with precisely $2m$ branch points, and there is no branching over the point at ∞. Since the total branching order here is $2m(n-1)$, we have $2m(n-1) = 2(n+g-1)$ by the Riemann–Hurwitz formula and hence $g = (m-1)(n-1)$. Just as in the nonsingular case, we can change the formula to put a branch point over ∞, either by taking the first product only up to $m-2$ (which will put a branch point $P_{m-1} = P_\infty$ over ∞) or by taking the second product only up to $m-1$ (which will put a branch point $Q_m = Q_\infty$ over ∞). This will change a bit the divisors of some functions and differentials, but will not make any essential change.

As in the case of nonsingular curves we can write a basis for the holomorphic differentials. If we think of z as the projection map onto the sphere, then we see that the divisor of the meromorphic differential dz is

$$\mathrm{div}(dz) = \frac{\prod_{i=0}^{m-1} P_i^{n-1} \prod_{i=1}^{m} Q_i^{n-1}}{\prod_{h=1}^{n} \infty_h^2},$$

where $z(P_i) = \lambda_i$ for every $0 \leq i \leq m-1$, $z(Q_i) = \mu_i$ for every $1 \leq i \leq m$, and the points ∞_h, $1 \leq h \leq n$, are the n points lying above the point at ∞. The divisor of the meromorphic function w is

$$\mathrm{div}(w) = \frac{\prod_{i=0}^{m-1} P_i \prod_{i=1}^{m} Q_i^{n-1}}{\prod_{h=1}^{n} \infty_h^m},$$

and thus the meromorphic differential dz/w has divisor $\prod_{i=0}^{m-1} P_i^{n-2} \prod_{h=1}^{n} \infty_h^{m-2}$ and is actually holomorphic. We will also find useful the fact that

$$\mathrm{div}\left(\frac{\prod_{t=1}^{m}(z-\mu_t)}{w} \right) = \frac{\prod_{i=1}^{m} Q_i}{\prod_{i=0}^{m-1} P_i}.$$

It thus follows that for $0 \leq k \leq n-2$, $\left(\prod_{t=1}^{m}(z-\mu_t)^k \right) dz/w^{k+1}$ (the product of dz/w by the kth power of $\left(\prod_{t=1}^{m}(z-\mu_t) \right)/w$) is a holomorphic differential with divisor

$$\prod_{i=0}^{m-1} P_i^{n-2-k} \prod_{i=1}^{m} Q_i^{k} \prod_{h=1}^{n} \infty_h^{m-2}.$$

In fact one can even multiply these $n-1$ differentials by a polynomial in z of degree at most $m-2$ and they remain holomorphic—we see in particular that multiplying each of these $n-1$ holomorphic differentials by the polynomials $(z-\lambda_i)^l$ or $(z-\mu_i)^l$ for some $0 \leq l \leq m-2$ leaves them holomorphic with divisors

$$P_i^{nl+n-2-k} \prod_{j\neq i} P_j^{n-2-k} \prod_{j=1}^{m} Q_j^{k} \prod_{h=1}^{n} \infty_h^{m-2-l}$$

and

$$\left(\prod_{j=0}^{m-1} P_j^{n-2-k} \right) Q_i^{nl+k} \prod_{j\neq i} Q_j^{k} \prod_{h=1}^{n} \infty_h^{m-2-l}.$$

respectively. It is now clear that for a given branch point P_i or Q_i we have exactly $(m-1)(n-1) = g$ holomorphic differentials, and checking the orders of these differentials at the given point we find that they are all distinct (as

$$\left\lfloor \frac{nl+n-2-k}{n} \right\rfloor = l, \qquad \left\{ \frac{nl+n-2-k}{n} \right\} = 1 - \frac{2+k}{n},$$

at a branch point P_i and

$$\left\lfloor \frac{nl+k}{n} \right\rfloor = l, \qquad \left\{ \frac{nl+k}{n} \right\} = \frac{k}{n},$$

at a branch point Q_i), which proves the following analog of Lemma 2.2.

Lemma 2.10. *The set of differentials* $(z-\lambda_i)^l \left(\prod_{t=1}^m (z-\mu_t)^k \right) dz/w^{k+1}$ *with k and l satisfying $0 \le k \le n-2$ and $0 \le l \le m-2$ form a basis for the differentials on the singular Z_n curve X, and this basis is adapted to the point P_i. Also, the set of differentials* $(z-\mu_i)^l \left(\prod_{t=1}^m (z-\mu_t)^k \right) dz/w^{k+1}$ *with k and l satisfying $0 \le k \le n-2$ and $0 \le l \le m-2$ form a basis for the differentials on the singular Z_n curve X, and this basis is adapted to the point Q_i.*

Checking the orders of these differentials at the given point P_i or Q_i proves the following analog of Lemma 2.3.

Lemma 2.11. *The gap sequence at any point P_i or Q_i is composed of all the numbers $1 \le t \le n(m-1)$ not divisible by n.*

Proof. Proposition 1.6 shows that the gap sequence at each such point is obtained by adding 1 to the orders of each of the differentials from Lemma 2.10. By doing so we find that these orders are either $nl+n-1-k$ or $nl+k+1$ with $0 \le k \le n-2$ and $0 \le l \le m-2$. Clearly any number $1 \le t \le n(m-1)$ not divisible by n is obtained this way (with l being $\lfloor t/n \rfloor$ and k being obtained either by $\{t/n\} = 1 - (k+1)/n$ in the first case or by $\{t/n\} = (k+1)/n$ in the second), and conversely, clearly these numbers are not divisible by n and are in the desired range. $\square$

2.3.2 Abel–Jacobi Images of Certain Divisors

It is also clear that as in the case of nonsingular curves the nth powers of the points P_i, $0 \le i \le m-1$, and Q_i, $1 \le i \le m$, are all equivalent divisors and thus the image of any branch point P_i or Q_i under the Abel–Jacobi map φ_R is a point of order n when R is any other branch point. This suggests that an analog of Lemma 2.4 holds as well, but this analog looks a bit different.

Lemma 2.12. *Let R be any of the $2m$ branch points of the projection map z. Then K_R is a point of order $2n$ in the jacobian variety. Furthermore, the image of any other branch point S by the Abel–Jacobi map φ_R with basepoint R is a point of order n in the jacobian variety.*

Proof. The second assertion was already proved in the last paragraph. As for the first, we have already remarked above that

$$\operatorname{div}\left(\frac{\prod_{t=1}^{m}(z-\mu_t)}{w}\right) = \frac{\prod_{i=1}^{m}Q_i}{\prod_{i=0}^{m-1}P_i},$$

so that the divisor $\prod_{i=0}^{m-1}P_i$ is equivalent to the divisor $\prod_{i=1}^{m}Q_i$. Let us assume for the notation that $R=P_0$. By looking at the divisors of the holomorphic differentials

$$(z-\lambda_0)^{m-2}\frac{dz}{w} \quad \text{and} \quad (z-\lambda_0)^{m-2}\left(\frac{\prod_{t=1}^{m}(z-\mu_t)}{w}\right)^{n-2}\frac{dz}{w},$$

we find that the divisors

$$P_0^{(m-2)n+n-2}\prod_{i=1}^{m-1}P_i^{n-2} \quad \text{and} \quad P_0^{(m-2)n}\prod_{i=1}^{m}Q_i^{n-2}$$

are canonical. It thus follows that

$$\varphi_{P_0}\left(\prod_{i=1}^{m-1}P_i^{n-2}\right) = \varphi_{P_0}\left(\prod_{i=1}^{m}Q_i^{n-2}\right) = -2K_{P_0},$$

and hence $-2K_{P_0}$ is a point of order n in the jacobian variety $J(X)$. This means that K_{P_0} itself is of order $2n$ in $J(X)$. Actually when n is even, this argument shows that $-2K_{P_0}$ is of order $n/2$ and thus K_{P_0} is of order n, but this will not make any difference in what follows, so we continue to say that K_{P_0} is of order $2n$ in any case. Clearly similar arguments hold when R is any other branch point. $\qquad\square$

The fact that the lemmas from the nonsingular case all have analogs in the singular case is not coincidental. We explain in Appendix A how the nonsingular and singular cases are both special cases of a more general family. In this context, Lemma 2.5 should also have an analog, and we say now that this analog is the fact that $\varphi_{P_0}\left(\prod_{i=1}^{m-1}P_i\right)$ equals $\varphi_{P_0}\left(\prod_{i=1}^{m}Q_i\right)$. In the presentation of the more general family in Appendix A it becomes clear why this is the proper analog for Lemma 2.5. As for now, we continue to treat the nonsingular and singular cases as two distinct families of Riemann surfaces.

As an example let us consider the singular curve

$$w^3 = (z-\lambda_0)(z-\lambda_1)(z-\lambda_2)(z-\mu_1)^2(z-\mu_2)^2(z-\mu_3)^2$$

(i.e., a singular Z_n curve with $n = 3$ and $m = 3$ without a branch point over ∞), which is a Riemann surface of genus $(3-1)(3-1) = 4$ branched over the 6 points λ_0, λ_1, λ_2, μ_1, μ_2 and μ_3. The two basic holomorphic differentials here are dz/w and $(z-\mu_1)(z-\mu_2)(z-\mu_3)dz/w^2$, and each of them can be multiplied by a linear function in z (of our choice) in order to obtain two more holomorphic differentials. For example, choosing the branch point P_0 gives us that

$$\frac{dz}{w}, \quad (z-\lambda_0)\frac{dz}{w}, \quad \frac{(z-\mu_1)(z-\mu_2)(z-\mu_3)dz}{w^2},$$

and

$$(z-\lambda_0)\frac{(z-\mu_1)(z-\mu_2)(z-\mu_3)dz}{w^2}$$

are four linearly independent holomorphic differentials with divisors

$$P_0 P_1 P_2 \infty_1 \infty_2 \infty_3, \quad P_0^4 P_1 P_2, \quad Q_1 Q_2 Q_3 \infty_1 \infty_2 \infty_3, \quad \text{and} \quad P_0^3 Q_1 Q_2 Q_3$$

respectively, and since the orders of these divisors at P_0 are all distinct, this is indeed a basis for the holomorphic differentials on X which is adapted to P_0 (as Lemma 2.10 shows). Therefore the gap sequence at P_0 is 1,2,4,5 (as Lemma 2.11 states it should be), whence we see that P_0 is a Weierstrass point. In addition we see from the fact that the divisor $P_0^4 P_1 P_2$ is canonical that

$$\varphi_{P_0}(P_1 P_2) = -2K_{P_0},$$

a fact which will be important later. It is clear that had we multiplied by $(z-\lambda_i)$ with another i or by $(z-\mu_i)$ we would have obtain bases adapted to one of the other five branch points P_1, P_2, Q_1, Q_2 and Q_3. This proves that their gap sequence is 1,2,4,5 as well and that they are all Weierstrass points (as Lemmas 2.10 and 2.11 have shown).

2.4 Non-Special Divisors of Degree g on Singular Z_n Curves

As in the nonsingular case, we are interested in non-special divisors of degree g whose support is in the set of branch points with P_0 removed.

2.4.1 An Example with $n = 3$ and $m = 3$

Before presenting the theorem characterizing these divisors on a general singular Z_n curve, we continue with the example of the special case with $n = 3$ and $m = 3$. We write a list of 31 divisors of degree $g = 4$ on this curve, and then show that they are all non-special. Throughout the following discussion we shall assume that

i and j are 1 and 2 and that k, l and m are 1, 2 and 3 such that $i \neq j$ and k, l and m are distinct. Now, the 31 divisors we speak of are those in the following 6 families: $P_1^2 P_2^2$ (1 such divisor), $P_i^2 Q_k^2$ (6 such divisors), $Q_k^2 Q_l^2$ (3 such divisors), $P_1 P_2 Q_k Q_l$ (3 such divisors), $P_i^2 P_j Q_k$ (6 such divisors) and $P_i Q_l Q_k^2$ (12 such divisors)—together 31 divisors.

Let us explain why the first three families of divisors are indeed non-special. We already saw that no holomorphic differential can vanish to order 2 at any of the branch points. It thus follows that if one of these divisors were special, then either $P_1^3 P_2^3$, $P_i^3 Q_k^3$ or $Q_k^3 Q_l^3$ would have been canonical (since $g = 4$, $2g - 2$ is 6). By examining the basis adapted to the point P_i, we see that the holomorphic differentials which vanish to order at least 3 at P_i are spanned by two differentials with divisors $P_i^4 P_0 P_j$ and $P_i^3 Q_1 Q_2 Q_3$. If we require the differential to vanish also at P_j, then we are left only with the differential with divisor $P_i^4 P_0 P_j$ (which is not $P_1^3 P_2^3$). If we also require vanishing at Q_k, then we are left only with the differential with divisor $P_i^3 Q_1 Q_2 Q_3$ (which is not $P_i^3 Q_k^3$). Similarly, examining the basis adapted to the point Q_k gives that the holomorphic differentials which vanish to order at least 3 at Q_k are spanned by two differentials with divisors $Q_k^4 Q_l Q_m$ and $Q_k^3 P_0 P_1 P_2$. If we require vanishing also at Q_l, then we are left only with the differential with divisor $Q_k^4 Q_l Q_m$ (which is not $Q_k^3 Q_l^3$), and if we require the differential to vanish also at P_i, then we are left only with the differential with divisor $Q_k^3 P_0 P_1 P_2$ (which is again not $P_i^3 Q_k^3$).

This argument also shows the non-specialty of the divisors in the last two families. Once again, since no holomorphic differential can vanish to order 2 at any of the branch points, we find that if such a divisor were special, then there would have been a canonical divisor whose divisor is a multiple of either $P_i^3 P_j Q_k$ or $P_i Q_l Q_k^3$. Examining the holomorphic differentials which vanish to order at least 3 at P_i again shows that the vanishing of such a differential at P_j as well implies that its divisor is $P_i^4 P_0 P_j$, and the vanishing of such a differential at Q_k as well implies that its divisor is $P_i^3 Q_1 Q_2 Q_3$, which cannot occur together. Similarly, the examination of the holomorphic differentials which vanish to order at least 3 at Q_k again shows that the vanishing of such a differential at Q_l as well implies that its divisor is $Q_k^4 Q_l Q_m$ and the vanishing of such a differential at P_i as well implies that its divisor is $Q_k^3 P_0 P_1 P_2$, which also cannot occur together.

It remains to see why the 3 divisors in the last remaining family, $P_1 P_2 Q_k Q_l$, are non-special. By examining the basis of the holomorphic differentials which is adapted to P_1 we find that those which vanish at P_1 are spanned by the three differentials with divisors $P_0 P_1 P_2 \infty_1 \infty_2 \infty_3$, $P_1^4 P_0 P_2$ and $P_1^3 Q_1 Q_2 Q_3$. From these, those which vanish also at P_2 are spanned by the differentials with divisors $P_0 P_1 P_2 \infty_1 \infty_2 \infty_3$ and $P_1^4 P_0 P_2$. These are the differentials dz/w and $(z - \lambda_1)dz/w$. This means that the space of differentials, which vanish at both P_1 and P_2, are those of the form $(az + b)dz/w$ with any complex a and b. Now, since dz/w does not vanish at Q_k, Q_l and Q_m (as we clearly see from its divisor), in order for the differential $(az + b)dz/w$ to vanish at Q_k and Q_l (so that we obtain a differential whose divisor is a multiple of $P_1 P_2 Q_k Q_l$) we must have both $a\mu_k + b = 0$ and $a\mu_l + b = 0$, which is of course impossible unless $a = b = 0$ since $\mu_k \neq \mu_l$. This proves that all of these 31 divisors are indeed non-special.

Comparing with the nonsingular case where all the non-special divisors "looked the same" in some sense, i.e., had the same form but with different choices of branch points, here these divisors are divided into 6 families which all have a different appearance. In order for this example to shed some light on the general situation, we need to see how they are indeed characterized by some property which does pick exactly these divisors from the set of all the divisors of degree g supported on the branch points. As in the nonsingular case, this characterization is based on some cardinality conditions. In order to understand them we have to write the divisors including the branch points which appear to the power 0 (i.e., do not appear at all). This means that our families are (in the same order): The first divisor is $P_1^2 P_2^2 Q_1^0 Q_2^0 Q_3^0$. The following 6 are $(P_i^2 Q_l^0 Q_m^0)(P_j^0 Q_k^2)$, and the next 3 are $Q_m^0(P_1^0 P_2^0 Q_k^2 Q_l^2)$. Then we have the 3 divisors $Q_m^0(P_1^1 P_2^1 Q_k^1 Q_l^1)$. After that the 6 divisors $(P_i^2 Q_l^0 Q_m^0)(P_j^1 Q_k^1)$, and finally the last 12 divisors are $Q_m^0(P_i^1 Q_l^1)(P_j^0 Q_k^2)$. The reason for the parentheses is explained in the next paragraph.

When looking at the families of divisors written in this way, we see that in each family 3 conditions hold: The number of P_s which appear to the 2nd power is one less than the number of Q_t which appear to the 0th power. The number of P_s which appear to the 1st power is the same as the number of Q_t which appear to the 1st power, and the number of P_s which appear to the 0th power is the same as the number of Q_t which appear to the 2nd power. The numbers of Q_t in the families (in this order) are $(3,0,0)$, $(2,0,1)$, $(1,0,2)$, $(1,2,0)$, $(2,1,0)$ and $(1,1,1)$. Since this exhausts the partitions of 3 points into 3 sets where the first one is nonempty, we find that every integral divisor of degree 4 which satisfies these conditions already appears in one of these 6 families (i.e., is one of these 31 divisors). In the notation of the general statement to follow, we denote the set of P_s which appear to the 2nd power by A, the set of P_s which appear to the 1st power by C_0, the set of P_s which appear to the 0th power (i.e., do not appear at all) by C_1, the set of Q_t which appear to the 0th power (i.e., do not appear at all) by B, the set of Q_t which appear to the 1st power by D_0, and the set of Q_t which appear to the 2nd power by D_1. Then the cardinality conditions say that $|B| = |A| + 1$ and $|D_j| = |C_j|$ with $j = 0, 1$ (recall that $|Y|$ is the cardinality of the set Y). The triples written, which characterize the families, are the cardinalities of B, D_0, and D_1, respectively. Note that as the sum of these cardinalities must be $m = 3$ and $|B| = |A| + 1 \geq 1$, these triples exhaust the possibilities. Hence our 6 families contain all the divisors which satisfy these cardinality conditions. The reader is strongly advised to make sure that this last paragraph is understood before going on.

2.4.2 Non-Special Divisors

We now return to our general exposition and explain how to choose an integral divisor of degree g with support in the set of branch points with P_0 removed. We recall that in the case of a nonsingular Z_n curve, we partitioned the set of $rn - 1$

branch points (P_0 was deleted) into n sets, the first of which we called A and the remaining ones we called C_j, $0 \leq j \leq n-2$, and we placed $r-1$ points in A and r points in each of the other sets. We then defined the divisor $\Delta = A^{n-1} \prod_{j=0}^{n-2} C_j^{n-2-j}$. We now have two sets of points, P_i, $1 \leq i \leq m-1$, and Q_i, $1 \leq i \leq m$. We partition the points P_i, $1 \leq i \leq m-1$, into n sets which we again call A and C_j, $0 \leq j \leq n-2$, and we also partition the points Q_i, $1 \leq i \leq m$, into n sets, which we now call B and D_j, $0 \leq j \leq n-2$. As for the cardinality conditions, we insist that $|B| = |A| + 1$ and $|D_j| = |C_j|$ for all $0 \leq j \leq n-2$. We note that the conditions on the C_j and the D_j alone suffice since A is the complement of the (disjoint) union of the C_j to the set of branch points P_i, $1 \leq i \leq m-1$ of cardinality $m-1$ and B is the complement of the (disjoint) union of the D_j to the set of branch points Q_i, $1 \leq i \leq m$ of cardinality m. We also note that some of the sets may indeed be empty. From such a given partition we form the integral divisor

$$\Delta = A^{n-1} B^0 \prod_{j=0}^{n-2} C_j^{n-2-j} D_j^{j+1}.$$

The first observation we make concerning this divisor Δ is that it is an integral divisor of degree g with support in the set of branch points with P_0 deleted. This follows from the fact that, for each $0 \leq j \leq n-2$, the factor $C_j^{n-2-j} D_j^{j+1}$ contributes $(n-2-j)|C_j| + (j+1)|D_j| = (n-1)|C_j|$ to the degree of Δ (recall that $|D_j| = |C_j|$), and the factor $A^{n-1} B^0 = A^{n-1}$ contributes $(n-1)|A|$. This combines to show that the degree of Δ is $(n-1)[|A| + \sum_{j=0}^{n-2} |C_j|]$, which equals $(n-1)(m-1) = g$ since A and the C_j with $0 \leq j \leq n-2$ are a partition of the set of branch points P_i, $1 \leq i \leq m-1$, which contains $m-1$ points.

We now prove the theorem about non-special divisors on a general singular Z_n curve of the type we now consider, which is the analog of Theorem 2.6.

Theorem 2.13. *Let A and C_j with $0 \leq j \leq n-2$ be a partition of the set of branch points P_i, $1 \leq i \leq m-1$, and B and D_j with $0 \leq j \leq n-2$ be a partition of the set of branch points Q_i, $1 \leq i \leq m$, which satisfy the cardinality conditions above, and let Δ be the integral divisor $A^{n-1} B^0 \prod_{j=0}^{n-2} C_j^{n-2-j} D_j^{j+1}$ of degree g as described above. Then $i(\Delta) = 0$, i.e., Δ is non-special.*

Proof. The proof is similar to the second proof we gave for Theorem 2.6. First note that also here we have a decomposition of the space $\Omega(1)$ of holomorphic differentials on X into a direct sum of subspaces $\Omega(1) = \bigoplus_{k=0}^{n-2} \Omega_k(1)$ where $\Omega_k(1)$ consists of all the differentials of the form $p_k(z) \left(\prod_{t=1}^{m} (z - \mu_t)^k \right) dz/w^{k+1}$ where p_k is a polynomial of degree at most $m-2$ (by Lemma 2.10). Since we have

$$\mathrm{Ord}_{P_i} \left[p_k(z) \left(\frac{\prod_{t=1}^{m} (z - \mu_t)}{w} \right)^k \frac{dz}{w} \right] = n\mathrm{Ord}_{\lambda_i} p_k + n - 2 - k$$

and

$$\mathrm{Ord}_{Q_i}\left[p_k(z)\left(\frac{\prod_{t=1}^{m}(z-\mu_t)}{w}\right)^k\frac{dz}{w}\right]=n\mathrm{Ord}_{\mu_i}p_k+k,$$

we find that if we fix a branch point R for a holomorphic differential ω whose decomposition into $\bigoplus_{k=0}^{n-2}\Omega_k(1)$ is $\sum_{k=0}^{n-2}\omega_k$, the numbers $\mathrm{Ord}_R\omega_k$ are all distinct (as are their residues modulo n). This means that Lemma 2.7 as stated for nonsingular curves holds here as well, and hence in order to show that our divisor Δ is non-special, it suffices to show that the spaces $\Omega_k(\Delta)$ with $0\le k\le n-2$ are all trivial. For the reader which already notices the similarity between the proof of this theorem and the proof of Theorem 2.6 we remark that the index k has been shifted by 1 between the two proofs, so some of the statements look slightly different because of that.

Fix $0\le k\le n-2$ and take a differential $\omega_k=p_k(z)\left(\prod_{t=1}^{m}(z-\mu_t)^k\right)dz/w^{k+1}$ in $\Omega_k(\Delta)$. The order of ω_k at a branch point P_i is $n-2-k$ if $p_k(\lambda_i)\ne0$ and is at least n if $p_k(\lambda_i)=0$, and the order of ω_k at a branch point Q_i is k if $p_k(\mu_i)\ne0$ and is at least n if $p_k(\mu_i)=0$. The assumption that ω_k is in $\Omega_k(\Delta)$ gives that $\mathrm{Ord}_{P_i}\omega_k\ge n-1$ for every branch point P_i which lies in A, $\mathrm{Ord}_{P_i}\omega_k\ge n-2-j$ for every $0\le j\le n-2$ and branch point P_i in C_j, and $\mathrm{Ord}_{Q_i}\omega_k\ge j+1$ for every $0\le j\le n-2$ and branch point Q_i in D_j. For a point P_i in C_j with $k\le j\le n-2$, the order $n-2-k$ suffices in any case, and for a point Q_i in B or in D_j with $0\le j\le k-1$ (this means only Q_i from B when $k=0$) the order k suffices as well. However, if P_i is in A or in C_j with $0\le j\le k-1$ (this means only P_i from A when $k=0$) or if Q_i is in D_j with $k\le j\le n-2$, then the order, $n-2-k$ for the former and k for the latter, cannot be the order of ω_k at this point, which means that every corresponding λ_i or μ_i must be a root of the polynomial p_k. Since we have $|D_j|=|C_j|$ for every $0\le j\le n-2$ and since A and C_j with $0\le j\le n-2$ form a partition of a set of cardinality $m-1$, we find that the set $A\cup\bigcup_{j=0}^{k-1}C_j\cup\bigcup_{j=k}^{n-2}D_j$ of necessary roots of p_k contains $m-1$ points; note that this holds for $k=0$ as well. However, since the degree of p_k cannot exceed $m-2$ and p_k must vanish at $m-1$ distinct values, we conclude that p_k must be the zero polynomial and we are done.　　　□

2.4.3 Characterizing All Non-Special Divisors

As in the nonsingular case, the converse of Theorem 2.13 also holds, and the proof is quite similar. We shall use an analog of Lemma 2.8, which again concerns partitions.

Lemma 2.14. *Assume that we have a partition of the $m-1$ branch points P_i with $1\le i\le m-1$ into n sets, one denoted by A and the others by C_j, $0\le j\le n-2$, and that we have a partition of the m branch points Q_i with $1\le i\le m$ into n sets, one denoted by B and the others by D_j, $0\le j\le n-2$. Assume also that*

$$(n-1)|A| + \sum_{j=0}^{n-3}(n-2-j)|C_j| + \sum_{j=0}^{n-2}(j+1)|D_j| = (n-1)(m-1).$$

Then either we have $|B| = |A| + 1$ *and* $|D_j| = |C_j|$ *for every* $0 \le j \le n-2$, *or the inequality*

$$|A| + \sum_{j=0}^{k-1}|C_j| \le |B| + \sum_{j=0}^{k-1}|D_j| - 2$$

holds for some $0 \le k \le n-2$ *(with $k=0$ giving the inequality $|A| \le |B| - 2$).*

Proof. If none of the inequalities hold, then, for every $0 \le k \le n-2$, we must have the reverse inequality

$$|A| + \sum_{j=0}^{k-1}|C_j| \ge |B| + \sum_{j=0}^{k-1}|D_j| - 1$$

(i.e., $|A| \ge |B| - 1$, $|A| + |C_0| \ge |B| + |D_0| - 1$, etc.). We sum these inequalities up and obtain that on the left-hand side $|A|$ appears in all the $n-1$ inequalities and the set C_j with $0 \le j \le n-3$ appears in exactly $n-2-j$ inequalities. Hence the left-hand side sums up to $(n-1)|A| + \sum_{j=0}^{n-3}(n-2-j)|C_j|$. On the other hand, we find that on the right-hand side the set $|B|$ appears in all of the $n-1$ inequalities and the set D_j with $0 \le j \le n-3$ appears in exactly $n-2-j$ inequalities, so that this side sums up to $(n-1)|B| + \sum_{j=0}^{n-3}(n-2-j)|D_j| - (n-1)$. This means that

$$(n-1)|A| + \sum_{j=0}^{n-3}(n-2-j)|C_j| \ge (n-1)|B| + \sum_{j=0}^{n-3}(n-2-j)|D_j| - (n-1).$$

If we add to both sides of the equation the expression $\sum_{j=0}^{n-2}(j+1)|D_j|$, then the left-hand side becomes the left-hand side of the assumption of the lemma, and the right-hand side becomes

$$(n-1)|B| + \sum_{j=0}^{n-3}(n-2-j+j+1)|D_j| - (n-1) = m(n-1) - (n-1) = g,$$

where $g = (m-1)(n-1)$ (since the sets B and D_j, $0 \le j \le n-2$ form a partition of a set of m points). From this we get the inequality

$$(n-1)|A| + \sum_{j=0}^{n-3}(n-2-j)|C_j| + \sum_{j=0}^{n-2}(j+1)|D_j| \ge (n-1)(m-1),$$

and in order for this to agree with the assumption of the lemma, all the inequalities must be equalities. Hence $|A| = |B| - 1$, and then, using exactly the same induction as we did in the nonsingular case, we obtain that $|D_l| = |C_l|$ for all $0 \le l \le n-3$. Since thus the union $A \cup \bigcup_{j=0}^{n-3} C_j$ is of cardinality one less than the union $B \cup \bigcup_{j=0}^{n-3} D_j$, and since the sets C_{n-2} and D_{n-2} are the complements of these unions in sets of cardinalities $m - 1$ and m respectively, we also conclude that $|D_{n-2}| = |C_{n-2}|$, and we are done.   □

We now prove the converse of Theorem 2.13, which is the analog of Theorem 2.9.

Theorem 2.15. *Every non-special divisor of degree g with support in the branch points distinct from P_0 is of the form* $\Delta = A^{n-1} B^0 \prod_{j=0}^{n-2} C_j^{n-2-j} D_j^{j+1}$ *where A and C_j, $0 \le j \le n - 2$, is a partition of the set of branch points P_i, $1 \le i \le m - 1$, and B and D_j, $0 \le j \le n - 2$, is a partition of the branch points Q_i, $1 \le i \le m$, such that $|B| = |A| + 1$ and $|D_j| = |C_j|$ for all $0 \le j \le n - 2$.*

Proof. The proof that no branch point can appear in Δ to power n or larger is exactly the same as in the proof of Theorem 2.9. We therefore call the set of branch points P_i appearing in Δ to the $(n - 1)$th power A and the set of branch points P_i appearing in Δ to the $(n - 2 - j)$th power C_j for every $0 \le j \le n - 3$ as before, and we also call the set of branch points Q_i appearing in Δ to the $(j + 1)$th power D_j for every $0 \le j \le n - 2$. Hence we find that Δ is indeed of the form $A^{n-1} B^0 \prod_{j=0}^{n-2} C_j^{n-2-j} D_j^{j+1}$ for partitions of the sets of branch points P_i, $1 \le i \le m - 1$, and Q_i, $1 \le i \le m$, into n sets each (with C_{n-2} and B the obvious complements). We then want to show that the cardinality conditions hold.

Let us now invoke Lemma 2.14. Since the degree of Δ is $g = (n - 1)(m - 1)$ the assumption of it is satisfied, and contradicting all the inequalities in it will allow the lemma to finish the proof. We thus assume that the kth inequality holds, and we take p_k to be the product of the linear expressions $z - \lambda_i$ for the points P_i from A and from C_j with $0 \le j \le k - 1$ (again this is only A if $k = 0$) and the linear expressions $z - \mu_i$ for the points Q_i from D_j with $k \le j \le n - 2$. As before we define ω_k to be $p_k(z) \left(\prod_{t=1}^{m} (z - \mu_t)^k \right) dz / w^{k+1}$. Since the sets B and D_j with $0 \le j \le k - 1$ complete the set of points Q_i which we have taken to a set of cardinality m, and since we have "replaced" these sets by sets whose total cardinality is less by at least 2 by the kth inequality, we find that the degree of p_k is at most $m - 2$. Hence ω_k is a holomorphic differential, and by considerations similar to those which we have seen in the proof of Theorem 2.13, we gather the following information. For every point P_i which lies in C_j with $k \le j \le n - 2$, the order $n - 2 - k$ which ω_k has at P_i suffices. For a point P_i in A or in C_j with $0 \le j \le k - 2$, the fact that $p_k(\lambda_i) = 0$ gives ω_k a high enough order as well. For every point Q_i which lies in B or in D_j with $0 \le j \le k - 1$, the order k which ω_k has at Q_i suffices, and for a point Q_i in D_j with $k \le j \le n - 2$, the fact that $p_k(\mu_i) = 0$ also gives ω_k a high enough order. Hence ω_k vanishes at every branch point to a high enough order, and we find that ω_k is a nonzero differential in $\Omega(\Delta)$ (and actually in $\Omega_k(\Delta)$). In particular $i(\Delta) \ge 1$ and Δ is special.

Since any of the inequalities in Lemma 2.14 implies that Δ is special, we find that if Δ is non-special, then none of these inequalities can hold. The fact that the degree of Δ is g yields the equality which is in the assumption of the lemma, and so applying the lemma assures us that we have $|B| = |A| + 1$ and $|D_j| = |C_j|$ for every $0 \le j \le n - 2$. This finishes the proof.   □

Note that up to changing every k here to $k+1$ and up to the fact that the cardinality conditions are different, the parts of the proofs of Theorems 2.13 and 2.15 which talk about the branch points P_i, $1 \leq i \leq m-1$, are exactly the same as the proofs of Theorems 2.6 and 2.9. This is, of course, not coincidental, but a result of the fact already mentioned that the nonsingular and singular cases are two special cases of one more general family of Riemann surfaces. This change of index from k to $k+1$ is more clear to the eye once one notices that here in the singular case the divisor ω_k was of the form $q_k(z)\mathrm{d}z/w^{k+1}$ where $q_k(z)$ is a polynomial in z (which is the product of $p_k(z)$ and $\prod_{t=1}^{m}(z-\mu_t)^k$). All this becomes clearer with the presentation of the more general family in Appendix A, as well as the general cardinality conditions which will become those of Lemma 2.8 in the special case of nonsingular curves and those of Lemma 2.14 in the special case of the singular curves described here.

We note that also in this case the fact that the specific point P_0 was excluded from the support plays no role in what we have done here, and therefore all of our statements in this section still hold also when P_0 appears in the support. We also recall that there is symmetry between the sets of points P_i, $0 \leq i \leq m-1$, and Q_i, $1 \leq i \leq m$, and hence between the sets A and C_j, $0 \leq j \leq n-2$, and the sets B and D_j, $0 \leq j \leq n-2$. The symmetry, which one must use when taking the basepoint later to be some point Q_i, is between A and D_{n-2}, between C_j and D_{n-1-j}, $0 \leq j \leq n-2$, and between B and C_{n-2}. Note that when adding P_0 into the deal, since it does not appear it should be added to C_{n-2}; hence, to the equalities $|C_{n-1-j}| = |D_{n-1-j}|$ we add the equality $|A| = |B \setminus Q_i|$ (we must take a point Q_i out now), and the equality $|C_{n-2} \cup P_0| = |D_{n-2}| + 1$, which completes the symmetry argument. Also in this case we have found all the non-special divisors of degree g supported on the branch points, since we claim that in any such divisor some branch point must not appear. Even more is true: we have already seen that the function $\left(\prod_{t=1}^{m}(z-\mu_t) \right)/w$ lies in $L(1/\prod_{i=0}^{m-1} P_i)$ (together with the constant functions), and clearly its reciprocal lies in $L(1/\prod_{i=1}^{m} Q_i)$ (again, together with the constant functions). Therefore both $r(1/\prod_{i=0}^{m-1} P_i)$ and $r(1/\prod_{i=1}^{m} Q_i)$ are at least 2, and by an argument similar to what we did in the singular case, we find that in any non-special divisor of degree g, at least one point P_i must not appear, and at least one point Q_i must not appear (this is again trivial for $n=2$ since then $g < m$—but it is now seen to be true in any case). However, as in the nonsingular case, also here the property of P_0 not appearing in the support of our divisors will be used in what follows (for the same reasons), so we continue assuming it.

2.5 Some Operators

In this section we define some operators on the set of integral divisors of degree g constructed from a partition of the set of branch points with P_0 deleted. We recall that in the nonsingular case we have written those divisors in the form $A^{n-1} \prod_{j=0}^{n-2} C_j^{n-2-j}$, where the set of branch points with P_0 omitted is the disjoint union of the sets A and C_j with $0 \leq j \leq n-2$ and the cardinality of A, is $r-1$ while the cardinality of the remaining sets is r. In the singular case with which we deal here we have written the divisors as $A^{n-1} B^0 \prod_{j=0}^{n-2} C_j^{n-2-j} D_j^{j+1}$, where the branch points P_i, $1 \leq i \leq m-1$, are the disjoint union of the sets A and C_j with $0 \leq j \leq n-2$, the branch points Q_i, $1 \leq i \leq m$, are the disjoint union of the sets B and D_j with $0 \leq j \leq n-2$, and we have insisted that $|A|+1 = |B|$ and that $|C_j| = |D_j|$ for all $0 \leq j \leq n-2$. We shall now define operators on these divisors.

2.5.1 Operators for the Nonsingular Case

We start with the definition of the operators in the nonsingular case.

Definition 2.16. For a divisor Δ of the form $A^{n-1} \prod_{j=0}^{n-2} C_j^{n-2-j}$, define

$$N(\Delta) = A^{n-1} \prod_{j=0}^{n-2} C_j^j.$$

For any branch point $R \in C_0$, define

$$T_R(\Delta) = (C_0 \setminus R)^{n-1} (A \cup R)^{n-2} \prod_{j=1}^{n-2} C_j^{j-1}.$$

Note that we have used the more compact notation $A \cup R$ and $C_0 \setminus R$ for the sets $A \cup \{R\}$ and $C_0 \setminus \{R\}$, and we shall continue to use this more compact notation. Now, it should be clear that for any non-special divisor Δ of degree g with support in the branch points with P_0 deleted, $N(\Delta)$ has the same form as Δ and so has $T_R(\Delta)$ for any $R \in C_0$, i.e., they are integral divisors of degree g with support in the set of branch points with P_0 deleted. They are also non-special since by Theorems 2.6 and 2.9 one simply checks the cardinality conditions, which for $N(\Delta)$ are obvious and for $T_R(\Delta)$ are also clear since $|C_0 \setminus R| = r-1$ and $|A \cup R| = r$. We emphasize that while the operator N can act on any non-special integral divisor of degree g with support in the set of branch points distinct from P_0 as defined above, the operator T_R for a given branch point R can act only on those divisors where the branch point R appears in C_0 (i.e., to the power $n-2$), and its image lies also in this set of divisors.

The following proposition demonstrates the role of the operators N and T_R from Definition 2.16.

Proposition 2.17. If $e = \varphi_{P_0}(\Delta) + K_{P_0}$ for $\Delta = A^{n-1} \prod_{j=0}^{n-2} C_j^{n-2-j}$, then we have $-e = \varphi_{P_0}(N(\Delta)) + K_{P_0}$. If $R \in C_0$, then

$$e + \varphi_{P_0}(R) = -(\varphi_{P_0}(T_R(\Delta)) + K_{P_0}).$$

Proof. Since in the nonsingular case Lemma 2.4 shows that K_{P_0} is a point of order 2 in the jacobian variety, it suffices to show that

$$\varphi_{P_0}(N(\Delta)) + \varphi_{P_0}(\Delta) = 0.$$

We now observe that by Definition 2.16 we have

$$\varphi_{P_0}(\Delta) + \varphi_{P_0}(N(\Delta)) = \varphi_{P_0}\left(A^{2n-2}\prod_{j=0}^{n-2}C_j^{n-2}\right).$$

By Lemma 2.4 this equals

$$\varphi_{P_0}\left(A^{n-2}\prod_{j=0}^{n-2}C_j^{n-2}\right) = (n-2)\varphi_{P_0}\left(\prod_{j=1}^{rn-1}P_i\right),$$

and vanishes by Lemma 2.5. In view of more general situations we remark that the vanishing of the last expression can also be proved using differentials. Explicitly, one uses the fact that the divisor $P_0^{n(r-2)+n-2}\prod_{j=1}^{rn-1}P_i^{n-2}$ is canonical as the divisor of the differential $(z-\lambda_0)^{r-2}dz/w$ and the fact that $-2K_{P_0} = 0$ here.

As for the second assertion, we first observe that since by Lemma 2.4 K_{P_0} is a point of order 2 it suffices to show that

$$\varphi_{P_0}(\Delta) + \varphi_{P_0}(R) + \varphi_{P_0}(T_R(\Delta)) = 0.$$

We first use Definition 2.16 to find that

$$\varphi_{P_0}(R) + \varphi_{P_0}(T_R(\Delta)) = \varphi_{P_0}\left(R(C_0\setminus R)^{n-1}(A\cup R)^{n-2}\prod_{j=1}^{n-2}C_j^{j-1}\right),$$

which clearly equals

$$\varphi_{P_0}\left(C_0^{n-1}A^{n-2}\prod_{j=1}^{n-2}C_j^{j-1}\right).$$

Adding $\varphi_{P_0}(\Delta)$ to the last expression gives

$$\varphi_{P_0}\left(C_0^{2n-3}A^{2n-3}\prod_{j=1}^{n-2}C_j^{n-3}\right),$$

which by Lemma 2.4 equals

$$\varphi_{P_0}\left(A^{n-3}\prod_{j=0}^{n-2}C_j^{n-3}\right) = (n-3)\varphi_{P_0}\left(\prod_{j=1}^{rn-1}P_i\right).$$

and vanishes by Lemma 2.5. Here the vanishing can also be seen by the fact that the divisor $P_0^{n(2r-2)+n-3}\prod_{j=1}^{rn-1}P_i^{n-3}$ is canonical as the divisor of the differential $(z-\lambda_0)^{2r-2}dz/w^2$ and $-2K_{P_0}=0$. This proves the proposition.   □

2.5.2 Operators for the Singular Case

We now make similar definitions for the singular case.

Definition 2.18. For $\Delta = A^{n-1}B^0\prod_{j=0}^{n-2}C_j^{n-2-j}D_j^{j+1}$, define

$$N(\Delta) = A^{n-1}B^0\prod_{j=0}^{n-2}C_j^j D_j^{n-1-j}.$$

For any branch point $R \in C_0$, define

$$T_R(\Delta) = (C_0\setminus R)^{n-1}D_0^0((A\cup R)^{n-2}B^1)\prod_{j=1}^{n-2}C_j^{j-1}D_j^{n-j},$$

and for $S \in B$, define

$$T_S(\Delta) = C_0^{n-1}(D_0\cup S)^0(A^{n-2}(B\setminus S)^1)\prod_{j=1}^{n-2}C_j^{j-1}D_j^{n-j}.$$

Also here it should be clear that for any non-special divisor Δ of degree g with support in the branch points with P_0 deleted, $N(\Delta)$ has the same form as Δ and so have $T_R(\Delta)$ for any $R \in C_0$ and $T_S(\Delta)$ for any $S \in B$. The general form is clear, and the verification of the cardinality conditions from Theorems 2.13 and 2.15 is again obvious for $N(\Delta)$ and very simple for $T_R(\Delta)$ and $T_S(\Delta)$, keeping track of the movement of the corresponding point R or S. Again we emphasize that while the operator N can act on any divisor Δ as defined above, the operators T_R and T_S for given branch points R and S can act only on an appropriate set of divisors (those with $R \in C_0$ or $S \in B$), and the images of these operators lie in the sets of divisors on which they can act. It is important to notice that for some divisors the set C_0 can now be empty, which means that no T_R can act on them. The set B, on the other hand, is never empty, so that on each divisor Δ at least one operator T_S can act.

The following proposition shows that the role of the operators N and T_R from Definition 2.18 is the same as their role in the nonsingular case, and also shows that the role of T_S is what it is expected to be.

Proposition 2.19. If $e = \varphi_{P_0}(\Delta) + K_{P_0}$ for $\Delta = A^{n-1}B^0\prod_{j=0}^{n-2}C_j^{n-2-j}D_j^{j+1}$, then we have $-e = \varphi_{P_0}(N(\Delta)) + K_{P_0}$. We also have that if $R \in C_0$, then

$$e + \varphi_{P_0}(R) = -(\varphi_{P_0}(T_R(\Delta)) + K_{P_0})$$

and if $S \in B$, then

$$e + \varphi_{P_0}(S) = -(\varphi_{P_0}(T_S(\Delta)) + K_{P_0}).$$

Proof. We are no longer in the situation where K_{P_0} is of order 2, but clearly it suffices to prove that

$$\varphi_{P_0}(\Delta) + \varphi_{P_0}(N(\Delta)) = -2K_{P_0}.$$

Using Definition 2.18 we find that the left-hand side equals

$$\varphi_{P_0}\left(A^{2n-2}B^0 \prod_{j=0}^{n-2} C_j^{n-2} D_j^n\right),$$

which equals

$$\varphi_{P_0}\left(A^{n-2} \prod_{j=0}^{n-2} C_j^{n-2}\right) = \varphi_{P_0}\left(P_0^{n(m-1)-2} \prod_{i=1}^{m-1} P_i^{n-2}\right)$$

by Lemma 2.12 and the fact that $\varphi_{P_0}(P_0)$ vanishes. Since the argument of φ_{P_0} is the divisor of the holomorphic differential $(z - \lambda_0)^{m-2} dz/w$, φ_{P_0} of this divisor equals $-2K_{P_0}$, as required.

For the other two statements we again note that K_{P_0} is no longer of order 2, but it suffices to prove that

$$\varphi_{P_0}(\Delta) + \varphi_{P_0}(R) + \varphi_{P_0}(T_R(\Delta)) = -2K_{P_0}$$

and

$$\varphi_{P_0}(\Delta) + \varphi_{P_0}(S) + \varphi_{P_0}(T_S(\Delta)) = -2K_{P_0}.$$

We once again use Definition 2.18 to see that

$$\varphi_{P_0}(R) + \varphi_{P_0}(T_R(\Delta)) = \varphi_{P_0}\left(R(C_0 \setminus R)^{n-1}(A \cup R)^{n-2}B^1 D_0^0 \prod_{j=1}^{n-2} C_j^{j-1} D_j^{n-j}\right),$$

which clearly equals

$$\varphi_{P_0}\left(C_0^{n-1} A^{n-2} B^1 D_0^0 \prod_{j=1}^{n-2} C_j^{j-1} D_j^{n-j}\right).$$

Similarly, we see that

$$\varphi_{P_0}(S) + \varphi_{P_0}(T_S(\Delta)) = \varphi_{P_0}\left(S C_0^{n-1} A^{n-2} (B \setminus S)^1 (D_0 \cup S)^0 \prod_{j=1}^{n-2} C_j^{j-1} D_j^{n-j}\right)$$

also gives the same expression

$$\varphi_{P_0}\left(C_0^{n-1}A^{n-2}B^1 D_0^0 \prod_{j=1}^{n-2} C_j^{j-1} D_j^{n-j}\right).$$

Now, adding $\varphi_{P_0}(\Delta)$ to the last expression gives

$$\varphi_{P_0}\left(C_0^{2n-3}A^{2n-3}B^1 D_0^1 \prod_{j=1}^{n-2} C_j^{n-3} D_j^{n+1}\right),$$

which equals

$$\varphi_{P_0}\left(A^{n-3}B \prod_{j=1}^{n-2} C_j^{n-3} D_j\right) = \varphi_{P_0}\left(P_0^{n(m-1)-3} \prod_{i=1}^{m-1} P_i^{n-3} \prod_{i=1}^{m} Q_i\right)$$

by Lemma 2.12 and the fact that $\varphi_{P_0}(P_0)$ vanishes. This finishes the proof of the proposition since this divisor is the divisor of the holomorphic differential $(z-\lambda_0)^{m-2}\left(\prod_{t=1}^{m}(z-\mu_t)\right)dz/w^2$ and hence its image by φ_{P_0} is $-2K_{P_0}$. $\square$

2.5.3 Properties of the Operators in Both Cases

We note that by combining the results of Proposition 2.17, we obtain the equality

$$\varphi_{P_0}(R) + \varphi_{P_0}(\Delta) + K_{P_0} = \varphi_{P_0}(N(T_R(\Delta))) + K_{P_0}$$

in the nonsingular case, and in the singular case, combining the results of Proposition 2.19, we obtain the equalities

$$\varphi_{P_0}(R) + \varphi_{P_0}(\Delta) + K_{P_0} = \varphi_{P_0}(N(T_R(\Delta))) + K_{P_0}$$

and

$$\varphi_{P_0}(S) + \varphi_{P_0}(\Delta) + K_{P_0} = \varphi_{P_0}(N(T_S(\Delta))) + K_{P_0}.$$

These equalities are very useful and will be applied many times in what follows.

We have actually defined $N(\Delta)$ and $T_R(\Delta)$ in Definition 2.16 and $N(\Delta)$, $T_R(\Delta)$ and $T_S(\Delta)$ in Definition 2.18 in order for Propositions 2.17 and 2.19 to hold. We will also be interested in examining the actions of these operators explicitly, i.e., how do they change the sets in the partitions. For this we write $N(\Delta)$ from Definition 2.16 as

$$A^{n-1} \prod_{j=0}^{n-2} C_{n-2-j}^{n-2-j}$$

(replacing the index j by $n-2-j$), which shows that in the nonsingular case the operator N fixes A and sends each C_j, $0 \leq j \leq n-2$, to C_{n-2-j}. In particular this means that N interchanges the sets C_j, $0 \leq j \leq n-2$, in pairs (except for $C_{(n-2)/2}$ when n is even, which is also fixed). As for the operator T_R, we write $T_R(\Delta)$ from Definition 2.16 as

$$(C_0 \setminus R)^{n-1} (A \cup R)^{n-2} \prod_{j=1}^{n-2} C_{n-1-j}^{n-2-j}$$

(now replacing j by $n-1-j$), so that the operator T_R of the nonsingular case mixes A and C_0, and sends each C_j, $1 \leq j \leq n-2$, to C_{n-1-j}. Therefore T_R interchanges the sets C_j, $1 \leq j \leq n-2$ in pairs (except for $C_{(n-1)/2}$ when n is odd, which is now fixed). The mixing of A and C_0 is, explicitly, to send the points of C_0 distinct from R to A and send A to C_0 while fixing R, which is again an interchange in a pair—the pair is now A and $C_0 \setminus R$. Therefore it does not leave the elements of every set together, but mixes the sets a little bit—now R is with the elements of A and not with the other elements of C_0. The most important observation we now have is that both operators are involutions (as they interchange sets in pairs).

As for the singular case, we write the divisor $N(\Delta)$ from Definition 2.18 as

$$A^{n-1} B^0 \prod_{j=0}^{n-2} C_{n-2-j}^{n-2-j} D_{n-2-j}^{j+1}$$

(replacing again the index j by $n-2-j$), so that in the singular case N fixes A and B and sends each C_j, $0 \leq j \leq n-2$ to C_{n-2-j} and each D_j, $0 \leq j \leq n-2$ to D_{n-2-j}. Again this means that N interchanges the sets C_j and D_j, $0 \leq j \leq n-2$ in pairs (still except for the fixed sets $C_{(n-2)/2}$ and $D_{(n-2)/2}$ when n is even). In order to examine the operators T_R and T_S we write $T_R(\Delta)$ and $T_S(\Delta)$ from Definition 2.18 as

$$(C_0 \setminus R)^{n-1} D_0^0 \left((A \cup R)^{n-2} B^1 \right) \prod_{j=1}^{n-2} C_{n-1-j}^{n-2-j} D_{n-1-j}^{j+1}$$

and

$$C_0^{n-1} (D_0 \cup S)^0 \left(A^{n-2} (B \setminus S)^1 \right) \prod_{j=1}^{n-2} C_{n-1-j}^{n-2-j} D_{n-1-j}^{j+1}$$

respectively (with j changed to $n-1-j$ again). Hence in the singular case, the operators T_R and T_S send each C_j, $1 \leq j \leq n-2$ to C_{n-1-j} and each D_j, $1 \leq j \leq n-2$ to D_{n-1-j}, so that they take these $2n-4$ sets and interchange them in pairs (with now $C_{(n-1)/2}$ and $D_{(n-1)/2}$ fixed when n is odd). As for their action on the remaining sets A, B, C_0 and D_0, our first observation is that T_R also sends B to D_0 and D_0 to B and that T_S also sends A to C_0 and C_0 to A (another pair for each). As for the mixing, we find that the mixing of A and C_0 by T_R is, explicitly, to send the points of C_0 distinct from R to A and send A to C_0 while fixing R. Similarly, the mixing of B and D_0 by T_S is to send the points of B distinct from S to D_0 and send D_0 to B while fixing S. In any case this is again an interchange in yet another pair: either A and

$C_0 \setminus R$ for T_R or $B \setminus S$ and D_0 for T_S. These operators again do not leave every set intact but do a little mixing, as after T_R we have that R is with the elements of A and not with the other elements of C_0, and after T_S we have that S is with the elements of D_0 and not with the other elements of B. In conclusion, all these operators are again involutions.

The reader should note that the parts in the action of the operators N and T_R in the singular case which concern the sets A and C_j, $0 \le j \le n-2$, are the same as in the nonsingular case. This continues the correspondence arising from the fact that these two cases are special cases of one bigger general family. The operator T_S appears only in the singular case simply because there are no branch points Q_i to be taken as S in the nonsingular case.

For reasons that will soon become clear, we will be interested, for a divisor Δ, in the branch points which do not appear in $N(\Delta)$. We would like to find the connection between these points for Δ and these points for divisors obtained from Δ by these operators. In the nonsingular case we have

$$\Delta = A^{n-1} \prod_{j=0}^{n-2} C_j^{n-2-j}, \quad T_R(\Delta) = (C_0 \setminus R)^{n-1} (A \cup R)^{n-2} \prod_{j=1}^{n-2} C_j^{j-1},$$

$$N(\Delta) = A^{n-1} \prod_{j=0}^{n-2} C_j^j, \quad \text{and} \quad N(T_R(\Delta)) = (C_0 \setminus R)^{n-1} \left(\prod_{j=1}^{n-2} C_j^{n-1-j} \right) (A \cup R)^0,$$

from which we can conclude that the only two branch points which do not appear in either $N(\Delta)$ or $N(T_R(\Delta))$ are P_0 and R. In the singular case we have

$$\Delta = A^{n-1} B^0 \prod_{j=0}^{n-2} C_j^{n-2-j} D_j^{j+1}, \quad N(\Delta) = A^{n-1} B^0 \prod_{j=0}^{n-2} C_j^j D_j^{n-1-j},$$

$$T_R(\Delta) = (C_0 \setminus R)^{n-1} D_0^0 (A \cup R)^{n-2} B^1 \prod_{j=1}^{n-2} C_j^{j-1} D_j^{n-j},$$

$$N(T_R(\Delta)) = (C_0 \setminus R)^{n-1} D_0^0 \left(\prod_{j=1}^{n-2} C_j^{n-1-j} D_j^j \right) (A \cup R)^0 B^{n-1},$$

$$T_S(\Delta) = C_0^{n-1} (D_0 \cup S)^0 A^{n-2} (B \setminus S)^1 \prod_{j=1}^{n-2} C_j^{j-1} D_j^{n-j},$$

and

$$N(T_S(\Delta)) = C_0^{n-1} (D_0 \cup S)^0 \left(\prod_{j=1}^{n-2} C_j^{n-1-j} D_j^j \right) A^0 (B \setminus S)^{n-1},$$

from which we conclude that the only branch points which do not appear in either $N(\Delta)$ or $N(T_R(\Delta))$ are P_0 and R and the only branch points which do not appear in either $N(\Delta)$ or $N(T_S(\Delta))$ are P_0 and S.

2.6 Theta Functions on Z_n Curves

In this section we obtain special results for theta functions which are associated to Z_n curves. These results apply equally to the nonsingular and singular cases, since they hold for general Z_n curves. They will also be applied later in Chapter 6 to the Z_n curves investigated there.

2.6.1 Non-Special Divisors as Characteristics for Theta Functions

The reason we have worked so hard in Sections 2.2 and 2.4 to characterize those divisors Δ such that $i(\Delta) = 0$ is that we now wish to consider certain theta functions on the curve. As we saw in Section 1.3 (specifically Theorem 1.8, the Riemann Vanishing Theorem), the function $f(P) = \theta(\varphi_{P_0}(P) - e, \Pi)$ is much more interesting when the point e, which always has a representation $e = \varphi_{P_0}(\Delta) + K_{P_0}$ with Δ an integral divisor of degree g by the Jacobi Inversion Theorem, has such a representation with $i(\Delta) = 0$ (otherwise f vanishes identically). This is why these divisors are the ones interesting us and the ones which play the main role in this book.

Recall that Lemma 2.4 in the nonsingular case and Lemma 2.12 in the singular case give that on a Z_n curve, after choosing a canonical homology basis and the dual basis for the holomorphic differentials, we have

$$\varphi_{P_0}(\Delta) + K_{P_0} = \Pi\frac{\varepsilon}{n} + I\frac{\varepsilon'}{n} + \Pi\frac{\delta}{2n} + I\frac{\delta'}{2n} = \frac{1}{2}\left(\Pi\frac{2\varepsilon+\delta}{n} + I\frac{2\varepsilon'+\delta'}{n}\right)$$

where ε, ε', δ and δ' are vectors in $\mathbb{Z}^g$ (ε and ε' come from $\varphi_{P_0}(\Delta)$ and δ and δ' come from K_{P_0}). Lemma 2.4 actually shows that in the nonsingular case δ and δ' are in $n\mathbb{Z}^g$, but this will make no difference in what follows, so we ignore it. It is also clear that changing ε or ε' by vectors in $n\mathbb{Z}^g$ or changing δ or δ' by vectors in $2n\mathbb{Z}^g$ does not change this value in $J(X)$. We call such an expression in $\mathbb{C}^g$ with a given choice of ε, ε', δ and δ' a *lift* of $\varphi_{P_0}(\Delta) + K_{P_0}$ to $\mathbb{C}^g$, and for a given lift we consider the theta function

$$\theta\begin{bmatrix}(2\varepsilon+\delta)/n \\ (2\varepsilon'+\delta')/n\end{bmatrix}(\varphi_{P_0}(P),\Pi) = \mathbf{e}(\cdot)\theta\left[\varphi_{P_0}(P) + \frac{1}{2}\left(\Pi\frac{2\varepsilon+\delta}{n} + I\frac{2\varepsilon'+\delta'}{n}\right),\Pi\right],$$

where the argument of the exponent is the linear function of $\varphi_{P_0}(P)$ (which depends on the characteristic) which comes from Eq. (1.3) (from which we also obtained the last equality). If we now write $e = \varphi_{P_0}(\Delta) + K_{P_0}$, then the Riemann Vanishing Theorem (Theorem 1.8) gives us that either the above function vanishes identically on the surface, or has g zeros (taking multiplicity into consideration) Q_i, $1 \leq i \leq g$, with $-e - \varphi_{P_0}\left(\prod_{i=1}^{g} Q_i\right) + K_{P_0}$. We shall denote the function $\theta\begin{bmatrix}(2\varepsilon+\delta)/n \\ (2\varepsilon'+\delta')/n\end{bmatrix}(\zeta,\Pi)$ by

$\theta[\Delta](\zeta,\Pi)$, knowing that at this point it depends on the lift of $\varphi_{P_0}(\Delta) + K_{P_0}$ to $\mathbb{C}^g$. The property exhibited in the following proposition, however, does not depend on the lift, and we later explain why the functions we will ultimately work with do not depend on any lifts.

Proposition 2.20. *Let Δ be a divisor of the form $A^{n-1} \prod_{j=0}^{n-2} C_j^{n-2-j}$ as in Section 2.2 or of the form $A^{n-1} B^0 \prod_{j=0}^{n-2} C_j^{n-2-j} D^{j+1}$ as in Section 2.4, and consider the theta function*

$$f(P) = \theta[\Delta](\varphi_{P_0}(P), \Pi)$$

for some lift of $\varphi_{P_0}(\Delta) + K_{P_0}$. Then the function $f(P)$ does not vanish identically on the surface and its divisor of zeros is the integral divisor of degree g denoted by $N(\Delta)$ in Definitions 2.16 and 2.18 respectively.

Proof. By Propositions 2.17 and 2.19 we find that if we denote the expression $\varphi_{P_0}(\Delta) + K_{P_0}$ by $-e$, then $e = \varphi_{P_0}(N(\Delta)) + K_{P_0}$ in the nonsingular and singular cases respectively. Now, since e is K_{P_0} plus the image of a non-special integral divisor Γ (which is $N(\Delta)$) of degree g by the Abel–Jacobi map φ_{P_0} and since $-e = [\Pi(2\varepsilon + n\delta)/n + I(2\varepsilon' + n\delta')/n]/2$, Corollary 1.11 implies that the function $\theta \begin{bmatrix} (2\varepsilon + \delta)/n \\ (2\varepsilon' + \delta')/n \end{bmatrix} (\varphi_{P_0}(P), \Pi)$ does not vanish identically and its zeros are the divisor Γ. Since this function is our f by definition and since $\Gamma = N(\Delta)$, this proves the proposition. $\square$

2.6.2 Quotients of Theta Functions with Characteristics Represented by Divisors

Our interest will be in the functions defined on the Z_n curve by

$$\frac{\theta^{en^2}[\Delta](\varphi_{P_0}(P), \Pi)}{\theta^{en^2}[\Xi](\varphi_{P_0}(P), \Pi)}$$

where Δ and Ξ are non-special integral divisors of degree g with support in the branch points distinct from P_0 (here, contrary to Sections 2.2 and 2.4, it is important that P_0 does not appear in the divisors Δ or Ξ) and e is an integer defined to be 1 if n is even and 2 if n is odd (i.e., $e = 2/\gcd(2,n)$). The most important remark is that these quotients are meromorphic functions on the Z_n curve. Since by Lemmas 2.4 and 2.12 the characteristics are $\begin{bmatrix} (2\varepsilon + \delta)/n \\ (2\varepsilon' + \delta')/n \end{bmatrix}$ and $\begin{bmatrix} (2\rho + \delta)/n \\ (2\rho' + \delta')/n \end{bmatrix}$, where the point $\Pi\varepsilon/n + I\varepsilon'/n$ of order n is $\varphi_{P_0}(\Delta)$, the point $\Pi\rho/n + I\rho'/n$ of order n is $\varphi_{P_0}(\Xi)$, and the point $\Pi\delta/2n + I\delta'/2n$ of order $2n$ is K_{P_0}, Eq. (1.6) shows that the quotient of the theta functions would already be single-valued on $J(X)$ taking the power to be n. This is so because $\alpha = 2\varepsilon + \delta$, $\alpha' = 2\varepsilon' + \delta'$, $\beta = 2\rho + \delta$ and $\beta' = 2\rho' + \delta'$, and the vectors $\alpha - \beta = 2(\varepsilon - \rho)$ and $\alpha' - \beta' = 2(\varepsilon' - \rho')$ are indeed in $2\mathbb{Z}^g$. Hence the composition is single-valued on X when the power is only n. However, we will be evaluating these functions at branch points, and since $\varphi_{P_0}(R)$ is a point of order n for any branch point R Eq. (1.7) with $\zeta = 0$ shows that with the power n^2 the evaluation of this quotient at a branch point is the same as evaluating another quotient at 0 (when writing a point of order n as $\Pi\gamma/2n + I\gamma'/2n$ we definitely have $\gamma \in 2\mathbb{Z}$, and we already saw that in our case $\alpha' - \beta' \in 2\mathbb{Z}$ as well).

Finally, recall that the functions we started with depended on the lift of the characteristics from $J(X)$ to $\mathbb{C}^g$ that we chose. Let us see what happens if we take a different lift. Eq. (1.4) shows that if we change the characteristic of the numerator by some vectors $2M$ and $2M'$ and the characteristic of the denominator by some vectors $2T$ and $2T'$ with M, M', T and T' in $\mathbb{Z}^g$ (which is the same as changing the lifts) then the quotient (with power 1) will be multiplied by $\mathbf{e}\left[(2\varepsilon + \delta)^t M'/2n - (2\rho + \delta)^t T'/2n\right]$. Hence if the power is n^2, this factor becomes $\mathbf{e}\left[n(\varepsilon^t M' + \delta^t M'/2 - \rho^t T' + \delta^t T'/2)\right] = \mathbf{e}\left[n\delta^t(M' - T')/2\right]$. Thus if n is even, this factor is 1 and our quotient is indeed independent of the choices of lifts of the characteristics from $J(X)$ to $\mathbb{C}^g$. If n is odd, however, this factor can be -1 (also in the nonsingular case where δ is known to belong to $n\mathbb{Z}^g$ by Lemma 2.4 it can be—multiplying again by the odd number n does not change that) and to get the desired independence we must take the square of this quotient, i.e., take the power to be $2n^2$. This is why we chose the power to be en^2, depending on the parity of n in this way, and we now know that our functions depend indeed only on the divisors (and P_0, of course), i.e., their images in $J(X)$, and not on any arbitrary choices of lifts to $\mathbb{C}^g$. We note (and the reader might check) that the independence of the en^2 powers of the quotients of the lifts from $J(X)$ to $\mathbb{C}$ holds for the theta constants (i.e., the values of the theta functions with characteristics at 0) as well, without the need to take quotients of two such constants. It is also good to notice that throughout this discussion we assumed that we keep a fixed lift of the vector K_{P_0} of Riemann constants, but the reader can verify that allowing this lift to vary causes no harm to the results.

2.6.3 Evaluating Quotients of Theta Functions at Branch Points

In order for the expressions we obtain by evaluating these quotients at branch points to have useful meaning we have to choose only points where neither the numerator nor the denominator vanish, and Proposition 2.20 shows that the points we should choose are only points which do not appear either in the support of $N(\Delta)$ or in the support of $N(\Xi)$. In particular, P_0 is a good point to substitute for any such Δ and Ξ, and this gives the evaluation of the quotient of theta functions (without the composition with φ_{P_0}) at 0. As we recall, these numbers, values of theta functions with characteristics at 0, are called *theta constants* of the Riemann surface X, and the Thomae formulae connects these constants with algebraic functions (evidently polynomials) of the parameters defining the surface X (in the case of a Z_n curve these are the numbers λ_i, and also the μ_i in the singular case). If now R is a point P_i with $i \neq 0$ which does not appear either in $N(\Delta)$ or in $N(\Xi)$ (i.e., R lies both in the sets $(C_0)_\Delta$ and $(C_0)_\Xi$), then Eq. (1.7) with $\zeta = 0$ translates to

$$\frac{\theta^{en^2}[\Delta](\varphi_{P_0}(R), \Pi)}{\theta^{en^2}[\Xi](\varphi_{P_0}(R), \Pi)} = \frac{\theta^{en^2}[\varphi_{P_0}(\Delta) + \varphi_{P_0}(R) + K_{P_0}](0, \Pi)}{\theta^{en^2}[\varphi_{P_0}(\Xi) + \varphi_{P_0}(R) + K_{P_0}](0, \Pi)},$$

and since by Proposition 2.17 in the nonsingular case and by Proposition 2.19 in the singular case we have

$$\varphi_{P_0}(\Delta) + \varphi_{P_0}(R) + K_{P_0} = -(\varphi_{P_0}(T_R(\Delta)) + K_{P_0})$$

and

$$\varphi_{P_0}(\Xi) + \varphi_{P_0}(R) + K_{P_0} = -(\varphi_{P_0}(T_R(\Xi)) + K_{P_0}),$$

the right-hand side of the last equation is simply the quotient

$$\frac{\theta^{en^2}[-(\varphi_{P_0}(T_R(\Delta)) + K_{P_0})](0, \Pi)}{\theta^{en^2}[-(\varphi_{P_0}(T_R(\Xi)) + K_{P_0})](0, \Pi)}.$$

Note that in this equality we have used the independence of the lifts: we transformed an equality which holds in $J(X)$ to an equality of theta quotients with the characteristics without knowing anything about their lifts to $\mathbb{C}^g$ (which are needed to define the theta quotients). We can do that since we have taken the power of the theta functions to be en^2. We now use Eq. (1.5) on both the numerator and denominator and conclude that

$$\frac{\theta^{en^2}[\Delta](\varphi_{P_0}(R), \Pi)}{\theta^{en^2}[\Xi](\varphi_{P_0}(R), \Pi)} = \frac{\theta^{en^2}[T_R(\Delta)](0, \Pi)}{\theta^{en^2}[T_R(\Xi)](0, \Pi)}. \qquad (2.1)$$

Similarly, if in the singular case S is a point Q_i which does not appear either in $N(\Delta)$ or in $N(\Xi)$ (i.e., S lies both in the sets B_Δ and B_Ξ) then Eq. (1.7) with $\zeta = 0$ translates to be

$$\frac{\theta^{en^2}[\Delta](\varphi_{P_0}(S), \Pi)}{\theta^{en^2}[\Xi](\varphi_{P_0}(S), \Pi)} = \frac{\theta^{en^2}[\varphi_{P_0}(\Delta) + \varphi_{P_0}(S) + K_{P_0}](0, \Pi)}{\theta^{en^2}[\varphi_{P_0}(\Xi) + \varphi_{P_0}(S) + K_{P_0}](0, \Pi)},$$

and since Proposition 2.19 gives

$$\varphi_{P_0}(\Delta) + \varphi_{P_0}(S) + K_{P_0} = -(\varphi_{P_0}(T_S(\Delta)) + K_{P_0})$$

and

$$\varphi_{P_0}(\Xi) + \varphi_{P_0}(S) + K_{P_0} = -(\varphi_{P_0}(T_S(\Xi)) + K_{P_0}),$$

the right-hand side of the last equation is simply the quotient

$$\frac{\theta^{en^2}[-(\varphi_{P_0}(T_S(\Delta))+K_{P_0})](0,\Pi)}{\theta^{en^2}[-(\varphi_{P_0}(T_S(\Xi))+K_{P_0})](0,\Pi)}$$

(using again the independence of the lift from $J(X)$ to $\mathbb{C}^g$). Applying Eq. (1.5) to both the numerator and denominator yields that

$$\frac{\theta^{en^2}[\Delta](\varphi_{P_0}(S),\Pi)}{\theta^{en^2}[\Xi](\varphi_{P_0}(S),\Pi)} = \frac{\theta^{en^2}[T_S(\Delta)](0,\Pi)}{\theta^{en^2}[T_S(\Xi)](0,\Pi)}. \qquad (2.2)$$

Since in all cases T_R is an involution, the divisor Ξ we should take in Eq. (2.1) for a divisor Δ and a point R in C_0 is $\Xi = T_R(\Delta)$, which would yield the equation

$$\frac{\theta^{en^2}[\Delta](\varphi_{P_0}(R),\Pi)}{\theta^{en^2}[T_R(\Delta)](\varphi_{P_0}(R),\Pi)} = \frac{\theta^{en^2}[T_R(\Delta)](0,\Pi)}{\theta^{en^2}[\Delta](0,\Pi)}. \qquad (2.3)$$

Similarly, the fact that in the singular case T_S is also an involution suggests that the divisor Ξ we should take in Eq. (2.2) for a divisor Δ and a point S in B is $\Xi = T_S(\Delta)$, and this would give that

$$\frac{\theta^{en^2}[\Delta](\varphi_{P_0}(S),\Pi)}{\theta^{en^2}[T_S(\Delta)](\varphi_{P_0}(S),\Pi)} = \frac{\theta^{en^2}[T_S(\Delta)](0,\Pi)}{\theta^{en^2}[\Delta](0,\Pi)}. \qquad (2.4)$$

2.6.4 Quotients of Theta Functions as Meromorphic Functions on Z_n Curves

For any divisors Δ and Ξ, Proposition 2.20 shows that the divisor of the meromorphic function

$$\frac{\theta^{en^2}[\Delta](\varphi_{P_0}(P),\Pi)}{\theta^{en^2}[\Xi](\varphi_{P_0}(P),\Pi)}$$

on X is $[N(\Delta)/N(\Xi)]^{en^2}$. In particular the divisor of

$$\frac{\theta^{en^2}[\Delta](\varphi_{P_0}(P),\Pi)}{\theta^{en^2}[T_R(\Delta)](\varphi_{P_0}(P),\Pi)}$$

is $[N(\Delta)/N(T_R(\Delta))]^{en^2}$ and in the singular case the divisor of

$$\frac{\theta^{en^2}[\Delta](\varphi_{P_0}(P),\Pi)}{\theta^{en^2}[T_S(\Delta)](\varphi_{P_0}(P),\Pi)}$$

is $\left[N(\Delta)/N(T_S(\Delta))\right]^{en^2}$. We also note that the remark made at the end of Section 2.5 gives that in the function

$$\frac{\theta^{en^2}[\Delta](\varphi_{P_0}(P),\Pi)}{\theta^{en^2}[T_R(\Delta)](\varphi_{P_0}(P),\Pi)}$$

there are only two branch points, P_0 and R, that we can substitute and get nontrivial information, and in the singular case it gives that in the function

$$\frac{\theta^{en^2}[\Delta](\varphi_{P_0}(P),\Pi)}{\theta^{en^2}[T_S(\Delta)](\varphi_{P_0}(P),\Pi)}$$

the two points P_0 and S are the only substitutions of branch points which yield nontrivial information. Another thing we note is that the divisors we get are all supported on branch points (inevitably) and that these branch points all appear to powers which are multiples of n (actually of n^2, but n suffices for what we want to say). Since these divisors are all of degree 0 (as powers of quotients of divisors of degree g), we see that this meromorphic function must be a constant multiple of a rational function of z. If, for every point P_i appearing to powers na_i in the numerator and nb_i in the denominator, we put a multiplier $(z-\lambda_i)^{a_i-b_i}$, and in the singular case, for every point Q_i appearing to powers nc_i in the numerator and nd_i in the denominator, we put a multiplier $(z-\mu_i)^{c_i-d_i}$, we get altogether a function whose divisor is the same as $\left[N(\Delta)/N(\Xi)\right]^{en^2}$, as we claimed.

The construction of the function in the last paragraph assumes that no branch point lies over ∞, as otherwise the expression $z-\lambda_i$ or $z-\mu_i$ for the relevant i is meaningless. On the other hand, the calculations we did for the theta quotients do not see whether the branch points dealt with lie over ∞ or not. Let us describe what is changed if there is a branch point over ∞ which appears in the divisors. Clearly, the divisors of the quotients of the theta functions do not change. Our first observation is that if no branch point over ∞ is involved, then the rational function of z we constructed indeed has neither a zero nor a pole at any point over ∞ (as at $z=\infty$ it attains the nonzero finite value 1). Therefore, if the branch point over ∞ appears in $N(\Delta)^{en^2}$ to the power na and in $N(\Xi)^{en^2}$ to the power nb, then by simply erasing the meaningless expression $(z-\infty)^{a-b}$ obtained here we have decreased the degree of the numerator by a and the degree of the denominator by b, so that the difference between the degrees of the numerator and denominator is $b-a$. Since every degree of the denominator contributes n to the order at the branch point over ∞ and every degree of the numerator contributes $-n$ to that order, we find that the order of the function thus constructed (with the $z-\infty$ multiplier omitted) has a zero of degree $na-nb$ at this branch point (and a pole of order equalling $nb-na$ if $na-nb$ is negative). This is consistent with the meaning of the divisor $\left[N(\Delta)/N(\Xi)\right]^{en^2}$. Therefore if there is a point over ∞ we do exactly the same construction and simply omit any meaningless expression we encounter. We summarize the above discussion as

Proposition 2.21. *For any two divisors Δ and Ξ, the function*

$$f(P) = \frac{\theta[\Delta]^{en^2}(\varphi_{P_0}(P),\Pi)}{\theta[\Xi]^{en^2}(\varphi_{P_0}(P),\Pi)}$$

is a well-defined meromorphic function on the Z_n curve X, which is independent of the lifts of $\varphi_{P_0}(\Delta)$, $\varphi_{P_0}(\Xi)$, and K_{P_0} from $J(X)$ to $\mathbb{C}^g$. This function has the same divisor $[N(\Delta)/N(\Xi)]^{en^2}$ as the function

$$\prod_Q (z - z(Q))^{en[v_Q(N(\Delta)) - v_Q(N(\Xi))]}$$

where the product runs over all the branch points on X and $v_Q(\Gamma)$ is the power to which the point Q appears in the divisor Γ, and where the expressions $z - \infty$ are omitted if appearing. Hence $f(P)$ is a nonzero constant multiple of this rational function of z.

Before continuing, let us look at two examples drawn from the nonsingular case. These examples will give us a taste of how the functions we have just constructed look. As we shall do in practice, we take the divisor Ξ to be $T_R(\Delta)$. For odd n, say $n = 7$, we have

$$\Delta = A^6 C_0^5 C_1^4 C_2^3 C_3^2 C_4^1 C_5^0, \quad N(\Delta) = A^6 C_5^5 C_4^4 C_3^3 C_2^2 C_1^1 C_0^0,$$

$$T_R(\Delta) = (C_0 \setminus R)^6 (A \cup R)^5 C_5^4 C_4^3 C_3^2 C_2^1 C_1^0, \quad \text{and} \quad N(T_R(\Delta)) = (C_0 \setminus R)^6 C_1^5 C_2^4 C_3^3 C_4^2 C_5^1,$$

and thus $N(\Delta)/N(T_R(\Delta)) = A^6 C_5^4 C_4^2/(C_0 \setminus R)^6 C_1^4 C_2^2$. Therefore the divisor of the meromorphic function we constructed is the 98th power of this divisor, and hence we have constructed the function

$$c \frac{\prod_{P_i \in A}(z - \lambda_i)^{84} \prod_{P_i \in C_5}(z - \lambda_i)^{56} \prod_{P_i \in C_4}(z - \lambda_i)^{28}}{\prod_{P_i \in (C_0 \setminus R)}(z - \lambda_i)^{84} \prod_{P_i \in C_1}(z - \lambda_i)^{56} \prod_{P_i \in C_2}(z - \lambda_i)^{28}},$$

where c is some nonzero complex constant. Note that even though the points of the set C_3 are neither zeros nor poles of this function, substituting them in the theta quotient is not informative and yields $0/0$. For an example of even n we take say $n = 8$, where we have

$$\Delta = A^7 C_0^6 C_1^5 C_2^4 C_3^3 C_4^2 C_5^1 C_6^0, \quad N(\Delta) = A^7 C_6^6 C_5^5 C_4^4 C_3^3 C_2^2 C_1^1 C_0^0,$$

$$T_R(\Delta) = (C_0 \setminus R)^7 (A \cup R)^6 C_6^5 C_5^4 C_4^3 C_3^2 C_2^1 C_1^0, \quad \text{and} \quad N(T_R(\Delta)) = (C_0 \setminus R)^7 C_1^6 C_2^5 C_3^4 C_4^3 C_5^2 C_6^1,$$

and therefore $N(\Delta)/N(T_R(\Delta)) = A^7 C_6^5 C_5^3 C_4/(C_0 \setminus R)^7 C_1^5 C_2^3 C_3$. The divisor of the meromorphic function that we constructed is the 64th power of this divisor, and hence the function we have constructed is

$$c \frac{\prod_{P_i \in A}(z - \lambda_i)^{56} \prod_{P_i \in C_6}(z - \lambda_i)^{40} \prod_{P_i \in C_5}(z - \lambda_i)^{24} \prod_{P_i \in C_4}(z - \lambda_i)^8}{\prod_{P_i \in (C_0 \setminus R)}(z - \lambda_i)^{56} \prod_{P_i \in C_1}(z - \lambda_i)^{40} \prod_{P_i \in C_2}(z - \lambda_i)^{24} \prod_{P_i \in C_3}(z - \lambda_i)^8},$$

where c is some nonzero complex constant. Here one really sees that every branch point except for P_0 and R is either a zero or a pole of this function. As the previous paragraph explains, if there is a branch point P_∞ over ∞ (in any case) and this point lies in A or in C_j then the function we have constructed is the one we have written (for the relevant case) with the factor $z - \infty$ simply omitted.

Chapter 3

Examples of Thomae Formulae

In this chapter we present some examples of Thomae formulae for the singular and nonsingular cases. These examples are intended to help the reader assimilate the general cases.

3.1 A Nonsingular Z_3 Curve with Six Branch Points

In this section we deal with the Riemann surface

$$w^3 = \prod_{i=0}^{5}(z - \lambda_i),$$

where $\lambda_i \neq \lambda_j$ for $i \neq j$. This is a nonsingular Z_n curve with $r = 2$, $n = 3$ and hence $en^2 = 18$, and the branch point on X which lies over λ_i is denoted as usual by P_i. The genus of this Riemann surface X is $(3-1)(2 \times 3 - 2)/2 = 4$, and we recall from Section 2.2 that the non-special divisors of degree $g = 4$ on X which are supported on the branch points distinct from P_0 are all the divisors of the form $P_i^2 P_j P_k$, which is $A^2 C_0^1 C_1^0$ with $A = \{P_i\}$, $C_0 = \{P_j, P_k\}$ and $C_1 = \{P_l, P_m\}$ (the complement). Throughout this entire section the indices i, j, k, l and m are always assumed to be 1, 2, 3, 4 and 5 in some order (i.e., being all distinct), so that the complement C_1 is indeed as noted. Using the explicit formulas for the operators N and T_R from Definition 2.16 in Section 2.5 we find that on $\Delta = P_i^2 P_j P_k$ only N, T_{P_j} and T_{P_k} can act, and that $N(\Delta) = P_i^2 P_l P_m$, $T_{P_j}(\Delta) = P_k^2 P_i P_j$ and $T_{P_k}(\Delta) = P_j^2 P_i P_k$.

H.M. Farkas and S. Zemel, *Generalizations of Thomae's Formula for Z_n Curves*, Developments in Mathematics 21, DOI 10.1007/978-1-4419-7847-9_3, © Springer Science+Business Media, LLC 2011

3.1.1 First Identities Between Theta Constants

Since near the end of Section 2.6 we saw that the "best" quotients to look at were those with a divisor Δ and the other divisor being $T_R(\Delta)$ for a permissible branch point R, we begin with the function

$$\frac{\theta^{18}[\Delta](\varphi_{P_0}(P),\Pi)}{\theta^{18}[T_{P_k}(\Delta)](\varphi_{P_0}(P),\Pi)} = \frac{\theta^{18}[P_i^2 P_j P_k](\varphi_{P_0}(P),\Pi)}{\theta^{18}[P_j^2 P_i P_k](\varphi_{P_0}(P),\Pi)}.$$

Since $N(\Delta) = P_i^2 P_l P_m$ and $N(T_{P_k}(\Delta)) = P_j^2 P_l P_m$, Proposition 2.21 gives us that this divisor of the function is the 18th power of P_i^2/P_j^2, i.e., P_i^{36}/P_j^{36}, and that this function is $c(z-\lambda_i)^{12}/(z-\lambda_j)^{12}$ where c is a nonzero complex constant.

If we now set $P = P_0$, we find that

$$\frac{\theta^{18}[P_i^2 P_j P_k](0,\Pi)}{\theta^{18}[P_j^2 P_i P_k](0,\Pi)} = c\frac{(\lambda_0 - \lambda_i)^{12}}{(\lambda_0 - \lambda_j)^{12}},$$

and therefore

$$\frac{\theta^{18}[P_i^2 P_j P_k](\varphi_{P_0}(P),\Pi)}{\theta^{18}[P_j^2 P_i P_k](\varphi_{P_0}(P),\Pi)} = \frac{(\lambda_0 - \lambda_j)^{12}}{(\lambda_0 - \lambda_i)^{12}} \times \frac{\theta^{18}[P_i^2 P_j P_k](0,\Pi)}{\theta^{18}[P_j^2 P_i P_k](0,\Pi)} \times \frac{(z-\lambda_i)^{12}}{(z-\lambda_j)^{12}}.$$

As already observed above, there is only one other substitution which is meaningful, and that is to set $P = P_k$. Under this substitution, we find that, on the one hand, the last equation gives us

$$\frac{\theta^{18}[P_i^2 P_j P_k](\varphi_{P_0}(P_k),\Pi)}{\theta^{18}[P_j^2 P_i P_k](\varphi_{P_0}(P_k),\Pi)} = \frac{(\lambda_0 - \lambda_j)^{12}}{(\lambda_0 - \lambda_i)^{12}} \times \frac{\theta^{18}[P_i^2 P_j P_k](0,\Pi)}{\theta^{18}[P_j^2 P_i P_k](0,\Pi)} \times \frac{(\lambda_k - \lambda_i)^{12}}{(\lambda_k - \lambda_j)^{12}},$$

and on the other hand, Eq. (2.3) gives us

$$\frac{\theta^{18}[P_i^2 P_j P_k](\varphi_{P_0}(P_k),\Pi)}{\theta^{18}[P_j^2 P_i P_k](\varphi_{P_0}(P_k),\Pi)} = \frac{\theta^{18}[P_j^2 P_i P_k](0,\Pi)}{\theta^{18}[P_i^2 P_j P_k](0,\Pi)},$$

recall that $T_{P_k}(P_i^2 P_j P_k) = P_j^2 P_i P_k$. Hence we find by combining the last two equations that

$$\frac{\theta^{36}[P_j^2 P_i P_k](0,\Pi)}{\theta^{36}[P_i^2 P_j P_k](0,\Pi)} = \frac{(\lambda_0 - \lambda_j)^{12}}{(\lambda_0 - \lambda_i)^{12}} \times \frac{(\lambda_k - \lambda_i)^{12}}{(\lambda_k - \lambda_j)^{12}},$$

and hence

$$\frac{\theta^{36}[P_i^2 P_j P_k](0,\Pi)}{(\lambda_0 - \lambda_i)^{12}(\lambda_k - \lambda_j)^{12}} = \frac{\theta^{36}[P_j^2 P_i P_k](0,\Pi)}{(\lambda_0 - \lambda_j)^{12}(\lambda_k - \lambda_i)^{12}}.$$

The above identity has been derived using the theta functions with the characteristic coming from $\Delta = P_i^2 P_j P_k$ and its image by T_{P_k}. There is another operator T_R that can act on Δ, the operator T_{P_j}, by which the image of Δ is $P_k^2 P_i P_j$. By starting with the function

$$\frac{\theta^{18}[\Delta](\varphi_{P_0}(P), \Pi)}{\theta^{18}[T_{P_j}(\Delta)](\varphi_{P_0}(P), \Pi)} = \frac{\theta^{18}[P_i^2 P_j P_k](\varphi_{P_0}(P), \Pi)}{\theta^{18}[P_k^2 P_i P_j](\varphi_{P_0}(P), \Pi)}$$

we obtain, using the same calculations with the roles of j and k interchanged, a corresponding result, which is

$$\frac{\theta^{36}[P_i^2 P_j P_k](0, \Pi)}{(\lambda_0 - \lambda_i)^{12}(\lambda_j - \lambda_k)^{12}} = \frac{\theta^{36}[P_k^2 P_i P_j](0, \Pi)}{(\lambda_0 - \lambda_k)^{12}(\lambda_j - \lambda_i)^{12}}.$$

We can, of course, combine these two results into one larger equation

$$\frac{\theta^{36}[P_i^2 P_j P_k](0, \Pi)}{(\lambda_0 - \lambda_i)^{12}(\lambda_k - \lambda_j)^{12}} = \frac{\theta^{36}[P_j^2 P_i P_k](0, \Pi)}{(\lambda_0 - \lambda_j)^{12}(\lambda_k - \lambda_i)^{12}} = \frac{\theta^{36}[P_k^2 P_i P_j](0, \Pi)}{(\lambda_0 - \lambda_k)^{12}(\lambda_i - \lambda_j)^{12}}. \quad (3.1)$$

The above formula is an identity between three terms, each one being a theta constant divided by a polynomial in the variables λ_i, $0 \leq i \leq 5$, defining the Riemann surface X. We could have started from any divisor Δ (i.e., any choice of the indices i, j, k, l, and m) and obtain a formula connecting the theta constant corresponding to Δ with two other theta constants, those corresponding to the two divisors $T_R(\Delta)$ where R is taken from $(C_0)_\Delta$ (which is of cardinality 2). It is clear from the form of Eq. (3.1) that starting from any other divisor appearing in it, i.e., from $T_R(\Delta)$ in place of Δ, we obtain exactly the same equality. We now prove

Proposition 3.1. *For a divisor Δ, define g_Δ to be the product of the 12th power of the difference between λ_0 and the z-value of the only branch point appearing in Δ to the 2nd power, and the 12th power of the difference between the z-values of the two branch points appearing in Δ to the 1st power. Then the quotient $\theta^{2en^2}[\Delta](0, \Pi)/g_\Delta$ is invariant under all the operators T_R.*

Proof. We show that the assertion of the proposition is simply Eq. (3.1). First we note that the denominator appearing under each theta constant in Eq. (3.1) can be constructed from the corresponding divisor in the way described in the proposition. It remains to prove that Eq. (3.1) gives the relations connecting all the divisors in a set closed under all the operators T_R. We know that the two divisors which appear with the divisor $\Delta = P_i^2 P_j P_k$ with which we started, are its images under the possible two operators T_R, namely T_{P_k} and T_{P_j}. As for the other two divisors, while the first one is taken by T_{P_k} back to Δ, and the second one is taken by T_{P_j} back to Δ (the operators T_R are involutions), there is one other operator T_R that can act on these divisors, which is T_{P_i} for both of them. Since T_{P_i} takes these two divisors to one another, the proposition is proved. □

3.1.2 The Thomae Formulae

The Thomae formulae are formulae that state that for every divisor Δ there is
a polynomial h_Δ in the parameters λ_i defining the Riemann surface such that
$\theta^{2en^2}[\Delta](0,\Pi)/h_\Delta$ is constant (i.e., independent of Δ). Proposition 3.1 is a result
in the right direction, but it gives many relations (one for each set closed under the
operators T_R), each one connecting exactly 3 theta constants. A priori there is no
connection between two different sets of 3 terms. We thus call Proposition 3.1 and
Eq. (3.1) a *PMT (Poor Man's Thomae)*. The full Thomae formulae will be obtained
by combining all the PMT formulae, (3.1) for different choices of Δ, into one larger
formula.

We remark that the PMT is an important step in the proof of the general case
as well. Unlike here however, where we obtained the PMT almost directly from
Eq. (2.3), in the general case we will need to do some manipulations on the results
obtained from Eq. (2.3) in order to obtain the PMT. As these manipulations will resemble the operations we apply, here and in general, to obtain the Thomae formulae
from the PMT, it is good to enter the subject with an example where the manipulations are needed only at one stage of the proof, and where they are simple at that
stage as well.

First, let us count how many formulas the PMT gives us, i.e., how many different numbers appear as a result of Eq. (3.1) when Δ runs over all the possible
divisors. Another way of looking at it is to ask how many minimal sets of divisors
that are closed under the operators T_R are there. The answer is given by taking the
number of our integral divisors and dividing it by 3. Now, since the first point P_i
can be chosen in 5 ways, and then we need choose two more points from the four
remaining points (which can be done in 6 ways), there are clearly 30 such divisors
(looking forward to the general case, 30 is $5!/(2!)^{3-1}(2-1)!$). Hence Eq. (3.1) is
in fact 10 different equalities, which can all give different values. We will now show
how to connect these equations. The main tool which connects values that appear
in two different formulae of PMT is Eq. (1.5), which states, in our language, that
$\theta[N(\Delta)](0,\Pi) = \theta[\Delta](0,\Pi)$ (actually the expression written here depends on the
lifts of the characteristics from $J(X)$ to $\mathbb{C}^g$, but the powers which we use do not).
This indeed helps, since Δ and $N(\Delta)$ do not appear in the same PMT. Therefore we
need to change the denominator g_Δ from Proposition 3.1 to something which allows
us to make these connections. We now present how this is done.

For $\Delta = P_i^2 P_j P_k$, $g_\Delta = (\lambda_0 - \lambda_i)^{12}(\lambda_j - \lambda_k)^{12}$, while $N(\Delta) = P_i^2 P_l P_m$ and thus
$g_{N(\Delta)} = (\lambda_0 - \lambda_i)^{12}(\lambda_l - \lambda_m)^{12}$. In order to combine them we should divide (3.1)
for a divisor Δ by $(\lambda_l - \lambda_m)^{12}$. We explain first why this is a "good" operation on
the PMT and then why this operation allows us to obtain the full Thomae formulae.
First, the new denominator is easily constructed from Δ: Proposition 3.1 shows us
how to construct g_Δ, and the extra multiplier $(\lambda_l - \lambda_m)^{12}$ can be described as the 12th
power of the difference between the z-values of the branch points not appearing in
Δ (which are not P_0, of course). Now, as this set of points, i.e., C_1, is invariant under
the operators T_R, looking at Eq. (3.1) as coming from $T_R(\Delta)$ and not Δ will tell us to

divide it by the same expression. If we denote the "new denominator" for Δ by h_Δ, then we have obtained a "new PMT", i.e., a formula with the same properties as the PMT, which states that

$$\frac{\theta^{36}[\Delta](0,\Pi)}{h_\Delta} = \frac{\theta^{36}[P_i^2 P_j P_k](0,\Pi)}{(\lambda_0 - \lambda_i)^{12}(\lambda_j - \lambda_k)^{12}(\lambda_l - \lambda_m)^{12}} \tag{3.2}$$

is invariant under the operators T_R.

We would not have passed from the old PMT in Eq. (3.1) to the new one in Eq. (3.2) if Eq. (3.2) did not have an advantage over Eq. (3.1). The advantage is clear once we examine Eq. (3.2) involving $N(\Delta)$ in comparison with the one involving Δ. When doing so we see, since $N(\Delta) = P_i^2 P_l P_m$ for this Δ, that the denominator under $\theta^{36}[N(\Delta)](0,\Pi)$ is $(\lambda_0 - \lambda_i)^{12}(\lambda_l - \lambda_m)^{12}(\lambda_j - \lambda_k)^{12}$ (either by the construction of the new denominator from the divisor, or simply by dividing Eq. (3.1) for $N(\Delta)$ by the missing expression $(\lambda_j - \lambda_k)^{12}$), which is the same as the denominator appearing under $\theta^{36}[\Delta](0,\Pi)$ in its corresponding Eq. (3.2). As we said before, the numerators are also the same, and hence Eq. (3.2) involving $N(\Delta)$ and the one involving Δ give the same constant. This can be expressed by the assertion that the expression in Eq. (3.2) is invariant not only under the operators T_R (the permissible ones at each point, of course), but also under the operator N, which means that this relation is constant on much larger sets of divisors.

We now prove that Eq. (3.2) is the Thomae formulae for our Z_3 curve, or equivalently

Theorem 3.2. *The expression appearing in Eq. (3.2) is independent of the divisor* Δ.

Proof. We already showed that this expression is invariant both under the operators T_R and under N. It remains to show that the only nonempty set of divisors which is closed under the operators T_R and N is the set containing all the divisors. For this we begin with $P_1^2 P_2 P_3$ and write a list of all the divisors in its orbit under the operators T_R and N. We do this in the form of an array, in which in each row appear 3 divisors that are connected by Eq. (3.1), and each new row begins with the N-image of a divisor already considered. Therefore in addition to the equality of the quotients corresponding to the items of that row, we also have the equality of them to the quotients corresponding to all the items preceding it. A new divisor whose N-image already appears in the array will appear in parentheses.

$$P_1^2P_2P_3 \quad P_2^2P_1P_3 \quad P_3^2P_1P_2$$
$$(P_1^2P_4P_5) \quad P_4^2P_1P_5 \quad P_5^2P_1P_4$$
$$(P_2^2P_4P_5) \quad P_4^2P_2P_5 \quad P_5^2P_2P_4$$
$$(P_3^2P_4P_5) \quad P_4^2P_3P_5 \quad P_5^2P_3P_4$$
$$(P_4^2P_2P_3) \quad P_2^2P_3P_4 \quad P_3^2P_2P_4$$
$$(P_5^2P_2P_3) \quad P_2^2P_3P_5 \quad P_3^2P_2P_5$$
$$(P_4^2P_1P_3) \quad P_1^2P_3P_4 \quad (P_3^2P_1P_4)$$
$$(P_5^2P_1P_3) \quad P_1^2P_3P_5 \quad (P_3^2P_1P_5)$$
$$(P_4^2P_1P_2) \quad (P_1^2P_2P_4) \quad (P_2^2P_1P_4)$$
$$(P_5^2P_1P_2) \quad (P_1^2P_2P_5) \quad (P_2^2P_1P_5)$$

Since we have 10 rows of 3 divisors each (and 15 divisors appearing without parentheses and 15 appear with them, as expected), we have obtained all the 30 nonspecial divisors of degree 4 supported on the branch points distinct from P_0 on X. This means that one can get from any divisor to any other divisor using these operators alone, which proves the theorem.   □

Theorem 3.2 states that the expression from Eq. (3.2) represents the Thomae formulae for the nonsingular Z_3 curve

$$w^3 = \prod_{i=0}^{5}(z - \lambda_i).$$

We now say a few words about the construction of the denominator h_Δ from the divisor Δ. This will help to understand the general construction. We already saw that the denominator h_Δ is the product of the 12th power of 3 differences: the one between λ_0 and the z-value of the branch point appearing in Δ to the 2nd power, the one between the z-values of the branch points appearing in Δ to the 1st power, and the one between the z-values of the branch points not appearing in Δ at all and which are not P_0. In the language of our sets, let us denote $A \cup P_0$ by C_{-1} (as we shall do in the general case to come—note that the cardinality condition on A in the general setting then becomes $|C_{-1}| = r$, as for the other sets C_j), and then h_Δ is simply the product of the 12th powers of all the differences between z-values of points which belong to the same set C_j (including $j = -1$!) of Δ. This is the correct, and in our opinion simplest, way to look at h_Δ.

3.1.3 Changing the Basepoint

Theorem 3.2 states that the quotient appearing in Eq. (3.2) is independent of the divisor Δ, which is a beautiful and symmetric formula once we know that h_Δ is constructed from Δ in an intrinsic manner. However, the careful reader may have noticed that there is still a choice we made, that may change the constant value of the quotient. This is the choice of the basepoint P_0, appearing implicitly in the fact that the characteristic of $\theta[\Delta]$ is $\varphi_{P_0}(\Delta) + K_{P_0}$. We now show that the constant value of the quotient is independent of the basepoint as well. This means that if we begin with another basepoint P_i and prove Theorem 3.2 for divisors whose support does not contain P_i and with the characteristic $\varphi_{P_i}(\Delta) + K_{P_i}$ for such divisors Δ, the constant value assured by Theorem 3.2 with basepoint P_i would bc the same as we obtained here with the basepoint P_0. We shall also give the "basepoint-independent" form of the Thomae formulae.

The tool to show that every choice of basepoint gives the same constant for the quotient is Corollary 1.13. By Lemma 2.4 the branch points satisfy the condition that $\varphi_{P_i}(P_j^n) = 0$ for every i and j with $n = 3$, and since all our divisors are integral, of degree $g = 4$, supported on the branch points (all of them—P_0 is no longer excluded as we allow any branch point to be the basepoint), and contain no branch point to the power $n = 3$ or higher, Corollary 1.13 can indeed by applied. Now, let us start with our divisors and basepoint P_0, and we want to show that the basepoint P_i for some $1 \le i \le 5$ gives the same constant. Choose a divisor Δ (with basepoint P_0) of the form $P_i^2 P_j P_k$ with the chosen point P_i appearing to the 2nd power, and then the divisor Γ_{P_i} from Corollary 1.13 is $P_0^2 P_j P_k$ and is non-special by Theorem 2.6 (which knows nothing about the basepoint P_0). It is clear that the sets C_0 and C_1 corresponding to Γ_{P_i} are the same as those corresponding to Δ, but A is now composed of P_0 instead of P_i. However, since for Γ_{P_i} the set C_{-1} is $A \cup P_i$ (and not P_0), we do get the same set C_{-1} as well, which also means that Eq. (3.2) gives the same expression for $h_{\Gamma_{P_i}}$ as for h_Δ. Now we can apply Corollary 1.13 to obtain the equality $\varphi_{P_i}(\Gamma_{P_i}) + K_{P_i} = \varphi_{P_0}(\Delta) + K_{P_0}$; this gives that the constant quotient obtained from the basepoint P_i (represented by the divisor Γ_{P_i}) is the same as the one from P_0 (represented by the divisor Δ), since these representatives have the same numerator and the same denominator.

Another thing that may change with the choice of the basepoint is the set of theta constants connected by the Thomae formulae, i.e., the set of characteristics one obtains for the theta function when going over all the divisors Theorems 2.6 and 2.9 give for any basepoint. However, we can use Proposition 1.14 to show that this is not the case. First, the divisors satisfy the assumptions of Proposition 1.14, and more importantly Lemma 2.5 gives the extra assumption, so that Proposition 1.14 is indeed applicable. Now, for any characteristic $\varphi_{P_0}(\Delta) + K_{P_0}$ obtained from the basepoint P_0 and any $1 \le i \le 5$, Proposition 1.14 gives that we have a divisor Γ_{P_i} such that $\varphi_{P_i}(\Gamma_{P_i}) + K_{P_i}$ gives the same characteristic, and since Γ_{P_i} is of the same form as Δ, Theorem 2.6 gives that Γ_{P_i} is non-special as well, and with P_i not appearing in its support. The converse follows either by an argument of symmetry or by the fact that the maps $\varphi_{P_i}(\cdot) + K_{P_i}$ are one-to-one on the relevant sets of divisors (as they are non-special—recall the remarks after Propositions 1.14 and 1.15) and the fact that there is the same number of divisors Γ with the basepoint P_i as divisors Δ with the basepoint P_0. Altogether the Thomae formulae from Theorem 3.2 with the 6 distinct branch points P_i, $0 \le i \le 5$, not only give the same constant, but involve the same theta constants. In order to demonstrate this we use the table we already have of the 30 divisors with basepoint P_0 appearing in the proof of Theorem 3.2. Choose also the basepoint P_5, and write the 30 divisors with basepoint P_5 in a table of the same form such that the entries give the same characteristics when applying $\varphi_{P_0}(\cdot) + K_{P_0}$ to those from the first table and $\varphi_{P_5}(\cdot) + K_{P_5}$ to those from the second. Note that in this table the rows are no longer invariant under the operators T_R (which can be defined for any basepoint). This gives (we omit the parentheses inside, which are not important for us here)

$$
\begin{array}{ccc}
P_4^2 P_0 P_1 & P_4^2 P_0 P_2 & P_4^2 P_0 P_3 \\
P_4^2 P_2 P_3 & P_1^2 P_2 P_3 & P_0^2 P_1 P_4 \\
P_4^2 P_1 P_3 & P_2^2 P_1 P_3 & P_0^2 P_2 P_4 \\
P_4^2 P_1 P_2 & P_3^2 P_1 P_2 & P_0^2 P_3 P_4 \\
P_1^2 P_0 P_4 & P_1^2 P_0 P_2 & P_1^2 P_0 P_3 \\
P_0^2 P_2 P_3 & P_3^2 P_1 P_4 & P_2^2 P_1 P_4 \\
P_2^2 P_0 P_4 & P_2^2 P_0 P_1 & P_2^2 P_0 P_3 \\
P_0^2 P_1 P_3 & P_3^2 P_2 P_4 & P_1^2 P_2 P_4 \\
P_3^2 P_0 P_4 & P_3^2 P_0 P_1 & P_3^2 P_0 P_2 \\
P_0^2 P_1 P_2 & P_2^2 P_3 P_4 & P_1^2 P_3 P_4.
\end{array}
$$

The reader can, as an exercise, use Proposition 1.14 to verify some of the entries in this table in both directions—either change the basepoint from P_0 to P_5 or change the basepoint from P_5 to P_0.

The discussion in the last two paragraphs suggests that we should perhaps adopt a different notation for the Thomae formulae in Theorem 3.2, a notation which will be indeed independent of the basepoint. The use of Δ is not good for this purpose, as we find that $\varphi_{P_0}(\Delta) + K_{P_0}$ equals $\varphi_{P_i}(\Gamma_{P_i}) + K_{P_i}$ rather than $\varphi_{P_i}(\Delta) + K_{P_i}$, and $\Gamma_{P_i} \neq \Delta$. Maintaining this notation has the property of giving the zeros of the theta function on the surface, but this presentation of the zeros of the theta function also depends on the basepoint we choose (Theorem 1.12 gives that $\varphi_Q(\Delta) + K_Q$ is independent of Q for divisors Δ of degree $g - 1$, and here our divisors have degree g), so that a basepoint-independent notation will lose this property anyway. Now, it is clear that if P_i appears to the 2nd power in Δ, then $P_0^2 \Delta = P_i^2 \Gamma_{P_i}$ and in this divisor (let us call it Ξ) P_0 and P_i play the same role (being in C_{-1}), and it is also clear that $\varphi_{P_0}(\Xi) = \varphi_{P_0}(\Delta)$ and $\varphi_{P_i}(\Xi) = \varphi_{P_i}(\Gamma_{P_i})$. We can thus replace the notation $\theta[\Delta]$ by $\theta[\Xi]$, where we define $\theta[\Xi]$ to be the theta function with characteristic $\varphi_{P_t}(\Xi) + K_{P_t}$ for any $0 \leq t \leq 5$ and this is independent of the choice of P_t by the second statement of Corollary 1.13. If we denote the expression for h_Δ appearing in Eq. (3.2) by h_Ξ (a more suitable notation since now both Ξ and h_Ξ are based on C_{-1}, rather than A), then we have actually proved

Theorem 3.3. *For every divisor $\Xi = C_{-1}^2 C_0^1 C_1^0$ where C_{-1}, C_0, and C_1 are disjoint and contain two branch points each (hence form a partition of the set of 6 branch points into $n = 3$ sets of cardinality $r = 2$), define $\theta[\Xi]$ to be the theta function with characteristic $\varphi_{P_t}(\Xi) + K_{P_t}$ for some $0 \leq t \leq 5$ (which is independent of the choice of P_t), and define h_Ξ to be the expression from Eq. (3.2). Then the quotient $\theta^{36}[\Xi](0, \Pi)/h_\Xi$ is independent of the divisor Ξ.*

Theorem 3.3 follows immediately from the discussion above and the fact that any such divisor Ξ comes in this way from any point in C_{-1} (and C_{-1} is not empty). Theorem 3.3 is the most "symmetric" Thomae formulae for our nonsingular Z_3 curve X, and can also give as a corollary the fact that all the Thomae formulae with the 6 different basepoints give the same constant. However, unlike Theorem 3.2 where no two divisors give the same characteristic, here the application M which takes such a divisor Ξ to $\Xi\left(\prod_{i=0}^5 P_i\right)/C_{-1}^3$ takes Ξ to a different divisor which represents (with any branch point as basepoint) the same characteristic as Ξ does. What we do have are $6!/(2!)^3 = 90$ such divisors Ξ, which split into 30 orbits of size 3 of M that give exactly the 30 characteristics.

We close this section by mentioning that the above steps are in large the steps of the proof in the general case. One uses Eq. (2.3) to obtain the first connections, and then "symmetrizes" these connections to obtain the PMT that are invariant under the operators T_R. Then one does some more manipulations to make the expression invariant also under N, and then one proves that this connects all the divisors. In the singular case the proof is similar, but the operators T_S also play their role. One should also notice that in the construction of h_Δ the omitted point P_0 (which is of course not in the support of any of these divisors) actually behaves as if it appears there to the 2nd power. This observation is understood when we look at the form of Theorem 3.3, and we shall see later that this is also the behavior we encounter in both the general nonsingular and singular cases.

3.2 A Singular Z_3 Curve with Six Branch Points

In this section we shall treat the case of the singular curve

$$w^3 = (z-\lambda_0)(z-\lambda_1)(z-\lambda_2)(z-\mu_1)^2(z-\mu_2)^2(z-\mu_3)^2,$$

where $\lambda_i \neq \lambda_j$ for $i \neq j$, $\mu_i \neq \mu_j$ for $i \neq j$ and $\lambda_i \neq \mu_j$ for any i and j. This is a singular Z_n curve with $m=3$ and $n=3$, hence $en^2 = 18$. We have already observed in Chapter 2 that in this case the genus is 4 and there are 31 divisors, which are divided into 6 families. The usual representation of the divisors in the singular case becomes in the case $n=3$ simply $\Delta = A^2 B^0 C_0^1 D_0^1 C_1^0 D_1^2$, and the division into families is according to the values of $|B| = |A| + 1$, $|D_0| = |C_0|$ and $|D_1| = |C_1|$ (recall the cardinality conditions!). In order to make our notation easier, we adopt conventions similar to Section 3.1 and assume throughout that i and j are always 1 and 2 and distinct and k, l and m are always 1, 2 and 3 and distinct (the m defining this singular Z_n curve is known to be 3, so no confusion with the new m will occur). The way of starting the proof in this case, i.e., obtaining the PMT, is more subtle here since divisors in different families do "look different" and other types of operators act on them.

3.2.1 First Identities between Theta Constants

We begin with the first divisor $\Delta = P_1^2 P_2^2$, whose set A is $\{P_1, P_2\}$, B is $\{Q_1, Q_2, Q_3\}$, and the others are empty. On this divisor no T_R can act, but N and all the operators T_S can, giving $N(\Delta) = P_1^2 P_2^2$ and $T_{Q_k}(\Delta) = P_1 P_2 Q_l Q_m$ for any $1 \leq k \leq 3$, where $\{l, m\}$ is the complement of $\{k\}$ in $\{1, 2, 3\}$. Therefore the "best" quotients to look at, as we saw near the end of Section 2.6, are the three quotients

$$\frac{\theta^{18}[\Delta](\varphi_{P_0}(P), \Pi)}{\theta^{18}[T_{Q_k}(\Delta)](\varphi_{P_0}(P), \Pi)} = \frac{\theta^{18}[P_1^2 P_2^2](\varphi_{P_0}(P), \Pi)}{\theta^{18}[P_1 P_2 Q_l Q_m](\varphi_{P_0}(P), \Pi)}$$

for the three possible values of k. We calculate that for any k the divisor $N(T_{Q_k}(\Delta))$ is $Q_l^2 Q_m^2$, so that Proposition 2.21 gives that the divisor of this quotient is the 18th power of $P_1^2 P_2^2 / Q_l^2 Q_m^2$, i.e., $P_1^{36} P_2^{36} / Q_l^{36} Q_m^{36}$, and hence that this quotient is the function

$$c_k \frac{(z - \lambda_1)^{12} (z - \lambda_2)^{12}}{(z - \mu_l)^{12} (z - \mu_m)^{12}},$$

where c_k is a nonzero complex constant.

As in the previous example, we can calculate the constants c_k by setting $P = P_0$, which gives

$$\frac{\theta^{18}[P_1^2 P_2^2](0, \Pi)}{\theta^{18}[P_1 P_2 Q_l Q_m](0, \Pi)} = c_k \frac{(\lambda_0 - \lambda_1)^{12} (\lambda_0 - \lambda_2)^{12}}{(\lambda_0 - \mu_l)^{12} (\lambda_0 - \mu_m)^{12}},$$

and thus we have

$$\frac{\theta^{18}[P_1^2 P_2^2](\varphi_{P_0}(P), \Pi)}{\theta^{18}[P_1 P_2 Q_l Q_m](\varphi_{P_0}(P), \Pi)} = \frac{(\lambda_0 - \mu_l)^{12} (\lambda_0 - \mu_m)^{12}}{(\lambda_0 - \lambda_1)^{12} (\lambda_0 - \lambda_2)^{12}} \times \frac{\theta^{18}[P_1^2 P_2^2](0, \Pi)}{\theta^{18}[P_1 P_2 Q_l Q_m](0, \Pi)}$$

$$\times \frac{(z - \lambda_1)^{12} (z - \lambda_2)^{12}}{(z - \mu_l)^{12} (z - \mu_m)^{12}}.$$

We already know that there is only one other meaningful substitution; this is to set $P = Q_k$. Under this substitution, we find that, on the one hand, the last equation gives us

$$\frac{\theta^{18}[P_1^2 P_2^2](\varphi_{P_0}(Q_k), \Pi)}{\theta^{18}[P_1 P_2 Q_l Q_m](\varphi_{P_0}(Q_k), \Pi)} = \frac{(\lambda_0 - \mu_l)^{12} (\lambda_0 - \mu_m)^{12}}{(\lambda_0 - \lambda_1)^{12} (\lambda_0 - \lambda_2)^{12}} \times \frac{\theta^{18}[P_1^2 P_2^2](0, \Pi)}{\theta^{18}[P_1 P_2 Q_l Q_m](0, \Pi)}$$

$$\times \frac{(\mu_k - \lambda_1)^{12} (\mu_k - \lambda_2)^{12}}{(\mu_k - \mu_l)^{12} (\mu_k - \mu_m)^{12}},$$

and, on the other hand, we obtain from Eq. (2.4) that

$$\frac{\theta^{18}[P_1^2 P_2^2](\varphi_{P_0}(Q_k), \Pi)}{\theta^{18}[P_1 P_2 Q_l Q_m](\varphi_{P_0}(Q_k), \Pi)} = \frac{\theta^{18}[P_1 P_2 Q_l Q_m](0, \Pi)}{\theta^{18}[P_1^2 P_2^2](0, \Pi)};$$

recall that $T_{Q_k}(P_1 P_2 Q_l Q_m) = P_1^2 P_2^2$, as T_{Q_k} is an involution. We can combine the last two equations together and find that

$$\frac{\theta^{36}[P_1 P_2 Q_l Q_m](0, \Pi)}{\theta^{36}[P_1^2 P_2^2](0, \Pi)} = \frac{(\lambda_0 - \mu_l)^{12}(\lambda_0 - \mu_m)^{12}}{(\lambda_0 - \lambda_1)^{12}(\lambda_0 - \lambda_2)^{12}} \times \frac{(\mu_k - \lambda_1)^{12}(\mu_k - \lambda_2)^{12}}{(\mu_k - \mu_l)^{12}(\mu_k - \mu_m)^{12}},$$

and hence

$$\frac{\theta^{36}[P_1^2 P_2^2](0, \Pi)}{(\lambda_0 - \lambda_1)^{12}(\lambda_0 - \lambda_2)^{12}(\mu_k - \mu_l)^{12}(\mu_k - \mu_m)^{12}}$$
$$= \frac{\theta^{36}[P_1 P_2 Q_l Q_m](0, \Pi)}{(\lambda_0 - \mu_l)^{12}(\lambda_0 - \mu_m)^{12}(\mu_k - \lambda_1)^{12}(\mu_k - \lambda_2)^{12}}.$$

The first thing we notice, comparing the two examples, is that the denominator under $\theta^{36}[P_1^2 P_2^2](0, \Pi)$ is different in these 3 equations, which means that in order to obtain the PMT (even involving this divisor alone) we have to do some manipulations on these equations. The way to merge these equalities into one big equation will be by making their left-hand sides equal, which is easily done by dividing the kth equation by $(\mu_l - \mu_m)^{12}$. Doing so we obtain the following equation (with the common left-hand sides appearing first):

$$\frac{\theta^{36}[P_1^2 P_2^2](0, \Pi)}{(\lambda_0 - \lambda_1)^{12}(\lambda_0 - \lambda_2)^{12}(\mu_1 - \mu_2)^{12}(\mu_1 - \mu_3)^{12}(\mu_2 - \mu_3)^{12}}$$
$$= \frac{\theta^{36}[P_1 P_2 Q_2 Q_3](0, \Pi)}{(\lambda_0 - \mu_2)^{12}(\lambda_0 - \mu_3)^{12}(\mu_1 - \lambda_1)^{12}(\mu_1 - \lambda_2)^{12}(\mu_2 - \mu_3)^{12}}$$
$$= \frac{\theta^{36}[P_1 P_2 Q_1 Q_3](0, \Pi)}{(\lambda_0 - \mu_1)^{12}(\lambda_0 - \mu_3)^{12}(\mu_2 - \lambda_1)^{12}(\mu_2 - \lambda_2)^{12}(\mu_1 - \mu_3)^{12}}$$
$$= \frac{\theta^{36}[P_1 P_2 Q_1 Q_2](0, \Pi)}{(\lambda_0 - \mu_1)^{12}(\lambda_0 - \mu_2)^{12}(\mu_3 - \lambda_1)^{12}(\mu_3 - \lambda_2)^{12}(\mu_1 - \mu_2)^{12}}. \tag{3.3}$$

3.2.2 The First Part of the Poor Man's Thomae

Let us now start with one of the divisors $P_1 P_2 Q_l Q_m$, where C_0 is $\{P_1, P_2\}$, D_0 is $\{Q_l, Q_m\}$, B is $\{Q_k\}$ and the other sets are empty. In the previous example we saw that starting with $T_R(\Delta)$ gives us the same equation as starting with Δ; but now this will no longer be the case and extra work must be done in order to obtain the PMT. On $\Delta = P_1 P_2 Q_l Q_m$, apart from N, the operators that can act are T_{Q_k} and the two operators T_{P_i} with i being 1 or 2. We already saw above that $N(\Delta) = Q_l^2 Q_m^2$, and we know that $T_{Q_k}(\Delta) = P_1^2 P_2^2$ (T_{Q_k} is an involution). T_{P_i}, however, takes Δ to the new divisor $P_j^2 P_i Q_k$ (recall that P_j is the branch point denoted P_i which is not P_0 or P_i). Taking the quotient

$$\frac{\theta^{18}[\Delta](\varphi_{P_0}(P),\Pi)}{\theta^{18}[T_{Q_k}(\Delta)](\varphi_{P_0}(P),\Pi)} = \frac{\theta^{18}[P_1P_2Q_lQ_m](\varphi_{P_0}(P),\Pi)}{\theta^{18}[P_1^2P_2^2](\varphi_{P_0}(P),\Pi)}$$

gives us the reciprocal of a function we already looked at, and therefore no new identities. On the other hand, consider the quotient

$$\frac{\theta^{18}[\Delta](\varphi_{P_0}(P),\Pi)}{\theta^{18}[T_{P_j}(\Delta)](\varphi_{P_0}(P),\Pi)} = \frac{\theta^{18}[P_1P_2Q_lQ_m](\varphi_{P_0}(P),\Pi)}{\theta^{18}[P_i^2P_jQ_k](\varphi_{P_0}(P),\Pi)}.$$

Since $N(\Delta) = Q_l^2 Q_m^2$ and $N(T_{P_j}(\Delta)) = P_i^2 Q_k^2$, the divisor of this function is, by Proposition 2.21, the 18th power of $Q_l^2 Q_m^2 / P_i^2 Q_k^2$, i.e., $Q_l^{36} Q_m^{36} / P_i^{36} Q_k^{36}$, and hence this function is

$$d_j \frac{(z-\mu_l)^{12}(z-\mu_m)^{12}}{(z-\lambda_i)^{12}(z-\mu_k)^{12}}$$

where d_j is a nonzero complex constant.

As usual, we calculate the value of the constant d_j by substituting $P = P_0$. This gives us the equality

$$\frac{\theta^{18}[P_1P_2Q_lQ_m](0,\Pi)}{\theta^{18}[P_i^2P_jQ_k](0,\Pi)} = d_j \frac{(\lambda_0-\mu_l)^{12}(\lambda_0-\mu_m)^{12}}{(\lambda_0-\lambda_i)^{12}(\lambda_0-\mu_k)^{12}},$$

and hence we have

$$\frac{\theta^{18}[P_1P_2Q_lQ_m](\varphi_{P_0}(P),\Pi)}{\theta^{18}[P_i^2P_jQ_k](\varphi_{P_0}(P),\Pi)} = \frac{(\lambda_0-\lambda_i)^{12}(\lambda_0-\mu_k)^{12}}{(\lambda_0-\mu_l)^{12}(\lambda_0-\mu_m)^{12}} \times \frac{\theta^{18}[P_1P_2Q_lQ_m](0,\Pi)}{\theta^{18}[P_i^2P_jQ_k](0,\Pi)}$$

$$\times \frac{(z-\mu_l)^{12}(z-\mu_m)^{12}}{(z-\lambda_i)^{12}(z-\mu_k)^{12}}.$$

Here the only other substitution, which is meaningful, is known to be setting $P = P_j$. Now, on the one hand, this substitution gives from the last equation that

$$\frac{\theta^{18}[P_1P_2Q_lQ_m](\varphi_{P_0}(P_j),\Pi)}{\theta^{18}[P_i^2P_jQ_k](\varphi_{P_0}(P_j),\Pi)} = \frac{(\lambda_0-\lambda_i)^{12}(\lambda_0-\mu_k)^{12}}{(\lambda_0-\mu_l)^{12}(\lambda_0-\mu_m)^{12}} \times \frac{\theta^{18}[P_1P_2Q_lQ_m](0,\Pi)}{\theta^{18}[P_i^2P_jQ_k](0,\Pi)}$$

$$\times \frac{(\lambda_j-\mu_l)^{12}(\lambda_j-\mu_m)^{12}}{(\lambda_j-\lambda_i)^{12}(\lambda_j-\mu_k)^{12}},$$

while Eq. (2.3) gives, on the other hand, that

$$\frac{\theta^{18}[P_1P_2Q_lQ_m](\varphi_{P_0}(P_j),\Pi)}{\theta^{18}[P_i^2P_jQ_k](\varphi_{P_0}(P_j),\Pi)} = \frac{\theta^{18}[P_i^2P_jQ_k](0,\Pi)}{\theta^{18}[P_1P_2Q_lQ_m](0,\Pi)}$$

(since $T_{P_j}(P_i^2P_jQ_k) = P_1P_2Q_lQ_m$, as T_{P_j} is an involution). We now combine the last two equations and obtain that

$$\frac{\theta^{36}[P_i^2 P_j Q_k](0,\Pi)}{\theta^{36}[P_1 P_2 Q_l Q_m](0,\Pi)} = \frac{(\lambda_0 - \lambda_i)^{12}(\lambda_0 - \mu_k)^{12}}{(\lambda_0 - \mu_l)^{12}(\lambda_0 - \mu_m)^{12}} \times \frac{(\lambda_j - \mu_l)^{12}(\lambda_j - \mu_m)^{12}}{(\lambda_j - \lambda_i)^{12}(\lambda_j - \lambda_k)^{12}},$$

and hence

$$\frac{\theta^{36}[P_1 P_2 Q_l Q_m](0,\Pi)}{(\lambda_0 - \mu_l)^{12}(\lambda_0 - \mu_m)^{12}(\lambda_1 - \lambda_2)^{12}(\lambda_j - \mu_k)^{12}}$$
$$= \frac{\theta^{36}[P_i^2 P_j Q_k](0,\Pi)}{(\lambda_0 - \lambda_i)^{12}(\lambda_0 - \mu_k)^{12}(\lambda_j - \mu_l)^{12}(\lambda_j - \mu_m)^{12}};$$

recall that $(\lambda_j - \lambda_i)^{12}$ is $(\lambda_1 - \lambda_2)^{12}$, independently of the choice of i and j. Adding in the previously obtained

$$\frac{\theta^{36}[P_1 P_2 Q_l Q_m](0,\Pi)}{(\lambda_0 - \mu_l)^{12}(\lambda_0 - \mu_m)^{12}(\lambda_1 - \mu_k)^{12}(\lambda_2 - \mu_k)^{12}}$$
$$= \frac{\theta^{36}[P_1^2 P_2^2](0,\Pi)}{(\lambda_0 - \lambda_1)^{12}(\lambda_0 - \lambda_2)^{12}(\mu_k - \mu_l)^{12}(\mu_k - \mu_m)^{12}},$$

again we have three equations, all involving $\theta^{36}[P_1 P_2 Q_l Q_m](0,\Pi)$ on their left-hand sides, but with different denominators. In order to merge them we divide the latter (older) equation by $(\lambda_1 - \lambda_2)^{12}$ and the jth new equation by $(\lambda_i - \mu_k)^{12}$. Doing so gives us the following equation:

$$\frac{\theta^{36}[P_1 P_2 Q_l Q_m](0,\Pi)}{(\lambda_0 - \mu_l)^{12}(\lambda_0 - \mu_m)^{12}(\lambda_1 - \lambda_2)^{12}(\lambda_1 - \mu_k)^{12}(\lambda_2 - \mu_k)^{12}}$$
$$= \frac{\theta^{36}[P_1^2 P_2^2](0,\Pi)}{(\lambda_0 - \lambda_1)^{12}(\lambda_0 - \lambda_2)^{12}(\mu_k - \mu_l)^{12}(\mu_k - \mu_m)^{12}(\lambda_1 - \lambda_2)^{12}}$$
$$= \frac{\theta^{36}[P_2^2 P_1 Q_k](0,\Pi)}{(\lambda_0 - \lambda_2)^{12}(\lambda_0 - \mu_k)^{12}(\lambda_1 - \mu_l)^{12}(\lambda_1 - \mu_m)^{12}(\lambda_2 - \mu_k)^{12}}$$
$$= \frac{\theta^{36}[P_1^2 P_2 Q_k](0,\Pi)}{(\lambda_0 - \lambda_1)^{12}(\lambda_0 - \mu_k)^{12}(\lambda_2 - \mu_l)^{12}(\lambda_2 - \mu_m)^{12}(\lambda_1 - \mu_k)^{12}}. \tag{3.4}$$

These are, of course, three sets of equalities between 4 expressions each—one set for every choice of k in the beginning.

Before combining these 4 sets of equalities, i.e., Eq. (3.4) with the different values of k and Eq. (3.3), with one another, we wish to see what other expressions can be obtained from the divisors involved. The divisors already appearing, which we have not taken as the starting point yet, are the divisors $\Delta = P_i^2 P_j Q_k$, where A is $\{P_i\}$, B is $\{Q_l, Q_m\}$, C_0 is $\{P_j\}$, D_0 is $\{Q_k\}$, and C_1 and D_1 are empty. On such a divisor the operators which can act are N, T_{P_j}, T_{Q_l} and T_{Q_m}. We already calculated that $N(\Delta) = P_i^2 Q_k^2$, and we already know that $T_{P_j}(\Delta) = P_1 P_2 Q_l Q_m$ (as T_{P_j} is an involution), but as for T_{Q_m}, it takes Δ to the divisor $P_j^2 P_i Q_l$, which already appeared. Similarly, T_{Q_l} takes Δ to $P_j^2 P_i Q_m$, which has also already appeared. The last two assertions are really the same in different notations (as l and m are symmetric in this case), but we have written them both to emphasize that if we start with a given such divisor, for example $P_2^2 P_1 Q_1$, then it can be taken either by T_{Q_3} to $P_1^2 P_2 Q_2$ or by T_{Q_2} to $P_1^2 P_2 Q_3$, which are two different divisors. We leave it as an exercise for the reader to see that starting with this type of divisor we get the following equality between 4 terms (again, the last two expressions are obtained from one another by interchanging l and m, but we have brought them both to illustrate the fact that starting from $P_2^2 P_1 Q_1$ we get equality between 4 terms):

$$\frac{\theta^{36}[P_i^2 P_j Q_k](0, \Pi)}{(\lambda_0 - \lambda_i)^{12}(\lambda_0 - \mu_k)^{12}(\lambda_j - \mu_l)^{12}(\lambda_j - \mu_m)^{12}(\mu_l - \mu_m)^{12}}$$

$$= \frac{\theta^{36}[P_1 P_2 Q_l Q_m](0, \Pi)}{(\lambda_0 - \mu_l)^{12}(\lambda_0 - \mu_m)^{12}(\lambda_1 - \lambda_2)^{12}(\lambda_j - \mu_k)^{12}(\mu_l - \mu_m)^{12}}$$

$$= \frac{\theta^{36}[P_j^2 P_i Q_l](0, \Pi)}{(\lambda_0 - \lambda_j)^{12}(\lambda_0 - \mu_l)^{12}(\mu_m - \lambda_i)^{12}(\lambda_j - \mu_l)^{12}(\mu_m - \mu_k)^{12}}$$

$$= \frac{\theta^{36}[P_j^2 P_i Q_m](0, \Pi)}{(\lambda_0 - \lambda_j)^{12}(\lambda_0 - \mu_m)^{12}(\mu_l - \lambda_i)^{12}(\lambda_j - \mu_m)^{12}(\mu_l - \mu_k)^{12}}.$$

We have not proven these formulae since, as we shall soon see, they add no extra information to what we obtain from combining Eqs. (3.3) and (3.4). However, the calculation of the action of the operators T_R and T_S in this discussion shows us that the set of divisors composed of these 3 families is closed under the actions of T_R and T_S.

We now want to merge the first four equalities of 4 expressions each, i.e., Eq. (3.3) and Eqs. (3.4) with k being 1, 2 and 3. We note that the first two theta constants in Eq. (3.4) (for any k) already appear in Eq. (3.3), but with different denominators. We would like them to have a common denominator. In order that this be the case we divide Eq. (3.3) by $(\lambda_1 - \lambda_2)^{12}$ and Eq. (3.4) by $(\mu_l - \mu_m)^{12}$ for every k. After that one sees that now the denominator under $\theta^{36}[P_1^2 P_2^2](0, \Pi)$ becomes the same expression (which is the one appearing in Eq. (3.5) below) in all four equations, and also that the new denominator under $\theta^{36}[P_1 P_2 Q_l Q_m](0, \Pi)$ becomes the same expression (which also appears in Eq. (3.5)) in Eq. (3.3) and in Eq. (3.4) with the corresponding k. Altogether we obtain the equality

$$\frac{\theta^{36}[P_1^2 P_2^2](0, \Pi)}{(\lambda_0 - \lambda_1)^{12}(\lambda_0 - \lambda_2)^{12}(\lambda_1 - \lambda_2)^{12}(\mu_1 - \mu_2)^{12}(\mu_1 - \mu_3)^{12}(\mu_2 - \mu_3)^{12}}$$

$$= \frac{\theta^{36}[P_1 P_2 Q_l Q_m](0, \Pi)}{(\lambda_0 - \mu_l)^{12}(\lambda_0 - \mu_m)^{12}(\mu_l - \mu_m)^{12}(\lambda_1 - \mu_k)^{12}(\lambda_2 - \mu_k)^{12}(\lambda_1 - \lambda_2)^{12}}$$

$$= \frac{\theta^{36}[P_i^2 P_j Q_k](0, \Pi)}{(\lambda_0 - \lambda_i)^{12}(\lambda_0 - \mu_k)^{12}(\lambda_i - \mu_k)^{12}(\lambda_j - \mu_l)^{12}(\lambda_j - \mu_m)^{12}(\mu_l - \mu_k)^{12}}, \quad (3.5)$$

which is an equality of 10 terms (as in the second expression we have 3 choices for k and in the third we have 3 choices for k and 2 choices for i and j). One can check that if we divide the equality of 4 terms obtained when starting with the divisor

$P_i^2 P_j Q_k$ by $(\lambda_i - \mu_k)^{12}$, then we get equalities that already appear in Eq. (3.5), so that we really get no extra information from these equalities. But it is good to know that the results obtained in this way are consistent with those we have.

We now prove

Proposition 3.4. *Equation* (3.5) *is part of the PMT for this singular Z_3 curve.*

Proof. We have already seen that the set of divisors appearing in Eq. (3.5) is closed under the operators T_R and T_S (and is a minimal such set). Let us denote, for every such divisor Δ, the denominator appearing under $\theta^{36}[\Delta](0,\Pi)$ in Eq. (3.5) by g_Δ. The content of Eq. (3.5) is that this quotient $\theta^{36}[\Delta](0,\Pi)/g_\Delta$ is invariant under T_R and T_S on the set of divisors composed of the families $P_1^2 P_2^2$, $P_1 P_2 Q_l Q_m$ and $P_i^2 P_j Q_k$. It remains to show how to construct g_Δ from the divisor Δ. We give an ad hoc proof for this, which will be replaced by a different one once we prove, later in this section, that Eq. (3.5) can be merged into the full Thomae formulae for our Z_3 curve X without needing to change the denominator. This proof will be used when proving the full Thomae formulae, so it is not a waste do to it here.

Let us now recall that $N(P_1^2 P_2^2)$ equals $P_1^2 P_2^2$, $N(P_1 P_2 Q_l Q_m)$ equals $Q_l^2 Q_m^2$, and $N(P_i^2 P_j Q_k)$ equals $P_i^2 Q_k^2$. One then sees that g_Δ can be constructed in the following way. Take the set of 2 branch points appearing in $N(\Delta)$ (and thus appearing to the 2nd power, as we saw) together with P_0, and take the product of the 12th powers of all the possible nonzero differences of z-values. In other words we have taken a set of three elements from the set $\{\lambda_0, \lambda_1, \lambda_2, \mu_1, \mu_2, \mu_3\}$, which leaves three additional numbers that are the z-values of the branch points not appearing in $N(\Delta)$ (and which are not P_0). Take now the product of the 12th powers of their nonzero differences, and multiply it by the first product. This gives us g_Δ, and therefore Eq. (3.5) means that $\theta^{36}[\Delta](0,\Pi)/g_\Delta$ with g_Δ thus constructed is invariant under T_R and T_S on this set of divisors. $\qquad\square$

3.2.3 Completing the Poor Man's Thomae

Since this set does not contain all the divisors, Proposition 3.4 gives that Eq. (3.5) is indeed a PMT and not a full Thomae formulae. Actually, it is only a partial PMT since it says nothing about the divisors in the other 3 families. Hence let us now see what happens when starting with divisors from the other families. We begin with the divisor $\Delta = P_i^2 Q_k^2$, with $A = \{P_i\}$, $B = \{Q_l, Q_m\}$, $C_1 = \{P_j\}$, $D_1 = \{Q_k\}$ and C_0 and D_0 empty. On Δ can act N, T_{Q_m} and T_{Q_l}, with images $P_i^2 P_j Q_k$, $P_i Q_l Q_k^2$ and $P_i Q_m Q_k^2$ respectively. Let us thus start with the quotient

$$\frac{\theta^{18}[\Delta](\varphi_{P_0}(P),\Pi)}{\theta^{18}[T_{Q_m}(\Delta)](\varphi_{P_0}(P),\Pi)} = \frac{\theta^{18}[P_i^2 Q_k^2](\varphi_{P_0}(P),\Pi)}{\theta^{18}[P_i Q_l Q_k^2](\varphi_{P_0}(P),\Pi)}.$$

Since $N(\Delta) = P_i^2 P_j Q_k$ and $N(T_{Q_m}(\Delta)) = P_j Q_k Q_l^2$, Proposition 2.21 gives us that the divisor of this function is the 18th power of P_i^2/Q_l^2, hence P_i^{36}/Q_l^{36}, and hence this function is the function $a_m(z - \lambda_i)^{12}/(z - \mu_l)^{12}$ where a_m is a nonzero complex constant. We calculate it by substituting $P = P_0$, which gives us

$$\frac{\theta^{18}[P_i^2 Q_k^2](0, \Pi)}{\theta^{18}[P_i Q_l Q_k^2](0, \Pi)} = a_m \frac{(\lambda_0 - \lambda_i)^{12}}{(\lambda_0 - \mu_l)^{12}},$$

and hence

$$\frac{\theta^{18}[P_i^2 Q_k^2](\varphi_{P_0}(P), \Pi)}{\theta^{18}[P_i Q_l Q_k^2](\varphi_{P_0}(P), \Pi)} = \frac{(\lambda_0 - \mu_l)^{12}}{(\lambda_0 - \lambda_i)^{12}} \times \frac{\theta^{18}[P_i^2 Q_k^2](0, \Pi)}{\theta^{18}[P_i Q_l Q_k^2](0, \Pi)} \times \frac{(z - \lambda_i)^{12}}{(z - \mu_l)^{12}}.$$

We proceed with the only other meaningful substitution $P = Q_m$ to get

$$\frac{\theta^{18}[P_i^2 Q_k^2](\varphi_{P_0}(Q_m), \Pi)}{\theta^{18}[P_i Q_l Q_k^2](\varphi_{P_0}(Q_m), \Pi)} = \frac{(\lambda_0 - \mu_l)^{12}}{(\lambda_0 - \lambda_i)^{12}} \times \frac{\theta^{18}[P_i^2 Q_k^2](0, \Pi)}{\theta^{18}[P_i Q_l Q_k^2](0, \Pi)} \times \frac{(\mu_m - \lambda_i)^{12}}{(\mu_m - \mu_l)^{12}}$$

from the last equation, and

$$\frac{\theta^{18}[P_i^2 Q_k^2](\varphi_{P_0}(Q_m), \Pi)}{\theta^{18}[P_i Q_l Q_k^2](\varphi_{P_0}(Q_m), \Pi)} = \frac{\theta^{18}[P_i Q_l Q_k^2](0, \Pi)}{\theta^{18}[P_i^2 Q_k^2](0, \Pi)}$$

from Eq. (2.4) (recall that $T_{Q_m}(P_i Q_l Q_k^2) = P_i^2 Q_k^2$, as T_{Q_m} is an involution). These combine to

$$\frac{\theta^{36}[P_i Q_l Q_k^2](0, \Pi)}{\theta^{36}[P_i^2 Q_k^2](0, \Pi)} = \frac{(\lambda_0 - \mu_l)^{12}}{(\lambda_0 - \lambda_i)^{12}} \times \frac{(\mu_m - \lambda_i)^{12}}{(\mu_m - \mu_l)^{12}},$$

and hence give

$$\frac{\theta^{36}[P_i^2 Q_k^2](0, \Pi)}{(\lambda_0 - \lambda_i)^{12}(\mu_m - \mu_l)^{12}} = \frac{\theta^{36}[P_i Q_l Q_k^2](0, \Pi)}{(\lambda_0 - \mu_l)^{12}(\mu_m - \lambda_i)^{12}}.$$

By starting with the other quotient, the one based on T_{Q_l}, we obtain, by interchanging the roles of l and m, the equation

$$\frac{\theta^{36}[P_i^2 Q_k^2](0, \Pi)}{(\lambda_0 - \lambda_i)^{12}(\mu_l - \mu_m)^{12}} = \frac{\theta^{36}[P_i Q_m Q_k^2](0, \Pi)}{(\lambda_0 - \mu_m)^{12}(\mu_l - \lambda_i)^{12}}$$

(here as well, l and m are just indices, but we write the two equations to emphasize that from a given divisor, say $\Delta = P_1^2 Q_1^2$, we can get two equalities, one relating it with $T_{Q_3}(\Delta) = P_1 Q_2 Q_1^2$, and the other with $T_{Q_2}(\Delta) = P_1 Q_3 Q_1^2$). It is clear now that the last two equations can be merged into one larger quality

$$\frac{\theta^{36}[P_i^2 Q_k^2](0, \Pi)}{(\lambda_0 - \lambda_i)^{12}(\mu_m - \mu_l)^{12}} = \frac{\theta^{36}[P_i Q_l Q_k^2](0, \Pi)}{(\lambda_0 - \mu_l)^{12}(\mu_m - \lambda_i)^{12}} = \frac{\theta^{36}[P_i Q_m Q_k^2](0, \Pi)}{(\lambda_0 - \mu_m)^{12}(\mu_l - \lambda_i)^{12}}.$$

$$(3.6)$$

Equation (3.6) actually represents 6 equations (we have 3 choices for k and 2 choices for i), each stating an equality between terms corresponding to 3 divisors. Of course, no divisor appears in two different Eqs. (3.6) (i.e., with difference choices of k and i), as easily seen from the way i and k appear in these divisors. We now want to prove

Proposition 3.5. *Equation* (3.6) *is the PMT for these 6 sets of 3 divisors each.*

Proof. We first have to show that this set of 3 divisors (for a given choice of k and i) is closed under the operators T_R and T_S. We already know that the only operators of this kind that can operate on $P_i^2 Q_k^2$ are T_{Q_m} and T_{Q_l}, which take it to the other two divisors, and it remains to check the other two divisors. On $\Delta = P_i Q_l Q_k^2$, i.e., with A being empty, $B = \{Q_m\}$, $C_0 = \{P_i\}$, $D_0 = \{Q_l\}$, $C_1 = \{P_j\}$ and $D_0 = \{Q_k\}$, only N, T_{P_i} and T_{Q_m} can operate. We already saw that $N(\Delta) = P_j Q_l Q_l^2$, and we know that $T_{Q_m}(\Delta) = P_i^2 Q_k^2$. Finally, one verifies that $T_{P_i}(\Delta) = P_i Q_m Q_k^2$, which is the third divisor appearing in Eq. (3.6). We leave it as an exercise to the reader to verify that when we start from the divisor $\Delta = P_i Q_l Q_k^2$ we obtain in the usual way (without any manipulations) exactly the corresponding Eq. (3.6). Clearly the case of the third divisor $P_i Q_m Q_k^2$ is symmetric to what we just did for $P_i Q_l Q_k^2$.

Let us again denote the denominator appearing under $\theta^{36}[\Delta](0, \Pi)$ in the corresponding Eq. (3.6) by g_Δ, so that Eq. (3.6) states that $\theta^{36}[\Delta](0, \Pi)/g_\Delta$ is invariant under T_R and T_S in these 6 sets of divisors. Again we have an ad hoc way to construct g_Δ from Δ, but we shall later see that the denominator h_Δ for the full Thomae formulae will be different for these divisors and for constructing it we shall need a different argument. Now, for a divisor Δ of the sort $P_i^2 Q_k^2$, g_Δ is the product of the 12th power of the difference between λ_0 and the z-value of the branch point P_i which appears in Δ (and thus appears to the 2nd power) and the 12th power of the difference between the z-values of the branch points Q_t which do not appear in Δ. On the other hand, for a divisor Δ of the sort $P_i Q_l Q_k^2$, g_Δ is the product of the 12th power of the difference between λ_0 and the z-value of the branch point Q_t which appears in Δ to the 1st power with the 12th power of the difference between the z-values of the branch point P_t appears in Δ to the 1st power and the branch point Q_t which does not appear in Δ at all. Thus Eq. (3.6) indeed states that $\theta^{36}[\Delta](0, \Pi)/g_\Delta$ is invariant under T_R and T_S in these 6 sets of divisors with g_Δ thus constructed, and hence it is indeed part of the PMT (or actually 6 PMT formulae). $\qquad\square$

Together with Proposition 3.4, Proposition 3.5 shows that we have almost finished writing the PMT for this singular Z_3 curve.

There is one family with which we have not yet dealt—the family containing the 3 divisors of the form $Q_k^2 Q_l^2$. Let us find the PMT for these divisors. Since $B = \{Q_m\}$, $C_1 = \{P_1, P_2\}$, $D_1 = \{Q_k, Q_l\}$ and the other sets are empty, we find that only two operators can act on $\Delta = Q_k^2 Q_l^2$ and they are N and T_{Q_m}. We have $N(\Delta) = P_1 P_2 Q_k Q_l$, and more importantly for the PMT we find that $T_{Q_m}(\Delta) = Q_k^2 Q_l^2 = \Delta$. Therefore this family consists of 3 sets that are closed under the operators T_R and T_S and each set contains one divisor. Therefore the PMT for these divisors is trivial, stating that when moving inside these one-divisor sets that are closed under the operators T_R

and T_S the quotient $\theta^{36}[\Delta](0,\Pi)/g_\Delta$ is left unchanged, whatever g_Δ may be. We shall call the PMT for this case the trivial formula with $g_\Delta = 1$ for each such Δ. Hence together with Propositions 3.4 and 3.5 we have found the PMT for this Z_3 curve—Eq. (3.5), the 6 Eqs. (3.6), and the 3 expressions $\theta^{36}[Q_k^2 Q_i^2](0,\Pi)/1$.

3.2.4 The Thomae Formulae

We now want to combine the PMT into the full Thomae formulae for this singular Z_3 curve. This will be done, as in the previous example, by using the fact that Eq. (1.5) shows that $\theta[N(\Delta)](0,\Pi) = \theta[\Delta](0,\Pi)$ (again, up to the proper power en^2). This is again helpful, as in many cases Δ and $N(\Delta)$ appear in different PMT formulas. First we notice that the divisor $P_i^2 Q_k^2$ appearing in the appropriate Eq. (3.6) and the divisor $P_i^2 P_j Q_k$ appearing in Eq. (3.5) are obtained from one another by the action of N. Therefore if we manipulate these two equations such that the denominator appearing under $\theta^{36}[P_i^2 P_j Q_k](0,\Pi)$ in Eq. (3.5) and the denominator appearing under $\theta^{36}[P_i^2 Q_k^2](0,\Pi)$ in Eq. (3.6) coincide, then we could merge them into a larger equation (actually, doing so for all k and i merges all the 7 equations, Eq. (3.5) and the 6 Eqs. (3.6), into one big equation). By examining these denominators we find that this merging can be done if Eq. (3.5) is not changed at all and Eq. (3.6) is divided by

$$(\lambda_0 - \mu_k)^{12}(\lambda_i - \mu_k)^{12}(\lambda_j - \mu_l)^{12}(\lambda_j - \mu_m)^{12}.$$

We do not show how to construct this extra denominator from $P_i^2 Q_k^2$ or from the other two divisors $P_i Q_l Q_k^2$ and $P_i Q_m Q_k^2$ appearing in the relevant Eq. (3.6) since we shall present the construction after obtaining the full Thomae formulae; this will be written when all the merging is done.

It remains to merge what we have called the 3 trivial PMT into the setup. This is done by noting that $Q_l^2 Q_m^2$ and the divisor $P_1 P_2 Q_l Q_m$ appearing in Eq. (3.5) are also obtained from one another by the action N. From here the merging is easy. Simply divide $\theta^{36}[Q_l^2 Q_m^2](0,\Pi)$ by the same denominator as the denominator appearing under $\theta^{36}[P_1 P_2 Q_l Q_m](0,\Pi)$ in Eq. (3.5). After doing so, we obtain the following equality, which includes the terms corresponding to all the divisors:

$$\frac{\theta^{36}[P_1^2 P_2^2](0, \Pi)}{(\lambda_0 - \lambda_1)^{12}(\lambda_0 - \lambda_2)^{12}(\lambda_1 - \lambda_2)^{12}(\mu_1 - \mu_2)^{12}(\mu_1 - \mu_3)^{12}(\mu_2 - \mu_3)^{12}}$$

$$= \frac{\theta^{36}[P_1 P_2 Q_l Q_m](0, \Pi)}{(\lambda_0 - \mu_l)^{12}(\lambda_0 - \mu_m)^{12}(\mu_l - \mu_m)^{12}(\lambda_1 - \mu_k)^{12}(\lambda_2 - \mu_k)^{12}(\lambda_1 - \lambda_2)^{12}}$$

$$= \frac{\theta^{36}[P_i^2 P_j Q_k](0, \Pi)}{(\lambda_0 - \lambda_i)^{12}(\lambda_0 - \mu_k)^{12}(\lambda_i - \mu_k)^{12}(\lambda_j - \mu_l)^{12}(\lambda_j - \mu_m)^{12}(\mu_l - \mu_m)^{12}}$$

$$= \frac{\theta^{36}[P_i^2 Q_k^2](0, \Pi)}{(\lambda_0 - \lambda_i)^{12}(\lambda_0 - \mu_k)^{12}(\lambda_i - \mu_k)^{12}(\lambda_j - \mu_l)^{12}(\lambda_j - \mu_m)^{12}(\mu_m - \mu_l)^{12}}$$

$$= \frac{\theta^{36}[P_i Q_l Q_k^2](0, \Pi)}{(\lambda_0 - \mu_l)^{12}(\lambda_0 - \mu_k)^{12}(\lambda_i - \mu_k)^{12}(\lambda_j - \mu_l)^{12}(\lambda_j - \mu_m)^{12}(\lambda_i - \mu_m)^{12}}$$

$$= \frac{\theta^{36}[Q_l^2 Q_m^2](0, \Pi)}{(\lambda_0 - \mu_l)^{12}(\lambda_0 - \mu_m)^{12}(\mu_l - \mu_m)^{12}(\lambda_1 - \mu_k)^{12}(\lambda_2 - \mu_k)^{12}(\lambda_1 - \lambda_2)^{12}}. \quad (3.7)$$

One may notice that the expression $(\lambda_i - \mu_m)^{12}(\lambda_j - \mu_m)^{12}$ appearing in the 5th line is actually $(\lambda_1 - \mu_m)^{12}(\lambda_2 - \mu_m)^{12}$ regardless of the choice of i and j.

Since Eq. (3.7) includes all the 6 families of divisors, we have actually finished proving

Theorem 3.6. *Equation* (3.7) *is the Thomae formulae for the singular Z_3 curve*

$$w^3 = (z - \lambda_0)(z - \lambda_1)(z - \lambda_2)(z - \mu_1)^2(z - \mu_2)^2(z - \mu_3)^2.$$

We have not said yet how to construct h_Δ from Δ in general, but for those who are satisfied with a family-dependent construction, one can define h_Δ as the denominator appearing under $\theta^{36}[\Delta](0, \Pi)$ in Eq. (3.7) (once one knows to which family Δ belongs, one can construct h_Δ accordingly) and then the proof of Theorem 3.6 is complete.

What we have really done is to define new denominators h_Δ, other than the g_Δ we had in the various formulas of the PMT, such that h_Δ will be invariant under N. For the divisor on the first line in Eq. (3.7) we had no problem, as N sends it to itself. For the divisors in the 2nd, 3rd, 4th and 6th lines (we now have the division into families listed in the lines of Eq. (3.7), and thus we speak about lines in this equation to represent the families of divisors) we simply made the h_Δ of the 4th and 6th lines to be the same as those of the 3rd and 2nd lines respectively, and this solved the problem. As for the 5th line, we have $N(P_i Q_l Q_k^2) = P_j Q_k Q_l^2$, and indeed if we interchange i and j and interchange k and l in the h_Δ of this line, then we get the same h_Δ. At the PMT stage these two divisors appear in Eq. (3.6) but with different choices of the values of i and k, and as for the second divisor that appears with $P_i Q_l Q_k^2$, which is $P_i Q_m Q_k^2$, its image by N is $P_j Q_k Q_m^2$ and appears in a third choice of i and k values for Eq. (3.6). The reader can check that in order to combine these parts of Eq. (3.6) by making the relevant denominators invariant under N, we should have divided it by $(\lambda_0 - \mu_k)^{12}(\lambda_j - \mu_l)^{12}(\lambda_j - \mu_m)^{12}$. This would yield the one combined equation

$$
\frac{\theta^{36}[P_i^2 Q_k^2](0, \Pi)}{(\lambda_0 - \lambda_i)^{12}(\lambda_0 - \mu_k)^{12}(\lambda_j - \mu_l)^{12}(\lambda_j - \mu_m)^{12}(\mu_m - \mu_l)^{12}}
$$
$$
= \frac{\theta^{36}[P_i Q_l Q_k^2](0, \Pi)}{(\lambda_0 - \mu_l)^{12}(\lambda_0 - \mu_k)^{12}(\lambda_j - \mu_l)^{12}(\lambda_1 - \mu_m)^{12}(\lambda_2 - \mu_m)^{12}},
$$

from which division by $(\lambda_i - \mu_k)^{12}$ gives the relevant part of the Thomae formulae in Eq. (3.7).

3.2.5 Relation with the General Singular Case

In order for this example to shed light on the general singular case, we have to find a general way of constructing the denominator h_Δ from the divisor Δ independently of the family to which Δ belongs. The reader is strongly advised to follow these calculations, as they make the Thomae formulae in the general singular case much more understandable. When describing the construction of g_Δ for Eq. (3.5) and Proposition 3.4 (which is the same as h_Δ) we actually gave the answer to the N-images of the divisors appearing there. As in the Thomae formulae (3.7) the denominators are invariant under N, we know the answer for the 1st, 4th and 6th lines—h_Δ is the product of the following expressions: the 12th powers of the differences between the z-values of any two distinct branch points, which are either P_0 or appear in Δ to the 2nd power, and the 12th powers of the differences between the z-values of any two distinct branch points that are not P_0 and do not appear in Δ at all.

In the language of our sets, let us denote $A \cup P_0$ by C_{-1} and B by D_{-1} (as we shall do in the general case—note that the cardinality condition on A and B becomes $|C_{-1}| = |D_{-1}|$, as for the other sets C_j and D_j). Then we see that h_Δ is the product of the 12th powers of the differences between the z-values of distinct points, which are either both in C_{-1}, both in D_1, or one in C_{-1} and one in D_1 (these are the points which appear to the 2nd power in Δ), multiplied by the product of the 12th powers of the differences between the z-values of distinct points that are either both in C_1, both in D_{-1}, or one in C_1 and one in D_{-1} (these are the points which do not appear in Δ at all, aside from P_0).

One can easily see that this construction does not hold for the other lines, so the best thing to do is to symmetrize it under N. Since N leaves $C_{-1} = A \cup P_0$ and $D_{-1} = B$ invariant and interchanges C_0 with C_1 and D_0 with D_1, we should multiply also by the symmetric expressions (this will not affect the answer for the 1st, 4th and 6th lines as their C_0 and D_0 are empty). This multiplies by the 12th powers of the differences between the z-values of distinct points, which are either both in C_0, both in D_0, one in C_{-1} and one in D_0, or one in C_0 and one in D_{-1}. One can check and see that what we have now finishes the construction not only for the 1st, 4th and 6th lines, but also for the 2nd and 3rd ones. For the 5th line it is not enough though, and the factors $(\lambda_i - \mu_k)^{12}$ and $(\lambda_j - \mu_l)^{12}$ are missing. These are the 12th powers of the differences between the z-values of branch points either one from C_0 and one from D_1, or one from C_1 and one from D_0. Multiplying by these expressions (which does nothing to the other lines as either C_0 and D_0 are empty or C_1 and D_1 are empty in any other line) gives the final answer. Altogether we see that h_Δ is the product of the 12th powers of the differences between the z-values of branch points, which are either in the same set C_j or D_j, or one in a set C_i and the other in a set D_j with $i \neq j$.

One should really make sure that the details of this argument are understood, especially this last statement, before going to the Thomae formulae in the general singular case and its proof. To make this easier for the reader, we give a full list, for each divisor Δ, of the sets C_{-1}, C_0, C_1, D_{-1}, D_0, and D_1 (in this order), and the denominator. We write the denominator as 6 pairs of numbers, and this represents the product of the 12th powers of the corresponding differences, and we have re-ordered the expressions from the Thomae formulae in Eq. (3.7) make the last statement clearer. The lines are ordered according to the lines in Eq. (3.7), or, explicitly, they correspond to the divisors $P_1^2P_2^2$, $P_1P_2Q_lQ_m$, $P_i^2P_jQ_k$, $P_i^2Q_k^2$, $P_iQ_lQ_k^2$, and $Q_l^2Q_m^2$, respectively.

	C_{-1}	C_0	C_1	D_{-1}	D_0	D_1	denominator
$P_0P_1P_2$	$\emptyset$	$\emptyset$	$Q_1Q_2Q_3$	$\emptyset$	$\emptyset$		$\lambda_0\lambda_1,\ \lambda_0\lambda_2,\ \lambda_1\lambda_2,\ \mu_1\mu_2,\ \mu_1\mu_3,\ \mu_2\mu_3$
P_0	P_1P_2	$\emptyset$	Q_k	Q_lQ_m	$\emptyset$		$\lambda_1\lambda_2,\ \mu_l\mu_m,\ \lambda_0\mu_l,\ \lambda_0\mu_m,\ \lambda_1\mu_k,\ \lambda_2\mu_k$
P_0P_i	P_j	$\emptyset$	Q_lQ_m	Q_k	$\emptyset$		$\lambda_0\lambda_i,\ \mu_l\mu_m,\ \lambda_0\mu_k,\ \lambda_i\mu_k,\ \lambda_j\mu_l,\ \lambda_j\mu_m$
P_0P_i	$\emptyset$	P_j	Q_lQ_m	$\emptyset$	Q_k		$\lambda_0\lambda_i,\ \mu_l\mu_m,\ \lambda_0\mu_k,\ \lambda_i\mu_k,\ \lambda_j\mu_l,\ \lambda_j\mu_m$
P_0	P_i	P_j	Q_m	Q_l	Q_k		$\lambda_0\mu_l,\ \lambda_0\mu_k,\ \lambda_i\mu_k,\ \lambda_j\mu_l,\ \lambda_j\mu_m,\ \lambda_i\mu_m$
P_0	$\emptyset$	P_1P_2	Q_k	$\emptyset$	Q_lQ_m		$\lambda_1\lambda_2,\ \mu_l\mu_m,\ \lambda_0\mu_l,\ \lambda_0\mu_m,\ \lambda_1\mu_k,\ \lambda_2\mu_k$.

This concludes the Thomae formulae and the way to construct the denominator h_Δ from Δ in this example of a singular Z_3 curve with $m = 3$.

In order to make this construction in the context of the general case, we now draw a matrix whose columns and rows are labeled by the sets C_j and D_j, $-1 \leq j \leq 1$, and with some of its entries containing numbers. When a number appears in an entry, whose row is labeled by a set Y and whose column is labeled by a set Z, this means that for $Y \neq Z$, one has to take the product of all the differences between the z-values of branch points one from Y and one from Z raised to the power determined by the number in the entry. If $Y = Z$, however, this means the product of all the differences between the z-values of distinct branch points from the set $Y = Z$, again to the power determined by the number in the entry. The expression for h_Δ represented by this matrix is the product of all these expressions.

	C_{-1}	D_{-1}	C_0	D_0	C_1	D_1	C_{-1}	D_{-1}
C_{-1}	12	0	0	12				
D_{-1}	0	12	12	0				
C_0			12	0	0	12		
D_0			0	12	12	0		
C_1					12	0	0	12
D_1					0	12	12	0

The reader is encouraged to verify that these two tables give the same expression for h_Δ we have in Eq. (3.7) for any non-special divisor Δ of degree $g = 4$ supported on the branch points distinct from P_0, hence they represent the Thomae formulae from Theorem 3.6. The reader can also verify that this is the case $n = 3$ of the pictorial description of the formula for h_Δ in Eq. (5.10) in Chapter 5.

3.2.6 Changing the Basepoint

As in the previous example, after obtaining the intrinsic construction of h_Δ from Δ we find that Theorem 3.6, which states that the value of the quotient appearing in Eq. (3.7) is independent of the divisor Δ, is a beautiful and symmetric formula. However, we still have the implicit dependence on the basepoint P_0. We show that here as well, starting with another basepoint gives the same constant value for the quotient in the Thomae formulae, and we give a notation that is independent of the basepoint. Lemma 2.12 gives that the branch points satisfy the condition that $\varphi_{P_i}(P_j^n) = 0$ for every i and j with $n = 3$ and similarly for $\varphi_{Q_k}(Q_l^n)$, $\varphi_{P_i}(Q_k^n)$ and $\varphi_{Q_k}(P_i^n)$, and since all the divisors are integral, of degree g, supported on the branch points (all of them—P_0 is no longer excluded, as we allow any branch point to be the basepoint), and contain no branch point to the power $n = 3$ or higher, we can apply Corollary 1.13.

Now, let us again start with our divisors and basepoint P_0, and we first want to show that the basepoints P_i for $i = 1$ or $i = 2$ gives the same constant. Let us choose $\Delta = P_1^2 P_2^2$ (with basepoint P_0—though we can replace this divisor by any other divisor in which P_i appears to the 2nd power), and then the divisor Γ_{P_i} from Corollary 1.13 is $P_0^2 P_j^2$ and is non-special by Theorem 2.13. It is clear that the sets C_0, C_1, D_0, and D_1 corresponding to Γ_{P_i} are empty (like those corresponding to Δ), and the set $B = D_{-1}$ remains $\{Q_1, Q_2, Q_3\}$ (like $B_\Delta = (D_{-1})_\Delta$), but now A is $\{P_0, P_j\}$ rather than $\{P_1, P_2\}$. Once again the fact that for Γ_{P_i} the set C_{-1} is $A \cup P_i$ (and not P_0) shows that we get the same set C_{-1}, which is composed of all the points P_t with $0 \leq t \leq 2$, and hence $h_{\Gamma_{P_i}}$ in the corresponding Eq. (3.7) is the same as our h_Δ. Now apply Corollary 1.13 to obtain $\varphi_{P_i}(\Gamma_{P_i}) + K_{P_i} = \varphi_{P_0}(\Delta) + K_{P_0}$, which gives that the constant quotient obtained from the basepoint P_i (represented by the divisor Γ_{P_i}) is the same as the one from P_0 (represented by the divisor Δ), since the numerators and denominators of these representatives are equal.

We would also like to see that when the basepoint is Q_k with $1 \leq k \leq 3$, then we also get the same constant. For this let us choose the divisor $\Delta = P_i^2 Q_k^2$ for $i = 1$ or $i = 2$ (with basepoint P_0—again we can choose any other divisor with Q_k appearing to the 2nd power) and then the divisor Γ_{Q_k} from Corollary 1.13 is $P_0^2 P_i^2$ and is non-special by Theorem 2.13. We claim that here as well the expression for $h_{\Gamma_{Q_k}}$ in the corresponding Eq. (3.7) is the same as our h_Δ. The sets are no longer the same and cannot be, as now the sets C_l, $-1 \leq l \leq 1$, are composed of points Q_t and the sets D_l, $-1 \leq l \leq 1$, are composed of points P_t, but we get the following equalities. C_0 and D_0 remain empty. C_{-1} is now $\{Q_k\}$ like $(D_1)_\Delta$, and D_{-1} is now $\{P_j\}$ like $(C_1)_\Delta$. Finally, C_1 is now $\{Q_l, Q_m\}$ like $(D_{-1})_\Delta$, and D_1 is now $\{P_0, P_i\}$ like $(C_{-1})_\Delta$. It is now easy to verify (and the reader is advised to do so) that the expression for $h_{\Gamma_{Q_k}}$ in Eq. (3.7) is the same as the one for h_Δ. Applying Corollary 1.13 again to obtain $\varphi_{Q_k}(\Gamma_{Q_k}) + K_{Q_k} = \varphi_{P_0}(\Delta) + K_{P_0}$ gives that the constant quotient obtained from the basepoint Q_k (represented by the divisor Γ_{Q_k}) is the same as the one from P_0 (represented by the divisor Δ), as stated.

As in the previous example, we would like to show that not only the constants are the same for all the possible basepoints, but so are the sets of characteristics one obtains for the theta function when going over all the divisors Theorems 2.13 and 2.15 give for any basepoint. Here we shall use Proposition 1.15 to show that this is the case. The divisors satisfy the assumptions of Proposition 1.15, and we also know that $P_0 P_1 P_2 / Q_1 Q_2 Q_3$ is principal, which gives the extra assumption and makes Proposition 1.15 applicable. Now, applying Proposition 1.15 shows that for any characteristic $\varphi_{P_0}(\Delta) + K_{P_0}$ obtained from the basepoint P_0 and any other branch point S (which is either P_1, P_2, Q_1, Q_2 or Q_3) there is a divisor Γ_S such that $\varphi_S(\Gamma_S) + K_S$ gives the same characteristic; by Theorem 2.13 this divisor Γ_S (which has the same form as Δ) is non-special as well (and we know that S does not appear in its support). In order to see the converse we can use an argument of symmetry or, alternatively, we can see that the maps $\varphi_S(\cdot) + K_S$ are one-to-one on the corresponding sets of divisors (again, they are non-special) and it is easy to see that there is the same number of divisors Γ with the basepoint S as divisors Δ with the basepoint P_0. Hence in addition to the fact that the Thomae formulae from Theorem 3.6 with the 6 distinct branch points P_0, P_1, P_2, Q_1, Q_2 and Q_3 give the same constant we see that they involve the same theta constants.

We now present the new notation for the Thomae formulae in Theorem 3.6 which will be indeed independent of the basepoint. Since it is clear that if P_i appears to the 2nd power in Δ, then $P_0^2 \Delta = P_i^2 \Gamma_{P_i}$, and if Q_k appears to the 2nd power in Δ, then $P_0^2 \Delta = Q_k^2 \Gamma_{Q_k}$, we shall again replace every divisor Δ with basepoint P_0 by $\Xi = P_0^2 \Delta$. We then replace again the notation $\theta[\Delta]$ by $\theta[\Xi]$, where $\theta[\Xi]$ is defined to be with characteristic $\varphi_S(\Xi) + K_S$ for any branch point S (either a P_i or a Q_k) and this is independent of the choice of S by the second statement of Corollary 1.13. If we again denote the expression for h_Δ appearing in Eq. (3.7) by h_Ξ (which is more suitable here as well), then we have nearly proved

Theorem 3.7. *For every divisor* $\Xi = C_{-1}^2 D_{-1}^0 C_0^1 D_0^1 C_1^0 D_1^2$ *where* C_{-1}, C_0, *and* C_1 *form a partition of the set* $\{P_0, P_1, P_2\}$ *and* D_{-1}, D_0, *and* D_1 *form a partition of the set* $\{Q_1, Q_2, Q_3\}$ *such that* $|D_j| = |C_j|$ *for* $-1 \leq j \leq 1$, *define* $\theta[\Xi]$ *to be the theta function with characteristic* $\varphi_S(\Xi) + K_S$ *for some branch point* S *(which is independent of the choice of* S), *and define* h_Ξ *to be the expression from Eq. (3.7).* *Then the quotient* $\theta^{36}[\Xi](0, \Pi)/h_\Xi$ *is independent of the divisor* Ξ.

The proof is not quite complete since the divisors Ξ we obtain from divisors Δ with basepoint P_0 are only divisors where C_{-1} and D_{-1} are not empty and C_{-1} contains P_0. We would like to remove this extra assumption, which breaks the lovely symmetry of Theorem 3.7. First we see that allowing the basepoint to be also P_1 or P_2 shows that Theorem 3.7 applies to any divisor Ξ with nonempty C_{-1} and D_{-1}. Then we note that choosing the basepoint to be some branch point Q_k, $1 \leq k \leq 3$, can give some divisors with empty C_{-1} and D_{-1}, provided that C_1 and D_1 are nonempty, which extends the validity of Theorem 3.7 also to this case. In the remaining case, which contains only the divisor $\Xi = P_0 P_1 P_2 Q_1 Q_2 Q_3$ with $C_0 = \{P_0, P_1, P_2\}$, $D_0 = \{Q_1, Q_2, Q_3\}$ and C_{-1}, D_{-1}, C_1 and D_1 empty, we shall have to use the operator M, which in this case multiplies Ξ by $P_0 P_1 P_2 / Q_1 Q_2 Q_3$ to give the divisor $\Upsilon = P_0^2 P_1^2 P_2^2$ (which has to satisfy $\varphi_S(\Xi) + K_S = \varphi_S(\Upsilon) + K_S$ for any branch point S as $P_0 P_1 P_2 / Q_1 Q_2 Q_3$ is a principal divisor). Since the only nontrivial expressions in h_Ξ are $[D_0, D_0]^{12}$ and $[C_0, C_0]^{12}$ and the only nontrivial expressions in h_Υ are $[C_{-1}, C_{-1}]^{12}$ and $[D_{-1}, D_{-1}]^{12}$, we find that $h_\Xi = h_\Upsilon$ and the quotients for Ξ and Υ are the same. This completes the proof of Theorem 3.7, which is the most "symmetric" Thomae formulae for our singular Z_3 curve X, and can also give as a corollary the fact that all the Thomae formulae with the 6 different basepoints give the same constant. Here as well, unlike Theorem 3.6 where no two divisors give the same characteristic, the operator M which takes such a divisor Ψ to $\Psi(P_0 P_1 P_2 / Q_1 Q_2 Q_3)(D_{-1}^3 / C_{-1}^3)$ takes Ψ to a different divisor which represents the same characteristic as Ψ does (with any branch point as basepoint). However, one sees that there are 93 divisors Ξ in Theorem 3.7 (1 of type $(3,0,0)$, 9 of type $(2,1,0)$, 9 of type $(2,0,1)$, 9 of type $(1,2,0)$, 36 of type $(1,1,1)$, 9 of type $(1,0,2)$, 1 of type $(0,3,0)$, 9 of type $(0,2,1)$, 9 of type $(0,1,2)$, and 1 of type $(0,0,3)$, where a type is characterized by the cardinality of C_{-1} and D_{-1}, the cardinality of C_0 and D_0, and the cardinality of C_1 and D_1), which split into 31 orbits of size 3 of M and give exactly the 31 characteristics.

3.3 A One-Parameter Family of Singular Z_n Curves with Four Branch Points

In this section we consider the family of singular Z_n curves with the equation

$$w^n = (z - \lambda_0)(z - \lambda_1)(z - \lambda)^{n-1}.$$

This is a singular Z_n curve with the given n and with $m = 2$, such that the branch point Q_2 lies over ∞. We shall denote the branch points P_0 and P_1 as usual, but the other branch points Q_1 and Q_2 will be denoted Q_λ and Q_∞.

3.3.1 Divisors and Operators

The genus of this Riemann surface X is $(n-1)(m-1) = n-1$. Since in this case the divisor of the function $(z - \lambda)/w$ is $Q_\lambda Q_\infty / P_0 P_1$, we see that all these Riemann surfaces are hyperelliptic. By Lemma 2.10 it is clear that the basis of the holomorphic differentials adapted to Q_λ and to Q_∞ is $(z-\lambda)^k dz/w^{k+1}$, $0 \le k \le n-2$. It is also clearly adapted to P_0 and P_1. Thus in the decomposition of the space $\Omega(1)$ into the direct sum of the spaces $\Omega_k(1)$ with $0 \le k \le n-2$, here we find that for each such k the space $\Omega_k(1)$ is 1-dimensional and is spanned by $(z-\lambda)^k dz/w^{k+1}$. Also, by Lemma 2.11 the gap sequences at any of the branch points is composed simply of the numbers from 1 to $n-1$, and hence the branch points are not Weierstrass points. One can easily find the $2g+2 = 2n$ Weierstrass points on this hyperelliptic Riemann surface, but since this information will not help us, we shall not work to obtain it.

We now find the non-special integral divisors Δ of degree $g = n-1$ with support in the set of branch points with P_0 deleted. These divisors are supported on the set $\{P_1, Q_\lambda, Q_\infty\}$. Theorem 2.13 gives us the set of these and Theorem 2.15 assures that this set contains all of them, and in this situation it is quite easy to write them all down. They are the n divisors $P_1^k Q_\lambda^{n-1-k}$, $0 \le k \le n-1$, and the n divisors $P_1^k Q_\infty^{n-1-k}$, $0 \le k \le n-1$, together $2n-1$ divisors since the two divisors with $k = n-1$ coincide. In terms of our representation of these divisors as $\Delta = A^{n-1} B^0 \prod_{j=0}^{n-2} C_j^{n-2-j} D_j^{j+1}$, we have for $\Delta = P_1^{n-1}$ that the set A is $\{P_1\}$, the set B is $\{Q_\lambda Q_\infty\}$ and the other sets are empty. For $\Delta = P_1^k Q_\lambda^{n-1-k}$ with $0 \le k \le n-2$, we have $B = \{Q_\infty\}$, $C_{n-2-k} = \{P_1\}$, $D_{n-2-k} = \{Q_\lambda\}$ and the rest empty. For $\Delta = P_1^k Q_\infty^{n-1-k}$ with $0 \le k \le n-2$, we have $B = \{Q_\lambda\}$, $C_{n-2-k} = \{P_1\}$, $D_{n-2-k} = \{Q_\infty\}$ and the rest empty. One easily sees that the cardinality conditions of Theorem 2.13 hold for all these divisors, and since B cannot be empty and there is only one P_t point to put ($P_t = P_1$), it is easy to see that this list exhausts all the possible divisors.

We shall now compute the actions of the operators from Definition 2.18 in Section 2.5 on these divisors. Recall that N fixes the sets A and B and sends the sets C_j and D_j, $0 \le j \le n-2$ to C_{n-2-j} and D_{n-2-j} respectively. Therefore we find that $N(P_1^{n-1}) = P_1^{n-1}$, while we have $N(P_1^k Q_\lambda^{n-1-k}) = P_1^{n-2-k} Q_\lambda^{k+1}$ and $N(P_1^k Q_\infty^{n-1-k}) = P_1^{n-2-k} Q_\infty^{k+1}$ for every $0 \le k \le n-2$. Note that for even n this means that the divisors $P_1^{(n-2)/2} Q_\lambda^{n/2}$ and $P_1^{(n-2)/2} Q_\infty^{n/2}$ are also invariant under N. As for the operators T_R (which is T_{P_1}) and T_S (which is either T_{Q_λ} or T_{Q_∞}), one sees that the only divisors on which T_{P_1} can act are $P_1^{n-2} Q_\lambda$ and $P_1^{n-2} Q_\infty$ (as they are the only ones with nonempty C_0), and that T_{P_1} sends one to the other. T_{Q_λ}, on the other hand, can act only on the divisors $P_1^k Q_\infty^{n-1-k}$ for $0 \le k \le n-1$, sending such a divisor with $k \le n-3$ to $P_1^{n-3-k} Q_\infty^{k+2}$ and sending P_1^{n-1} and $P_1^{n-2} Q_\infty$ to one another. Similarly, T_{Q_∞} can act only on the divisors $P_1^k Q_\lambda^{n-1-k}$, sending such a divisor with $k \le n-3$ to $P_1^{n-3-k} Q_\lambda^{k+2}$ and P_1^{n-1} and $P_1^{n-2} Q_\lambda$ to one another. The reader can easily check these assertions.

Having said all this we now warn the reader that the method of proof we use for this example, which in fact preceded the general theory, is not going to be the same as the one used for the general singular case. Our previous examples were meant to motivate the general case and make it more palatable. Here our intention is to indicate other techniques. What we shall do is in place of taking the quotients

$$\frac{\theta[\Delta](\varphi_{P_0}(P), \Pi)}{\theta[T_S(\Delta)](\varphi_{P_0}(P), \Pi)}$$

(as we saw, $T_R = T_{P_1}$ does not provide much information), we take the quotients

$$\frac{\theta[\Delta](\varphi_{P_0}(P), \Pi)}{\theta[N(T_S(\Delta))](\varphi_{P_0}(P), \Pi)}.$$

To see this we calculate

$$N(T_{Q_\lambda}(P_1^k Q_\infty^{n-1-k})) = N(P_1^{n-3-k} Q_\infty^{k+2}) = P_1^{k+1} Q_\infty^{n-2-k}, \quad 0 \le k \le n-3$$

and

$$N(T_{Q_\lambda}(P_1^{n-2} Q_\infty)) = N(P_1^{n-1}) = P_1^{n-1}, \quad N(T_{Q_\lambda}(P_1^{n-1})) = N(P_1^{n-2} Q_\infty) = Q_\infty^{n-1}$$

for $k = n-2$ and $k = n-1$ respectively. Similarly, we have

$$N(T_{Q_\infty}(P_1^k Q_\lambda^{n-1-k})) = N(P_1^{n-3-k} Q_\lambda^{k+2}) = P_1^{k+1} Q_\lambda^{n-2-k}, \quad 0 \le k \le n-3$$

and

$$N(T_{Q_\infty}(P_1^{n-2} Q_\lambda)) = N(P_1^{n-1}) = P_1^{n-1}, \quad N(T_{Q_\infty}(P_1^{n-1})) = N(P_1^{n-2} Q_\lambda) = Q_\lambda^{n-1}$$

for $k = n-2$ and $k = n-1$ respectively. Thus these operations take k to $k+1$ modulo n. The reader should remember this when looking at the quotients introduced. Note that we will have to split into cases according to the parity of n, and in fact this will also occur in the general case (both nonsingular and singular). We remark that while in the case of even n here we will obtain the same conclusion as in the general case, in the case of odd n here we shall get Thomae formulae with the powers $2n^2$ of the theta constants, instead of $4n^2$. The denominators here as well are the same as in the general case (again, with the powers halved), which means that in the case of odd n the results we obtain in this example are a bit stronger than what we shall prove in the general singular case. This is an additional reason for the presentation.

3.3.2 First Identities Between Theta Constants

Let us define, for every $1 \le k \le n-1$, the function

$$f_k(P) = \frac{\theta^{en^2}[P_1^{k-1} Q_\lambda^{n-k}](\varphi_{P_0}(P), \Pi)}{\theta^{en^2}[P_1^k Q_\lambda^{n-1-k}](\varphi_{P_0}(P), \Pi)}$$

and for $k = 0$ the function

$$f_0(P) = \frac{\theta^{en^2}[P_1^{n-1}](\varphi_{P_0}(P), \Pi)}{\theta^{en^2}[Q_\lambda^{n-1}](\varphi_{P_0}(P), \Pi)};$$

recall that $e = 1$ for even n and $e = 2$ for odd n. As usual, these are well-defined meromorphic functions on the Riemann surface X. Note that the divisor appearing in the denominator is indeed the NT_{Q_∞}-image of the divisor appearing in the numerator for every $0 \le k \le n - 1$. Note also that the definition for $k = 0$ is the natural extension, as the non-integral divisor $P_1^{-1}Q_\lambda^n$ is equivalent to the integral divisor P_1^{n-1} via the meromorphic function $(z - \lambda_1)/(z - \lambda)$. Now, by looking at the divisors $N(P_1^{k-1}Q_\lambda^{n-k}) = P_1^{n-1-k}Q_\lambda^k$ and $N(P_1^k Q_\lambda^{n-k-1}) = P_1^{n-2-k}Q_\lambda^{k+1}$ (and also $N(P_1^{n-1}) = P_1^{n-1}$ and $N(Q_\lambda^{n-1}) = P_1^{n-2}Q_\lambda$ for $k = 0$), applying Proposition 2.21 gives us here that the divisor of f_k is simply $P_1^{en^2}/Q_\lambda^{en^2}$ and therefore that the functions f_k we have constructed are $c_k(z - \lambda_1)^{en}/(z - \lambda)^{en}$ where c_k, $0 \le k \le n - 1$ are nonzero complex constants.

We begin to deal with these functions in the usual way. We substitute $P = P_0$ and obtain for $k \ge 1$

$$\frac{\theta^{en^2}[P_1^{k-1}Q_\lambda^{n-k}](0, \Pi)}{\theta^{en^2}[P_1^k Q_\lambda^{n-1-k}](0, \Pi)} = c_k \frac{(\lambda_0 - \lambda_1)^{en}}{(\lambda_0 - \lambda)^{en}},$$

and thus

$$\frac{\theta^{en^2}[P_1^{k-1}Q_\lambda^{n-k}](\varphi_{P_0}(P), \Pi)}{\theta^{en^2}[P_1^k Q_\lambda^{n-1-k}](\varphi_{P_0}(P), \Pi)} = \frac{(\lambda_0 - \lambda)^{en}}{(\lambda_0 - \lambda_1)^{en}} \times \frac{\theta^{en^2}[P_1^{k-1}Q_\lambda^{n-k}](0, \Pi)}{\theta^{en^2}[P_1^k Q_\lambda^{n-1-k}](0, \Pi)} \times \frac{(z - \lambda_1)^{en}}{(z - \lambda)^{en}},$$

while for $k = 0$

$$\frac{\theta^{en^2}[P_1^{n-1}](0, \Pi)}{\theta^{en^2}[Q_\lambda^{n-1}](0, \Pi)} = c_0 \frac{(\lambda_0 - \lambda_1)^{en}}{(\lambda_0 - \lambda)^{en}},$$

and thus

$$\frac{\theta^{en^2}[P_1^{n-1}](\varphi_{P_0}(P), \Pi)}{\theta^{en^2}[Q_\lambda^{n-1}](\varphi_{P_0}(P), \Pi)} = \frac{(\lambda_0 - \lambda)^{en}}{(\lambda_0 - \lambda_1)^{en}} \times \frac{\theta^{en^2}[P_1^{n-1}](0, \Pi)}{\theta^{en^2}[Q_\lambda^{n-1}](0, \Pi)} \times \frac{(z - \lambda_1)^{en}}{(z - \lambda)^{en}}.$$

We now substitute $P = Q_\infty$ (i.e., take $z \to \infty$) and obtain

$$\frac{\theta^{en^2}[P_1^{k-1}Q_\lambda^{n-k}](\varphi_{P_0}(Q_\infty), \Pi)}{\theta^{en^2}[P_1^k Q_\lambda^{n-1-k}](\varphi_{P_0}(Q_\infty), \Pi)} = \frac{(\lambda_0 - \lambda)^{en}}{(\lambda_0 - \lambda_1)^{en}} \times \frac{\theta^{en^2}[P_1^{k-1}Q_\lambda^{n-k}](0, \Pi)}{\theta^{en^2}[P_1^k Q_\lambda^{n-1-k}](0, \Pi)}$$

for $k \ge 1$ and

$$\frac{\theta^{en^2}[P_1^{n-1}](\varphi_{P_0}(Q_\infty), \Pi)}{\theta^{en^2}[Q_\lambda^{n-1}](\varphi_{P_0}(Q_\infty), \Pi)} = \frac{(\lambda_0 - \lambda)^{en}}{(\lambda_0 - \lambda_1)^{en}} \times \frac{\theta^{en^2}[P_1^{n-1}](0, \Pi)}{\theta^{en^2}[Q_\lambda^{n-1}](0, \Pi)}$$

for $k = 0$. Here we cannot use Eq. (2.4), but we can use the preceding Eq. (2.2), which yields (apply T_{Q_∞} to each divisor and then apply N and use Eq. (1.5) in its form $\theta[N(\Delta)](0,\Pi) = \theta[\Delta](0,\Pi)$):

$$\frac{\theta^{en^2}[P_1^{k-1}Q_\lambda^{n-k}](\varphi_{P_0}(Q_\infty),\Pi)}{\theta^{en^2}[P_1^kQ_\lambda^{n-1-k}](\varphi_{P_0}(Q_\infty),\Pi)} = \frac{\theta^{en^2}[P_1^{n-2-k}Q_\lambda^{k+1}](0,\Pi)}{\theta^{en^2}[P_1^{n-3-k}Q_\lambda^{k+2}](0,\Pi)} = \frac{\theta^{en^2}[P_1^kQ_\lambda^{n-1-k}](0,\Pi)}{\theta^{en^2}[P_1^{k+1}Q_\lambda^{n-2-k}](0,\Pi)}$$

for $1 \le k \le n-3$,

$$\frac{\theta^{en^2}[P_1^{n-3}Q_\lambda^2](\varphi_{P_0}(Q_\infty),\Pi)}{\theta^{en^2}[P_1^{n-2}Q_\lambda](\varphi_{P_0}(Q_\infty),\Pi)} = \frac{\theta^{en^2}[Q_\lambda^{n-1}](0,\Pi)}{\theta^{en^2}[P_1^{n-1}](0,\Pi)} = \frac{\theta^{en^2}[P_1^{n-2}Q_\lambda](0,\Pi)}{\theta^{en^2}[P_1^{n-1}](0,\Pi)}$$

for $k = n-2$,

$$\frac{\theta^{en^2}[P_1^{n-2}Q_\lambda](\varphi_{P_0}(Q_\infty),\Pi)}{\theta^{en^2}[P_1^{n-1}](\varphi_{P_0}(Q_\infty),\Pi)} = \frac{\theta^{en^2}[P_1^{n-1}](0,\Pi)}{\theta^{en^2}[P_1^{n-2}Q_\lambda](0,\Pi)} = \frac{\theta^{en^2}[P_1^{n-1}](0,\Pi)}{\theta^{en^2}[Q_\lambda^{n-1}](0,\Pi)}$$

for $k = n-1$, and

$$\frac{\theta^{en^2}[P_1^{n-1}](\varphi_{P_0}(Q_\infty),\Pi)}{\theta^{en^2}[Q_\lambda^{n-1}](\varphi_{P_0}(Q_\infty),\Pi)} = \frac{\theta^{en^2}[P_1^{n-2}Q_\lambda](0,\Pi)}{\theta^{en^2}[P_1^{n-3}Q_\lambda^2](0,\Pi)} = \frac{\theta^{en^2}[Q_\lambda^{n-1}](0,\Pi)}{\theta^{en^2}[P_1Q_\lambda^{n-2}](0,\Pi)}$$

for $k = 0$. As a consequence we obtain

$$L_k = \frac{\theta^{2en^2}[P_1^kQ_\lambda^{n-k-1}](0,\Pi)}{\theta^{en^2}[P_1^{k+1}Q_\lambda^{n-2-k}](0,\Pi)\theta^{en^2}[P_1^{k-1}Q_\lambda^{n-k}](0,\Pi)} = \frac{(\lambda_0-\lambda)^{en}}{(\lambda_0-\lambda_1)^{en}}$$

for all $1 \le k \le n-2$,

$$L_{n-1} = \frac{\theta^{2en^2}[P_1^{n-1}](0,\Pi)}{\theta^{en^2}[Q_\lambda^{n-1}](0,\Pi)\theta^{en^2}[P_1^{n-2}Q_\lambda](0,\Pi)} = \frac{(\lambda_0-\lambda)^{en}}{(\lambda_0-\lambda_1)^{en}}$$

for $k = n-1$ and

$$L_0 = \frac{\theta^{2en^2}[Q_\lambda^{n-1}](0,\Pi)}{\theta^{en^2}[P_1Q_\lambda^{n-2}](0,\Pi)\theta^{en^2}[P_1^{n-1}](0,\Pi)} = \frac{(\lambda_0-\lambda)^{en}}{(\lambda_0-\lambda_1)^{en}}$$

for $k = 0$. Using these equalities we shall prove one part of the Thomae formulae for this case.

Before doing this we observe that we could have proceeded differently by basically interchanging the roles of Q_λ and Q_∞. We could have defined for $1 \le k \le n-1$ the function

$$g_k(P) = \frac{\theta^{en^2}[P_1^{k-1}Q_\infty^{n-k}](\varphi_{P_0}(P),\Pi)}{\theta^{en^2}[P_1^kQ_\infty^{n-1-k}](\varphi_{P_0}(P),\Pi)},$$

and for $k = 0$ the function

$$g_0(P) = \frac{\theta^{en^2}[P_1^{n-1}](\varphi_{P_0}(P), \Pi)}{\theta^{en^2}[Q_\infty^{n-1}](\varphi_{P_0}(P), \Pi)}.$$

These are also meromorphic functions on X where the divisor appearing in the denominator is indeed the NT_{Q_λ}-image of the divisor appearing in the numerator for every $0 \le k \le n-1$, and $P_1^{-1}Q_\infty^n$ is equivalent to P_1^{n-1} via the function $z - \lambda_1$. Here we see that the divisor of g_k is $P_1^{en^2}/Q_\infty^{en^2}$, and hence that the functions g_k we have constructed are $d_k(z-\lambda_1)^{en}$ where d_k, $0 \le k \le n-1$ are nonzero complex constants.

Again we substitute $P = P_0$ to obtain for $k \ge 1$

$$\frac{\theta^{en^2}[P_1^{k-1}Q_\infty^{n-k}](0, \Pi)}{\theta^{en^2}[P_1^kQ_\infty^{n-1-k}](0, \Pi)} = d_k(\lambda_0 - \lambda_1)^{en},$$

and hence

$$\frac{\theta^{en^2}[P_1^{k-1}Q_\infty^{n-k}](\varphi_{P_0}(P), \Pi)}{\theta^{en^2}[P_1^kQ_\infty^{n-1-k}](\varphi_{P_0}(P), \Pi)} = \frac{\theta^{en^2}[P_1^{k-1}Q_\infty^{n-k}](0, \Pi)}{\theta^{en^2}[P_1^kQ_\infty^{n-1-k}](0, \Pi)} \times \frac{(z-\lambda_1)^{en}}{(\lambda_0 - \lambda_1)^{en}},$$

while for $k = 0$

$$\frac{\theta^{en^2}[P_1^{n-1}](0, \Pi)}{\theta^{en^2}[Q_\infty^{n-1}](0, \Pi)} = d_0(\lambda_0 - \lambda_1)^{en},$$

and hence

$$\frac{\theta^{en^2}[P_1^{n-1}](\varphi_{P_0}(P), \Pi)}{\theta^{en^2}[Q_\infty^{n-1}](\varphi_{P_0}(P), \Pi)} = \frac{\theta^{en^2}[P_1^{n-1}](0, \Pi)}{\theta^{en^2}[Q_\infty^{n-1}](0, \Pi)} \times \frac{(z-\lambda_1)^{en}}{(\lambda_0 - \lambda_1)^{en}}.$$

When substituting $P = Q_\lambda$ we now obtain

$$\frac{\theta^{en^2}[P_1^{k-1}Q_\infty^{n-k}](\varphi_{P_0}(Q_\lambda), \Pi)}{\theta^{en^2}[P_1^kQ_\infty^{n-1-k}](\varphi_{P_0}(Q_\lambda), \Pi)} = \frac{\theta^{en^2}[P_1^{k-1}Q_\infty^{n-k}](0, \Pi)}{\theta^{en^2}[P_1^kQ_\infty^{n-1-k}](0, \Pi)} \times \frac{(\lambda-\lambda_1)^{en}}{(\lambda_0 - \lambda_1)^{en}}$$

for $k \ge 1$ and

$$\frac{\theta^{en^2}[P_1^{n-1}](\varphi_{P_0}(Q_\lambda), \Pi)}{\theta^{en^2}[Q_\infty^{n-1}](\varphi_{P_0}(Q_\lambda), \Pi)} = \frac{\theta^{en^2}[P_1^{n-1}](0, \Pi)}{\theta^{en^2}[Q_\infty^{n-1}](0, \Pi)} \times \frac{(\lambda-\lambda_1)^{en}}{(\lambda_0 - \lambda_1)^{en}}$$

for $k = 0$. Also here we use Eq. (2.2) since Eq. (2.4) is not applicable, and this yields (now apply T_{Q_λ} to each divisor and then apply N)

$$\frac{\theta^{en^2}[P_1^{k-1}Q_\infty^{n-k}](\varphi_{P_0}(Q_\lambda), \Pi)}{\theta^{en^2}[P_1^kQ_\infty^{n-1-k}](\varphi_{P_0}(Q_\lambda), \Pi)} = \frac{\theta^{en^2}[P_1^{n-2-k}Q_\infty^{k+1}](0, \Pi)}{\theta^{en^2}[P_1^{n-3-k}Q_\infty^{k+2}](0, \Pi)} = \frac{\theta^{en^2}[P_1^kQ_\infty^{n-1-k}](0, \Pi)}{\theta^{en^2}[P_1^{k+1}Q_\infty^{n-2-k}](0, \Pi)}$$

for $1 \le k \le n-3$,

$$\frac{\theta^{en^2}[P_1^{n-3}Q_\infty^2](\varphi_{P_0}(Q_\lambda),\Pi)}{\theta^{en^2}[P_1^{n-2}Q_\infty](\varphi_{P_0}(Q_\lambda),\Pi)} = \frac{\theta^{en^2}[Q_\infty^{n-1}](0,\Pi)}{\theta^{en^2}[P_1^{n-1}](0,\Pi)} = \frac{\theta^{en^2}[P_1^{n-2}Q_\infty](0,\Pi)}{\theta^{en^2}[P_1^{n-1}](0,\Pi)}$$

for $k = n-2$,

$$\frac{\theta^{en^2}[P_1^{n-2}Q_\infty](\varphi_{P_0}(Q_\lambda),\Pi)}{\theta^{en^2}[P_1^{n-1}](\varphi_{P_0}(Q_\lambda),\Pi)} = \frac{\theta^{en^2}[P_1^{n-1}](0,\Pi)}{\theta^{en^2}[P_1^{n-2}Q_\infty](0,\Pi)} = \frac{\theta^{en^2}[P_1^{n-1}](0,\Pi)}{\theta^{en^2}[Q_\infty^{n-1}](0,\Pi)}$$

for $k = n-1$, and

$$\frac{\theta^{en^2}[P_1^{n-1}](\varphi_{P_0}(Q_\lambda),\Pi)}{\theta^{en^2}[Q_\infty^{n-1}](\varphi_{P_0}(Q_\lambda),\Pi)} = \frac{\theta^{en^2}[P_1^{n-2}Q_\infty](0,\Pi)}{\theta^{en^2}[P_1^{n-3}Q_\infty^2](0,\Pi)} = \frac{\theta^{en^2}[Q_\infty^{n-1}](0,\Pi)}{\theta^{en^2}[P_1Q_\infty^{n-2}](0,\Pi)}$$

for $k = 0$. From this we deduce

$$M_k = \frac{\theta^{2en^2}[P_1^kQ_\infty^{n-k-1}](0,\Pi)}{\theta^{en^2}[P_1^{k+1}Q_\infty^{n-2-k}](0,\Pi)\theta^{en^2}[P_1^{k-1}Q_\infty^{n-k}](0,\Pi)} = \frac{(\lambda-\lambda_1)^{en}}{(\lambda_0-\lambda_1)^{en}}$$

for all $1 \leq k \leq n-2$,

$$M_{n-1} = \frac{\theta^{2en^2}[P_1^{n-1}](0,\Pi)}{\theta^{en^2}[Q_\infty^{n-1}](0,\Pi)\theta^{en^2}[P_1^{n-2}Q_\infty](0,\Pi)} = \frac{(\lambda-\lambda_1)^{en}}{(\lambda_0-\lambda_1)^{en}}$$

for $k = n-1$ and

$$M_0 = \frac{\theta^{2en^2}[Q_\infty^{n-1}](0,\Pi)}{\theta^{en^2}[P_1Q_\infty^{n-2}](0,\Pi)\theta^{en^2}[P_1^{n-1}](0,\Pi)} = \frac{(\lambda-\lambda_1)^{en}}{(\lambda_0-\lambda_1)^{en}}$$

for $k = 0$. These equalities will allow us to prove the other part the Thomae formulae for this case, and then we combine them.

This is the point where we have to split the cases according to the parity of n. The ideas in the even n and odd n cases are similar, but the details are very different. We thus present the two cases fully.

3.3.3 Even n

We begin with even n, hence $e = 1$. We use again Eq. (1.5) in the form of the identity $\theta[N(\Delta)](0,\Pi) = \theta[\Delta](0,\Pi)$ (up to the proper power, as usual), to see that $L_{(n-2)/2}$ satisfies

$$L_{(n-2)/2} = \frac{\theta^{2n^2}[P_1^{(n-2)/2}Q_\lambda^{n/2}](0,\Pi)}{\theta^{2n^2}[P_1^{(n-4)/2}Q_\lambda^{(n+2)/2}](0,\Pi)} = \frac{\theta^{2n^2}[P_1^{(n-2)/2}Q_\lambda^{n/2}](0,\Pi)}{\theta^{2n^2}[P_1^{n/2}Q_\lambda^{(n-2)/2}](0,\Pi)} = \frac{(\lambda_0-\lambda)^n}{(\lambda_0-\lambda_1)^n}.$$

We now multiply

$$L_{(n-4)/2}L_{(n-2)/2} = \frac{\theta^{n^2}[P_1^{(n-2)/2}Q_\lambda^{n/2}](0,\Pi)}{\theta^{n^2}[P_1^{(n-6)/2}Q_\lambda^{(n+4)/2}](0,\Pi)} = \frac{(\lambda_0-\lambda)^{2n}}{(\lambda_0-\lambda_1)^{2n}}$$

and square to obtain

$$\frac{\theta^{2n^2}[P_1^{(n-2)/2}Q_\lambda^{n/2}](0,\Pi)}{\theta^{2n^2}[P_1^{(n-6)/2}Q_\lambda^{(n+4)/2}](0,\Pi)} = \frac{(\lambda_0-\lambda)^{4n}}{(\lambda_0-\lambda_1)^{4n}}.$$

Similarly, we have

$$L_{(n-2)/2}L_{n/2} = \frac{\theta^{n^2}[P_1^{(n-2)/2}Q_\lambda^{n/2}](0,\Pi)}{\theta^{n^2}[P_1^{(n+2)/2}Q_\lambda^{(n-4)/2}](0,\Pi)} = \frac{(\lambda_0-\lambda)^{2n}}{(\lambda_0-\lambda_1)^{2n}}$$

and the square

$$\frac{\theta^{2n^2}[P_1^{(n-2)/2}Q_\lambda^{n/2}](0,\Pi)}{\theta^{2n^2}[P_1^{(n+2)/2}Q_\lambda^{(n-4)/2}](0,\Pi)} = \frac{(\lambda_0-\lambda)^{4n}}{(\lambda_0-\lambda_1)^{4n}}.$$

Let us rewrite the last two equalities as

$$\frac{\theta^{2n^2}[P_1^{(n-2)/2}Q_\lambda^{n/2}](0,\Pi)}{(\lambda_0-\lambda)^{4n}} = \frac{\theta^{2n^2}[P_1^{(n-6)/2}Q_\lambda^{(n+4)/2}](0,\Pi)}{(\lambda_0-\lambda_1)^{4n}}$$

and

$$\frac{\theta^{2n^2}[P_1^{(n-2)/2}Q_\lambda^{n/2}](0,\Pi)}{(\lambda_0-\lambda)^{4n}} = \frac{\theta^{2n^2}[P_1^{(n+2)/2}Q_\lambda^{(n-4)/2}](0,\Pi)}{(\lambda_0-\lambda_1)^{4n}}.$$

We now prove

Proposition 3.8. *The following equalities hold: For $1 \leq j \leq (n-4)/2$ we have*

$$\frac{\theta^{2n^2}[P_1^{(n-2j)/2}Q_\lambda^{(n+2j-2)/2}](0,\Pi)}{(\lambda_0-\lambda)^{4jn}} = \frac{\theta^{2n^2}[P_1^{(n-2j-4)/2}Q_\lambda^{(n+2j+2)/2}](0,\Pi)}{(\lambda_0-\lambda_1)^{4jn}},$$

and for $j = (n-2)/2$ we have

$$\frac{\theta^{2n^2}[P_1 Q_\lambda^{n-2}](0,\Pi)}{(\lambda_0-\lambda)^{2n(n-2)}} = \frac{\theta^{2n^2}[P_1^{n-1}](0,\Pi)}{(\lambda_0-\lambda_1)^{2n(n-2)}}.$$

Moreover, we have

$$\frac{\theta^{2n^2}[P_1^{(n+2j-4)/2}Q_\lambda^{(n-2j+2)/2}](0,\Pi)}{(\lambda_0-\lambda)^{4jn}} = \frac{\theta^{2n^2}[P_1^{(n+2j)/2}Q_\lambda^{(n-2j-2)/2}](0,\Pi)}{(\lambda_0-\lambda_1)^{4jn}}$$

for all $1 \leq j \leq (n-2)/2$.

Proof. The proof is by induction on j, the first case $j = 1$ being the equations obtained above. Now, assume the result is true for some $1 \leq j \leq (n-4)/2$, which means that

$$\frac{\theta^{2n^2}[P_1^{(n-2j)/2}Q_\lambda^{(n+2j-2)/2}](0,\Pi)}{\theta^{2n^2}[P_1^{(n-2j-4)/2}Q_\lambda^{(n+2j+2)/2}](0,\Pi)} = \frac{(\lambda_0 - \lambda)^{4jn}}{(\lambda_0 - \lambda_1)^{4jn}}$$

and

$$\frac{\theta^{2n^2}[P_1^{(n+2j-4)/2}Q_\lambda^{(n-2j+2)/2}](0,\Pi)}{\theta^{2n^2}[P_1^{(n+2j)/2}Q_\lambda^{(n-2j-2)/2}](0,\Pi)} = \frac{(\lambda_0 - \lambda)^{4jn}}{(\lambda_0 - \lambda_1)^{4jn}}.$$

Recall that

$$\frac{\theta^{2n^2}[P_1^{(n-2j-2)/2}Q_\lambda^{(n+2j)/2}](0,\Pi)\theta^{2n^2}[P_1^{(n-2j-4)/2}Q_\lambda^{(n+2j+2)/2}](0,\Pi)}{\theta^{2n^2}[P_1^{(n-2j)/2}Q_\lambda^{(n+2j-2)/2}](0,\Pi)\theta^{2n^2}[P_1^{(n-2j-6)/2}Q_\lambda^{(n+2j+4)/2}](0,\Pi)} = \frac{(\lambda_0 - \lambda)^{4n}}{(\lambda_0 - \lambda_1)^{4n}}$$

for $1 \leq j \leq (n-6)/2$ (both being equal to $L_{(n-2j-2)/2}^2 L_{(n-2j-4)/2}^2$) and

$$L_1^2 L_0^2 = \frac{\theta^{2n^2}[P_1 Q_\lambda^{n-2}](0,\Pi)\theta^{2n^2}[Q_\lambda^{n-1}](0,\Pi)}{\theta^{2n^2}[P_1^2 Q_\lambda^{n-3}](0,\Pi)\theta^{2n^2}[P_1^{n-1}](0,\Pi)} = \frac{(\lambda_0 - \lambda)^{4n}}{(\lambda_0 - \lambda_1)^{4n}}$$

for $j = (n-4)/2$, and also

$$\frac{\theta^{2n^2}[P_1^{(n+2j-2)/2}Q_\lambda^{(n-2j)/2}](0,\Pi)\theta^{2n^2}[P_1^{(n+2j)/2}Q_\lambda^{(n-2j-2)/2}](0,\Pi)}{\theta^{2n^2}[P_1^{(n+2j-4)/2}Q_\lambda^{(n-2j+2)/2}](0,\Pi)\theta^{2n^2}[P_1^{(n+2j+2)/2}Q_\lambda^{(n-2j-4)/2}](0,\Pi)} = \frac{(\lambda_0 - \lambda)^{4n}}{(\lambda_0 - \lambda_1)^{4n}}$$

for all j (both being equal to $L_{(n+2j-2)/2}^2 L_{(n+2j)/2}^2$). Multiplication gives

$$\frac{\theta^{2n^2}[P_1^{(n-2j-2)/2}Q_\lambda^{(n+2j)/2}](0,\Pi)}{\theta^{2n^2}[P_1^{(n-2j-6)/2}Q_\lambda^{(n+2j+4)/2}](0,\Pi)} = \frac{(\lambda_0 - \lambda)^{4(j+1)n}}{(\lambda_0 - \lambda_1)^{4(j+1)n}}$$

for $1 \leq j \leq (n-6)/2$ and

$$\frac{\theta^{2n^2}[P_1 Q_\lambda^{n-2}](0,\Pi)}{\theta^{2n^2}[P_1^{n-1}](0,\Pi)} = \frac{(\lambda_0 - \lambda)^{2n(n-2)}}{(\lambda_0 - \lambda_1)^{2n(n-2)}}$$

for $j = (n-4)/2$, and, for the other expressions,

$$\frac{\theta^{2n^2}[P_1^{(n+2j-2)/2}Q_\lambda^{(n-2j)/2}](0,\Pi)}{\theta^{2n^2}[P_1^{(n+2j+2)/2}Q_\lambda^{(n-2j-4)/2}](0,\Pi)} = \frac{(\lambda_0 - \lambda)^{4(j+1)n}}{(\lambda_0 - \lambda_1)^{4(j+1)n}}$$

for all j, which easily gives the needed identities for $j+1$ for all j. $\square$

Since the quotients M_k are the same but with Q_λ replaced by Q_∞ and with the rational expression in the parameters replaced by $(\lambda - \lambda_1)^n/(\lambda_0 - \lambda_1)^n$, the same considerations prove

Proposition 3.9. *The following equalities hold: For* $1 \le j \le (n-4)/2$ *we have*

$$\frac{\theta^{2n^2}[P_1^{(n-2j)/2}Q_\infty^{(n+2j-2)/2}](0,\Pi)}{(\lambda - \lambda_1)^{4jn}} = \frac{\theta^{2n^2}[P_1^{(n-2j-4)/2}Q_\infty^{(n+2j+2)/2}](0,\Pi)}{(\lambda_0 - \lambda_1)^{4jn}},$$

and for $j = (n-2)/2$ *we have*

$$\frac{\theta^{2n^2}[P_1 Q_\infty^{n-2}](0,\Pi)}{(\lambda - \lambda_1)^{2n(n-2)}} = \frac{\theta^{2n^2}[P_1^{n-1}](0,\Pi)}{(\lambda_0 - \lambda_1)^{2n(n-2)}}.$$

Moreover, we have

$$\frac{\theta^{2n^2}[P_1^{(n+2j-4)/2}Q_\infty^{(n-2j+2)/2}](0,\Pi)}{(\lambda - \lambda_1)^{4jn}} = \frac{\theta^{2n^2}[P_1^{(n+2j)/2}Q_\infty^{(n-2j-2)/2}](0,\Pi)}{(\lambda_0 - \lambda_1)^{4jn}}$$

for all $1 \le j \le (n-2)/2$.

We note that in both Propositions 3.8 and 3.9 each part can be proved from the other by using $\theta[N(\Delta)](0,\Pi) = \theta[\Delta](0,\Pi)$ from Eq. (1.5). Therefore for every j the expressions appearing in the two parts of Proposition 3.8 are equal, and so are the expressions appearing in the two parts of Proposition 3.9 (actually for $j = 1$ and for $j = (n-2)/2$ the two parts of both Propositions 3.8 and 3.9 have a common expression).

3.3.4 An Example with $n = 10$

In order for the path we take from here to be clearer, we work out an example of $n = 10$, where Proposition 3.8 yields the following sequence of equalities: $j = 1$ gives

$$\frac{\theta^{200}[P_1^4 Q_\lambda^5](0,\Pi)}{(\lambda_0 - \lambda)^{40}} = \frac{\theta^{200}[P_1^2 Q_\lambda^7](0,\Pi)}{(\lambda_0 - \lambda_1)^{40}}, \quad \frac{\theta^{200}[P_1^4 Q_\lambda^5](0,\Pi)}{(\lambda_0 - \lambda)^{40}} = \frac{\theta^{200}[P_1^6 Q_\lambda^3](0,\Pi)}{(\lambda_0 - \lambda_1)^{40}};$$

$j = 2$ gives

$$\frac{\theta^{200}[P_1^3 Q_\lambda^6](0,\Pi)}{(\lambda_0 - \lambda)^{80}} = \frac{\theta^{200}[P_1 Q_\lambda^8](0,\Pi)}{(\lambda_0 - \lambda_1)^{80}}, \quad \frac{\theta^{200}[P_1^5 Q_\lambda^4](0,\Pi)}{(\lambda_0 - \lambda)^{80}} = \frac{\theta^{200}[P_1^7 Q_\lambda^2](0,\Pi)}{(\lambda_0 - \lambda_1)^{80}};$$

$j = 3$ gives

$$\frac{\theta^{200}[P_1^2 Q_\lambda^7](0,\Pi)}{(\lambda_0 - \lambda)^{120}} = \frac{\theta^{200}[Q_\lambda^9](0,\Pi)}{(\lambda_0 - \lambda_1)^{120}}, \quad \frac{\theta^{200}[P_1^6 Q_\lambda^3](0,\Pi)}{(\lambda_0 - \lambda)^{120}} = \frac{\theta^{200}[P_1^8 Q_\lambda](0,\Pi)}{(\lambda_0 - \lambda_1)^{120}};$$

and $j = 4$ gives

$$\frac{\theta^{200}[P_1 Q_\lambda^8](0,\Pi)}{(\lambda_0 - \lambda)^{160}} = \frac{\theta^{200}[P_1^9](0,\Pi)}{(\lambda_0 - \lambda_1)^{160}}, \quad \frac{\theta^{200}[P_1^7 Q_\lambda^2](0,\Pi)}{(\lambda_0 - \lambda)^{160}} = \frac{\theta^{200}[P_1^9](0,\Pi)}{(\lambda_0 - \lambda_1)^{160}}.$$

In addition, before proving Proposition 3.8 we already saw that the quotient $L_{(n-2)/2}$ (which is L_4 here) gives

$$\frac{\theta^{200}[P_1^4 Q_\lambda^5](0,\Pi)}{(\lambda_0 - \lambda)^{10}} = \frac{\theta^{200}[P_1^3 Q_\lambda^6](0,\Pi)}{(\lambda_0 - \lambda_1)^{10}} = \frac{\theta^{200}[P_1^5 Q_\lambda^4](0,\Pi)}{(\lambda_0 - \lambda_1)^{10}}.$$

The way to obtain the Thomae formulae for the case $n = 10$ from these equalities is as follows. We divide the equation for $j = 1$ by $(\lambda_0 - \lambda)^{120}$, the equation for $j = 2$ by $(\lambda_0 - \lambda)^{160}$, the equation for $j = 3$ by $(\lambda_0 - \lambda_1)^{40}$, and the equation for $j = 4$ by $(\lambda_0 - \lambda_1)^{80}$. After that we can combine the equations for $j = 1$ and $j = 3$ to

$$\frac{\theta^{200}[Q_\lambda^9](0,\Pi)}{(\lambda_0 - \lambda_1)^{160}} = \frac{\theta^{200}[P_1^2 Q_\lambda^7](0,\Pi)}{(\lambda_0 - \lambda_1)^{40}(\lambda_0 - \lambda)^{120}} = \frac{\theta^{200}[P_1^4 Q_\lambda^5](0,\Pi)}{(\lambda_0 - \lambda)^{160}}$$

$$= \frac{\theta^{200}[P_1^6 Q_\lambda^3](0,\Pi)}{(\lambda_0 - \lambda_1)^{40}(\lambda_0 - \lambda)^{120}} = \frac{\theta^{200}[P_1^8 Q_\lambda](0,\Pi)}{(\lambda_0 - \lambda_1)^{160}},$$

and the equations for $j = 2$ and $j = 4$ to

$$\frac{\theta^{200}[P_1^5 Q_\lambda^4](0,\Pi)}{(\lambda_0 - \lambda)^{240}} = \frac{\theta^{200}[P_1^7 Q_\lambda^2](0,\Pi)}{(\lambda_0 - \lambda_1)^{80}(\lambda_0 - \lambda)^{160}} = \frac{\theta^{200}[P_1^9](0,\Pi)}{(\lambda_0 - \lambda_1)^{240}}$$

$$= \frac{\theta^{200}[P_1 Q_\lambda^8](0,\Pi)}{(\lambda_0 - \lambda_1)^{80}(\lambda_0 - \lambda)^{160}} = \frac{\theta^{200}[P_1^3 Q_\lambda^6](0,\Pi)}{(\lambda_0 - \lambda)^{240}}.$$

In order to combine these two equations into one identity we recall that the equation of L_4 contains the theta constants corresponding to $P_1^3 Q_\lambda^6$ and $P_1^5 Q_\lambda^4$ from the identity from $j = 2$ and $j = 4$, and also the theta constants corresponding to $P_1^4 Q_\lambda^5$ from the one from $j = 1$ and $j = 3$; hence it can be the merging tool. We thus divide the equation of L_4 by $(\lambda_0 - \lambda)^{240}$, the identity from $j = 2$ and $j = 4$ by $(\lambda_0 - \lambda_1)^{10}$ and the one from $j = 1$ and $j = 3$ by $(\lambda_0 - \lambda)^{90}$, and then these equations can be merged into

$$\frac{\theta^{200}[P_1^9](0,\Pi)}{(\lambda_0-\lambda_1)^{250}} = \frac{\theta^{200}[Q_\lambda^9](0,\Pi)}{(\lambda_0-\lambda_1)^{160}(\lambda_0-\lambda)^{90}} = \frac{\theta^{200}[P_1 Q_\lambda^8](0,\Pi)}{(\lambda_0-\lambda_1)^{90}(\lambda_0-\lambda)^{160}}$$

$$= \frac{\theta^{200}[P_1^2 Q_\lambda^7](0,\Pi)}{(\lambda_0-\lambda_1)^{40}(\lambda_0-\lambda)^{210}} = \frac{\theta^{200}[P_1^3 Q_\lambda^6](0,\Pi)}{(\lambda_0-\lambda_1)^{10}(\lambda_0-\lambda)^{240}} = \frac{\theta^{200}[P_1^4 Q_\lambda^5](0,\Pi)}{(\lambda_0-\lambda)^{250}}$$

$$= \frac{\theta^{200}[P_1^5 Q_\lambda^4](0,\Pi)}{(\lambda_0-\lambda_1)^{10}(\lambda_0-\lambda)^{240}} = \frac{\theta^{200}[P_1^6 Q_\lambda^3](0,\Pi)}{(\lambda_0-\lambda_1)^{40}(\lambda_0-\lambda)^{210}} = \frac{\theta^{200}[P_1^7 Q_\lambda^2](0,\Pi)}{(\lambda_0-\lambda_1)^{90}(\lambda_0-\lambda)^{160}}$$

$$= \frac{\theta^{200}[P_1^8 Q_\lambda](0,\Pi)}{(\lambda_0-\lambda_1)^{160}(\lambda_0-\lambda)^{90}},$$

after some reordering of the quotients. On the other hand, Proposition 3.9 and the equation for M_4 show that we can change every Q_λ to Q_∞ and $\lambda_0-\lambda$ to $\lambda_1-\lambda$, and hence obtain

$$\frac{\theta^{200}[P_1^9](0,\Pi)}{(\lambda_0-\lambda_1)^{250}} = \frac{\theta^{200}[Q_\infty^9](0,\Pi)}{(\lambda_0-\lambda_1)^{160}(\lambda_1-\lambda)^{90}} = \frac{\theta^{200}[P_1 Q_\infty^8](0,\Pi)}{(\lambda_0-\lambda_1)^{90}(\lambda_1-\lambda)^{160}}$$

$$= \frac{\theta^{200}[P_1^2 Q_\infty^7](0,\Pi)}{(\lambda_0-\lambda_1)^{40}(\lambda_1-\lambda)^{210}} = \frac{\theta^{200}[P_1^3 Q_\infty^6](0,\Pi)}{(\lambda_0-\lambda_1)^{10}(\lambda_1-\lambda)^{240}} = \frac{\theta^{200}[P_1^4 Q_\infty^5](0,\Pi)}{(\lambda_1-\lambda)^{250}}$$

$$= \frac{\theta^{200}[P_1^5 Q_\infty^4](0,\Pi)}{(\lambda_0-\lambda_1)^{10}(\lambda_1-\lambda)^{240}} = \frac{\theta^{200}[P_1^6 Q_\infty^3](0,\Pi)}{(\lambda_0-\lambda_1)^{40}(\lambda_1-\lambda)^{210}} = \frac{\theta^{200}[P_1^7 Q_\infty^2](0,\Pi)}{(\lambda_0-\lambda_1)^{90}(\lambda_1-\lambda)^{160}}$$

$$= \frac{\theta^{200}[P_1^8 Q_\infty](0,\Pi)}{(\lambda_0-\lambda_1)^{160}(\lambda_1-\lambda)^{90}}.$$

Since the expressions appearing first in these two equations are the same, all these expressions are equal, which gives us the Thomae formulae for this singular Z_{10} curve.

3.3.5 Thomae Formulae for Even n

After seeing this example, we are ready to state and prove the Thomae formulae for this family of singular Z_n curves for any even n. The expressions for the denominators h_Δ are as follows. For the divisor $\Delta = P_1^k Q_\lambda^{n-1-k}$, $0 \le k \le n-1$, we define h_Δ to be

$$(\lambda_0-\lambda_1)^{n(n/2-k-1)^2}(\lambda_0-\lambda)^{n(k+1)(n-1-k)} \tag{3.8}$$

and for the divisor $\Delta = P_1^k Q_\infty^{n-1-k}$, $0 \le k \le n-1$, we define h_Δ to be

$$(\lambda_0-\lambda_1)^{n(n/2-k-1)^2}(\lambda_1-\lambda)^{n(k+1)(n-1-k)} \tag{3.9}$$

(note that for the divisor $\Delta = P_1^{n-1}$ Eqs. (3.8) and (3.9) coincide to define h_Δ to be $(\lambda_0-\lambda_1)^{n^3/4}$). With these definitions we have

Theorem 3.10. *The quotient* $\theta^{2n^2}[\Delta](0,\Pi)/h_\Delta$ *where* h_Δ *is defined by Eqs.* (3.8) *and* (3.9) *is independent of* Δ.

Proof. First assume that 4 divides n. Then for every $1 \leq i \leq n/4$ divide the $(2i-1)$th equation of Proposition 3.8 by

$$(\lambda_0 - \lambda_1)^{4n(i-1)^2}(\lambda_0 - \lambda)^{n^3/4 - 4ni^2}.$$

Thus for $1 \leq i \leq (n-4)/4$ the denominators appearing under the expressions $\theta^{2n^2}[P_1^{(n-4i-2)/2}Q_\lambda^{(n+4i)/2}](0,\Pi)$ and $\theta^{2n^2}[P_1^{(n+4i-2)/2}Q_\lambda^{(n-4i)/2}](0,\Pi)$ in both the $(2i-1)$th and $(2i+1)$th equations become the same expression

$$(\lambda_0 - \lambda_1)^{4ni^2}(\lambda_0 - \lambda)^{n^3/4 - 4ni^2},$$

and for $i = n/4$ the corresponding denominator appears under $\theta^{2n^2}[P_1^{n-1}](0,\Pi)$ in the $[(n-2)/2]$th equation. This gives the equality

$$\frac{\theta^{2n^2}[P_1^{n-1}](0,\Pi)}{(\lambda_0 - \lambda_1)^{n^3/4}} = \frac{\theta^{2n^2}[P_1 Q_\lambda^{n-2}](0,\Pi)}{(\lambda_0 - \lambda_1)^{n(n-4)^2/4}(\lambda_0 - \lambda)^{2n(n-2)}} = \cdots$$

$$= \frac{\theta^{2n^2}[P_1^{(n-4i-2)/2}Q_\lambda^{(n+4i)/2}](0,\Pi)}{(\lambda_0 - \lambda_1)^{4ni^2}(\lambda_0 - \lambda)^{n^3/4 - 4ni^2}} = \cdots = \frac{\theta^{2n^2}[P_1^{(n-2)/2}Q_\lambda^{n/2}]}{(\lambda_0 - \lambda)^{n^3/4}} = \cdots$$

$$= \frac{\theta^{2n^2}[P_1^{(n+4i-2)/2}Q_\lambda^{(n-4i)/2}](0,\Pi)}{(\lambda_0 - \lambda_1)^{4ni^2}(\lambda_0 - \lambda)^{n^3/4 - 4ni^2}} = \cdots = \frac{\theta^{2n^2}[P_1^{n-3}Q_\lambda^2](0,\Pi)}{(\lambda_0 - \lambda_1)^{n(n-4)^2/4}(\lambda_0 - \lambda)^{2n(n-2)}}.$$

On the other hand, for every $1 \leq i \leq (n-4)/4$, divide the $2i$th equation of Proposition 3.8 by

$$(\lambda_0 - \lambda_1)^{4ni(i-1)}(\lambda_0 - \lambda)^{n^2(n-4)/4 - 4ni(i+1)}.$$

Then we find that for every $1 \leq i \leq (n-8)/4$ the denominators appearing under $\theta^{2n^2}[P_1^{(n-4i-4)/2}Q_\lambda^{(n+4i+2)/2}](0,\Pi)$ and $\theta^{2n^2}[P_1^{(n+4i)/2}Q_\lambda^{(n-4i-2)/2}](0,\Pi)$ in both the $2i$th and $(2i+2)$th equations become the same expression

$$(\lambda_0 - \lambda_1)^{4ni(i+1)}(\lambda_0 - \lambda)^{n^2(n-4)/4 - 4ni(i+1)},$$

and for the value $i = (n-4)/4$ the corresponding denominator appears under $\theta^{2n^2}[Q_\lambda^{n-1}](0,\Pi)$ and $\theta^{2n^2}[P_1^{n-2}Q_\lambda](0,\Pi)$. Here this gives the two equalities

$$\frac{\theta^{2n^2}[Q_\lambda^{n-1}](0,\Pi)}{(\lambda_0 - \lambda_1)^{n^2(n-4)/4}} = \cdots = \frac{\theta^{2n^2}[P_1^{(n-4i-4)/2}Q_\lambda^{(n+4i+2)/2}](0,\Pi)}{(\lambda_0 - \lambda_1)^{4ni(i+1)}(\lambda_0 - \lambda)^{n^2(n-4)/4 - 4ni(i+1)}}$$

$$= \cdots = \frac{\theta^{2n^2}[P_1^{(n-4)/2}Q_\lambda^{(n+2)/2}]}{(\lambda_0 - \lambda)^{n^2(n-4)/4}}$$

and

$$\frac{\theta^{2n^2}[P_1^{n/2}Q_\lambda^{(n-2)/2}]}{(\lambda_0-\lambda)^{n^2(n-4)/4}}=\dots=\frac{\theta^{2n^2}[P_1^{(n+4i)/2}Q_\lambda^{(n-4i-2)/2}](0,\Pi)}{(\lambda_0-\lambda_1)^{4ni(i+1)}(\lambda_0-\lambda)^{n^2(n-4)/4-4ni(i+1)}}$$

$$=\dots=\frac{\theta^{2n^2}[P_1^{n-2}Q_\lambda](0,\Pi)}{(\lambda_0-\lambda_1)^{n^2(n-4)/4}}.$$

The expressions in the last two equalities are in fact all equal since the operator N takes any divisor appearing in a theta constant in one of them to a divisor whose theta constant lies in the other equality over the same denominator, but we can ignore it as we now merge all three equalities together.

Now, the identity for $L_{(n-2)/2}$ shows us that

$$\frac{\theta^{2n^2}[P_1^{(n-2)/2}Q_\lambda^{n/2}](0,\Pi)}{(\lambda_0-\lambda)^n}=\frac{\theta^{2n^2}[P_1^{(n-4)/2}Q_\lambda^{(n+2)/2}](0,\Pi)}{(\lambda_0-\lambda_1)^n}=\frac{\theta^{2n^2}[P_1^{n/2}Q_\lambda^{(n-2)/2}](0,\Pi)}{(\lambda_0-\lambda_1)^n},$$

which contains exactly one theta constant from each of the equalities above. Divide the last equation by $(\lambda_0-\lambda)^{n^3/4-n}$, and then divide the two equations preceding it by $(\lambda_0-\lambda_1)^n(\lambda_0-\lambda)^{n^2-n}$. This allows us to merge all four equations into one. When checking the denominators under the theta constants, after doing so we find that the denominator under the expressions $\theta^{2n^2}[P_1^{(n-4i-2)/2}Q_\lambda^{(n+4i)/2}](0,\Pi)$ and $\theta^{2n^2}[P_1^{(n+4i-2)/2}Q_\lambda^{(n-4i)/2}](0,\Pi)$ from the first equality remains

$$(\lambda_0-\lambda_1)^{4ni^2}(\lambda_0-\lambda)^{n^3/4-4ni^2},$$

while the denominator under $\theta^{2n^2}[P_1^{(n-4i-4)/2}Q_\lambda^{(n+4i+2)/2}](0,\Pi)$ from the second equality and under $\theta^{2n^2}[P_1^{(n+4i)/2}Q_\lambda^{(n-4i-2)/2}](0,\Pi)$ from the third equality is now

$$(\lambda_0-\lambda_1)^{n(2i+1)^2}(\lambda_0-\lambda)^{n^3/4-n(2i+1)^2}$$

(as we have $4ni(i+1)+n=n(2i+1)^2$ and $n^2(n-4)/4+n^2=n^3/4$). The reader can verify that in the modified identity for $L_{(n-2)/2}$ any denominator appearing under some theta constant coincides with the denominator appearing under the same theta constant in the corresponding modified equality, which does the merging. When examining the divisors in the usual form $P_1^kQ_\lambda^{n-1-k}$ we see that

$$P_1^kQ_\lambda^{n-1-k}=\begin{cases}P_1^{(n-4i-4)/2}Q_\lambda^{(n+4i+2)/2}, i=(n-2k-4)/4 & k\le(n-4)/2\text{ even}\\P_1^{(n+4i)/2}Q_\lambda^{(n-4i-2)/2}, i=(2k-n)/4 & k\ge n/2\text{ even}\\P_1^{(n-4i-2)/2}Q_\lambda^{(n+4i)/2}, i=(n-2k-2)/4 & k\le(n-2)/2\text{ odd}\\P_1^{(n+4i-2)/2}Q_\lambda^{(n-4i)/2}, i=(2k+2-n)/4 & k\ge(n-2)/2\text{ odd}.\end{cases}$$

The reader can easily verify that $0\le i\le n/4$ in any case, and, once we substitute the corresponding value of i in the denominators we have obtained, and take into account that

$$\frac{n^3}{4} - n\left(\frac{n}{2} - k - 1\right)^2 = n(k+1)(n-1-k),$$

it becomes evident that the denominator appearing under $\theta^{2n^2}[P_1^k Q_\lambda^{n-1-k}](0,\Pi)$ is the one from Eq. (3.8). The multiple definition for $k = (n-2)/2$ gives, of course, $i = 0$ in both cases and the same denominator, so there is no problem here. This proves the relation for the divisors $P_1^k Q_\lambda^{n-1-k}$ in the case where 4 divides n.

Now assume that 4 does not divide n. In this case we divide the $(2i-1)$th equation with $1 \le i \le (n-2)/4$ of Proposition 3.8 by

$$(\lambda_0 - \lambda_1)^{4n(i-1)^2}(\lambda_0 - \lambda)^{n(n-2)^2/4 - 4ni^2}.$$

We get that after doing so, for every $1 \le i \le (n-6)/4$ the denominators appearing under $\theta^{2n^2}[P_1^{(n-4i-2)/2}Q_\lambda^{(n+4i)/2}](0,\Pi)$ and $\theta^{2n^2}[P_1^{(n+4i-2)/2}Q_\lambda^{(n-4i)/2}](0,\Pi)$ in the $(2i-1)$th and $(2i+1)$th equations become the same expression

$$(\lambda_0 - \lambda_1)^{4ni^2}(\lambda_0 - \lambda)^{n(n-2)^2/4 - 4ni^2};$$

for the value $i = (n-2)/4$ the denominators appearing under the theta constants $\theta^{2n^2}[Q_\lambda^{n-1}](0,\Pi)$ and $\theta^{2n^2}[P_1^{n-2}Q_\lambda](0,\Pi)$ follow the same rule. This combines to give the equality

$$\frac{\theta^{2n^2}[Q_\lambda^{n-1}](0,\Pi)}{(\lambda_0 - \lambda_1)^{n(n-2)^2/4}} = \cdots = \frac{\theta^{2n^2}[P_1^{(n-4i-2)/2}Q_\lambda^{(n+4i)/2}](0,\Pi)}{(\lambda_0 - \lambda_1)^{4ni^2}(\lambda_0 - \lambda)^{n(n-2)^2/4 - 4ni^2}} = \cdots$$

$$= \frac{\theta^{2n^2}[P_1^{(n-2)/2}Q_\lambda^{n/2}]}{(\lambda_0 - \lambda)^{n(n-2)^2/4}} = \cdots = \frac{\theta^{2n^2}[P_1^{(n+4i-2)/2}Q_\lambda^{(n-4i)/2}](0,\Pi)}{(\lambda_0 - \lambda_1)^{4ni^2}(\lambda_0 - \lambda)^{n(n-2)^2/4 - 4ni^2}} = \cdots$$

$$= \frac{\theta^{2n^2}[P_1^{n-2}Q_\lambda](0,\Pi)}{(\lambda_0 - \lambda_1)^{n(n-2)^2/4}}.$$

As for the other equations, for every $1 \le i \le (n-2)/4$, divide the $2i$th equation of Proposition 3.8 by

$$(\lambda_0 - \lambda_1)^{4ni(i-1)}(\lambda_0 - \lambda)^{n(n^2-4)/4 - 4ni(i+1)}.$$

Then we find that for every $1 \le i \le (n-6)/4$ the denominators appearing under $\theta^{2n^2}[P_1^{(n-4i-4)/2}Q_\lambda^{(n+4i+2)/2}](0,\Pi)$ and $\theta^{2n^2}[P_1^{(n+4i)/2}Q_\lambda^{(n-4i-2)/2}](0,\Pi)$ in both the $2i$th and $(2i+2)$th equations become the same expression

$$(\lambda_0 - \lambda_1)^{4ni(i+1)}(\lambda_0 - \lambda)^{n(n^2-4)/4 - 4ni(i+1)},$$

and for $i = (n-2)/2$ the same applies for the denominator under the expression $\theta^{2n^2}[P_1^{n-1}](0,\Pi)$. Together this gives the equality

$$\frac{\theta^{2n^2}[P_1^{n/2}Q_\lambda^{(n-2)/2}]}{(\lambda_0-\lambda)^{n(n^2-4)/4}} = \cdots = \frac{\theta^{2n^2}[P_1^{(n+4i)/2}Q_\lambda^{(n-4i-2)/2}](0,\Pi)}{(\lambda_0-\lambda_1)^{4ni(i+1)}(\lambda_0-\lambda)^{n(n^2-4)/4-4ni(i+1)}} = \cdots$$

$$= \frac{\theta^{2n^2}[P_1^{n-3}Q_\lambda](0,\Pi)}{(\lambda_0-\lambda_1)^{n(n-2)(n-6)/4}(\lambda_0-\lambda)^{2n(n-2)}} = \frac{\theta^{2n^2}[P_1^{n-1}](0,\Pi)}{(\lambda_0-\lambda_1)^{n(n^2-4)/4}}$$

$$= \frac{\theta^{2n^2}[P_1 Q_\lambda^{n-2}](0,\Pi)}{(\lambda_0-\lambda_1)^{n(n-2)(n-6)/4}(\lambda_0-\lambda)^{2n(n-2)}} = \cdots$$

$$= \frac{\theta^{2n^2}[P_1^{(n-4i-4)/2}Q_\lambda^{(n+4i+2)/2}](0,\Pi)}{(\lambda_0-\lambda_1)^{4ni(i+1)}(\lambda_0-\lambda)^{n(n^2-4)/4-4ni(i+1)}} = \cdots = \frac{\theta^{2n^2}[P_1^{(n-4)/2}Q_\lambda^{(n+2)/2}]}{(\lambda_0-\lambda)^{n(n^2-4)/4}},$$

here already one long equality and not two separate ones.

Again we take the identity for $L_{(n-2)/2}$ to obtain

$$\frac{\theta^{2n^2}[P_1^{(n-2)/2}Q_\lambda^{n/2}](0,\Pi)}{(\lambda_0-\lambda)^n} = \frac{\theta^{2n^2}[P_1^{(n-4)/2}Q_\lambda^{(n+2)/2}](0,\Pi)}{(\lambda_0-\lambda_1)^n} = \frac{\theta^{2n^2}[P_1^{n/2}Q_\lambda^{(n-2)/2}](0,\Pi)}{(\lambda_0-\lambda_1)^n},$$

and then divide the last equation by the expression $(\lambda_0-\lambda)^{n(n^2-4)/4}$, the one preceding it by $(\lambda_0-\lambda_1)^n$, and the remaining one by $(\lambda_0-\lambda)^{n^2-n}$. We can then merge the three equations into one. We now check the denominators under the theta constants after these operations. Then we find that the denominator under $\theta^{2n^2}[P_1^{(n-4i-2)/2}Q_\lambda^{(n+4i)/2}](0,\Pi)$ and $\theta^{2n^2}[P_1^{(n+4i-2)/2}Q_\lambda^{(n-4i)/2}](0,\Pi)$ from the first equality is now

$$(\lambda_0-\lambda_1)^{4ni^2}(\lambda_0-\lambda)^{n^3/4-4ni^2},$$

while the denominator under the expressions $\theta^{2n^2}[P_1^{(n-4i-4)/2}Q_\lambda^{(n+4i+2)/2}](0,\Pi)$ and $\theta^{2n^2}[P_1^{(n+4i)/2}Q_\lambda^{(n-4i-2)/2}](0,\Pi)$ from the second equality is now

$$(\lambda_0-\lambda_1)^{n(2i+1)^2}(\lambda_0-\lambda)^{n^3/4-n(2i+1)^2}$$

(again we use $4ni(i+1)+n=n(2i+1)^2$, and we have $n(n-2)^2/4+n^2-n=n^3/4$ and $n(n^2-4)/4+n=n^3/4$—the last power of $\lambda_0-\lambda$ was obtained by adding and subtracting n). From these theta constants we know that three appear in the identity for $L_{(n-2)/2}$, and after the manipulations we did we find that also the denominators coincide. Therefore we can merge all these equalities into one. As for writing the divisors in the usual form, here we find that

$$P_1^k Q_\lambda^{n-1-k} = \begin{cases} P_1^{(n-4i-2)/2}Q_\lambda^{(n+4i)/2}, i=(n-2k-2)/4 & k \le (n-2)/2 \text{ even} \\ P_1^{(n+4i-2)/2}Q_\lambda^{(n-4i)/2}, i=(2k+2-n)/4 & k \ge (n-2)/2 \text{ even} \\ P_1^{(n-4i-4)/2}Q_\lambda^{(n+4i+2)/2}, i=(n-2k-4)/4 & k \le (n-4)/2 \text{ odd} \\ P_1^{(n+4i)/2}Q_\lambda^{(n-4i-2)/2}, i=(2k-n)/4 & k \ge n/2 \text{ odd}. \end{cases}$$

Once again it can be easily verified that $0 \leq i \leq (n-2)/4$ in any case and that we have under $\theta^{2n^2}[P_1^k Q_\lambda^{n-1-k}](0,\Pi)$ exactly the denominator from Eq. (3.8), once the corresponding value of i is taken. We have again used the equality

$$\frac{n^3}{4} - n\left(\frac{n}{2} - k - 1\right)^2 = n(k+1)(n-1-k).$$

The multiple definition here is for $k = (n-2)/2$ again, and still gives $i = 0$ in both cases and the same denominator and causes no problem. This concludes for the divisors $P_1^k Q_\lambda^{n-1-k}$ where 4 does not divide n.

This proves the theorem for the divisors $P_1^k Q_\lambda^{n-1-k}$, $0 \leq k \leq n-1$. Note however that Proposition 3.9 says that everything we have said about the divisors $P_1^k Q_\lambda^{n-1-k}$ is correct for the divisors $P_1^k Q_\infty^{n-1-k}$ as well once every multiplier $\lambda_0 - \lambda$ is changed to $\lambda_1 - \lambda$, and this action takes the expressions from Eq. (3.8) to the corresponding expressions in Eq. (3.9). Since the same holds for the identity for $M_{n-2/2}$, we find that the same arguments prove the assertion of the theorem for the divisors $P_1^k Q_\infty^{n-1-k}$, $0 \leq k \leq n-1$. Since $k = n-1$ gives the same divisor P_1^{n-1} and the same denominator in both sets, the theorem is proved.   ☐

Theorem 3.10 is the Thomae formula for the family of compact Riemann surfaces whose algebraic equation is

$$w^n = (z - \lambda_0)(z - \lambda_1)(z - \lambda)^{n-1}$$

for even n.

3.3.6 Odd n

We now turn our attention to odd values of n, hence $e = 2$. We use again the proper power of the identity $\theta[N(\Delta)](0,\Pi) = \theta[\Delta](0,\Pi)$ from Eq. (1.5) to see that $L_{(n-3)/2}$ and $L_{(n-1)/2}$ satisfy

$$L_{(n-3)/2} = \frac{\theta^{2n^2}[P_1^{(n-3)/2} Q_\lambda^{(n+1)/2}](0,\Pi)}{\theta^{2n^2}[P_1^{(n-5)/2} Q_\lambda^{(n+3)/2}](0,\Pi)} = \frac{(\lambda_0 - \lambda)^{2n}}{(\lambda_0 - \lambda_1)^{2n}}$$

and

$$L_{(n-1)/2} = \frac{\theta^{2n^2}[P_1^{(n-1)/2} Q_\lambda^{(n-1)/2}](0,\Pi)}{\theta^{2n^2}[P_1^{(n+1)/2} Q_\lambda^{(n-3)/2}](0,\Pi)} = \frac{(\lambda_0 - \lambda)^{2n}}{(\lambda_0 - \lambda_1)^{2n}}$$

(as $\theta^{2n^2}[P_1^{(n-3)/2} Q_\lambda^{(n+1)/2}](0,\Pi)$ and $\theta^{2n^2}[P_1^{(n-1)/2} Q_\lambda^{(n-1)/2}](0,\Pi)$ cancel with one another). We write the last two equalities as

$$\frac{\theta^{2n^2}[P_1^{(n-3)/2}Q_\lambda^{(n+1)/2}](0,\Pi)}{(\lambda_0-\lambda)^{2n}} = \frac{\theta^{2n^2}[P_1^{(n-5)/2}Q_\lambda^{(n+3)/2}](0,\Pi)}{(\lambda_0-\lambda_1)^{2n}}$$

and

$$\frac{\theta^{2n^2}[P_1^{(n-1)/2}Q_\lambda^{(n-1)/2}](0,\Pi)}{(\lambda_0-\lambda)^{2n}} = \frac{\theta^{2n^2}[P_1^{(n+1)/2}Q_\lambda^{(n-3)/2}](0,\Pi)}{(\lambda_0-\lambda_1)^{2n}}.$$

The proposition we now prove is

Proposition 3.11. *The following equalities hold: For $1 \leq j \leq (n-3)/2$ we have*

$$\frac{\theta^{2n^2}[P_1^{(n-2j-1)/2}Q_\lambda^{(n+2j-1)/2}](0,\Pi)}{(\lambda_0-\lambda)^{2jn}} = \frac{\theta^{2n^2}[P_1^{(n-2j-3)/2}Q_\lambda^{(n+2j+1)/2}](0,\Pi)}{(\lambda_0-\lambda_1)^{2jn}},$$

and for $j=(n-1)/2$ we have

$$\frac{\theta^{2n^2}[Q_\lambda^{n-1}](0,\Pi)}{(\lambda_0-\lambda)^{2n(n-1)}} = \frac{\theta^{2n^2}[P_1^{n-1}](0,\Pi)}{(\lambda_0-\lambda_1)^{2n(n-1)}}.$$

Moreover, we have

$$\frac{\theta^{2n^2}[P_1^{(n+2j-3)/2}Q_\lambda^{(n-2j+1)/2}](0,\Pi)}{(\lambda_0-\lambda)^{2jn}} = \frac{\theta^{2n^2}[P_1^{(n+2j-1)/2}Q_\lambda^{(n-2j-1)/2}](0,\Pi)}{(\lambda_0-\lambda_1)^{2jn}}$$

for all $1 \leq j \leq (n-1)/2$.

Proof. Again we use induction on j, where the case $j=1$ is just the equations that we have obtained above. As for the induction step, assume the result is true for some $1 \leq j \leq (n-3)/2$, i.e., that

$$\frac{\theta^{2n^2}[P_1^{(n-2j-1)/2}Q_\lambda^{(n+2j-1)/2}](0,\Pi)}{\theta^{2n^2}[P_1^{(n-2j-3)/2}Q_\lambda^{(n+2j+1)/2}](0,\Pi)} = \frac{(\lambda_0-\lambda)^{2jn}}{(\lambda_0-\lambda_1)^{2jn}}$$

and

$$\frac{\theta^{2n^2}[P_1^{(n+2j-3)/2}Q_\lambda^{(n-2j+1)/2}](0,\Pi)}{\theta^{2n^2}[P_1^{(n+2j-1)/2}Q_\lambda^{(n-2j-1)/2}](0,\Pi)} = \frac{(\lambda_0-\lambda)^{2jn}}{(\lambda_0-\lambda_1)^{2jn}}.$$

We now look at the equality obtained from the expressions for $L_{(n-2j-3)/2}$ and $L_{(n+2j-1)/2}$. The first gives us

$$\frac{\theta^{4n^2}[P_1^{(n-2j-3)/2}Q_\lambda^{(n+2j+1)/2}](0,\Pi)}{\theta^{2n^2}[P_1^{(n-2j-1)/2}Q_\lambda^{(n+2j-1)/2}](0,\Pi)\theta^{2n^2}[P_1^{(n-2j-5)/2}Q_\lambda^{(n+2j+3)/2}](0,\Pi)} = \frac{(\lambda_0-\lambda)^{2n}}{(\lambda_0-\lambda_1)^{2n}}$$

for $j \leq (n-5)/2$ and

$$\frac{\theta^{4n^2}[Q_\lambda^{n-1}](0,\Pi)}{\theta^{2n^2}[P_1 Q_\lambda^{n-2}](0,\Pi)\theta^{2n^2}[P_1^{n-1}](0,\Pi)} = \frac{(\lambda_0 - \lambda)^{2n}}{(\lambda_0 - \lambda_1)^{2n}}$$

for $j = (n-3)/2$ (as it involves L_0), and the second gives

$$\frac{\theta^{4n^2}[P_1^{(n+2j-1)/2} Q_\lambda^{(n-2j-1)/2}](0,\Pi)}{\theta^{2n^2}[P_1^{(n+2j-3)/2} Q_\lambda^{(n-2j+1)/2}](0,\Pi)\theta^{2n^2}[P_1^{(n+2j+1)/2} Q_\lambda^{(n-2j-3)/2}](0,\Pi)} = \frac{(\lambda_0 - \lambda)^{2n}}{(\lambda_0 - \lambda_1)^{2n}}$$

for all j. Multiplying the corresponding elements gives

$$\frac{\theta^{2n^2}[P_1^{(n-2j-3)/2} Q_\lambda^{(n+2j+1)/2}](0,\Pi)}{\theta^{2n^2}[P_1^{(n-2j-5)/2} Q_\lambda^{(n+2j+3)/2}](0,\Pi)} = \frac{(\lambda_0 - \lambda)^{2(j+1)n}}{(\lambda_0 - \lambda_1)^{2(j+1)n}}$$

for $j \le (n-5)/2$ and

$$\frac{\theta^{2n^2}[Q_\lambda^{n-1}](0,\Pi)}{\theta^{2n^2}[P_1^{n-1}](0,\Pi)} = \frac{(\lambda_0 - \lambda)^{n(n-1)}}{(\lambda_0 - \lambda_1)^{n(n-1)}}$$

for $j = (n-3)/2$, and, on the other hand, also

$$\frac{\theta^{2n^2}[P_1^{(n+2j-1)/2} Q_\lambda^{(n-2j-1)/2}](0,\Pi)}{\theta^{2n^2}[P_1^{(n+2j+1)/2} Q_\lambda^{(n-2j-3)/2}](0,\Pi)} = \frac{(\lambda_0 - \lambda)^{2(j+1)n}}{(\lambda_0 - \lambda_1)^{2(j+1)n}}$$

for all j. From this the needed identities for $j+1$ for all j follow immediately.   □

The same considerations with the quotients M_k, which are the same but with Q_λ replaced by Q_∞ and the rational expression in the parameters replaced by the rational expression $(\lambda - \lambda_1)^n/(\lambda_0 - \lambda_1)^n$, prove

Proposition 3.12. *The following equalities hold: For* $1 \le j \le (n-3)/2$ *we have*

$$\frac{\theta^{2n^2}[P_1^{(n-2j-1)/2} Q_\infty^{(n+2j-1)/2}](0,\Pi)}{(\lambda - \lambda_1)^{2jn}} = \frac{\theta^{2n^2}[P_1^{(n-2j-3)/2} Q_\infty^{(n+2j+1)/2}](0,\Pi)}{(\lambda_0 - \lambda_1)^{2jn}},$$

and for $j = (n-1)/2$ *we have*

$$\frac{\theta^{2n^2}[Q_\infty^{n-1}](0,\Pi)}{(\lambda - \lambda_1)^{2n(n-1)}} = \frac{\theta^{2n^2}[P_1^{n-1}](0,\Pi)}{(\lambda_0 - \lambda_1)^{2n(n-1)}}.$$

Moreover, we have

$$\frac{\theta^{2n^2}[P_1^{(n+2j-3)/2} Q_\infty^{(n-2j+1)/2}](0,\Pi)}{(\lambda - \lambda_1)^{2jn}} = \frac{\theta^{2n^2}[P_1^{(n+2j-1)/2} Q_\infty^{(n-2j-1)/2}](0,\Pi)}{(\lambda_0 - \lambda_1)^{2jn}}$$

for all $1 \le j \le (n-1)/2$.

Also here we note (as we noted previously for the case of even n) that in both Propositions 3.11 and 3.12 each part can be proved from the other by using $\theta[N(\Delta)](0, \Pi) = \theta[\Delta](0, \Pi)$ from Eq. (1.5). Thus for every j, the expressions appearing in the two parts of Proposition 3.11 are equal, and so are expressions appearing in the two parts of Proposition 3.12 (here only for $j = (n-1)/2$ the two parts have common expressions).

3.3.7 An Example with $n = 9$

We once again begin with an example to make the way clearer. We look at $n = 9$, where Proposition 3.11 yields the following sequence of equalities: $j = 1$ gives

$$\frac{\theta^{162}[P_1^3 Q_\lambda^5](0, \Pi)}{(\lambda_0 - \lambda)^{18}} = \frac{\theta^{162}[P_1^2 Q_\lambda^6](0, \Pi)}{(\lambda_0 - \lambda_1)^{18}}, \quad \frac{\theta^{162}[P_1^4 Q_\lambda^4](0, \Pi)}{(\lambda_0 - \lambda)^{18}} = \frac{\theta^{162}[P_1^5 Q_\lambda^3](0, \Pi)}{(\lambda_0 - \lambda_1)^{18}};$$

$j = 2$ gives

$$\frac{\theta^{162}[P_1^2 Q_\lambda^6](0, \Pi)}{(\lambda_0 - \lambda)^{36}} = \frac{\theta^{162}[P_1 Q_\lambda^7](0, \Pi)}{(\lambda_0 - \lambda_1)^{36}}, \quad \frac{\theta^{162}[P_1^5 Q_\lambda^3](0, \Pi)}{(\lambda_0 - \lambda)^{36}} = \frac{\theta^{162}[P_1^6 Q_\lambda^2](0, \Pi)}{(\lambda_0 - \lambda_1)^{36}};$$

$j = 3$ gives

$$\frac{\theta^{162}[P_1 Q_\lambda^7](0, \Pi)}{(\lambda_0 - \lambda)^{54}} = \frac{\theta^{162}[Q_\lambda^8](0, \Pi)}{(\lambda_0 - \lambda_1)^{54}}, \quad \frac{\theta^{162}[P_1^6 Q_\lambda^2](0, \Pi)}{(\lambda_0 - \lambda)^{54}} = \frac{\theta^{162}[P_1^7 Q_\lambda](0, \Pi)}{(\lambda_0 - \lambda_1)^{54}};$$

and $j = 4$ gives

$$\frac{\theta^{162}[Q_\lambda^8](0, \Pi)}{(\lambda_0 - \lambda)^{72}} = \frac{\theta^{162}[P_1^8](0, \Pi)}{(\lambda_0 - \lambda_1)^{72}}, \quad \frac{\theta^{162}[P_1^7 Q_\lambda](0, \Pi)}{(\lambda_0 - \lambda)^{72}} = \frac{\theta^{162}[P_1^8](0, \Pi)}{(\lambda_0 - \lambda_1)^{72}}.$$

The way to obtain the Thomae formulae for the case $n = 9$ from these equalities is as follows. We divide the equation for $j = 1$ by $(\lambda_0 - \lambda)^{162}$, the one for $j = 2$ by $(\lambda_0 - \lambda_1)^{18}(\lambda_0 - \lambda)^{126}$, the one for $j = 3$ by $(\lambda_0 - \lambda_1)^{54}(\lambda_0 - \lambda)^{72}$, and the one for $j = 4$ by $(\lambda_0 - \lambda_1)^{108}$. After that we can combine the four equations to

$$\frac{\theta^{162}[P_1^4 Q_\lambda^4](0, \Pi)}{(\lambda_0 - \lambda)^{180}} = \frac{\theta^{162}[P_1^5 Q_\lambda^3](0, \Pi)}{(\lambda_0 - \lambda_1)^{18}(\lambda_0 - \lambda)^{162}} = \frac{\theta^{162}[P_1^6 Q_\lambda^2](0, \Pi)}{(\lambda_0 - \lambda_1)^{54}(\lambda_0 - \lambda)^{126}}$$

$$= \frac{\theta^{162}[P_1^7 Q_\lambda](0, \Pi)}{(\lambda_0 - \lambda_1)^{108}(\lambda_0 - \lambda)^{72}} = \frac{\theta^{162}[P_1^8](0, \Pi)}{(\lambda_0 - \lambda_1)^{180}} = \frac{\theta^{162}[Q_\lambda^8](0, \Pi)}{(\lambda_0 - \lambda_1)^{108}(\lambda_0 - \lambda)^{72}}$$

$$= \frac{\theta^{162}[P_1 Q_\lambda^7](0, \Pi)}{(\lambda_0 - \lambda_1)^{54}(\lambda_0 - \lambda)^{126}} = \frac{\theta^{162}[P_1^2 Q_\lambda^6](0, \Pi)}{(\lambda_0 - \lambda_1)^{18}(\lambda_0 - \lambda)^{162}} = \frac{\theta^{162}[P_1^3 Q_\lambda^5](0, \Pi)}{(\lambda_0 - \lambda)^{180}}.$$

We can get another equality by changing every Q_λ to Q_∞ and $\lambda_0 - \lambda$ to $\lambda_1 - \lambda$ as Proposition 3.12 shows, and this gives

$$\frac{\theta^{162}[P_1^4 Q_\infty^4](0,\Pi)}{(\lambda_1 - \lambda)^{180}} = \frac{\theta^{162}[P_1^5 Q_\infty^3](0,\Pi)}{(\lambda_0 - \lambda_1)^{18}(\lambda_1 - \lambda)^{162}} = \frac{\theta^{162}[P_1^6 Q_\infty^2](0,\Pi)}{(\lambda_0 - \lambda_1)^{54}(\lambda_1 - \lambda)^{126}}$$

$$= \frac{\theta^{162}[P_1^7 Q_\infty](0,\Pi)}{(\lambda_0 - \lambda_1)^{108}(\lambda_1 - \lambda)^{72}} = \frac{\theta^{162}[P_1^8](0,\Pi)}{(\lambda_0 - \lambda_1)^{180}} = \frac{\theta^{162}[Q_\infty^8](0,\Pi)}{(\lambda_0 - \lambda_1)^{108}(\lambda_1 - \lambda)^{72}}$$

$$= \frac{\theta^{162}[P_1 Q_\infty^7](0,\Pi)}{(\lambda_0 - \lambda_1)^{54}(\lambda_1 - \lambda)^{126}} = \frac{\theta^{162}[P_1^2 Q_\infty^6](0,\Pi)}{(\lambda_0 - \lambda_1)^{18}(\lambda_1 - \lambda)^{162}} = \frac{\theta^{162}[P_1^3 Q_\infty^5](0,\Pi)}{(\lambda_1 - \lambda)^{180}}.$$

The expression appearing in the middle of each one of these two equations is the same, which means that all these expressions are equal. This gives us the Thomae formulae for this singular Z_9 curve.

3.3.8 Thomae Formulae for Odd n

Bearing this example in mind, we now state and prove the Thomae formulae for this family of singular Z_n curves for any odd n. The denominators h_Δ are the following expressions. For the divisor $\Delta = P_1^k Q_\lambda^{n-1-k}$, $0 \leq k \leq n-1$, we define h_Δ to be

$$(\lambda_0 - \lambda_1)^{n(n-2k-1)(n-2k-3)/4}(\lambda_0 - \lambda)^{n(k+1)(n-1-k)}, \tag{3.10}$$

and for the divisor $\Delta = P_1^k Q_\infty^{n-1-k}$, $0 \leq k \leq n-1$, we define h_Δ to be

$$(\lambda_0 - \lambda_1)^{n(n-2k-1)(n-2k-3)/4}(\lambda_1 - \lambda)^{n(k+1)(n-1-k)}; \tag{3.11}$$

for $\Delta = P_1^{n-1}$ Eqs. (3.10) and (3.11) are easily seen to coincide and the value of h_Δ here is $(\lambda_0 - \lambda_1)^{n(n^2-1)/4}$. The result we have here is

Theorem 3.13. *The quotient* $\theta^{2n^2}[\Delta](0,\Pi)/h_\Delta$ *where* h_Δ *is defined by Eqs. (3.10) and (3.11) is independent of* Δ.

Proof. For every $1 \leq j \leq (n-1)/2$, divide the jth equation of Proposition 3.11 by

$$(\lambda_0 - \lambda_1)^{nj(j-1)}(\lambda_0 - \lambda)^{n(n^2-1)/4-nj(j+1)}.$$

This gives us that for every $1 \leq j \leq (n-3)/2$ the denominators appearing under $\theta^{2n^2}[P_1^{(n-2j-3)/2} Q_\lambda^{(n+2j+1)/2}](0,\Pi)$ and $\theta^{2n^2}[P_1^{(n+2j-1)/2} Q_\lambda^{(n-2j-1)/2}](0,\Pi)$ in both the jth and $(j+1)$th equations become the same expression

$$(\lambda_0 - \lambda_1)^{nj(j+1)}(\lambda_0 - \lambda)^{n(n^2-1)/4-nj(j+1)},$$

and for $j = (n-1)/2$ the corresponding denominator appears under the expression $\theta^{2n^2}[P_1^{n-1}](0,\Pi)$. All this combines to give

$$\frac{\theta^{2n^2}[P_1^{(n-1)/2}Q_\lambda^{(n-1)/2}]}{(\lambda_0-\lambda)^{n(n^2-1)/4}} = \cdots = \frac{\theta^{2n^2}[P_1^{(n+2j-1)/2}Q_\lambda^{(n-2j-1)/2}](0,\Pi)}{(\lambda_0-\lambda_1)^{nj(j+1)}(\lambda_0-\lambda)^{n(n^2-1)/4-nj(j+1)}}$$

$$= \cdots = \frac{\theta^{2n^2}[P_1^{n-2}Q_\lambda](0,\Pi)}{(\lambda_0-\lambda_1)^{n(n-1)(n-3)/4}(\lambda_0-\lambda)^{n(n-1)}} = \frac{\theta^{2n^2}[P_1^{n-1}](0,\Pi)}{(\lambda_0-\lambda_1)^{n(n^2-1)/4}}$$

$$= \frac{\theta^{2n^2}[Q_\lambda^{n-1}](0,\Pi)}{(\lambda_0-\lambda_1)^{n(n-1)(n-3)/4}(\lambda_0-\lambda)^{n(n-1)}} = \cdots$$

$$= \frac{\theta^{2n^2}[P_1^{(n-2j-3)/2}Q_\lambda^{(n+2j+1)/2}](0,\Pi)}{(\lambda_0-\lambda_1)^{nj(j+1)}(\lambda_0-\lambda)^{n(n^2-1)/4-nj(j+1)}} = \cdots = \frac{\theta^{2n^2}[P_1^{(n-3)/2}Q_\lambda^{(n+1)/2}]}{(\lambda_0-\lambda)^{n(n^2-1)/4}}.$$

We now want to write these divisors in the usual form. Here we have

$$P_1^k Q_\lambda^{n-1-k} = \begin{cases} P_1^{(n-2j-3)/2}Q_\lambda^{(n+2j+1)/2}, & j=(n-2k-3)/2 \quad k\le(n-3)/2 \\ P_1^{(n+2j-1)/2}Q_\lambda^{(n-2j-1)/2}, & j=(2k+1-n)/2 \quad k\ge(n-1)/2, \end{cases}$$

with no need to split into even and odd k cases. It is clear that $0 \le j \le (n-1)/2$ in any case, and that the denominator appearing under $\theta^{2n^2}[P_1^k Q_\lambda^{n-1-k}](0,\Pi)$ is exactly the expression from Eq. (3.10); substitute the corresponding value of j, and use the equality

$$n\frac{n^2-1}{4} - n\frac{(n-2k-1)(n-2k-3)}{4} = n(k+1)(n-1-k).$$

This finishes proving the theorem for every odd n, but just for the divisors $P_1^k Q_\lambda^{n-1-k}$, $0 \le k \le n-1$. Now, Proposition 3.12 says that everything said about the divisors $P_1^k Q_\lambda^{n-1-k}$ holds also for $P_1^k Q_\infty^{n-1-k}$ after replacing every $\lambda_0-\lambda$ to $\lambda_1-\lambda$, an action which takes the expressions from Eq. (3.10) to the corresponding expressions in Eq. (3.11). This proves the theorem also for the divisors $P_1^k Q_\infty^{n-1-k}$, $0 \le k \le n-1$, and since $k=n-1$ gives the same divisor P_1^{n-1} and the same denominator in both cases, we are done. □

Theorem 3.13 is the Thomae formula for the family of compact Riemann surfaces whose algebraic equation is

$$w^n = (z-\lambda_0)(z-\lambda_1)(z-\lambda)^{n-1}$$

for odd n.

3.3.9 Changing the Basepoint

Theorems 3.10 and 3.13 give us the Thomae formulae for these Z_n curves depending on the parity of n, but they also depend a priori on the choice of P_0 as the basepoint throughout. There are 3 other possible branch points, P_1, Q_λ, and Q_∞, and for each of them we can prove Thomae formulae (again, depending on the parity of n), but the divisors will be different as now P_0 is allowed to appear in them while the new basepoint is not. It is quite easy to verify at this stage that when the new basepoint is P_1 the $2n-1$ divisors are $P_0^k Q_\lambda^{n-1-k}$, $0 \le k \le n-1$ and $P_0^k Q_\infty^{n-1-k}$, $0 \le k \le n-1$ with the two divisors with $k = n-1$ coinciding. When the new basepoint is Q_λ, they are $Q_\infty^k P_j^{n-1-k}$, $0 \le k \le n-1$, with j being 0 or 1, and with the two divisors with $k = n-1$ coinciding. Finally, when the new basepoint is Q_∞, they are $Q_\lambda^k P_j^{n-1-k}$, with $0 \le k \le n-1$, with j being 0 or 1, and with the two divisors with $k = n-1$ coinciding. Now, for each basepoint and the appropriate set of divisors one can prove Thomae formulae, and in principal obtain 4 different constants. We now show that this is not the case.

We begin with changing the basepoint to P_1. The action of the operator N is to take the divisor $P_0^k Q_\lambda^{n-1-k}$ with $0 \le k \le n-2$ to $P_0^{n-2-k} Q_\lambda^{k+1}$, the divisor $P_0^k Q_\infty^{n-1-k}$ with $0 \le k \le n-2$ to $P_0^{n-2-k} Q_\infty^{k+1}$, and to leave P_0^{n-1} invariant. The operator T_{P_0} takes $P_0^{n-2} Q_\lambda$ and $P_0^{n-2} Q_\infty$ to one another. The operator T_{Q_∞} takes $P_0^k Q_\lambda^{n-1-k}$ with $0 \le k \le n-3$ to $P_0^{n-3-k} Q_\lambda^{k+2}$ and takes $P_0^{n-2} Q_\lambda$ and P_0^{n-2} to one another. Finally the operator T_{Q_λ} takes $P_0^k Q_\infty^{n-1-k}$ with $0 \le k \le n-3$ to $P_0^{n-3-k} Q_\infty^{k+2}$ and takes $P_0^{n-2} Q_\infty$ and P_0^{n-2} to one another. Now, the Thomae formulae with the basepoint P_1 is proved with the denominators h_Δ which we now write. When n is even and $\Delta = P_0^k Q_\lambda^{n-1-k}$, $0 \le k \le n-1$, we define h_Δ to be

$$(\lambda_0 - \lambda_1)^{n(n/2-k-1)^2} (\lambda_1 - \lambda)^{n(k+1)(n-1-k)},$$

and for the divisor $\Delta = P_0^k Q_\infty^{n-1-k}$, $0 \le k \le n-1$, we define h_Δ to be

$$(\lambda_0 - \lambda_1)^{n(n/2-k-1)^2} (\lambda_0 - \lambda)^{n(k+1)(n-1-k)}.$$

The two expressions coincide for $k = n-1$ to $h_\Delta = (\lambda_0 - \lambda_1)^{n^3/4}$ for $\Delta = P_0^{n-1}$. When n is odd and $\Delta = P_0^k Q_\lambda^{n-1-k}$, $0 \le k \le n-1$, we define h_Δ to be

$$(\lambda_0 - \lambda_1)^{n(n-2k-1)(n-2k-3)/4} (\lambda_1 - \lambda)^{n(k+1)(n-1-k)},$$

and for the divisor $\Delta = P_0^k Q_\infty^{n-1-k}$, $0 \le k \le n-1$, we define h_Δ to be

$$(\lambda_0 - \lambda_1)^{n(n-2k-1)(n-2k-3)/4} (\lambda_0 - \lambda)^{n(k+1)(n-1-k)}.$$

Again for $k = n-1$ the two expressions coincide and give for $\Delta = P_0^{n-1}$ the expression $h_\Delta = (\lambda_0 - \lambda_1)^{n(n^2-1)/4}$. This can be done either by proving Theorems 3.10 and

3.13 again with the basepoint P_1 or by simply using the obvious symmetry we have and interchange 0 and 1 throughout. Now, we have seen that the comparison of the constant obtained from the basepoint P_0 with the one obtained from the basepoint P_1 is easiest if we choose the divisor Δ (with the basepoint P_0) to contain P_1 to the $(n-1)$th power, hence $\Delta = P_1^{n-1}$. Since P_0^n, P_1^n, Q_λ^n, and Q_∞^n are all equivalent, we can apply Corollary 1.13. The divisor $\Gamma_{P_1} = P_0^{n-1}$ was seen to be non-special, and we see from Theorems 3.10 and 3.13 that the denominator h_Δ with basepoint P_0 and the denominator $h_{\Gamma_{P_1}}$ with basepoint P_1 coincide. Since $\varphi_{P_0}(\Delta) + K_{P_0} = \varphi_{P_1}(\Gamma_{P_1}) + K_{P_1}$ by Corollary 1.13, we find by the usual argument that the constant quotient obtained from the basepoint P_1 (represented by the divisor Γ_{P_1}) is the same as the one from P_0 (represented by the divisor Δ).

We now turn to the basepoint Q_λ. Here the operator N takes $Q_\infty^k P_j^{n-1-k}$, with j being 0 or 1 and $0 \le k \le n-2$, to $Q_\infty^{n-2-k} P_j^{k+1}$ and leaves Q_∞^{n-1} invariant. The operator T_{Q_∞} takes $Q_\infty^{n-2} P_0$ and $Q_\infty^{n-2} P_1$ to one another. As for the remaining two operators, we have that T_{P_j}, with j being 0 or 1, takes $Q_\infty^k P_i^{n-1-k}$, with $0 \le k \le n-3$ and i being 0 or 1 and distinct from j, to $Q_\infty^{n-3-k} P_i^{k+2}$, and takes $Q_\infty^{n-2} P_i$ (with the same i) and Q_∞^{n-2} to one another. We can now state the Thomae formulae with the basepoint Q_λ, which is proved with the following denominators h_Δ (j is again either 0 or 1). When n is even and $\Delta = Q_\infty^k P_j^{n-1-k}$, $0 \le k \le n-1$, we define h_Δ to be

$$(\lambda_0 - \lambda_1)^{n(n/2-k-1)^2} (\lambda - \lambda_j)^{n(k+1)(n-1-k)};$$

again for $k = n-1$ the expressions obtained from the two values of j coincide and $h_\Delta = (\lambda_0 - \lambda_1)^{n^3/4}$ for $\Delta = Q_\infty^{n-1}$. When n is odd and the divisor is $\Delta = Q_\infty^k P_j^{n-1-k}$, $0 \le k \le n-1$, we define h_Δ to be

$$(\lambda_0 - \lambda_1)^{n(n-2k-1)(n-2k-3)/4} (\lambda - \lambda_j)^{n(k+1)(n-1-k)};$$

once again for $k = n-1$ the expressions obtained from the two values of j coincide and $h_\Delta = (\lambda_0 - \lambda_1)^{n(n^2-1)/4}$ for $\Delta = Q_\infty^{n-1}$. The proof is similar, but the calculations are a bit different. For example, the quotients

$$\frac{\theta^{en^2}[Q_\infty^{k-1} P_j^{n-k}](\varphi_{Q_\lambda}(P), \Pi)}{\theta^{en^2}[Q_\infty^k P_j^{n-1-k}](\varphi_{Q_\lambda}(P), \Pi)}$$

for $1 \le k \le n-1$, and the quotient

$$\frac{\theta^{en^2}[Q_\infty^{n-1}](\varphi_{Q_\lambda}(P), \Pi)}{\theta^{en^2}[P_j^{n-1}](\varphi_{Q_\lambda}(P), \Pi)}$$

for $k = 0$, with j being 0 or 1, are the functions $c_{jk}/(z - \lambda_j)^{en}$. Here there is no obvious symmetry between this branch point and the points P_0 and P_1, though in a few paragraphs, when we connect this example to the general case, we collect the information needed to complete the symmetry argument that spares us from the need of reproving Theorems 3.10 and 3.13 from the start. It is an excellent exercise for the reader to verify this result. Now, by the usual reasoning, the divisor Δ we take in order to make the comparison of the constants obtained from the basepoints P_0 and Q_λ is Q_λ^{n-1}, and Corollary 1.13 gives us that $\Gamma_{Q_\lambda} = P_0^{n-1}$. For this (non-special) divisor it is easy to see from Theorems 3.10 and 3.13 that the denominator h_Δ with basepoint P_0 and the denominator $h_{\Gamma_{Q_\lambda}}$ with basepoint Q_λ coincide (both of them with $k = 0$). Corollary 1.13 gives us $\varphi_{P_0}(\Delta) + K_{P_0} = \varphi_{Q_\lambda}(\Gamma_{Q_\lambda}) + K_{Q_\lambda}$, from which we find that the constant quotient obtained from the basepoint Q_λ (represented by the divisor Γ_{Q_λ}) is also the same as the one from P_0 (represented by the divisor Δ), as needed.

We conclude with the last branch point Q_∞. The action of the operators is exactly the same as with the basepoint Q_λ, but with Q_∞ replaced by Q_λ throughout. The Thomae formulae with the basepoint Q_∞ is proved with the following denominators h_Δ, with again i and j being 0 and 1 and distinct. When n is even and $\Delta = Q_\lambda^k P_j^{n-1-k}$, $0 \leq k \leq n - 1$, we define h_Δ to be

$$(\lambda_0 - \lambda_1)^{n(n/2-k-1)^2}(\lambda - \lambda_i)^{n(k+1)(n-1-k)};$$

as before for $k = n - 1$ the expressions obtained from the two values of j coincide and $h_\Delta = (\lambda_0 - \lambda_1)^{n^3/4}$ for $\Delta = Q_\lambda^{n-1}$. When n is odd and the divisor is $\Delta = Q_\lambda^k P_j^{n-1-k}$, $0 \leq k \leq n - 1$, we define h_Δ to be

$$(\lambda_0 - \lambda_1)^{n(n-2k-1)(n-2k-3)/4}(\lambda - \lambda_i)^{n(k+1)(n-1-k)};$$

as always for $k = n - 1$ the expressions obtained from the two values of j coincide and $h_\Delta = (\lambda_0 - \lambda_1)^{n(n^2-1)/4}$ for $\Delta = Q_\lambda^{n-1}$. The proof is again similar but not exactly the same, as the quotients

$$\frac{\theta^{en^2}[Q_\lambda^{k-1}P_j^{n-k}](\varphi_{Q_\infty}(P), \Pi)}{\theta^{en^2}[Q_\lambda^k P_j^{n-1-k}](\varphi_{Q_\infty}(P), \Pi)}$$

for $1 \leq k \leq n - 1$, and the quotient

$$\frac{\theta^{en^2}[Q_\lambda^{n-1}](\varphi_{Q_\infty}(P), \Pi)}{\theta^{en^2}[P_j^{n-1}](\varphi_{Q_\infty}(P), \Pi)}$$

for $k = 0$ (j is 0 or 1 again), are now the functions $c_{jk}(z - \lambda)^{en}/(z - \lambda_j)^{en}$ (recall that substituting $P = Q_\infty$ means taking the limit $z \to \infty$, but this leads only to finite nonzero limits as we had with the functions f_k with the basepoint P_0). Again a symmetry argument can give this result, using the information in the next paragraph. The interested reader can learn more by verifying this result. We now compare the constants obtained from the basepoints P_0 and Q_∞ by taking the divisor $\Delta = Q_\infty^{n-1}$, for which Corollary 1.13 gives $\Gamma_{Q_\infty} = P_0^{n-1}$. The denominator h_Δ with basepoint P_0 and the denominator $h_{\Gamma_{Q_\infty}}$ with basepoint Q_∞ in Theorems 3.10 and 3.13 are once again the same expression (again both of them are with $k = 0$), and since we have $\varphi_{P_0}(\Delta) + K_{P_0} = \varphi_{Q_\infty}(\Gamma_{Q_\infty}) + K_{Q_\infty}$ by Corollary 1.13, we conclude that the constant quotient obtained from the basepoint Q_∞ (represented by the divisor Γ_{Q_∞}) is the same as the one from P_0 (represented by the divisor Δ). This finishes showing that all four constants are equal.

Once again we would also like to show that all four basepoints involve the same theta constants, i.e., the same characteristics. This will be done using Proposition 1.15, which is applicable since we know that $P_0P_1/Q_\lambda Q_\infty$ is a principal divisor. It is a bit harder to see that all the divisors we have are of the form described in Proposition 1.15, but denoting B by D_{-1} and examining the description of these divisors in the beginning of this section shows that this is indeed the case. As in the previous examples, Proposition 1.15 gives us that every characteristic obtained from the basepoint P_0 is also obtained from any other basepoint and vice versa (here we even know the number of these characteristics—it is $2n - 1$), but since we can write all the divisors explicitly, we state that the equalities Proposition 1.15 gives us are

$$\varphi_{P_0}(P_1^k Q_\lambda^{n-1-k}) + K_{P_0} = \varphi_{P_1}(P_0^{n-2-k} Q_\infty^{k+1}) + K_{P_1}$$
$$= \varphi_{Q_\lambda}(P_0^{n-1-k} Q_\infty^k) + K_{Q_\lambda} = \varphi_{Q_\infty}(P_1^{k+1} Q_\lambda^{n-2-k}) + K_{Q_\infty}$$

and

$$\varphi_{P_0}(P_1^k Q_\infty^{n-1-k}) + K_{P_0} = \varphi_{P_1}(P_0^{n-2-k} Q_\lambda^{k+1}) + K_{P_1}$$
$$= \varphi_{Q_\lambda}(P_1^{k+1} Q_\infty^{n-2-k}) + K_{Q_\lambda} = \varphi_{Q_\infty}(P_0^{n-1-k} Q_\lambda^k) + K_{Q_\infty}$$

for every $0 \leq k \leq n - 2$ and

$$\varphi_{P_0}(P_1^{n-1}) + K_{P_0} = \varphi_{P_1}(P_0^{n-1}) + K_{P_1} = \varphi_{Q_\lambda}(Q_\infty^{n-1}) + K_{Q_\lambda} = \varphi_{Q_\infty}(Q_\lambda^{n-1}) + K_{Q_\infty}$$

for the one remaining divisor. One easily sees that all the divisors appearing here are non-special, and the reader can verify the above equalities using the principality of the divisor $P_0P_1/Q_\lambda Q_\infty$ and moving nth powers of branch points in the right directions.

The above suggests that we shall ultimately have in this example a notation which is independent of the basepoint. We do have such a notation, and it is based (as usual) on divisors of the form $\Xi = P_0^{n-1}\Delta$ and similar forms, but these divisors do not have at this point a general form suited for all of them and suited for writing the denominators h_Ξ nicely. Therefore we shall not write these expressions here, and we just remark that they shall ultimately appear as the special case of the general Theorem 5.8 below. We remark once again that for odd n the result here is slightly improved by the fact that all the powers can be cut in half (so that the numerators are always theta constants raised to the $2n^2$ power, as when n is even, rather than to the $4n^2$ power).

3.3.10 Relation with the General Singular Case

We now describe how this example relates to the general singular case. Since λ appears to the power $n-1$ and there is branching over ∞ as if it appears to the power $n-1$ as well (this means that we need $n-1$ to complete the degree of the polynomial $(z-\lambda_0)(z-\lambda_1)(z-\lambda)^{n-1}$, which is w^n, to a multiple of n), we should treat this curve as the curve of the algebraic equation

$$w^n = (z-\lambda_0)(z-\lambda_1)(z-\mu_1)^{n-1}(z-\mu_2)^{n-1},$$

where μ_1 is λ and ∞ plays the role of μ_2. As the reader can check, we can prove Thomae formulae for this curve (which is simply the singular case with $m=2$) just as we did in this section, but with the following changes. Every λ is changed to μ_1, and since we treat μ_2 as a finite number, substituting Q_2 in the functions f_k yields the extra factor $(\mu_2-\lambda_1)^{en}/(\mu_2-\mu_1)^{en}$, which multiplies the expression $(\lambda_0-\mu_1)^{en}/(\lambda_0-\lambda_1)^{en}$ appearing in the quotients L_k. Similarly, the functions g_k are $d_k(z-\lambda_1)^{en}/(z-\mu_2)^{en}$, which means that the extra factor $(\lambda_0-\mu_2)^{en}/(\mu_1-\mu_2)^{en}$ must multiply the expression $(\mu_1-\lambda_1)^{en}/(\lambda_0-\lambda_1)^{en}$ appearing in the quotients M_k. Therefore each expression $\lambda_0-\lambda_1$ must be accompanied by $\mu_1-\mu_2$ to the same power, each expression $\lambda_0-\mu_1$ must be accompanied by $\lambda_1-\mu_2$ to the same power, and each $\lambda_1-\mu_1$ must be accompanied by $\lambda_0-\mu_2$ to the same power.

Let us now show how to write the denominator h_Δ (for Δ a divisor with basepoint P_0) in an intrinsic way, similar to what we do in the general singular case. This is important, since the method of proof here was not the same and we want to make sure all our results are consistent. We again denote the set $A\cup P_0$ by C_{-1} and B by D_{-1}. Now, for the divisor $P_1^k Q_1^{n-1-k}$ with $0\le k\le n-2$, we find that the already existing expressions, $\lambda_0-\lambda_1$ and $\lambda_0-\mu_1$, are the difference of the z-values of P_0 from the set C_{-1} and P_1 from the set C_{n-2-k}, and the difference of the z-values of P_0 from the set C_{-1} and Q_1 from the set D_{n-2-k}, respectively. The new factors, $\mu_1-\mu_2$ and $\lambda_1-\mu_2$, are the difference of the z-values of Q_2 from the set D_{-1} and Q_1 from the set D_{n-2-k}, and the difference of the z-values of Q_2 from the set D_{-1} and P_1 from the set C_{n-2-k}, respectively. As for the divisor $P_1^k Q_2^{n-1-k}$ with $0\le k\le n-2$, the already existing expressions, $\lambda_0-\lambda_1$ and $\lambda_1-\mu_1$, are the difference of the z-values of P_0 from the set C_{-1} and P_1 from the set C_{n-2-k}, and the difference of the z-values of Q_1 from the set D_{-1} and P_1 from the set C_{n-2-k}, respectively. The new factors, $\mu_1-\mu_2$ and $\lambda_0-\mu_2$, are the difference of the z-values of Q_1 from the set D_{-1} and Q_2 from the set D_{n-2-k}, and the difference of the z-values of P_0 from the set C_{-1} and Q_2 from the set D_{n-2-k}, respectively. For the remaining divisor P_1^{n-1}, the existing expression $\lambda_0-\lambda_1$ is the difference of the z-values of the points P_0 and P_1 from C_{-1}, and the new factor $\mu_1-\mu_2$ is the difference of the z-values of the points Q_0 and Q_1 from D_{-1}.

In order to verify consistency, the reader might want to take a glimpse at Eqs. (5.11) and (5.10) below, which give the denominators for the Thomae formulae in the general singular case. The comparison shows us that all the expressions from the last paragraph appear there, and the (many) other expressions appearing there contribute nothing here since most of the sets C_j and D_j, $-1 \leq j \leq n-2$ are empty. The powers are also correct, except for the fact that h_Δ from Eq. (5.10) is the square of h_Δ from Theorem 3.13, which is consistent with the fact that in Theorem 3.13 it appears under $\theta^{2n^2}[\Delta](0, \Pi)$ and in Theorem 5.5 (of the general case below) it appears under $\theta^{4n^2}[\Delta](0, \Pi)$. This means that squaring the result of Theorem 3.13 (and thus weakening it a bit) gives the special case $m = 2$ of Theorem 5.5, while Theorem 3.10 is exactly the special case $m = 2$ of Theorem 5.6. Therefore the results in this section are slightly stronger (in the odd n case only), but consistent with the general singular ones.

3.4 Nonsingular Z_n Curves with $r = 1$ and Small n

When we defined the nonsingular Z_n curves we stated that the proof works "nicely" only for Z_n curves with equation

$$w^n = \prod_{i=0}^{rn-1} (z - \lambda_i)$$

(we assume for convenience that there is no branch point over ∞), and $r \geq 2$. In this section we present the case $r = 1$, explain what goes wrong there, and also demonstrate how the tool of basepoint change helps to finish the proof. We prove the Thomae formulae for $r = 1$ and $n = 4$ in the same way as we did in the previous examples (and as we do later for general nonsingular Z_n curves with $r \geq 2$) and show that the case $r = 1$ and $n = 3$ is trivial. The case $n = 2$ (and $r = 1$) is excluded, since then $g = 0$. When $n \geq 5$ we show that our proof does not suffice to obtain the full Thomae formulae and we need the tool of basepoint change to obtain it.

3.4.1 The Set of Divisors as a Principal Homogenous Space for S_{n-1}

All the general considerations we have done for nonsingular curves until now hold for the case $r = 1$ with no problem. The genus is $g = (n-1)(n-2)/2$ (whence we automatically assume $n \geq 3$, as curves of genus 0 do not interest us here), and Theorems 2.6 and 2.9 show that the non-special divisors supported on the branch points with P_0 deleted are of the form $\prod_{j=0}^{n-2} C_j^{n-2-j}$ where the sets C_j, $0 \leq j \leq n-2$, are of cardinality $r = 1$ (A is omitted, being of cardinality $r - 1 = 0$ and hence empty). Since in each set C_j there is only one branch point, which we shall denote by $P_{i_{j+1}}$, this means that our divisors are all of the form $\prod_{j=1}^{n-1} P_{i_j}^{n-1-j}$, where we have replaced the index j by $j+1$, and where the numbers i_j, $1 \leq j \leq n-1$, are the numbers from 1 to $n-1$ in some order. Hence there are exactly $(n-1)!$ such divisors. A useful way of looking at the set of divisors in this case is as a principal homogenous space for the group S_{n-1}, i.e., a set on which this group acts freely and transitively, where the action of the element $\sigma \in S_{n-1}$ is through the indices, i.e., to take $\prod_{j=1}^{n-1} P_{i_j}^{n-1-j}$ to $\prod_{j=1}^{n-1} P_{i_{\sigma^{-1}(j)}}^{n-1-j}$. We emphasize that this representation of the set of divisors is valuable only in this specific case of a nonsingular Z_n curve with $r=1$, and any attempt to do similar things in other cases will meet difficulties.

Let us now look at the operators as defined in Section 2.5. First, all of our divisors are such that there is exactly one operator T_R which can act—the one where R is the unique point P_{i_1} in the set C_0 of cardinality 1. Therefore we can (again—for this case and for this case only) define a single operator T which will act on Δ as the only permissible operator T_R. Now, for a divisor $\Delta = \prod_{j=1}^{n-1} P_{i_j}^{n-1-j}$ we have that $N(\Delta)$ is $\prod_{j=1}^{n-1} P_{i_j}^{j-1}$ and $T(\Delta) = T_{P_{i_1}}(\Delta)$ is $P_{i_1}^{n-2} \prod_{j=2}^{n-1} P_{i_j}^{j-2}$, by Definition 2.16 translated to this language. However, it is perhaps more convenient to write these expressions as $N(\Delta) = \prod_{j=1}^{n-1} P_{i_{n-j}}^{n-1-j}$ (replacing j by $n-j$) and $T(\Delta) = P_{i_1}^{n-2} \prod_{j=2}^{n-1} P_{i_{n+1-j}}^{n-1-j}$ (replacing j by $n+1-j$). This is the case since now it is easier to see that the operator N interchanges the index i_j, $1 \le j \le n-1$, with the index i_{n-j}, and the operator T fixes the index i_1 and interchanges the index i_j, $2 \le j \le n-1$, with the index i_{n+1-j}. Therefore the action of N is the same as the action of the element
$$\begin{pmatrix} 1 & 2 & \dots & n-2 & n-1 \\ n-1 & n-2 & \dots & 2 & 1 \end{pmatrix}$$
of S_{n-1}, and the action of T is the same as the action of the element $\begin{pmatrix} 1 & 2 & \dots & n-2 & n-1 \\ 1 & n-1 & \dots & 3 & 2 \end{pmatrix}$. We shall identify N and T with these elements (both of order 2, as expected) of S_{n-1} throughout this section.

We shall see that the proof of the Thomae formulae in the general case (singular and nonsingular) goes in the following stages. We obtain the PMT which is invariant under the operators T_R (and also T_S in the singular case), we then modify it to be invariant under N, and then we prove that using the operators T_R (and T_S in the singular case) and N we can get from any divisor to any other divisor. What fails in the nonsingular case with $r=1$ and $n \ge 5$ is only the last stage. The assertion of this the last stage is equivalent, in our description, to the statement that N and T generate the group S_{n-1}, and we later see that for $n \ge 5$ they do not.

3.4.2 The Case $n = 4$

We begin with the case $n = 4$, where we do prove the full Thomae formulae in the usual way, and it is not trivial. In this case the genus is $(4-1)(4-2)/2 = 3$ and there are $3! = 6$ divisors which are of the form $P_{i_1}^2 P_{i_2}^1 P_{i_3}^0$ with i_1, i_2 and i_3 being the numbers 1, 2 and 3 in any order. This set of divisors is a principal homogenous space for the group S_3. Here N is the permutation interchanging 1 and 3 and T is the permutation interchanging 1 and 2.

Let us use the notation where i, j, and k are the numbers 1, 2, and 3 in any order, and start with the divisor $\Delta = P_i^2 P_j$. We have $T(\Delta) = T_{P_i}(\Delta) = P_i^2 P_k$, and we consider the function

$$\frac{\theta^{16}[\Delta](\varphi_{P_0}(P), \Pi)}{\theta^{16}[T(\Delta)](\varphi_{P_0}(P), \Pi)} = \frac{\theta^{16}[P_i^2 P_j](\varphi_{P_0}(P), \Pi)}{\theta^{16}[P_i^2 P_k](\varphi_{P_0}(P), \Pi)}.$$

Since $N(\Delta) = P_k^2 P_j$ and $N(T(\Delta)) = P_j^2 P_k$, Proposition 2.21 shows that the divisor of this function is the 16th power of P_k/P_j, i.e., P_k^{16}/P_j^{16}, and hence that this function is $c(z - \lambda_k)^4/(z - \lambda_j)^4$ where c is a nonzero complex constant. As usual, we set $P = P_0$, which yields

$$\frac{\theta^{16}[P_i^2 P_j](0, \Pi)}{\theta^{16}[P_i^2 P_k](0, \Pi)} = c\frac{(\lambda_0 - \lambda_k)^4}{(\lambda_0 - \lambda_j)^4},$$

and thus

$$\frac{\theta^{16}[P_i^2 P_j](\varphi_{P_0}(P), \Pi)}{\theta^{16}[P_i^2 P_k](\varphi_{P_0}(P), \Pi)} = \frac{(\lambda_0 - \lambda_j)^4}{(\lambda_0 - \lambda_k)^4} \times \frac{\theta^{16}[P_i^2 P_j](0, \Pi)}{\theta^{16}[P_i^2 P_k](0, \Pi)} \times \frac{(z - \lambda_k)^4}{(z - \lambda_j)^4}.$$

Now, since $T = T_{P_i}$, the only reasonable substitution is $P = P_i$, which, on the one hand, yields

$$\frac{\theta^{16}[P_i^2 P_j](\varphi_{P_0}(P_i), \Pi)}{\theta^{16}[P_i^2 P_k](\varphi_{P_0}(P_i), \Pi)} = \frac{(\lambda_0 - \lambda_j)^4}{(\lambda_0 - \lambda_k)^4} \times \frac{\theta^{16}[P_i^2 P_j](0, \Pi)}{\theta^{16}[P_i^2 P_k](0, \Pi)} \times \frac{(\lambda_i - \lambda_k)^4}{(\lambda_i - \lambda_j)^4}$$

from the last equation, and, on the other hand, we have

$$\frac{\theta^{16}[P_i^2 P_j](\varphi_{P_0}(P_i), \Pi)}{\theta^{16}[P_i^2 P_k](\varphi_{P_0}(P_i), \Pi)} = \frac{\theta^{16}[P_i^2 P_k](0, \Pi)}{\theta^{16}[P_i^2 P_j](0, \Pi)}$$

from Eq. (2.3) (since $T_{P_i}(P_i^2 P_k) = P_i^2 P_j$ as T_{P_i} is an involution). We therefore obtain

$$\frac{\theta^{32}[P_i^2 P_k](0, \Pi)}{\theta^{32}[P_i^2 P_j](0, \Pi)} = \frac{(\lambda_0 - \lambda_j)^4}{(\lambda_0 - \lambda_k)^4} \times \frac{(\lambda_i - \lambda_k)^4}{(\lambda_i - \lambda_j)^4},$$

which gives us

$$\frac{\theta^{32}[P_i^2 P_j](0, \Pi)}{(\lambda_0 - \lambda_k)^4(\lambda_i - \lambda_j)^4} = \frac{\theta^{32}[P_i^2 P_k](0, \Pi)}{(\lambda_0 - \lambda_j)^4(\lambda_i - \lambda_k)^4}. \tag{3.12}$$

Since the operator T_{P_i} is the only operator T_R which can act on $P_i^2 P_j$ and $P_i^2 P_k$ and it takes them to one another, we find that Eq. (3.12) is the PMT for this Z_4 curve. Indeed, if we define for every divisor Δ the expression g_Δ to be the product of the 4th power of the difference between λ_0 and the z-value of the branch point which does not appear in Δ (and is not P_0) and the 4th power of the difference between the z-values of the branch points which do appear in Δ then by what we have just said Eq. (3.12) states that $\theta^{32}[\Delta](0,\Pi)/g_\Delta$ is invariant under the operators T_R (or actually T in our case). It is easy to see that starting from the other divisor $P_i^2 P_k$ gives us exactly the same Eq. (3.12).

As usual, the Thomae formulae is obtained by using the fact that Eq. (1.5) gives $\theta[N(\Delta)](0,\Pi) = \theta[\Delta](0,\Pi)$ to the proper power, and thus we would like to replace the denominators g_Δ by more symmetric denominators h_Δ which are also invariant under N. For this we observe that for the divisor $\Delta = P_i^2 P_j$ we have the denominator $g_\Delta = (\lambda_0 - \lambda_k)^4 (\lambda_i - \lambda_j)^4$, and thus for $N(\Delta) = P_k^2 P_j$ we have the denominator $g_{N(\Delta)} = (\lambda_0 - \lambda_i)^4 (\lambda_k - \lambda_j)^4$. We notice that for the divisor $T(\Delta) = P_i^2 P_k$ we obtain $N(T(\Delta)) = P_j^2 P_k$, which means that looking at the PMT in Eq. (3.12) as coming from $T(\Delta)$ the expression we would have examined instead of $g_{N(\Delta)}$ would have been $g_{N(T(\Delta))} = (\lambda_0 - \lambda_i)^4 (\lambda_j - \lambda_k)^4$, which equals $g_{N(\Delta)}$. Therefore we divide Eq. (3.12) by this common expression and obtain, by defining $h_\Delta = g_\Delta g_{N(\Delta)}$, that

$$\frac{\theta^{32}[\Delta](0,\Pi)}{h_\Delta} = \frac{\theta^{32}[P_i^2 P_j](0,\Pi)}{(\lambda_0 - \lambda_k)^4 (\lambda_i - \lambda_j)^4 (\lambda_0 - \lambda_i)^4 (\lambda_k - \lambda_j)^4} \tag{3.13}$$

is invariant both under T and under N. Under T it is invariant since we got Eq. (3.13) from Eq. (3.12), and under N since both the numerator and the denominator are invariant under N.

We now prove

Theorem 3.14. *Equation* (3.13) *is the Thomae formulae for the nonsingular* Z_4 *curve* X.

Proof. We already know that Eq. (3.13) is invariant under T and N. It thus suffices to show that with T and N one can get from any divisor to any other. Since the set of divisors is a principal homogenous space for S_3, we are reduced to showing that N and T generate the group S_3. But N is the permutation interchanging 1 and 3 and T is the permutation interchanging 1 and 2, and these are known to generate S_3. Since there are only 6 divisors, we can also write this explicitly. Let us start with the divisor $\Delta = P_1^2 P_2$, and then we have

$$\Delta = P_1^2 P_2, \quad T(\Delta) = P_1^2 P_3, \quad NT(\Delta) = P_2^2 P_3,$$

$$N(\Delta) = P_3^2 P_2, \quad TN(\Delta) = P_3^2 P_1, \quad NTN(\Delta) = TNT(\Delta) = P_2^2 P_1$$

(an easy calculation of permutations shows that $NTN = TNT$ in this case). This proves the theorem. □

If we denote, as usual, $A \cup P_0 = \{P_0\}$ by C_{-1}, we find that h_Δ is the product of the 4th power of four differences, three of them being the differences between the z-values of the points in neighboring sets (C_j and C_{j+1} for j being -1, 0, and 1) and the fourth being the difference between the z-values of the point in C_{-1} and the point in C_2. The reader can check that this is consistent with the result we shall obtain for general nonsingular Z_4 curves, Theorem 4.7 and Eq. (4.8) with $n = 4$ in Chapter 4 (recall that we cannot take differences of z-values of distinct points from the same set when all the sets are of cardinality $r = 1$).

3.4.3 Changing the Basepoint for $n = 4$

As in the previous examples, Theorem 3.14 shows that the quotient in Eq. (3.13) is independent of Δ, but still depends on our choice of P_0 as the basepoint. Therefore, allowing the basepoint to change we obtain 4 constants, one for each basepoint P_i with $0 \leq i \leq 3$, which in principal can be different. We now want to show that all these constants are in fact equal, and obtain a basepoint-independent expression for the Thomae formulae. Even though Lemma 2.4 gives us the needed condition on the branch points and all the divisors we have are supported on the branch points and contain no point to order $n = 4$ or higher, we cannot use Corollary 1.13 in the manner we have always done. This is since all of our divisors have empty A and Corollary 1.13 needs a point appearing in Δ to the power $n - 1 = 3$ to work. However, we have another tool which gives results of the nature of Corollary 1.13, and this is Proposition 1.14, which is easily seen to be applicable to the situation we have here by Lemma 2.5. We choose some other branch point P_i with $1 \leq i \leq 3$, and in order for the writing to be simple (i.e., for us to apply M only once) we choose a divisor $\Delta = P_i^2 P_j$ (with the basepoint P_0) where the chosen point P_i appears to the 2nd power. Then Proposition 1.14 gives that $\varphi_{P_0}(\Delta) + K_{P_0} = \varphi_{P_i}(\Gamma_{P_i}) + K_{P_i}$ where $\Gamma_{P_i} = P_j^2 P_k$ (with basepoint P_i) is integral of degree $g = 3$ and supported on the branch points distinct from P_i, and is again non-special by Theorem 2.6.

In order for this to connect the Thomae formulae with basepoint P_0 to the one with basepoint P_i we have to show that $h_{\Gamma_{P_i}}$ from Eq. (3.13) with basepoint P_i is the same as our h_Δ. For this we notice that for the divisor Γ_{P_i} we have the sets $C_{-1} = \{P_i\}$ like $(C_0)_\Delta$, $C_0 = \{P_j\}$ like $(C_1)_\Delta$, $C_1 = \{P_k\}$ like $(C_2)_\Delta$ and $C_2 = \{P_0\}$ like $(C_{-1})_\Delta$ (as we remarked after Proposition 1.14, we get a "rotation" of the sets). From this it is easy to see, using the construction of h_Δ from Δ and of $h_{\Gamma_{P_i}}$ from Γ_{P_i}, that $h_{\Gamma_{P_i}} = h_\Delta$. Therefore we can still deduce that the constant quotient obtained from the basepoint P_i (represented by the divisor Γ_{P_i}) is the same as the one from P_0 (represented by the divisor Δ), as we wanted. We could have done this with any other divisor instead of Δ, but we chose this one to present explicitly the action of M, which turns out to be very important for the nonsingular case with $r = 1$.

Since we have already applied Proposition 1.14, let us extract everything we can from it. As in the previous examples, we want not only the constants from the different basepoints to be the same, but also the characteristics, and we do it as before. Proposition 1.14 shows that the 6 divisors with basepoint P_0 give the same characteristics as divisors of any other basepoint, and either by symmetry or by the fact that different non-special divisors with the same basepoint give different characteristics and the fact that there are the same number 6 of divisors for every basepoint, we obtain the desired result. We now write the correspondence of the divisors which give the same characteristics explicitly:

$$\varphi_{P_0}(P_1^2P_2) + K_{P_0} = \varphi_{P_1}(P_2^2P_3) + K_{P_1} = \varphi_{P_2}(P_3^2P_0) + K_{P_2} = \varphi_{P_3}(P_0^2P_1) + K_{P_3},$$

$$\varphi_{P_0}(P_1^2P_3) + K_{P_0} = \varphi_{P_1}(P_3^2P_2) + K_{P_1} = \varphi_{P_2}(P_0^2P_1) + K_{P_2} = \varphi_{P_3}(P_2^2P_0) + K_{P_3},$$

$$\varphi_{P_0}(P_2^2P_1) + K_{P_0} = \varphi_{P_1}(P_3^2P_0) + K_{P_1} = \varphi_{P_2}(P_1^2P_3) + K_{P_2} = \varphi_{P_3}(P_0^2P_2) + K_{P_3},$$

$$\varphi_{P_0}(P_2^2P_3) + K_{P_0} = \varphi_{P_1}(P_0^2P_2) + K_{P_1} = \varphi_{P_2}(P_3^2P_1) + K_{P_2} = \varphi_{P_3}(P_1^2P_0) + K_{P_3},$$

$$\varphi_{P_0}(P_3^2P_1) + K_{P_0} = \varphi_{P_1}(P_2^2P_0) + K_{P_1} = \varphi_{P_2}(P_0^2P_3) + K_{P_2} = \varphi_{P_3}(P_1^2P_2) + K_{P_3},$$

and

$$\varphi_{P_0}(P_3^2P_2) + K_{P_0} = \varphi_{P_1}(P_0^2P_3) + K_{P_1} = \varphi_{P_2}(P_1^2P_0) + K_{P_2} = \varphi_{P_3}(P_2^2P_1) + K_{P_3}.$$

One can easily verify that all the divisors appearing here are non-special by Theorem 2.6, and that this list exhausts all of them with all the possible basepoints. It is also worth mentioning once more that since the action of M on divisors with basepoint P_0 (and thus on divisors with any basepoint) leaves the denominator h_Δ invariant, we know that every basepoint change (which can be looked at as successive applications of M) leaves it invariant. Therefore our results are indeed consistent and we do not get unwanted extra relations.

All this suggests that here also we can find a basepoint-independent notation for the Thomae formulae, and this is done as follows. We again look at $\Xi = P_0^3\Delta$ where Δ is a divisor with basepoint P_0, but here no matter which Δ and other basepoint P_i we choose, the divisor $\Upsilon = P_i^3\Gamma_{P_i}$ does not equal Ξ. However, it can be obtained from Ξ by applications of M. Since we saw that these applications leave the characteristic (which are now $\varphi_{P_i}(\Xi) + K_{P_i}$ for any $0 \le t \le 3$ and is independent of t by the second statement of Corollary 1.13) and the denominator from Eq. (3.13) (which we shall denote by h_Ξ rather than h_Δ, and it is more suitable here as well as Ξ also includes the set C_{-1}) invariant, we find that this quotient is invariant under M as well. We can define as before $\theta[\Xi]$ as the theta function with this characteristic, and altogether we have

Theorem 3.15. *For any divisor* $\Xi = P_i^3 P_j^2 P_k^1 P_l^0$ *with i, j, k, and l being 0, 1, 2, and 3 in some order define* $\theta[\Xi]$ *to be the theta function with characteristic* $\varphi_{P_t}(\Xi) + K_{P_t}$ *for some* $0 \le t \le 3$ *(which is independent of the choice of* P_t*) and define* h_Ξ *as in Eq. (3.13). Then the quotient* $\theta^{32}[\Xi](0,\Pi)/h_\Xi$ *is independent of the choice of the divisor* Ξ.

Theorem 3.15 is the most "symmetric" Thomae formulae for the nonsingular Z_4 curve X, and can give as a corollary the fact that all the Thomae formulae with the 4 different basepoints give the same constant. As usual, unlike Theorem 3.14 where no two divisors give the same characteristic, here the application M takes Ξ to a different divisor which represents (with any branch point as basepoint) the same characteristic as Ξ does. Here we have $4! = 24$ such divisors Ξ, which split into the 6 orbits of size 4 of M listed above, and these give the 6 characteristics.

3.4.4 The Case $n = 3$

Before turning to the more problematic cases $n \geq 5$, let us first look at the case $n = 3$. Here we find that the genus is $(3 - 1)(3 - 2)/2 = 1$, and there are only $2! = 2$ divisors, which are of the form $P_{i_1}^1 P_{i_2}^0$, with i_1 and i_2 being 1 and 2 and distinct. Explicitly, these are P_1 and P_2. The operators satisfy, where i and j are 1 and 2 and distinct, that $N(P_i) = P_j$ and $T(P_i) = T_{P_i}(P_i) = P_i$ (i.e., T is the trivial permutation in S_2 and N is the nontrivial one). This means that we cannot even begin the process of applying Proposition 2.21, but we do not need to; there are only two divisors and N takes one to the other. A simple use of Eq. (1.5) in its form $\theta[N(\Delta)](0, \Pi) = \theta[\Delta](0, \Pi)$ gives

Theorem 3.16. $\theta^{36}[P_1](0, \Pi) = \theta^{36}[P_2](0, \Pi)$ is the Thomae formulae for this nonsingular Z_3 curve.

This result is quite trivial, but as it is the Thomae formulae for a Z_n curve, we stated it as a theorem. In the general nonsingular case with $n = 3$ Theorem 4.6 and Eq. (4.7) show that the denominator h_Δ contains only differences of z-values of distinct points from the same set, and here there are no such differences as all the sets are of cardinality $r = 1$ (including $C_{-1} = A \cup P_0 = \{P_0\}$). Therefore Theorem 3.16 is consistent with the result for general nonsingular Z_3 curves, Theorem 4.6 with $n = 3$ in Chapter 4. We remark here that the same argument actually gives the slightly stronger result that these two theta constants are equal when taken only to the 18th power, but Theorem 3.16 was formulated this way to be a special case of Theorem 4.6.

Also in this trivial case, Theorem 3.16 shows that the two theta constants are equal, but still changing the basepoint may give, in principal, another value. Hence we have 3 constants (one for each basepoint P_i with $0 \leq i \leq 2$), which we wish to compare and then write the Thomae formulae in a basepoint-independent manner. Here as well we cannot use Corollary 1.13 in the way we always did (even though Lemma 2.4 gives us the needed condition on the branch points and the divisors we have are supported on the branch points and contain no point to order $n = 3$ or higher) since the set A is always empty, but we can proceed as we did in the case $n = 4$. Proposition 1.14 is again applicable (by what we have said before, by Lemma 2.5, and by the fact that the divisors are of the form assumed in Proposition 1.14— this of course holds for every n), and for another branch point P_i with i being 1 or 2, we choose the divisor $\Delta = P_i$ (the same P_i appearing to the 1st power) with the basepoint P_0 (again in order to apply M only once). Then Proposition 1.14 gives that $\varphi_{P_0}(\Delta) + K_{P_0} = \varphi_{P_i}(\Gamma_{P_i}) + K_{P_i}$ where $\Gamma_{P_i} = P_j$ with basepoint P_i is integral of genus $g = 1$ and supported on the branch points distinct from P_i (j is the other number 1 or 2 which is not i), and is again non-special by Theorem 2.6.

The connection of the Thomae formulae with basepoint P_0 to the one with basepoint P_i is very easy, as all the denominators (h_Δ, $h_{\Gamma_{P_i}}$, etc.) are 1. Therefore the constant quotient (or simply theta constants) obtained from the basepoint P_i (represented by the divisor Γ_{P_i}) is the same as the one from P_0 (represented by the divisor

Δ). We could have, of course, done this with the other divisor P_j, but we have chosen P_i in order to apply M and not M^2. As in the case $n = 4$ we shall use Proposition 1.14 to show that the 2 divisors with basepoint P_0 give the same characteristics as divisors of any other basepoint and vice versa, using the same considerations as always. Writing these results explicitly gives

$$\varphi_{P_0}(P_1) + K_{P_0} = \varphi_{P_1}(P_2) + K_{P_1} = \varphi_{P_2}(P_0) + K_{P_2}$$

and

$$\varphi_{P_0}(P_2) + K_{P_0} = \varphi_{P_1}(P_0) + K_{P_1} = \varphi_{P_2}(P_1) + K_{P_2},$$

which shows we get the "rotation" of the sets as we remarked after Proposition 1.14. One easily sees that these 6 divisors, 2 with each of the 3 basepoints, are exactly all the non-special divisors we consider. Using again the divisor $\Xi = P_0^2 \Delta$ where Δ is a divisor with basepoint P_0, the divisor $\Upsilon = P_i^2 \Gamma_{P_i}$ again never equals Ξ no matter which Δ and other basepoint P_i we choose, but is obtained from it by applications of M. We write the characteristic as $\varphi_{P_t}(\Xi) + K_{P_t}$ for any $0 \le t \le 2$ (which is again independent of t by the second statement of Corollary 1.13), and as the denominator h_Ξ is 1 for every Ξ (as it was for every Δ) we can define $\theta[\Xi]$ as always to be the theta function with this characteristic and obtain the basepoint-independent notation for the Thomae formulae in the form

Theorem 3.17. *For any divisor* $\Xi = P_i^2 P_j^1 P_k^0$ *with distinct i, j and k between 0 and 2 define* $\theta[\Xi]$ *to be the theta function with characteristic* $\varphi_{P_t}(\Xi) + K_{P_t}$ *for some* $0 \le t \le 2$ *(which is independent of the choice of P_t) and define $h_\Xi = 1$. Then the quotient* $\theta^{36}[\Xi](0,\Pi)/h_\Xi$ *(which is simply $\theta[\Xi](0,\Pi)$) is independent of the choice of the divisor Ξ.*

Theorem 3.17 is still a quite trivial result, as there are only 6 such divisors Ξ. Among those, $P_0^2 P_1$, $P_1^2 P_2$ and $P_2^2 P_0$ are equivalent and give the same theta function, and also $P_0^2 P_2$, $P_2^2 P_1$ and $P_1^2 P_0$ are equivalent and give the same theta function. Therefore Theorem 3.17 actually connects the values of two theta constants that are connected by the operator N and hence Theorem 3.17 is nothing more than Eq. (1.5) raised to the 36th power. This is actually what Theorem 3.16 was, but Theorem 3.17 is the more "symmetric" form of the (trivial) Thomae formulae for the nonsingular Z_3 curve X. As we remarked after Theorem 3.16, Theorem 3.17 also holds when the powers of the theta constants is only 18. Theorem 3.17 has as a corollary the fact that the (trivial) Thomae formulae with the 3 different basepoints give the same constant. In this case we have 6 divisors Ξ in Theorem 3.17, divided into 2 orbits of size 3 of M as listed above, and this gives the 2 characteristics.

3.4.5 The Problem with $n \ge 5$

More interesting is the case where $n \ge 5$. Here we can begin our process and follow quite a few steps of the general proof without encountering any problems. We get the

PMT which is invariant under T (which will relate only sets of 2 divisors each since there is only one operator T and it is an involution), and we construct the formula (which again looks like the general one when substituting $r = 1$) that is invariant under N and T. Therefore what we prove is that the quotient $\theta^{2en^2}[\Delta](0, \Pi)/h_\Delta$ is constant on the orbits of the subgroup generated by N and T in S_{n-1} inside the set of the non-special integral divisors of degree g supported on the set of branch points distinct from P_0. Since Proposition 3.18 below will show that for $n \geq 5$ the elements N and T do not generate all of S_{n-1}, this leaves us with still a Poor man's Thomae formulae, which will be proved to be indeed the full Thomae formulae only after the basepoint change is applied.

We now prove

Proposition 3.18. *For $n \geq 4$ the elements N and T generate a subgroup G of order $2(n-1)$ in S_{n-1}.*

Proof. We recall that we have identified the operators N and T with elements of S_{n-1}, namely

$$N = \begin{pmatrix} 1 & 2 & \dots & n-2 & n-1 \\ n-1 & n-2 & \dots & 2 & 1 \end{pmatrix}, \quad T = \begin{pmatrix} 1 & 2 & \dots & n-2 & n-1 \\ 1 & n-1 & \dots & 3 & 2 \end{pmatrix}.$$

Hence NT identifies with the element $\begin{pmatrix} 1 & 2 & 3 & \dots & n-2 & n-1 \\ n-1 & 1 & 2 & \dots & n-3 & n-2 \end{pmatrix}$ of S_{n-1}, which has order $n - 1$. When writing an element in G in its reduced form as a word in the generators N and T we can start either with N or T, and after writing one generator we must either write the other generator or stop (since both N and T are involutions). Since TN is the inverse of NT, it is also of order $n - 1$, and this also gives that the word $(TN)^k$ is the same as $(NT)^{n-1-k}$ for $1 \leq k \leq n-2$, and the word $(TN)^k T$ is the same as $(NT)^{n-2-k}N$ for $0 \leq k \leq n-2$. This leaves us with the words starting with N, and since $(NT)^{n-1} = Id$ there are exactly $2(n-1)$ such words (the partial words of $(NT)^{n-1}$ of length from 0 to $2n-3$). Since NT is of order $n - 1$ and no less and T does not lie in the group generated by NT (this is clear), no more relations exist between these words and they represent G faithfully. Thus $|G| = 2(n-1)$ and the proposition is proved. □

The place where we have used the fact that $n \geq 4$ is when we stated that T does not lie in the group generated by NT; this is not true when $n = 3$ as then T is trivial (and we do get a group of order 2 and not 4, which happens to be all of S_2). We just mention that this shows that the group G is isomorphic, for $n \geq 4$, to the dihedral group D_{n-1}, which is indeed of order $2(n-1)$ (and for $n = 3$ to the cyclic group of order 2).

Proposition 3.18 explains why in the case $n = 4$ we obtained the full Thomae formulae in Theorem 3.14, while for $n \geq 5$ we do not yet do so. When $n = 4$ we have $2 \times 3 = 6 = 3!$, so that N and T generate a subgroup of S_3 of the same size as S_3 and thus generate all of S_3 (as we saw in the proof of Theorem 3.14). Similarly, for $n = 3$ the group G is all of S_2 and we got full (trivial, but full) Thomae formulae in Theorem 3.16. On the other hand, for $n \geq 5$, $2(n-1)$ is strictly smaller than $(n-1)!$, so that N and T generate only a subgroup (of index $(n-2)!/2$) in S_{n-1}. Hence what we prove in this case is at best only a PMT, which states that the quotient $\theta^{2en^2}[\Delta](0,\Pi)/h_\Delta$ is constant on each of the $(n-2)!/2$ orbits of the group G, but can, in principal, attain different values on different orbits. We will have to let the basepoint vary to obtain that all these orbits of G give the same constant value, and it does not suffice to work with only one fixed basepoint.

3.4.6 The Case $n = 5$

In order to demonstrate this, we shall do in detail the case $n = 5$. In this case there are $4! = 24$ divisors that are all of the form $\Delta = P_i^3 P_j^2 P_k^1 P_l^0$ where i, j, k, and l are 1, 2, 3, and 4 in any order. For such a divisor Δ we have $N(\Delta) = P_l^3 P_k^2 P_j^1 P_i^0$ and $T(\Delta) = T_{P_i}(\Delta) = P_i^3 P_l^2 P_k^1 P_j^0$. We consider, as usual, the quotient

$$\frac{\theta^{50}[\Delta](\varphi_{P_0}(P),\Pi)}{\theta^{50}[T(\Delta)](\varphi_{P_0}(P),\Pi)} = \frac{\theta^{50}[P_i^3 P_j^2 P_k](\varphi_{P_0}(P),\Pi)}{\theta^{50}[P_i^3 P_l^2 P_k](\varphi_{P_0}(P),\Pi)},$$

and since $N(\Delta) = P_l^3 P_k^2 P_j$ and $N(T(\Delta)) = P_j^3 P_k^2 P_l$, Proposition 2.21 shows that the divisor of this function is the 50th power of P_l^2/P_j^2, i.e., P_l^{100}/P_j^{100}, and that this function is $c(z-\lambda_l)^{20}/(z-\lambda_j)^{20}$, where c is a nonzero complex constant. By putting $P = P_0$ we have

$$\frac{\theta^{50}[P_i^3 P_j^2 P_k](0,\Pi)}{\theta^{50}[P_i^3 P_l^2 P_k](0,\Pi)} = c\frac{(\lambda_0-\lambda_l)^{20}}{(\lambda_0-\lambda_j)^{20}},$$

and thus

$$\frac{\theta^{50}[P_i^3 P_j^2 P_k](\varphi_{P_0}(P),\Pi)}{\theta^{50}[P_i^3 P_l^2 P_k](\varphi_{P_0}(P),\Pi)} = \frac{(\lambda_0-\lambda_j)^{20}}{(\lambda_0-\lambda_l)^{20}} \times \frac{\theta^{50}[P_i^3 P_j^2 P_k](0,\Pi)}{\theta^{50}[P_i^3 P_l^2 P_k](0,\Pi)} \times \frac{(z-\lambda_l)^{20}}{(z-\lambda_j)^{20}}.$$

Since $T = T_{P_i}$ the other substitution we make is $P = P_i$. This gives, from the last equation

$$\frac{\theta^{50}[P_i^3 P_j^2 P_k](\varphi_{P_0}(P_i),\Pi)}{\theta^{50}[P_i^3 P_l^2 P_k](\varphi_{P_0}(P_i),\Pi)} = \frac{(\lambda_0-\lambda_j)^{20}}{(\lambda_0-\lambda_l)^{20}} \times \frac{\theta^{50}[P_i^3 P_j^2 P_k](0,\Pi)}{\theta^{50}[P_i^3 P_l^2 P_k](0,\Pi)} \times \frac{(\lambda_i-\lambda_l)^{20}}{(\lambda_i-\lambda_j)^{20}},$$

and from Eq. (2.3)

$$\frac{\theta^{50}[P_i^3 P_l^2 P_k](\varphi_{P_0}(P_i),\Pi)}{\theta^{50}[P_i^3 P_l^2 P_k](\varphi_{P_0}(P_i),\Pi)} = \frac{\theta^{50}[P_i^3 P_l^2 P_k](0,\Pi)}{\theta^{50}[P_i^3 P_j^2 P_k](0,\Pi)},$$

so that together we have

$$\frac{\theta^{100}[P_i^3 P_l^2 P_k](0, \Pi)}{\theta^{100}[P_i^3 P_j^2 P_k](0, \Pi)} = \frac{(\lambda_0 - \lambda_j)^{20}}{(\lambda_0 - \lambda_l)^{20}} \times \frac{(\lambda_i - \lambda_l)^{20}}{(\lambda_i - \lambda_j)^{20}}$$

or

$$\frac{\theta^{100}[P_i^3 P_j^2 P_k](0, \Pi)}{(\lambda_0 - \lambda_l)^{20}(\lambda_i - \lambda_j)^{20}} = \frac{\theta^{100}[P_i^3 P_l^2 P_k](0, \Pi)}{(\lambda_0 - \lambda_j)^{20}(\lambda_i - \lambda_l)^{20}}. \tag{3.14}$$

As in the case $n = 4$, it is clear that starting with $T(\Delta) = P_i^3 P_l^2 P_k$ gives the same Eq. (3.14) and that the smallest nonempty sets that are closed under T are of size 2. We define g_Δ to be the product of the 20th power of the difference between λ_0 and the z-value of the branch point not appearing in Δ (which is not P_0), and the 20th power of the difference between the z-values of the branch point appearing to the 3rd power in Δ, and the branch point appearing to the 2nd power in Δ. Then Eq. (3.14) states that $\theta^{100}[\Delta](0, \Pi)/g_\Delta$ is invariant under T. This is the PMT for this Z_5 curve.

When wanting to replace g_Δ by a denominator h_Δ which is invariant under N, we find that the correcting multipliers $g_{N(\Delta)} = (\lambda_0 - \lambda_i)^{20}(\lambda_l - \lambda_k)^{20}$ (starting from Δ) and $g_{N(T(\Delta))} = (\lambda_0 - \lambda_i)^{20}(\lambda_j - \lambda_k)^{20}$ (starting from $T(\Delta)$) are no longer the same expression. However, if we put in the additional multiplier, which is the 20th power of the difference between the z-values of the branch point appearing to the 2nd power in Δ and the branch point appearing to the 1st power in Δ, then the "extra multiplier" $g_{N(\Delta)}$ will be multiplied by $(\lambda_j - \lambda_k)^{20}$, the "extra multiplier" $g_{N(T(\Delta))}$ will be multiplied by $(\lambda_l - \lambda_k)^{20}$, and the multipliers become the same expression. When examining the equation corresponding to $N(\Delta)$, we find that apart from the (clearly N-invariant) expression $g_\Delta g_{N(\Delta)}$ that we have in the new denominator, the expression $(\lambda_j - \lambda_k)^{20}$ is also invariant under N. This is so, since λ_j appears in C_1 and λ_j appears in C_2, N interchanges C_1 with C_2, and the power 20 is even. We therefore divide Eq. (3.14) by $(\lambda_0 - \lambda_i)^{20}(\lambda_l - \lambda_k)^{20}(\lambda_j - \lambda_k)^{20}$, and obtain

Theorem 3.19. *The expression*

$$\frac{\theta^{100}[\Delta](0, \Pi)}{h_\Delta} = \frac{\theta^{100}[P_i^3 P_j^2 P_k](0, \Pi)}{(\lambda_0 - \lambda_l)^{20}(\lambda_i - \lambda_j)^{20}(\lambda_0 - \lambda_i)^{20}(\lambda_l - \lambda_k)^{20}(\lambda_j - \lambda_k)^{20}}$$

is invariant under T and N.

There is an alternative description of the procedure for the construction of h_Δ which we present now. This will be the language to be used in Chapter 4. We begin by writing our divisor Δ in the form $C_0^3 C_1^2 C_2^1 C_3^0$ (recalling that in our case there is no set A since its cardinality when $r = 1$ is zero). In this notation we write $T(\Delta)$ as $\widetilde{C_0}^3 \widetilde{C_1}^2 \widetilde{C_2}^1 \widetilde{C_3}^0$ with

$$C_0 = \{P_i\}, \quad C_1 = \{P_j\}, \quad C_2 = \{P_k\}, \quad C_3 = \{P_l\},$$

$$\widetilde{C_0} = \{P_i\}, \quad \widetilde{C_1} = \{P_l\}, \quad \widetilde{C_2} = \{P_k\}, \quad \widetilde{C_3} = \{P_j\},$$

and let $C_{-1} = \widetilde{C_{-1}} = \{P_0\}$. Using the notation

$$[Y, Z] = \prod_{S \in Y, R \in Z} (z(S) - z(R))$$

from Definition 4.1 below for disjoint sets Y and Z of branch points (which will be formally introduced in Chapter 4) we write the PMT in Eq. (3.14) as

$$\frac{\theta^{100}[C_0^3 C_1^2 C_2](0, \Pi)}{[C_{-1}, C_3]^{20} [C_0, C_1]^{20}} = \frac{\theta^{100}[\widetilde{C_0}^3 \widetilde{C_1}^2 \widetilde{C_2}](0, \Pi)}{[\widetilde{C_{-1}}, \widetilde{C_3}]^{20} [\widetilde{C_0}, \widetilde{C_1}]^{20}}.$$

This is a nice way of seeing the PMT as $\theta^{100}[\Delta](0, \Pi)/g_\Delta$ being invariant under T with g_Δ being constructed from Δ in a very clear way. However, we prefer having an expression where the denominator h_Δ is invariant also under the operator N. Since $N(C_0^3 C_1^2 C_2) = C_3^3 C_2^2 C_1$ we find that such a symmetrization needs to have the property that for every expression appearing in it, the expression, obtained by replacing any C_j with $0 \le j \le 3$ by C_{3-j} (and C_{-1} remaining fixed), should appear as well and to the same power. This means that for invariance under N we certainly have to multiply the left-hand side of the equation by $[C_{-1}, C_0]^{20} [C_2, C_3]^{20}$, and compensate by multiplying the right-hand side by the same expression, which is $[\widetilde{C_{-1}}, \widetilde{C_0}]^{20} [\widetilde{C_1}, \widetilde{C_2}]^{20}$. The equation now becomes

$$\frac{\theta^{100}[C_0^3 C_1^2 C_2](0, \Pi)}{[C_{-1}, C_3]^{20} [C_0, C_1]^{20} [C_{-1}, C_0]^{20} [C_2, C_3]^{20}} = \frac{\theta^{100}[\widetilde{C_0}^3 \widetilde{C_1}^2 \widetilde{C_2}](0, \Pi)}{[\widetilde{C_{-1}}, \widetilde{C_0}]^{20} [\widetilde{C_0}, \widetilde{C_1}]^{20} [\widetilde{C_{-1}}, \widetilde{C_3}]^{20} [\widetilde{C_1}, \widetilde{C_2}]^{20}},$$

which is no longer symmetric. In order to reobtain the symmetry we thus multiply the left-hand side by $[C_1, C_2]^{20}$ and the right-hand side by the same thing, which is now $[\widetilde{C_2}, \widetilde{C_3}]^{20}$. This yields the equality

$$\frac{\theta^{100}[C_0^3 C_1^2 C_2](0, \Pi)}{[C_{-1}, C_3]^{20} [C_0, C_1]^{20} [C_{-1}, C_0]^{20} [C_2, C_3]^{20} [C_1, C_2]^{20}}$$
$$= \frac{\theta^{100}[\widetilde{C_0}^3 \widetilde{C_1}^2 \widetilde{C_2}](0, \Pi)}{[\widetilde{C_{-1}}, \widetilde{C_0}]^{20} [\widetilde{C_0}, \widetilde{C_1}]^{20} [\widetilde{C_{-1}}, \widetilde{C_3}]^{20} [\widetilde{C_1}, \widetilde{C_2}]^{20} [\widetilde{C_2}, \widetilde{C_3}]^{20}},$$

and the reader can observe that it is both symmetric and invariant under N. This is an alternative description of the derivation of h_Δ, which actually says the same thing in a slightly different language. Note that in each denominator appear pairs of sets C_j with difference 1 between their indices, and in one pair this difference is 4. This is an easy way to remember the construction of h_Δ from Δ, and also shows the formula we obtained is the special case with $n = 5$ and $r = 1$ of the general formula for nonsingular Z_n curves with odd n in Theorem 4.6 and Eq. (4.7). We remark again that Theorem 3.19 does not yet give us the full Thomae formulae.

3.4.7 The Orbits for $n = 5$

Let us now see in detail why Theorem 3.19 does not give the full Thomae formulae yet. Let us start with a divisor Δ, and then we know from Theorem 3.19 that the value $\theta^{100}[\Delta](0, \Pi)/h_\Delta$ is constant on the set

$$\Delta, \quad N(\Delta), \quad TN(\Delta), \quad NTN(\Delta),$$

$$T(\Delta), \quad NT(\Delta), \quad TNT(\Delta), \quad NTNT(\Delta) = TNTN(\Delta),$$

and the actions of N and T add no additional divisors to this set. This is so because $TNTN = NTNT$: they are clearly inverses, and we saw in the proof of Proposition 3.18 that TN is of order 4 in S_4. Thus $TNTN$ is of order 2 and hence is equal to its inverse. This is indeed a set of 8 divisors, as Proposition 3.18 predicts, and thus it covers only one third of the whole set of 24 divisors. This means that the set of 24 divisors splits into 3 sets of 8 divisors each, namely the 3 orbits of the group G of order 8, and that $\theta^{100}[\Delta](0, \Pi)/h_\Delta$ is constant on every set. However, at this stage we do not know how to compare the values it attains on different sets. We now write the 3 sets explicitly, and we organize them according to the last list (i.e., start with some Δ, writing $N(\Delta)$ right after it and $T(\Delta)$ below it, etc.): One set is composed of

$$P_1^3 P_2^2 P_3, \quad P_4^3 P_3^2 P_2, \quad P_4^3 P_1^2 P_2, \quad P_3^3 P_2^2 P_1,$$

$$P_1^3 P_4^2 P_3, \quad P_2^3 P_3^2 P_4, \quad P_2^3 P_1^2 P_4, \quad P_3^3 P_4^2 P_1,$$

the second set is composed of

$$P_2^3 P_1^2 P_3, \quad P_4^3 P_3^2 P_1, \quad P_4^3 P_2^2 P_1, \quad P_3^3 P_1^2 P_2,$$

$$P_2^3 P_4^2 P_3, \quad P_1^3 P_3^2 P_4, \quad P_1^3 P_2^2 P_4, \quad P_3^3 P_4^2 P_2,$$

and the third set is composed of

$$P_1^3 P_3^2 P_2, \quad P_4^3 P_2^2 P_3, \quad P_4^3 P_1^2 P_3, \quad P_3^3 P_2^2 P_1,$$

$$P_1^3 P_4^2 P_2, \quad P_3^3 P_2^2 P_4, \quad P_3^3 P_1^2 P_4, \quad P_2^3 P_4^2 P_1.$$

The reader can check that these are indeed 3 disjoint sets of 8 divisors each, which are the orbits of G whose union is all the 24 divisors have on X.

Theorem 3.19 gives that with the basepoint P_0 there are 3 values (which we have not yet compared) such that each of the 24 quotients equals one of these values. Since we have 5 options for the basepoints, in principal we can have 15 values, and each of the 120 quotients equals one of these values that corresponds to the "right" basepoint. We now show that the process of basepoint change not only gives that the 3 values from the basepoint P_0 are all the values obtained from the other basepoint, but also that these values are in fact equal. Once again Corollary 1.13 cannot help us, and we shall use Proposition 1.14 for all these purposes.

3.4.8 Changing the Basepoint for $n = 5$

Let us start with the first assertion. Start with our divisors and with the basepoint P_0, and choose one other branch point P_i with $1 \leq i \leq 4$. We shall show that each of the 3 values obtained from the basepoint P_0 equals a value which comes from the basepoint P_i. Choose such a value, i.e., a set of 8 divisors from the 3 sets written above (an orbit of G), and choose a divisor $\Delta = P_i^3 P_j^2 P_k$ in this set such that the chosen point P_i appears in it to the 3rd power (one can see that in each set each branch point appears to any given power between 0 and 3 in exactly 2 divisors, so this can be done). Again we do this choice so that we shall need only one application of M when applying Proposition 1.14, and thus the writing will be simpler. Now, Proposition 1.14 shows that $\varphi_{P_0}(\Delta) + K_{P_0} = \varphi_{P_i}(\Gamma_{P_i}) + K_{P_i}$ where $\Gamma_{P_i} = P_j^3 P_k^2 P_l$ (where l is the 4th number between 1 and 4 which is not i, j or k), and the divisor Γ_{P_i} is integral, of degree $g = 6$, supported on the branch points distinct from P_i, and non-special by Theorem 2.6.

Once again we examine also the denominator h_Δ from Theorem 3.19 and compare it with the expression giving $h_{\Gamma_{P_i}}$ with basepoint P_i in order to make this a connection between the quotients. For this we notice that for the divisor Γ_{P_i} we have the sets $C_{-1} = \{P_i\}$ like $(C_0)_\Delta$, $C_0 = \{P_j\}$ like $(C_1)_\Delta$, $C_1 = \{P_k\}$ like $(C_2)_\Delta$, $C_2 = \{P_l\}$ like $(C_3)_\Delta$ and $C_2 = \{P_0\}$ like $(C_{-1})_\Delta$. This again demonstrates what the "rotation" M does to the sets, and it makes it easy to see, using the construction of h_Δ from Δ and of $h_{\Gamma_{P_i}}$ from Γ_{P_i}, that $h_{\Gamma_{P_i}} = h_\Delta$ (just follow the differences between the indices of the sets appearing in each pair). As always we obtain that the constant quotient obtained from the basepoint P_i (represented by the divisor Γ_{P_i}) is the same as the one from P_0 (represented by the divisor Δ), which here means that every value attained by a quotient with the basepoint P_i is already attained by a quotient with the basepoint P_0, as we wanted. As before, the application of Proposition 1.14 gives that the theta constants (or equivalently characteristics) are the same for every basepoint P_i, $0 \leq i \leq 4$, and since the application of Proposition 1.14 can always be obtained by successive applications of M and we saw that M keeps the denominator invariant, we do get the same value of quotient for any two divisors Δ with basepoint P_0 and its Γ_{P_i} with basepoint P_i for any $1 \leq i \leq 4$.

All this shows that there are indeed only three values for all the quotients obtained from all the different basepoints. In order to show that these three constants are the same, we shall have to make use of the divisors with some basepoint other than P_0. Let us choose P_4. The three sets of divisors corresponding to the basepoint P_4 are as follows: One set is composed of

$$P_1^3 P_2^2 P_3, \quad P_0^3 P_3^2 P_2, \quad P_0^3 P_1^2 P_2, \quad P_3^3 P_2^2 P_1,$$

$$P_1^3 P_0^2 P_3, \quad P_2^3 P_3^2 P_0, \quad P_2^3 P_1^2 P_0, \quad P_3^3 P_0^2 P_1,$$

the second set is composed of

$$P_2^3 P_1^2 P_3, \quad P_0^3 P_3^2 P_1, \quad P_0^3 P_2^2 P_1, \quad P_3^3 P_1^2 P_2,$$

$$P_2^3 P_0^2 P_3, \quad P_1^3 P_3^2 P_0, \quad P_1^3 P_2^2 P_0, \quad P_3^3 P_0^2 P_2,$$

and the third set is composed of

$$P_1^3 P_3^2 P_2, \quad P_0^3 P_2^2 P_3, \quad P_0^3 P_1^2 P_3, \quad P_2^3 P_3^2 P_1,$$

$$P_1^3 P_0^2 P_2, \quad P_3^3 P_2^2 P_0, \quad P_3^3 P_1^2 P_0, \quad P_3^3 P_0^2 P_1.$$

We have written the divisors in each set by the same logic we have written the divisors in the sets with basepoint P_0, i.e., with the actions of N and T as before. The order of the sets is chosen such that a basepoint change from P_0 to P_4 which is done using one application of M takes a divisor from the jth set with basepoint P_0 to a divisor from the jth set here, with the basepoint P_4. The reader can verify that $P_4^3 P_3^2 P_2$ and $P_4^3 P_1^2 P_2$, from the first set with basepoint P_0, are taken by M to $P_3^3 P_2^2 P_1$ and $P_1^3 P_2^2 P_3$, from the first set with basepoint P_4, respectively. Also, $P_4^3 P_3^2 P_1$ and $P_4^3 P_2^2 P_1$, from the second set with basepoint P_0, are taken by M to $P_3^3 P_1^2 P_2$ and $P_2^3 P_1^2 P_3$, from the second set with basepoint P_4, respectively. Finally, $P_4^3 P_2^2 P_3$ and $P_4^3 P_1^2 P_3$, from the third set with basepoint P_0, are taken by M to $P_2^3 P_3^2 P_1$ and $P_1^3 P_3^2 P_2$, from the third set with basepoint P_4, respectively.

We have done the basepoint change from P_0 to P_4, but only on divisors where this is done by a single application of M. However, Proposition 1.14 is not limited to these basepoint changes, and we obtain the comparison of the three constants by using it for more divisors. Let us choose the first set of divisors with basepoint P_0, and apply the basepoint change to P_4 on all of them. Proposition 1.14 gives us that this yields

$$P_0^3 P_1^2 P_2(M^4), \quad P_3^3 P_2^2 P_1(M^1), \quad P_1^3 P_2^2 P_3(M^1), \quad P_0^3 P_3^2 P_2(M^4),$$

$$P_3^3 P_2^2 P_0(M^2), \quad P_1^3 P_0^2 P_2(M^3), \quad P_3^3 P_0^2 P_2(M^3), \quad P_1^3 P_2^2 P_0(M^2),$$

where in parentheses we have written after each divisor the power of M used in the application of Proposition 1.14 for obtaining it. We see that while the first four divisors lie in the first set of the basepoint P_4, the following two lie in the third set, and the last two lie in the second one. Since we have already seen that the application of M leaves the denominator invariant, this indeed gives us the desired connection also between the quotients, making the three constants (which, as we saw, are the same three constants for any choice of basepoint) equal. Of course, we could have done it using any of the other two sets of divisors with basepoint P_0, and obtain the same conclusion, which is

Theorem 3.20. *The quotient in Theorem 3.19 is the Thomae formulae for the non-singular Z_5 curve X.*

What allowed us to strengthen Theorem 3.19 to Theorem 3.20 was the fact that the basepoint change added more operations, i.e., elements in $S_{n-1} = S_4$ which are not in G, which also leave the quotient from Theorem 3.19 invariant. In our situation of $n = 5$ it suffices to add one such relation, since the index of G (whose cardinality is 8) in S_4 (whose cardinality is 24) is 3 and prime, meaning that there are no inter-mediate subgroups between G and S_4. In the general case we shall have to prove that the elements we add generate all of S_{n-1}, or, alternatively, look at specific elements that are known to generate S_{n-1} together with G. We now demonstrate this with the specific element we shall later take in the general case, and choose the initial divisor Δ such that all the intermediate calculations are already listed above. We take the divisor Δ to be $P_4^3 P_1^2 P_3$ from the third set of the basepoint P_0, and apply the base-point change to P_4 using a single application of M to obtain $P_1^3 P_3^2 P_2$ from the third set of the basepoint P_4. Applying now the element NT of the group G (we do this with divisors with the basepoint P_4, so that this divisor is actually $P_1^3 P_3^2 P_2^1 P_0^0$, but Theorem 3.19 holds equally for the basepoint P_4) yields the divisor $P_3^3 P_2^2 P_0$, which is also from the third set of the basepoint P_4; we saw that this was the image of the basepoint change by M^2 of the divisor $P_1^3 P_4^2 P_3$ from the first set of the basepoint P_0. This means that this operator, $M^{-2}NTM$, gives another element of S_4 (namely the transposition τ_{12}, as one can see from its action) which is not in G and which leaves the quotient invariant. This element $M^{-2}NTM$ turns out to be the transposition τ_{12} in the general case as well and generates with G all of S_{n-1}, as we shall prove in Chapter 4.

Note that aside from Theorem 3.20, we have obtained that we get the same con-stant for all the 5 possible choices of basepoint P_i, $0 \le i \le 4$. We also obtain, using Proposition 1.14, that the 24 divisors with basepoint P_0 give the same characteristics as divisors arising from any other choice of basepoint, and by the usual argument we conclude that the set of characteristics we get from each basepoint is the same set. As in the previous examples, we would like now to express the Thomae formulae in a basepoint-independent notation. The usual system, which worked in the other examples, can serve us here as well. We take, for a divisor Δ with basepoint P_0, the divisor $\Xi = P_0^4 \Delta$ we used above for the basepoint change, but recall that $P_i^4 \Gamma_{P_i}$ does not equal Ξ for any choice of P_i but is only obtained from it using applications of M. We define $\theta[\Xi]$ as the theta function with characteristic $\varphi_{P_t}(\Xi) + K_{P_t}$ for some branch point P_t, $0 \le t \le 4$ (which is independent of t by the second statement in Corollary 1.13), and we denote the expression for h_Δ from Theorem 3.19 by h_Ξ (again Ξ expresses the set C_{-1} and hence this notation is more suitable). All this gives us

Theorem 3.21. *For any divisor* $\Xi = P_i^4 P_j^3 P_k^2 P_l^1 P_m^0$, *with* i, j, k, l, *and* m *distinct between being 0 and 4, define* $\theta[\Xi]$ *to be the theta function with characteristic* $\varphi_{P_t}(\Xi) + K_{P_t}$ *for some* $0 \le t \le 4$ *(which is independent of the choice of* P_t*) and define* h_Ξ *as in Theorem 3.19. Then the quotient* $\theta^{100}[\Xi](0,\Pi)/h_\Xi$ *is independent of* Δ.

Theorem 3.21 is the more "symmetric" way to express the Thomae formulae we obtained for the nonsingular Z_5 curve X, and contains Theorem 3.20 and the fact that all the basepoints give the same constant as corollaries. This seems like a circu-lar claim, but one can prove (and it is even somewhat easier) Theorem 3.21 directly from Theorem 3.19, without using Theorem 3.20. This is done using a certain em-bedding of $S_{n-1} = S_4$ (and hence G) in $S_n = S_5$, considering M as another element of $S_n = S_5$, and then showing that M and G generate $S_n = S_5$. We shall present this in the case of general n and $r = 1$ in Chapter 4, and even remark that M and NT suffice to generate all of S_n and we do not need all three elements M, N and T for this. We just note, toward the general case, that this means that we now have an action of the operators N and T on these divisors Ξ by leaving the basepoint invariant. Explicitly, for Ξ as in Theorem 3.21, we have $N(\Xi) = P_i^4 P_m^3 P_l^2 P_k^1 P_j^0$ and $T(\Xi) = P_i^4 P_j^3 P_m^2 P_l^1 P_k^0$. In addition we have the operator M which as we have seen acts on these divisors. We shall discuss this and what happens with $r \geq 2$ from this aspect when we do the general case in Chapter 4.

Chapter 4

Thomae Formulae for Nonsingular Z_n Curves

In this chapter we shall present a proof of the Thomae formula for the general non-singular Z_n curve associated to the equation

$$w^n = \prod_{i=0}^{rn-1} (z - \lambda_i),$$

where $\lambda_i \neq \lambda_j$ for $i \neq j$. With the examples of Chapter 3 as our guide we can continue along with general statements, and distinguishing between even and odd values of n at the necessary stage. We shall later have to split the cases $r \geq 2$ and $r = 1$, but meanwhile we make no assumption on r.

4.0.1 A Useful Notation

In order to simplify the notation in some of the formulae below we make the following

Definition 4.1. For a subset Y of the set of branch points we define

$$[z, Y] = \prod_{S \in Y} (z - z(S)) = \prod_{P_i \in Y} (z - \lambda_i),$$

which is a meromorphic function on X. For a branch point R and such a set Y we define

$$[R, Y] = \prod_{S \in Y, S \neq R} (z(R) - z(S)) = \prod_{P_i \in Y, P_i \neq R} (\sigma - \lambda_i)$$

(where $\sigma = z(R)$), excluding R from the product. This is a nonzero complex number. For two disjoint subsets Y and Z of branch points we define

H.M. Farkas and S. Zemel, *Generalizations of Thomae's Formula for Z_n Curves*,
Developments in Mathematics 21, DOI 10.1007/978-1-4419-7847-9_4,
© Springer Science+Business Media, LLC 2011

$$[Y,Z] = \prod_{S\in Y, R\in Z} (z(S) - z(R)) = \prod_{P_i \in Y, P_j \in Z} (\lambda_i - \lambda_j),$$

also a nonzero complex number. Finally, for one such set Y we define

$$[Y,Y] = \prod_{P_i, P_j \in Y, i<j} (z(P_i) - z(P_j)) = \prod_{P_i, P_j \in Y, i<j} (\lambda_i - \lambda_j),$$

again a nonzero complex number.

Note that in the expression defining $[Y,Y]$ in Definition 4.1, the difference between any two distinct branch points in Y is taken exactly once (in the order given). Also note that if Y contains P_i, then $[P_i, Y] = [P_i, Y \setminus P_i]$, or equivalently, if Y does not contain P_i, then $[P_i, Y] = [P_i, Y \cup P_i]$. We agree that if the product is empty (i.e., one of the sets is empty, $Y = \{P_i\}$ in $[P_i, Y]$, or if $|Y| = 1$ in $[Y,Y]$), then the relevant symbol (of brackets) is simply the number 1 (or the constant function 1 in the case $[z, \emptyset]$).

The careful reader can see, when examining Definition 4.1, that changing the ordering of the points inside the expressions may change their value, but only by a sign. However, in all the expressions we shall obtain, these brackets will appear to an even power (some multiple of the even number en), and hence the different choices are irrelevant. Therefore we shall be able to replace $[Y,Z]$ with $[Z,Y]$ freely if needed, and if U, V, and W are disjoint sets, then we can replace $[U,W][V,W]$ by $[U \cup V, W]$ freely (as it is also easy to see from Definition 4.1 that they are equal up to sign). Less trivial, but very useful, is the following

Lemma 4.2. *Up to sign we have $[Y,Y][Z,Z][Y,Z] = [Y \cup Z, Y \cup Z]$ for two disjoint sets Y and Z of branch points, and $[R,W][W \setminus R, W \setminus R] = [W,W]$ for any branch point R which belongs to the set W.*

Proof. By Definition 4.1, the expression $[Y \cup Z, Y \cup Z]$ is the product of all the differences $z(S) - z(R)$ with $S \in Y \cup Z$, $R \in Y \cup Z$, $R \neq S$ and such that every pair is taken exactly once (i.e., from each two pairs (S,R) and (R,S) exactly one is taken). We partition the set of these pairs into 3 subsets: first the set of pairs of points from Y, second the set of pairs of points from Z, and third the set $Y \times Z$ of pairs (S,R) with $S \in Y$ and $R \in Z$. This covers all the pairs since the "partners" of the pairs in $Z \times Y$ are already taken. Now, the product of the first pairs gives $[Y,Y]$, the product of the second pairs gives $[Z,Z]$, and the product of the third pairs gives $[Y,Z]$. Since we used the same order on the branch points, the $[Y,Y]$ and $[Z,Z]$ parts appear in $[Y \cup Z, Y \cup Z]$ in the proper order, but maybe some pairs that appear in $[Y,Z]$ appeared in $[Y \cup Z, Y \cup Z]$ as if coming from $[Z,Y]$ (this will happen whenever we have $P_j \in Z$ and $P_i \in Y$ with $j < i$). This establishes the first equality up to sign, as claimed.

The second equality is a special case of the first one. Take $Y = \{R\}$ and $Z = W \setminus R$. Then $Y \cup Z = W$, $[Y,Y] = 1$ and contributes nothing, and $[Y,Z] = [R,W]$ as explained above. Thus the first equality becomes the second one in this setting.   □

Corollary 4.3. *Assume that* Y *and* Z *form a pair of disjoint sets of branch points. Assume that* V *and* W *form another pair of disjoint sets of branch points, and that* $Y \cup Z = V \cup W$. *Then we have* $[Y,Y][Z,Z][Y,Z] = [V,V][W,W][V,W]$ *up to sign.*

Proof. If we denote $Y \cup Z = V \cup W$ by U, then Lemma 4.2 shows that the two expressions both equal $[U,U]$ up to sign, whence the corollary. $\square$

As already mentioned, we shall only be working with the expressions from Definition 4.1 raised to even powers (en and its multiples). Hence the specific order of the branch points which is used to define $[Y,Y]$ in Definition 4.1 will not matter, so that our results will not depend on that order. Another good thing to notice is that this means that all the "equalities up to sign" in Lemma 4.2 and Corollary 4.3 will essentially be exact equalities in all our applications.

4.1 The Poor Man's Thomae Formulae

Let us now start by constructing what we call the PMT for a general nonsingular Z_n curve.

4.1.1 First Identities Between Theta Constants

As in the examples in Chapter 3, the way to obtain the PMT involving the theta constant $\theta[\Delta](0,\Pi)$ for a non-special integral divisor $\Delta = A^{n-1} \prod_{j=0}^{n-2} C_j^{n-2-j}$ of degree g with support in the branch points distinct from P_0 passes through the quotient

$$f(P) = \frac{\theta[\Delta]^{en^2}(\varphi_{P_0}(P),\Pi)}{\theta[\Xi]^{en^2}(\varphi_{P_0}(P),\Pi)}$$

for some other such divisor Ξ. The considerations near the end of Section 2.6 show again that the "best" divisors to take for Ξ are $T_R(\Delta)$ for some branch point $R \in C_0$. In this case Proposition 2.21 shows that this function is the function

$$c \prod_Q (z - z(Q))^{en[v_Q(N(\Delta)) - v_Q(N(T_R(\Delta)))]}$$

where c is a nonzero complex constant. Definition 2.16 shows that

$$N(\Delta) = A^{n-1} \prod_{j=0}^{n-2} C_j^j, \quad T_R(\Delta) = (C_0 \setminus R)^{n-1}(A \cup R)^{n-2} \prod_{j=1}^{n-2} C_j^{j-1},$$

and

$$N(T_R(\Delta)) = (C_0 \setminus R)^{n-1} \Big(\prod_{j=1}^{n-2} C_j^{n-1-j} \Big) (A \cup R)^0.$$

Thus comparing the powers to which a branch point from each set appears in $N(\Delta)$ and $N(T_R(\Delta))$ we obtain the following explicit expressions for the functions in Proposition 2.21. For odd n it gives

$$c \frac{[z,A]^{2n(n-1)} \prod_{j=(n+1)/2}^{n-2} [z,C_j]^{2n(2j-(n-1))}}{[z,C_0 \setminus R]^{2n(n-1)} \prod_{j=1}^{(n-3)/2} [z,C_j]^{2n(n-1-2j)}}$$

(with the powers of $[z,C_{(n-1)/2}]$ canceling), and for even n it gives

$$c \frac{[z,A]^{n(n-1)} \prod_{j=n/2}^{n-2} [z,C_j]^{n(2j-(n-1))}}{[z,C_0 \setminus R]^{n(n-1)} \prod_{j=1}^{(n-2)/2} [z,C_j]^{n(n-1-2j)}}$$

(with all the sets appearing).

Actually, the only difference between the even and odd n in this case is (except for the value of e) "what happens in the middle sets". When n is odd the set $C_{(n-1)/2}$ does not appear in the divisor of this function (so that the branch points in it are regular values of the theta quotient), and the "neighboring" sets $C_{(n+1)/2}$ and $C_{(n-3)/2}$ appear to the $4n$th power. However, when n is even all the sets appear in the divisor of this function and the "middle sets" $C_{n/2}$ and $C_{(n-2)/2}$ appear to the nth power. We saw this in the examples of $n = 7$ and $n = 8$ at the end of Section 2.6. This difference is not significant at this point, if we notice that in any case we have that the power to which $[z,A]$ appears is $en(n-1)$, the power to which $[z,C_0 \setminus R]$ appears is $-en(n-1)$, and the power to which $[z,C_j]$ with $1 \le j \le n-2$ appears is $en(2j-(n-1))$ (compare the numbers $v_Q(N(\Delta))$ and $v_Q(N(T_R(\Delta)))$ for any such Q). We conclude that the explicit form of Proposition 2.21 with $\Xi = T_R(\Delta)$ is

$$f(P) = c[z,A]^{en(n-1)} [z,C_0 \setminus R]^{-en(n-1)} \prod_{j=1}^{n-2} [z,C_j]^{en(2j-(n-1))},$$

independently of the parity of n. We write this equation in the more convenient form

$$\frac{\theta[\Delta]^{en^2}(\varphi_{P_0}(P), \Pi)}{\theta[T_R(\Delta)]^{en^2}(\varphi_{P_0}(P), \Pi)} = f(P) = c \frac{[z,A]^{en(n-1)} \prod_{j=1}^{n-2} [z,C_j]^{enj}}{[z,C_0 \setminus R]^{en(n-1)} \prod_{j=1}^{n-2} [z,C_j]^{en(n-1-j)}},$$

which is not "reduced" like the expressions for odd and even n but has the advantage of being independent of the parity of n. This form is based directly on the form of the divisors $N(\Delta)$ and $T_R(\Delta)$.

As in the example, we can find the constant c by substituting $P = P_0$, which gives us

$$\frac{\theta[\Delta]^{en^2}(0,\Pi)}{\theta[T_R(\Delta)]^{en^2}(0,\Pi)} = c\frac{[P_0,A]^{en(n-1)}\prod_{j=1}^{n-2}[P_0,C_j]^{enj}}{[P_0,C_0\setminus R]^{en(n-1)}\prod_{j=1}^{n-2}[P_0,C_j]^{en(n-1-j)}},$$

and thus

$$f(P) = \frac{\theta[\Delta]^{en^2}(0,\Pi)}{\theta[T_R(\Delta)]^{en^2}(0,\Pi)} \times \frac{[P_0,C_0\setminus R]^{en(n-1)}\prod_{j=1}^{n-2}[P_0,C_j]^{en(n-1-j)}}{[P_0,A]^{en(n-1)}\prod_{j=1}^{n-2}[P_0,C_j]^{enj}}$$

$$\times \frac{[z,A]^{en(n-1)}\prod_{j=1}^{n-2}[z,C_j]^{enj}}{[z,C_0\setminus R]^{en(n-1)}\prod_{j=1}^{n-2}[z,C_j]^{en(n-1-j)}}.$$

Now we substitute $P = R$, which, on the one hand, gives us

$$f(R) = \frac{\theta[\Delta]^{en^2}(0,\Pi)}{\theta[T_R(\Delta)]^{en^2}(0,\Pi)} \times \frac{[P_0,C_0\setminus R]^{en(n-1)}\prod_{j=1}^{n-2}[P_0,C_j]^{en(n-1-j)}}{[P_0,A]^{en(n-1)}\prod_{j=1}^{n-2}[P_0,C_j]^{enj}}$$

$$\times \frac{[R,A]^{en(n-1)}\prod_{j=1}^{n-2}[R,C_j]^{enj}}{[R,C_0\setminus R]^{en(n-1)}\prod_{j=1}^{n-2}[R,C_j]^{en(n-1-j)}},$$

and on the other hand, Eq. (2.3) gives us

$$f(R) = \frac{\theta[\Delta]^{en^2}(\varphi_{P_0}(R),\Pi)}{\theta[T_R(\Delta)]^{en^2}(\varphi_{P_0}(R),\Pi)} = \frac{\theta[T_R(\Delta)]^{en^2}(0,\Pi)}{\theta[\Delta]^{en^2}(0,\Pi)}.$$

Hence we can combine the two equations and obtain

$$\frac{\theta[T_R(\Delta)]^{2en^2}(0,\Pi)}{\theta[\Delta]^{2en^2}(0,\Pi)} = \frac{[P_0,C_0\setminus R]^{en(n-1)}\prod_{j=1}^{n-2}[P_0,C_j]^{en(n-1-j)}}{[P_0,A]^{en(n-1)}\prod_{j=1}^{n-2}[P_0,C_j]^{enj}}$$

$$\times \frac{[R,A]^{en(n-1)}\prod_{j=1}^{n-2}[R,C_j]^{enj}}{[R,C_0\setminus R]^{en(n-1)}\prod_{j=1}^{n-2}[R,C_j]^{en(n-1-j)}},$$

which is equivalent to

$$\frac{\theta[\Delta]^{2en^2}(0,\Pi)}{[P_0,A]^{en(n-1)}[R,C_0\setminus R]^{en(n-1)}\prod_{j=1}^{n-2}[P_0,C_j]^{enj}[R,C_j]^{en(n-1-j)}}$$

$$= \frac{\theta[T_R(\Delta)]^{2en^2}(0,\Pi)}{[P_0,C_0\setminus R]^{en(n-1)}[R,A]^{en(n-1)}\prod_{j=1}^{n-2}[P_0,C_j]^{en(n-1-j)}[R,C_j]^{enj}}. \quad (4.1)$$

Equation (4.1), whose proof is just a calculation, illustrates the motivation for defining the operators T_R and N. Its proof shows that by properly defining T_R and N we obtain the first relations between the theta constants on X (those expressed in Eq. (4.1)) using simple and elementary calculations. The PMT will be obtained from Eq. (4.1) by symmetrizing the denominators to eliminate the dependence on the point

R. We shall later see, as we saw in the examples in Chapter 3, that obtaining the full Thomae formulae from the PMT is done using similar ideas.

4.1.2 Symmetrization over R and the Poor Man's Thomae

Equation (4.1) gives, for a divisor Δ and a branch point $R \in C_0$, a relation between $\theta[\Delta](0,\Pi)$ and $\theta[T_R(\Delta)](0,\Pi)$ (together with the polynomials in the variables λ_i defining the nonsingular Z_n curve X). We notice that it has the property, which we would later like to preserve in all the equations we have, that starting with the divisor $T_R(\Delta)$ and the branch point R (which also gives a relation between these theta constants) gives exactly the same relation. This will be easier to see after we set up some notation. Recall that the sets A and C_j, $0 \le j \le n-2$ depend on the divisor with which we work, so we assume that they are the sets corresponding to Δ, and denote the sets corresponding to $T_R(\Delta)$ by adding a tilde to the notation. Now, Definition 2.16 shows that

$$T_R(\Delta) = (C_0 \setminus R)^{n-1}(A \cup R)^{n-2}\prod_{j=1}^{n-2} C_j^{j-1},$$

and hence we have $\widetilde{A} = C_0 \setminus R$, $\widetilde{C_0} = A \cup R$, and $\widetilde{C_j} = C_{n-i-j}$ for $1 \le j \le n-2$. Therefore, writing the denominator under $\theta[T_R(\Delta)](0,\Pi)$ in Eq. (4.1) using the sets with the tilde gives

$$[P_0,\widetilde{A}]^{en(n-1)}[R,\widetilde{C_0} \setminus R]^{en(n-1)}\prod_{j=1}^{n-2}[P_0,\widetilde{C_j}]^{enj}[R,\widetilde{C_j}]^{en(n-1-j)};$$

note that we have changed the multiplication index j to $n-1-j$. This is exactly the denominator appearing under $\theta[\Delta](0,\Pi)$ in Eq. (4.1) with a tilde on all the sets. This shows that starting with $T_R(\Delta)$ and R instead of Δ and R (which means putting a tilde where there is not and removing it from any place where it appears) would give exactly the same Eq. (4.1).

The crucial property that the PMT must have is that the denominator appearing under the theta constant corresponding to a divisor Δ must be constructed from Δ alone. This is, of course, not the case in Eq. (4.1), as one clearly sees that the denominator appearing under $\theta[\Delta](0,\Pi)$ there depends on the branch point R chosen from C_0. We must now find a way to eliminate this dependence on R. We shall do this by dividing Eq. (4.1) by suitable expressions, preserving the property that we already have—the equation must be the same no matter from which divisor appearing in it we begin. We already saw that Eq. (4.1) has this property, and we do not want to lose it.

For this we want to take the expressions depending on R in this denominator, which are $[R,C_0 \setminus R]^{en(n-1)} = [R,C_0]^{en(n-1)}$ and $[R,C_j]^{en(n-1-j)}$ for $1 \le j \le n-2$, and we multiply them (i.e., divide Eq. (4.1)) by appropriate expressions that will give expressions independent of R. By Lemma 4.2 we find that multiplying $[R,C_0]$ by $[C_0 \setminus R, C_0 \setminus R]$ would give $[C_0, C_0]$ (all taken to an even power) which is independent of R, and it is clear that multiplying $[R,C_j]$ with $1 \leq j \leq n-2$ by $[C_0 \setminus R, C_j]$ would give $[C_0, C_j]$ (all taken to an even power) which is also independent of R. Therefore our first try would be to divide Eq. (4.1) by

$$[C_0 \setminus R, C_0 \setminus R]^{en(n-1)} \prod_{j=1}^{n-2} [C_0 \setminus R, C_j]^{en(n-1-j)}.$$

However, we must check the effect of this action on the other side of Eq. (4.1). Writing these sets in the notation with the tilde, we find that the expression by which we divided Eq. (4.1) is

$$[\widetilde{A}, \widetilde{A}]^{en(n-1)} \prod_{j=1}^{n-2} [\widetilde{A}, \widetilde{C}_j]^{enj};$$

note that we have replaced the multiplication index j by $n-1-j$. Therefore, in order not to lose the property that starting with $T_R(\Delta)$ and R gives us the same equation as starting with Δ and R, we must divide Eq. (4.1) also by

$$[A,A]^{en(n-1)} \prod_{j=1}^{n-2} [A,C_j]^{enj},$$

which in the notation of the sets with the tilde equals

$$[\widetilde{C_0} \setminus R, \widetilde{C_0} \setminus R]^{en(n-1)} \prod_{j=1}^{n-2} [\widetilde{C_0} \setminus R, \widetilde{C}_j]^{en(n-1-j)}$$

(replacing j with $n-1-j$ again). Clearly the product of these two expressions looks the same in the notation with the tilde as in the notation without it (a fact which is related to T_R being an involution), and this turns out to be the expression by which dividing Eq. (4.1) gives the PMT, as we now show.

Let us denote the set $A \cup P_0$ by C_{-1}. Then, together with the "merging" anticipated in the previous paragraph, we find that for any $1 \leq j \leq n-2$, $[P_0, C_j]^{enj}$ from Eq. (4.1) and $[A, C_j]^{enj}$ from the new multiplier can be merged into $[C_{-1}, C_j]^{enj}$. Moreover, Lemma 4.2 shows that the $en(n-1)$th powers of $[P_0, A] = [P_0, C_{-1}]$ and $[A,A] = [C_{-1} \setminus P_0, C_{-1} \setminus P_0]$ can be merged into $[C_{-1}, C_{-1}]^{en(n-1)}$. All this happens in the denominator appearing under $\theta[\Delta](0, \Pi)$, while under $\theta[T_R(\Delta)](0, \Pi)$ the same thing happens with a tilde on all the sets, where the set $\widetilde{C_{-1}}$ is defined to be $\widetilde{A} \cup P_0 = C_0 \setminus R \cup P_0$. This shows that the property that starting with $T_R(\Delta)$ and R gives us the same equation as starting with Δ and R is preserved. The formula thus becomes

$$\frac{\theta[\Delta]^{2en^2}(0,\Pi)}{[C_{-1},C_{-1}]^{en(n-1)}[C_0,C_0]^{en(n-1)}\prod_{j=1}^{n-2}[C_{-1},C_j]^{enj}[C_0,C_j]^{en(n-1-j)}}$$

$$= \frac{\theta[T_R(\Delta)]^{2en^2}(0,\Pi)}{[\widetilde{C_{-1}},\widetilde{C_{-1}}]^{en(n-1)}[\widetilde{C_0},\widetilde{C_0}]^{en(n-1)}\prod_{j=1}^{n-2}[\widetilde{C_{-1}},\widetilde{C_j}]^{enj}[\widetilde{C_0},\widetilde{C_j}]^{en(n-1-j)}}. \qquad (4.2)$$

The meaning of Eq. (4.2) is the content of the following

Proposition 4.4. *For a divisor* $\Delta = A^{n-1}\prod_{j=0}^{n-2}C_j^{n-2-j}$, *define*

$$g_\Delta = [C_{-1},C_{-1}]^{en(n-1)}[C_0,C_0]^{en(n-1)}\prod_{j=1}^{n-2}[C_{-1},C_j]^{enj}[C_0,C_j]^{en(n-1-j)},$$

where $C_{-1} = A \cup P_0$. *Then the quotient* $\theta^{2en^2}[\Delta](0,\Pi)/g_\Delta$ *is invariant under all the operators* T_R *and is the PMT for the nonsingular* Z_n *curve* X.

Proof. It is clear how to construct g_Δ from Δ and that the construction depends only on Δ. Our notation of the sets with the tilde in Eq. (4.2) shows that this equation states that

$$\frac{\theta^{2en^2}[\Delta](0,\Pi)}{g_\Delta} = \frac{\theta^{2en^2}[T_R(\Delta)](0,\Pi)}{g_{T_R(\Delta)}}.$$

Hence applying T_R does not change the quotient. Since Δ and R were arbitrary, except for the demand that R lies in $(C_0)_\Delta$, of course, the proposition is clear. $\qquad\square$

   Proposition 4.4 shows that Eq. (4.2) is no longer an equality between two terms, but can be merged into an equality of many more terms—one term for each divisor to which one can get from Δ using successive applications of the operators T_R. This can also be seen from Eq. (4.2) itself. The denominator under $\theta[\Delta](0,\Pi)$ is independent of the branch point $R \in C_0$, and hence for the same divisor Δ the r Eqs. (4.2) (one for each $R \in C_0$) can be merged into one long equality of $r+1$ terms (one is the common expression $\theta[\Delta](0,\Pi)/g_\Delta$, and the other r expressions are $\theta[T_R(\Delta)](0,\Pi)/g_{T_R(\Delta)}$ for the branch points R from C_0, together $r+1$ quotients). This equality of $r+1$ terms comes from a given divisor Δ (and the various points $R \in C_0$), but starting with another divisor, say $T_R(\Delta)$ for some $R \in C_0$, we obtain another equality of $r+1$ terms, involving two terms from the first equality and $r-1$ other terms (which belong to the divisors $T_Q(T_R(\Delta))$ for $Q \in A$). This allows us to broaden the equality to include much more (when r is large) than only $r+1$ terms.

4.1.3 Reduced Formulae

Before we explain how to go on to the full Thomae formulae, we remark that for a given minimal set of divisors that is closed under the action of the operators T_R

(on which the quotient from Proposition 4.4 is constant by the proposition), the quotients are not reduced. This means that there are expressions which appear under $\theta[\Delta](0,\Pi)$ for each divisor Δ in this set. This occurs for the following reason. Fix a divisor Δ in the set, and define for the moment the set G to be $C_{-1} \cup C_0$. Then it is clear that $\widetilde{G} = \widetilde{C_{-1}} \cup \widetilde{C_0}$ is the same set as G (being the union of the sets $C_0 \setminus R \cup P_0$ and $A \cup R$), and hence G is the same set of branch points for every divisor in this set of divisors. Then we see that the expressions $[G,C_j]^{en\min\{j,n-1-j\}}$ for $1 \le j \le n-2$, which are the expressions $[\widetilde{G},\widetilde{C_j}]^{en\min\{j,n-1-j\}}$ for $1 \le j \le n-2$ when replacing j by $n-1-j$, appear both in the product of $[C_{-1},C_j]^{enj}$ and $[C_0,C_j]^{en(n-1-j)}$ in the left-hand side of Eq. (4.2) and in the product of $[\widetilde{C_{-1}},\widetilde{C_j}]^{enj}$ and $[\widetilde{C_0},\widetilde{C_j}]^{en(n-1-j)}$ in the right-hand side of Eq. (4.2). Since the tilde only interchanges between C_j and C_{n-1-j} for every such j, these expressions thus appear under every theta constant in this set of divisors. We warn the reader that this does not hold for the expressions $[C_{-1},C_{-1}]^{en(n-1)}$ and $[C_0,C_0]^{en(n-1)}$—even though in Eq. (4.2) for a given divisor Δ and branch point R there are common expressions there (i.e., $[A,A]^{en(n-1)}$ and $[C_0 \setminus R, C_0 \setminus R]^{en(n-1)}$, the latter clearly depending on R), letting the point R vary and multiple applications of the operators T_R mix C_{-1} and C_0 more and more and there is no part of this expression which appears under all the theta constants in the set.

These common expressions essentially appear since when we merged the expressions for the function $f(P)$ as a rational function of z for odd n and even n we used unreduced quotients. One could have, of course, started with the reduced expressions for $f(P)$ appearing in the beginning of the section and worked on the odd and even n cases separately. We leave as an exercise for the reader to verify that applying the same calculations to the reduced expressions gives, instead of Eq. (4.1), the equation

$$\frac{\theta[\Delta]^{4n^2}(0,\Pi)}{([P_0,A][R,C_0 \setminus R])^{2n(n-1)} \prod_{j=(n+1)/2}^{n-2}[P_0,C_j]^{2n(2j-(n-1))} \prod_{j=1}^{(n-3)/2}[R,C_j]^{2n(n-1-2j)}}$$
$$= \frac{\theta[T_R(\Delta)]^{4n^2}(0,\Pi)}{([P_0,C_0 \setminus R][R,A])^{2n(n-1)} \prod_{j=1}^{(n-3)/2}[P_0,C_j]^{2n(n-1-2j)} \prod_{j=(n+1)/2}^{n-2}[R,C_j]^{2n(2j-(n-1))}}$$

for odd n and the equation

$$\frac{\theta[\Delta]^{2n^2}(0,\Pi)}{([P_0,A][R,C_0 \setminus R])^{n(n-1)} \prod_{j=n/2}^{n-2}[P_0,C_j]^{n(2j-(n-1))} \prod_{j=1}^{(n-2)/2}[R,C_j]^{n(n-1-2j)}}$$
$$= \frac{\theta[T_R(\Delta)]^{2n^2}(0,\Pi)}{([P_0,C_0 \setminus R][R,A])^{n(n-1)} \prod_{j=1}^{(n-2)/2}[P_0,C_j]^{n(n-1-2j)} \prod_{j=n/2}^{n-2}[R,C_j]^{n(2j-(n-1))}}$$

for even n. Then the symmetrization over R replaces Eq. (4.2) by its reduced form and one can reprove Proposition 4.4 with

$$g_\Delta = ([C_{-1}, C_{-1}][C_0, C_0])^{2n(n-1)} \prod_{j=(n+1)/2}^{n-2} [C_{-1}, C_j]^{2n(2j-(n-1))} \prod_{j=1}^{(n-3)/2} [C_0, C_j]^{2n(n-1-2j)}$$

(4.3)

when n is odd, and

$$g_\Delta = ([C_{-1}, C_{-1}][C_0, C_0])^{n(n-1)} \prod_{j=n/2}^{n-2} [C_{-1}, C_j]^{n(2j-(n-1))} \prod_{j=1}^{(n-2)/2} [C_0, C_j]^{n(n-1-2j)}$$

(4.4)

when n is even.

It is not hard to see that Eqs. (4.3) and (4.4) are obtained from Eq. (4.2) by canceling the common factor, i.e., by multiplying Eq. (4.2) (and hence dividing g_Δ) by the expression

$$\prod_{j=1}^{n-2} [G, C_j]^{en \min\{j, n-1-j\}} = \prod_{j=1}^{n-2} [\widetilde{G}, \widetilde{C}_j]^{en \min\{j, n-1-j\}},$$

which equals

$$\prod_{j=1}^{(n-1)/2} [C_{-1}, C_j]^{2nj} [C_0, C_j]^{2nj} \prod_{j=(n+1)/2}^{n-2} [C_{-1}, C_j]^{2n(n-1-j)} [C_0, C_j]^{2n(n-1-j)}$$

(and the same with the tilde) when n is odd and

$$\prod_{j=1}^{(n-2)/2} [C_{-1}, C_j]^{nj} [C_0, C_j]^{nj} \prod_{j=n/2}^{n-2} [C_{-1}, C_j]^{n(n-1-j)} [C_0, C_j]^{n(n-1-j)}$$

(and the same with the tilde) when n is even (note that in the odd n case taking the first product until $(n-3)/2$ and the second from $(n-1)/2$ gives the same result).

Another way to see the whole reduced process in comparison with the unreduced one, in view of future cases, is as follows. In all the unreduced expressions we have some expressions appearing to the power $en(n-1)$, and the others, which are based on the index j, appear either to the power enj or to the power $en(n-1-j)$. The reduced expressions are obtained by leaving the expressions appearing to the power $en(n-1)$ as they are, but changing the other powers as follows. Every power enj is replaced by $en(2j-(n-1))$, and every power $en(n-1-j)$ is replaced by $en(n-1-2j)$. Then the product over j is taken only over those j such that the corresponding power is positive. This means that in the expressions of the first kind the product is taken over $(n+1)/2 \le j \le n-2$ for odd n and over $n/2 \le j \le n-2$ for even n, and in the expressions of the second kind the product is taken over $1 \le j \le (n-3)/2$ for odd n and over $1 \le j \le (n-2)/2$ for even n. The value $j = (n-1)/2$ for odd n does not appear since both $2j - (n-1)$ and $n-1-2j$ vanish then, hence each of the products for odd n can include this value without changing anything. In any case, the Thomae formulae can now be obtained either from the reduced Eqs. (4.3) and (4.4) or from Proposition 4.4 and Eq. (4.2) themselves.

Our goal now is to improve the PMT of Proposition 4.4 in the following way: We wish to preserve the property that the equality is between terms of the form $\theta[\Delta](0,\Pi)/h_\Delta$ where the denominator h_Δ is constructed from Δ in a certain way, and also preserve the property that the quotient is invariant under the operators T_R. The additional property we now require is that the quotient also be invariant under the operator N. Since by Eq. (1.5) we know that $\theta[N(\Delta)](0,\Pi) = \theta[\Delta](0,\Pi)$, this is equivalent to simply demanding that the denominator h_Δ which we construct should satisfy $h_{N(\Delta)} = h_\Delta$. We remind the reader that this is the way we worked in the example in Section 3.1 (and actually in all the examples in Chapter 3, except for the one in Section 3.3), but still before we prove the general case we do the example of $n = 5$, which will illuminate the process of the general case ($n = 3$ is too small and avoids the complications we wish to present).

4.2 Example with $n = 5$ and General r

For the case $n = 5$ (hence $e = 2$) and arbitrary r we find that the PMT, i.e., Proposition 4.4 and Eq. (4.2), states that the expression

$$\frac{\theta^{100}[\Delta](0,\Pi)}{[C_{-1},C_{-1}]^{40}[C_0,C_0]^{40}[C_{-1},C_1]^{10}[C_0,C_1]^{30}[C_{-1},C_2]^{20}[C_0,C_2]^{20}[C_{-1},C_3]^{30}[C_0,C_3]^{10}}$$

(4.5)

(where Δ is $A^4 C_0^3 C_1^2 C_2^1 C_3^0$ since $n = 5$) is invariant under the operators T_R, and the denominator was denoted by g_Δ. The reduced PMT in Eq. (4.3) states the same for the expression

$$\frac{\theta^{100}[\Delta](0,\Pi)}{[C_{-1},C_{-1}]^{40}[C_0,C_0]^{40}[C_{-1},C_3]^{20}[C_0,C_1]^{20}},$$

but since we proved in detail the unreduced formula, we shall continue with it. Note that this reduced PMT gives back Eq. (3.14) from Section 3.4 when $r = 1$ since the expressions $[C_{-1},C_{-1}]$ and $[C_0,C_0]$ are $[Y,Y]$ for sets Y of cardinality 1, hence are 1. Indeed, the process for working with the reduced PMT in this case is the one we used in Section 3.4 with $n = 5$. We shall continue to denote the sets corresponding to $T_R(\Delta)$ by adding the tilde, so that

$$\widetilde{A} = C_0 \setminus R, \quad \widetilde{C_{-1}} = \widetilde{A} \cup P_0 = C_0 \setminus R \cup P_0, \quad \widetilde{C_0} = A \cup R,$$

$$\widetilde{C_1} = C_3, \quad \widetilde{C_2} = C_2, \quad \text{and} \quad \widetilde{C_3} = C_1$$

as in Section 3.4. Our objective now is to make the denominator invariant under the operator N. Here the construction will be a bit more complicated than it was in the nonsingular example in Chapter 3, where it was immediate as to how to proceed.

We recall from Definition 2.16 that $N(\Delta) = A^{n-1} \prod_{j=0}^{n-2} C_j^j$, and in our case of $n = 5$ it is simply $A^4 C_3^3 C_2^2 C_1^1 C_0^0$. This means that the denominator $g_{N(\Delta)}$ appearing under $\theta^{100}[N(\Delta)](0,\Pi)$ in its PMT is obtained from g_Δ by replacing C_0 by C_3, C_1 by C_2, C_2 by C_1 and C_3 by C_0 (and fixing every A and thus every C_{-1}). Therefore $g_{N(\Delta)}$ equals

$$[C_{-1},C_{-1}]^{40}[C_3,C_3]^{40}[C_{-1},C_2]^{10}[C_3,C_2]^{30}[C_{-1},C_1]^{20}[C_3,C_1]^{20}[C_{-1},C_0]^{30}[C_3,C_0]^{10}.$$

Now, in order to replace the denominator g_Δ in Eq. (4.5) by another denominator h_Δ which is also invariant under N we first multiply it (or equivalently divide Eq. (4.5)) by the "missing elements" which appear in $g_{N(\Delta)}$ and not in g_Δ. We shall add the missing elements in a specific order, and the reason for this specific order will be clear as we proceed.

4.2.1 Correcting the Expressions Involving C_{-1}

We begin with the expression $[C_{-1},C_1]^{10}$. The expression $[C_{-1},C_1]$ appears in both g_Δ and $g_{N(\Delta)}$, but to different powers, and this is the difference which needs correction. Now, since we divide Eq. (4.5) also for $T_R(\Delta)$ by the same expression, we write this in the notation of the sets with the tilde and find that we have divided Eq. (4.5) by $[\widetilde{C_0 \setminus R \cup P_0},\widetilde{C_3}]^{10}$. This means that to keep the invariance of the whole expression under the operators T_R, we must divide Eq. (4.5) also by this expression with the tilde omitted. However, there are two problems with this expression: first it depends on R, and second, by applying another operator T_Q to $T_R(\Delta)$ the sets C_{-1} and C_0 get even more mixed. This means that the expression invariant under the operators T_R, by which we need to divide Eq. (4.5), is

$$[C_{-1},C_1]^{10}[C_0,C_1]^{10}[C_{-1},C_3]^{10}[C_0,C_3]^{10}.$$

By putting $G = C_{-1} \cup C_0$ once again we find that $[C_{-1},C_1][C_0,C_1] = [G,C_1]$ and $[C_{-1},C_3][C_0,C_3] = [G,C_3]$. Since up to sign we also have that $[\widetilde{G},\widetilde{C_1}] = [G,C_3]$ and $[\widetilde{G},\widetilde{C_3}] = [G,C_1]$ (we already observed that $\widetilde{G} = G$) and the power 10 is even, we find that this expression is indeed invariant under the operators T_R, i.e., is equal to the expression obtained from it by putting a tilde on all the sets. This gives the new denominator

$$[C_{-1},C_{-1}]^{40}[C_0,C_0]^{40}[C_{-1},C_1]^{20}[C_0,C_1]^{40}[C_{-1},C_2]^{20}[C_0,C_2]^{20}[C_{-1},C_3]^{40}[C_0,C_3]^{20}.$$

In order to continue the comparison we note that the expression by which we multiplied $g_{N(\Delta)}$ is

$$[C_{-1},C_2]^{10}[C_3,C_2]^{10}[C_{-1},C_0]^{10}[C_3,C_0]^{10},$$

giving the new denominator

$$[C_{-1},C_{-1}]^{40}[C_3,C_3]^{40}[C_{-1},C_2]^{20}[C_3,C_2]^{40}[C_{-1},C_1]^{20}[C_3,C_1]^{20}[C_{-1},C_0]^{40}[C_3,C_0]^{20}.$$

Our next objective is to fix the expression $[C_{-1}, C_0]^{40}$, which appears in $g_{N(\Delta)}$ but not in g_Δ. Note that we should indeed do this correction after the previous one, as the last correction changed this power from 30 to 40. This means that had we done it the other way around we would have corrected the 30th power, and then needed to go back and correct the other 10. These considerations explain why we work in this specific order. Now, in the notation of the sets with the tilde, this expression is $[\widetilde{C_0} \setminus R \cup P_0, \widetilde{A} \cup R]^{40}$, and depends on R. We of course need something independent of R, and hence it seems that we should take something like the "least common multiple" of these expressions as R varies. However, applying another operator T_Q on $T_R(\Delta)$ mixes C_{-1} and C_0 more. This means that the proper correction at this stage should be multiplying the denominator (or dividing the Eq. (4.5)) by

$$[C_{-1}, C_{-1}]^{40} [C_0, C_0]^{40} [C_{-1}, C_0]^{40},$$

which by Corollary 4.3 is invariant under the operators T_R because we have that the unions $\widetilde{C_{-1}} \cup \widetilde{C_0}$ and $C_{-1} \cup C_0$ are $\widetilde{G}$ and G, respectively, and they are equal. Therefore the last expression is the same as the one obtained from it by putting a tilde on all the sets since the power 40 is even. This gives now the denominator

$$[C_{-1}, C_{-1}]^{80} [C_0, C_0]^{80} [C_{-1}, C_0]^{40} [C_{-1}, C_1]^{20} [C_0, C_1]^{40} [C_{-1}, C_2]^{20} [C_0, C_2]^{20}$$
$$\times [C_{-1}, C_3]^{40} [C_0, C_3]^{20}.$$

We note that the expression by which we multiplied $g_{N(\Delta)}$ is

$$[C_{-1}, C_{-1}]^{40} [C_3, C_3]^{40} [C_{-1}, C_3]^{40},$$

giving the new denominator

$$[C_{-1}, C_{-1}]^{80} [C_3, C_3]^{80} [C_{-1}, C_3]^{40} [C_{-1}, C_2]^{20} [C_3, C_2]^{40} [C_{-1}, C_1]^{20} [C_3, C_1]^{20}$$
$$\times [C_{-1}, C_0]^{40} [C_3, C_0]^{20}.$$

Examining the last two new denominators shows that all the expressions involving the set C_{-1} now appear in both of them to the same powers, which means that we have completed correcting these expressions.

4.2.2 Correcting the Expressions Not Involving C_{-1}

Next we turn to correct the expressions $[C_3, C_3]^{80}$, $[C_3, C_2]^{40}$ and $[C_3, C_1]^{20}$, which appear in $g_{N(\Delta)}$ but not in g_Δ. Again note that the powers 80 and 40 were originally 40 and 30 respectively, so this correction should indeed be done after the two others. In the notation with the tilde these expressions are $[\widetilde{C_1}, \widetilde{C_1}]^{80}$, $[\widetilde{C_1}, \widetilde{C_2}]^{40}$ and $[\widetilde{C_1}, \widetilde{C_3}]^{20}$, so that we need to divide Eq. (4.5) also by $[C_1, C_1]^{80}$ and by $[C_1, C_2]^{40}$ ($[C_1, C_3]^{20} = [C_3, C_1]^{20}$ appears in the original expressions). Therefore the proper expression by which we need to divide Eq. (4.5) now is

$$[C_3,C_3]^{80}[C_3,C_2]^{40}[C_3,C_1]^{20}[C_1,C_1]^{80}[C_1,C_2]^{40},$$

which is invariant under the operators T_R since up to sign we have

$$[\widetilde{C_3},\widetilde{C_3}] = [C_1,C_1], \quad [\widetilde{C_1},\widetilde{C_1}] = [C_3,C_3], \quad [\widetilde{C_3},\widetilde{C_2}] = [C_1,C_2],$$

$$[\widetilde{C_1},\widetilde{C_2}] = [C_3,C_2], \quad \text{and} \quad [\widetilde{C_3},\widetilde{C_1}] = [C_1,C_3],$$

and the powers 20, 40 and 80 are even. This gives the denominator

$$[C_{-1},C_{-1}]^{80}[C_0,C_0]^{80}[C_{-1},C_0]^{40}[C_{-1},C_1]^{20}[C_0,C_1]^{40}[C_{-1},C_2]^{20}[C_0,C_2]^{20}$$
$$\times [C_{-1},C_3]^{40}[C_0,C_3]^{20}[C_3,C_3]^{80}[C_3,C_2]^{40}[C_3,C_1]^{20}[C_1,C_1]^{80}[C_1,C_2]^{40},$$

and since $g_{N(\Delta)}$ is thus multiplied by

$$[C_0,C_0]^{80}[C_0,C_1]^{40}[C_0,C_2]^{20}[C_2,C_2]^{80}[C_2,C_1]^{40},$$

this gives the denominator

$$[C_{-1},C_{-1}]^{80}[C_3,C_3]^{80}[C_{-1},C_3]^{40}[C_{-1},C_2]^{20}[C_3,C_2]^{40}[C_{-1},C_1]^{20}[C_3,C_1]^{20}$$
$$\times [C_{-1},C_0]^{40}[C_3,C_0]^{20}[C_0,C_0]^{80}[C_0,C_1]^{40}[C_0,C_2]^{20}[C_2,C_2]^{80}[C_2,C_1]^{40}.$$

Now we are nearly finished: the only difference is that the denominator for Δ contains $[C_1,C_1]^{80}$ and that for $N(\Delta)$ contains $[C_2,C_2]^{80}$. Therefore we divide yet again Eq. (4.5) by $[C_2,C_2]^{80}$, and since this expression also equals $[\widetilde{C_2},\widetilde{C_2}]^{80}$, it is invariant under the operators T_R and is a good expression to divide Eq. (4.5) by. This means that we obtain the denominator

$$h_\Delta = [C_{-1},C_{-1}]^{80}[C_0,C_0]^{80}[C_{-1},C_0]^{40}[C_{-1},C_1]^{20}[C_0,C_1]^{40}[C_{-1},C_2]^{20}[C_0,C_2]^{20}$$
$$\times [C_{-1},C_3]^{40}[C_0,C_3]^{20}[C_3,C_3]^{80}[C_3,C_2]^{40}[C_3,C_1]^{20}[C_1,C_1]^{80}[C_1,C_2]^{40}[C_2,C_2]^{80},$$

and since this means that the denominator for $N(\Delta)$ was multiplied by $[C_1,C_1]^{80}$, we find that

$$h_{N(\Delta)} = [C_{-1},C_{-1}]^{80}[C_3,C_3]^{80}[C_{-1},C_3]^{40}[C_{-1},C_2]^{20}[C_3,C_2]^{40}[C_{-1},C_1]^{20}[C_3,C_1]^{20}$$
$$\times [C_{-1},C_0]^{40}[C_3,C_0]^{20}[C_0,C_0]^{80}[C_0,C_1]^{40}[C_0,C_2]^{20}[C_2,C_2]^{80}[C_2,C_1]^{40}[C_1,C_1]^{80}$$

and equals h_Δ. Since we were careful throughout the process to preserve the invariance under T_R (starting from Eq. (4.5) and verifying it after each stage), we find that $\theta^{100}[\Delta](0,\Pi)/h_\Delta$ with this h_Δ is still invariant under the operators T_R. Since we already know by Eq. (1.5) that $\theta[N(\Delta)](0,\Pi) = \theta[\Delta](0,\Pi)$ and since we saw that $h_{N(\Delta)} = h_\Delta$, this quotient is also invariant under N.

Before writing the final Thomae formulae, we want to make a remark about the last two stages. In the last stage we needed to compensate $[C_2,C_2]^{80}$, which came from the stage preceding it. Merging these two stages together we see that the expression by which we divided Eq. (4.5) was

$$[C_3,C_3]^{80}[C_3,C_2]^{40}[C_3,C_1]^{20}[C_1,C_1]^{80}[C_1,C_2]^{40}[C_2,C_2]^{80}.$$

This expression can be described as follows. Take all the expressions $[C_j,C_j]^{80}$ with $1 \leq j \leq 3$, all the expressions $[C_j,C_{j+1}]^{40}$ with $1 \leq j \leq 2$, and the expression $[C_j,C_{j+2}]^{20}$ with $j = 1$, and multiply them all. This means the product of all the expressions $[C_j,C_i]$ with $j \leq i$, which do not involve C_0 or C_{-1}, and to powers depending only on the difference $i - j$. Similarly, for $N(\Delta)$ the expression by which we divided was

$$[C_0,C_0]^{80}[C_0,C_1]^{40}[C_0,C_2]^{20}[C_2,C_2]^{80}[C_2,C_1]^{40}[C_1,C_1]^{80},$$

which has a similar description. Here we multiply all the expressions $[C_j,C_j]^{80}$ with $0 \leq j \leq 2$, all the expressions $[C_j,C_{j+1}]^{40}$ with $0 \leq j \leq 1$, and the expression $[C_j,C_{j+2}]^{20}$ for $j = 0$. Therefore we get the product of all the expressions $[C_j,C_i]$ with $j \leq i$, which do not involve C_3 or C_{-1}, and to powers depending only on $i - j$. We saw that partial expressions (like the one not including $[C_2,C_2]^{80}$ at first) do not suffice, and we need all the sets. This is why when constructing h_Δ in the general case one can merge all these stages to one by this definition (as one can see in detail in Appendix A).

4.2.3 Reduction and the Thomae Formulae for $n = 5$

Now, this last equation could have been the Thomae formulae for this case, but we would like our Thomae formulae to be reduced. We see that in the current expression for h_Δ every expression $[C_i,C_j]$ appears at least to the 20th power. This means that there is a "common divisor" for all of the h_Δ, which is

$$\left[\bigcup_{j=-1}^{3} C_j, \bigcup_{j=-1}^{3} C_j \right]^{20} = \prod_{0 \leq i < j \leq 5r-1} (\lambda_i - \lambda_j)^{20}$$

and is clearly invariant under everything. Therefore we shall now multiply Eq. (4.5) by this expression, and obtain that $\theta^{100}[\Delta](0,\Pi)/h_\Delta$ with

$$h_\Delta = [C_{-1},C_{-1}]^{60}[C_0,C_0]^{60}[C_{-1},C_0]^{20}[C_0,C_1]^{20}[C_{-1},C_3]^{20}$$
$$\times [C_3,C_3]^{60}[C_3,C_2]^{20}[C_1,C_1]^{60}[C_1,C_2]^{20}[C_2,C_2]^{60} \qquad (4.6)$$

is invariant under the operators T_R and N. This is the main part in the proof of our ultimate goal here,

Theorem 4.5. *The expression $\theta^{100}[\Delta](0,\Pi)/h_\Delta$ with h_Δ from Eq. (4.6) is independent of the divisor Δ, and is the Thomae formulae for the nonsingular Z_5 curve X.*

Note that Theorem 4.5 contains Theorem 3.19 as its special case with $r = 1$, since the description appearing in the paragraph following Theorem 3.19 gives exactly Eq. (4.6) with $r = 1$ (again, removing the expressions $[Y,Y]$ for sets Y with cardinality 1).

We have not yet proved Theorem 4.5, as we have not shown that one can get from any divisor to any other divisor using successive operations of T_R and N. In the case $r = 1$, however, we did prove it—it is the basepoint dependent version of Theorem 3.21 with the basepoint P_0. The proof for $r \geq 2$ really uses the assumption $r \geq 2$ (unused until now, but necessary if one does not want to use the basepoint change yet, as seen in Section 3.4), and we leave it for the general case in Section 4.4. We just want to note that we can describe h_Δ of Eq. (4.6) as the product of

$$\prod_{j=-1}^{3} [C_j,C_j]^{60}, \quad \prod_{j=-1}^{2} [C_j,C_{j+1}]^{20},$$

and $[C_j,C_{j+4}]^{20}$ with $j = -1$.

We wish to say a word about this reduction at the end. The need for this reduction essentially comes from the fact that we have started with Eq. (4.5), which comes from the unreduced PMT in Eq. (4.2) and not from the reduced one in Eq. (4.3). Now, Eq. (4.3) with $n = 5$ states that

$$\frac{\theta[\Delta]^{100}(0,\Pi)}{[C_{-1},C_{-1}]^{40}[C_0,C_0]^{40}[C_{-1},C_3]^{20}[C_0,C_1]^{20}}$$

is invariant under the operators T_R. We leave it as an exercise to the reader to see that starting from this equation one can get directly to the reduced Thomae formulae by the following stages. First divide it by

$$[C_{-1},C_{-1}]^{20}[C_0,C_0]^{20}[C_{-1},C_0]^{20}$$

(which comes initially from $[C_{-1},C_0]^{20}$), then by

$$[C_3,C_3]^{60}[C_3,C_2]^{20}[C_1,C_1]^{60}[C_1,C_2]^{20}$$

(which comes from $[C_3,C_3]^{60}$ and $[C_2,C_3]^{20}$), and finally by $[C_2,C_2]^{60}$. This yields Eq. (4.6) and Theorem 4.5 as stated, without needing the reduction. The reader may notice that this is the same process we did for $r = 1$ in Section 3.4, again without the expressions $[Y,Y]$ for sets Y with cardinality 1. This process is simpler, but the advantage of showing the longer and unreduced process is the fact that it demonstrates the idea of the order of the stages better. For general n (either odd or even) the stages should be done in the proper order also when starting with the reduced PMT in Eqs. (4.3) and (4.4) and not only with the unreduced one in Eq.

(4.2). Hence it is important to demonstrate them in the example, for which we chose to work with the unreduced PMT here. For more details we refer to Appendix A.

4.3 Invariance also under N

In this section we shall obtain the Thomae formulae for a general nonsingular Z_n curve, even though the proof that this is indeed the full Thomae formulae will be completed only in Section 4.4. What we do prove in this section is that the quotient $\theta^{2en^2}[\Delta](0,\Pi)/h_\Delta$, where h_Δ is some polynomial depending on the divisor Δ in the parameters λ_i, $0 \le i \le nr - 1$, is invariant not only under all the operators T_R, but also under N. This is done, as in the example of $n = 5$ in Section 4.2, by dividing the PMT from Proposition 4.4 and Eq. (4.2) (or alternatively Eqs. (4.3) and (4.4)) by certain invariant expressions, such that replacing the denominator g_Δ by the new denominator h_Δ will still give quotients invariant under the operators T_R, but now also under N. Since Eq. (1.5) shows us that $\theta[N(\Delta)](0,\Pi) = \theta[\Delta](0,\Pi)$ the last statement is equivalent to saying that h_Δ is invariant under N, i.e., that $h_{N(\Delta)} = h_\Delta$ for every divisor Δ.

As we have seen in the example of $n = 5$ in Section 4.2, it turns out that it is possible to do "corrections" to the PMT (either to the reduced Eqs. (4.3) and (4.4) depending on the parity of n, or to the unreduced Eq. (4.2) in both odd and even n, but even in the latter the corrections depend on the parity of n) in order to obtain the desired expression h_Δ. We shall not give all the details here, and turn directly to the description of the denominator h_Δ, although the interested reader can find the details of its construction in Appendix A. The form of h_Δ depends on the parity of n, although by taking some power of the expressions we obtain one could "merge" the two cases into one (at the cost of writing a formula which is a bit weaker), as we explain in Section 4.4.

4.3.1 The Description of h_Δ for Odd n

We first present the case of odd n. An intuitive way of describing h_Δ is pictorial. This description is based on a matrix with n rows and $n + (n - 3)/2$ columns. The rows are labeled by the sets C_j, $-1 \le j \le n - 2$, in increasing order, and the columns are labeled also by the sets C_j, $-1 \le j \le n - 2$, in increasing order, but followed by C_j, $-1 \le j \le (n - 7)/2$, in increasing order. The reader may, if more convenient, treat the columns as if labeled by the sets C_j, $-1 \le j \le (3n - 7)/2$, in increasing order, and assume that $C_j = C_{j-n}$ for $j \ge n - 1$. For example, for $n = 9$ we have 9 rows and 12 columns, with the last 3 columns being labeled C_{-1}, C_0, and C_1 (which can be considered as C_8, C_9, and C_{10}).

In this matrix we consider all the diagonals of maximal length, i.e., the one running from (C_{-1}, C_{-1}) to (C_{n-2}, C_{n-2}), then the one running from (C_{-1}, C_0) to (C_{n-2}, C_{-1}) (or equivalently (C_{n-2}, C_{n-1})), and ending with the one running from $(C_{-1}, C_{(n-5)/2})$ to $(C_{n-2}, C_{(n-7)/2})$ (or equivalently $(C_{n-2}, C_{(3n-7)/2})$). Note that, at least in the notation with the sets C_j with $j \geq n-1$, each diagonal contains the pairs of sets (C_i, C_{i+k}) for some fixed k (with i running from -1 to $n-2$). We call this diagonal the kth diagonal, with $0 \leq k \leq (n-3)/2$. Now, for each such k we take all the elements on the kth diagonal, raise them to the power

$$2n \frac{(n-2k)^2-1}{4} = 2n \left(\frac{n-1}{2}-k\right)\left(\frac{n+1}{2}-k\right) = 2n\left(\frac{n^2-1}{4} - k(n-k)\right),$$

and take the product of all of them. The powers decrease with k, starting from $2n(n^2-1)/4$ for $k=0$ and ending with $4n$ for $k=(n-3)/2$, and they are all even integers (as they are divisible by $2n$). This means that as a formula we can write (again, allowing sets of the form C_j with $j \geq n-1$)

$$h_\Delta = \prod_{k=0}^{(n-3)/2} \prod_{i=-1}^{n-2} [C_i, C_{i+k}]^{2n[(n^2-1)/4-k(n-k)]}.$$

In order to see it in the pictorial description, we give here the matrix for $n=9$, where the double vertical line marks the point to the right of which lie the columns labeled by "sets with two names".

360	216	108	36								
	360	216	108	36							
		360	216	108	36						
			360	216	108	36					
				360	216	108	36				
					360	216	108	36			
						360	216	108	36		
							360	216	108	36	
								360	216	108	36

In order to prove that this h_Δ gives the desired Thomae formulae, we should first obtain a formula for it which is easier to compare with g_Δ, i.e., a formula which expresses h_Δ only with the usual sets C_j with $-1 \leq j \leq n-2$. In order to do so we see that for a fixed $1 \leq k \leq (n-3)/2$ (for $k=0$ nothing needs to be changed) we have that for $i \geq n-1-k$ the expression $[C_i, C_{i+k}]$ means $[C_i, C_{i+k-n}]$, and thus (up to sign) equals $[C_{i+k-n}, C_i]$ or equivalently $[C_j, C_{j+n-k}]$ for $j = i+k-n$ (which satisfies $-1 \leq j \leq k-2$). This means that for this k the product over i becomes

$$\prod_{j=-1}^{n-2-k} [C_j, C_{j+k}]^{2n[(n^2-1)/4-k(n-k)]} \prod_{j=-1}^{k-2} [C_j, C_{j+n-k}]^{2n[(n^2-1)/4-k(n-k)]},$$

where we have changed the first multiplication index to j as well. Since it is clear that the power remains the same when we replace k with $n-k$, we now see that the second product is the same as the first one but corresponding to $n-k$ rather than k, and since $1 \le k \le (n-3)/2$ we have $(n+3)/2 \le n-k \le n-1$. This means that we can write

$$h_\Delta = \prod_{k=0}^{(n-3)/2} \prod_{j=-1}^{n-2-k} [C_j, C_{j+k}]^{2n[(n^2-1)/4-k(n-k)]}$$

$$\times \prod_{k=(n+3)/2}^{n-1} \prod_{j=-1}^{n-2-k} [C_j, C_{j+k}]^{2n[(n^2-1)/4-k(n-k)]}.$$

The two values that are missing now, $k = (n-1)/2$ and $k = (n+1)/2$, can be added with no effect, since the corresponding power, which we already saw that also equals $2n[(n-1)/2-k][(n+1)/2-k]$, vanishes for these values of k. This gives the more compact formula

$$h_\Delta = \prod_{k=0}^{n-1} \prod_{j=-1}^{n-2-k} [C_j, C_{j+k}]^{2n[(n^2-1)/4-k(n-k)]}. \qquad (4.7)$$

Before turning to proving the properties of this h_Δ, we want to check what the expression for h_Δ in Eq. (4.7) is, explicitly, for small values of odd n, and compare it with the examples. We start with $n = 3$, where $2n = 6$, and k can be either 0, 1 or 2. However, the power

$$2n \left(\frac{n-1}{2} - k \right) \left(\frac{n+1}{2} - k \right) = 6(1-k)(2-k)$$

vanishes for $k = 1$ and $k = 2$, leaving only $k = 0$ with the power 12. This can also be seen in the pictorial description, since the number of columns is $n + (n-3)/2 = 3$ like the number $n = 3$ of rows, which means that there is only one long diagonal (which corresponds to $k = 0$). Note that in the case $r = 2$ this gives exactly Eq. (3.2) from Section 3.1, which means that the combination of Theorems 4.6 and 4.8 below contains Theorem 3.2 from Section 3.1 as the special case with $n = 3$ and $r = 2$. Also in the case $r = 1$ we have a special case already written, which is the expression from Theorem 3.16 from Section 3.4 with denominator 1. It is indeed the special case of Eq. (4.7) with $n = 3$ and $r = 1$ since the expressions for $k = 0$ are of the sort $[Y, Y]$ with sets Y of cardinality $r = 1$. Hence Theorem 3.16 is the special case of Theorem 4.6 with $n = 3$ and $r = 1$, which gives the Thomae formulae since $n = 3$ is small enough. As for connections with previous works, Theorems 4.6 and 4.8 contain one of the main results of [EiF] as the special case with $n = 3$.

As for $n = 5$, we have $2n = 10$ and k varies from 0 to 4. The power

$$2n \left(\frac{n-1}{2} - k \right) \left(\frac{n+1}{2} - k \right) = 10(3-k)(2-k)$$

vanishes for $k = 2$ and $k = 3$ and gives 60 for $k = 0$ and 20 for $k = 1$ and $k = 4$, the latter referring only to the expression $[C_{-1}, C_3]$. As for the pictorial description, the number of columns is $n + (n-3)/2 = 6$, one more than the number $n = 5$ or rows, which gives two long diagonals, the zeroth for $k = 0$ and the first for $k = 1$ and $k = 4$. Therefore it is clear that Eq. (4.7) gives Eq. (4.6) as its special case with $n = 5$, and also the expression appearing in Theorem 3.19 as its special case with $n = 5$ and $r = 1$. Therefore Theorem 4.6 below contains Theorem 4.5 as the special case with $n = 5$ and Theorem 3.19 the special case with $n = 5$ and $r = 1$, and once we prove Theorem 4.8 below Theorem 4.5 will give the Thomae formulae for general nonsingular Z_5 curves with $r \geq 2$. As we said before, Theorem 3.21 gives the Thomae formulae for $n = 5$ and $r = 1$.

4.3.2 N-Invariance for Odd n

We now prove

Theorem 4.6. *The expression* $\theta^{4n^2}[\Delta](0, \Pi)/h_\Delta$ *with* h_Δ *from Eq. (4.7) is invariant under the action of the operators N and T_R.*

Proof. We begin with the invariance under N, which by Eq. (1.5) in its form $\theta[N(\Delta)](0, \Pi) = \theta[\Delta](0, \Pi)$ reduces to verifying that $h_{N(\Delta)} = h_\Delta$. We recall from Definition 2.16 that the operator N fixes C_{-1} and interchanges C_j with $0 \leq j \leq n-2$ by C_{n-2-j}. Since the action of N is different on C_{-1} than on the other sets, we need to treat the expressions involving C_{-1} separately from the others. First we claim that for each $0 \leq k \leq n-1$, the corresponding product with $0 \leq j \leq n-2-k$ (excluding $j = -1$) in Eq. (4.7) is invariant under N. This is because N takes $[C_j, C_{j+k}]$ to $[C_{n-2-j}, C_{n-2-j-k}]$, which equals (up to sign again) $[C_i, C_{i+k}]$ for $i = n-2-j-k$, and when j runs through the numbers from 0 to $n-2-k$ so does i (in decreasing order, but this changes nothing). Therefore indeed applying N to this product leaves it invariant for each $0 \leq k \leq n-1$.

It remains to deal with the expressions containing C_{-1}, i.e., to prove that the expression

$$\prod_{k=0}^{n-1} [C_{-1}, C_{k-1}]^{2n[(n^2-1)/4 - k(n-k)]}$$

is invariant under N. Now, for the expression with $k = 0$ it is clear (N fixes C_{-1}). For any other permissible value of k, we see that N takes $[C_{-1}, C_{k-1}]$ (up to the usual sign, which we can ignore) to $[C_{-1}, C_{n-1-k}]$, which we write as $[C_{-1}, C_{(n-k)-1}]$. This means that on the product with $1 \leq k \leq n-1$, the action of N simply takes k to $n-k$, and since the power is invariant under this action and as k runs from 1 to $n-1$ so does $n-k$, we find that this product is invariant under N. This finishes the proof that the expression for h_Δ in Eq. (4.7) is invariant under N, or that $h_{N(\Delta)} = h_\Delta$.

The important ingredient in the proof of the invariance under the operators T_R is of course the PMT, Proposition 4.4. We shall use it with the reduced expression for

g_Δ, the one appearing in Eq. (4.3) (n is odd). We write

$$\frac{\theta^{2n^2}[\Delta](0,\Pi)}{h_\Delta} = \frac{\theta^{2n^2}[\Delta](0,\Pi)}{g_\Delta} \bigg/ \frac{h_\Delta}{g_\Delta},$$

and since the (reduced) PMT in Proposition 4.4 gives us that the quotient in the numerator is invariant under the operators T_R, it remains to show that the quotient in the denominator is also invariant under the operators T_R. This is indeed much easier to verify as this expression no longer involves the theta constants, but only expressions which we have already written explicitly. We therefore fix a divisor Δ (which determines the sets C_j, $-1 \leq j \leq n-2$) and a branch point $R \in C_0$. We adopt again the notation using the sets with the tilde to denote the sets corresponding to the divisor $T_R(\Delta)$, and we now show that $h_\Delta/g_\Delta = h_{T_R(\Delta)}/g_{T_R(\Delta)}$, or equivalently that $h_\Delta/h_{T_R(\Delta)} = g_\Delta/g_{T_R(\Delta)}$.

Let us, for this purpose, split the expression for h_Δ from Eq. (4.7) into the product of two expressions, s_Δ and t_Δ, where t_Δ is composed of all the expressions containing either C_{-1} or C_0, and s_Δ is the product of all the rest. This means that s_Δ contains all the elements in the products with $j \geq 1$ and t_Δ contains the elements with $j = -1$ and $j = 0$. Explicitly,

$$h_\Delta = s_\Delta t_\Delta, \quad s_\Delta = \prod_{k=0}^{n-3} \prod_{j=1}^{n-2-k} [C_j, C_{j+k}]^{2n[(n^2-1)/4-k(n-k)]},$$

and

$$t_\Delta = \prod_{k=0}^{n-2} [C_0, C_k]^{2n[(n^2-1)/4-k(n-k)]} \prod_{k=-1}^{n-2} [C_{-1}, C_k]^{2n[(n^2-1)/4-(k+1)(n-1-k)]},$$

where in the elements with $j = -1$ in t_Δ we have replaced k by $k+1$. We do this splitting since now we claim that $s_\Delta = s_{T_R(\Delta)}$. In order to see this we recall that we have $\widetilde{C}_j = C_{n-1-j}$ for every $1 \leq j \leq n-2$, which means (using considerations similar to those used to prove the N-invariance) that for each $0 \leq k \leq n-3$, the corresponding product over $1 \leq j \leq n-2-k$ in the expression defining s_Δ remains the same when putting a tilde on all the sets. Indeed, for $j \geq 1$ we see that the expression $[\widetilde{C}_j, \widetilde{C}_{j+k}]$ is $[C_{n-1-j}, C_{n-1-j-k}]$, which equals (up to sign) $[C_i, C_{i+k}]$ with $i = n-1-j-k$, and when j runs through the numbers from 1 to $n-2-k$ so does i (again in decreasing order). Hence in the quotient $h_\Delta/h_{T_R(\Delta)} = s_\Delta t_\Delta/s_{T_R(\Delta)} t_{T_R(\Delta)}$ we can cancel s_Δ with $s_{T_R(\Delta)}$, hence this quotient equals $t_\Delta/t_{T_R(\Delta)}$. It remains to show that $t_\Delta/t_{T_R(\Delta)} = g_\Delta/g_{T_R(\Delta)}$.

We now recall that the set $G = C_{-1} \cup C_0$ equals $\widetilde{G} = \widetilde{C_{-1}} \cup \widetilde{C_0}$, meaning that writing the expression for t_Δ with the set G will help us to see what further cancelations we can do in the quotient $t_\Delta/t_{T_R(\Delta)}$. Now, for this we first compare the powers to which $[C_0, C_k]$ and $[C_{-1}, C_k]$ appear in the expression for t_Δ. Since

$$(k+1)(n-1-k) = k(n-k) + (n-1-2k)$$

we find that we have to split the products on k into 3 parts: one with $k = -1$ and $k = 0$, one with $1 \le k \le (n-3)/2$, and one with $(n+1)/2 \le k \le n-2$; we omit the value $k = (n-1)/2$ since then both expressions $k(n-k)$ and $(k+1)(n-1-k)$ equal $(n^2-1)/4$, hence both corresponding powers vanish and this value can be ignored. Dealing with the part with $k = -1$ and $k = 0$ we find that the expressions $[C_{-1}, C_{-1}]$, $[C_0, C_0]$ and $[C_{-1}, C_0]$ all appear to a power at least $2n(n^2 - 4n + 3)/4$, and by Lemma 4.2 we find that we have the equality $[C_{-1}, C_{-1}][C_0, C_0][C_{-1}, C_0] = [G, G]$ (up to the usual sign). As for $1 \le k \le (n-3)/2$, the power to which $[C_{-1}, C_k]$ appears is smaller than the power to which $[C_0, C_k]$ appears (recall that the expressions $k(n-k)$ and $(k+1)(n-1-k)$ appear with a minus sign), while for $(n+1)/2 \le k \le n-2$ the power to which $[C_0, C_k]$ appears is smaller than the power to which $[C_{-1}, C_k]$ appears. In any case we have $[C_{-1}, C_k][C_0, C_k] = [G, C_k]$ (up to the usual sign). This means that we can write the expression for t_Δ as

$$[G, G]^{2n(n^2 - 4n + 3)/4} ([C_{-1}, C_{-1}][C_0, C_0])^{2n(n-1)}$$

$$\times \prod_{k=1}^{(n-3)/2} [G, C_k]^{2n[(n^2-1)/4 - (k+1)(n-1-k)]} [C_0, C_k]^{2n(n-1-2k)}$$

$$\times \prod_{k=(n+1)/2}^{n-2} [G, C_k]^{2n[(n^2-1)/4 - k(n-k)]} [C_{-1}, C_k]^{2n(2k - (n-1))}.$$

Let us now put all the expressions involving the set G together, and write them in a way that will make it easier to see how they cancel in the quotient $t_\Delta / t_{T_R(\Delta)}$. We recall that for every $1 \le j \le n-2$, the product $[G, C_j][G, C_{n-1-j}]$ is the same as the expression obtained from it by putting a tilde on all the sets (because $\widetilde{G} = G$, $\widetilde{C}_j = C_{n-1-j}$ and $\widetilde{C_{n-1-j}} = C_j$), meaning that it makes a good expression to work with for our purposes. Moreover, if we focus on the values $(n+1)/2 \le k \le n-2$, then we note that $1 \le n-1-k \le (n-3)/2$ and that the power to which $[G, C_{n-1-k}]$ appears is $2n$ times the complement with respect to $(n^2-1)/4$ of

$$[(n-1-k)+1][n-1-(n-1-k)] = k(n-k).$$

This means that we can reorder the last expression for t_Δ and write it as

$$[G, G]^{2n(n^2 - 4n + 3)/4} \prod_{k=(n+1)/2}^{n-2} ([G, C_k][G, C_{n-1-k}])^{2n[(n^2-1)/4 - k(n-k)]}$$

$$\times ([C_{-1}, C_{-1}][C_0, C_0])^{2n(n-1)} \prod_{k=1}^{(n-3)/2} [C_0, C_k]^{2n(n-1-2k)} \prod_{k=(n+1)/2}^{n-2} [C_{-1}, C_k]^{2n(2k - (n-1))}.$$

Note that the second line is exactly the reduced expression for g_Δ in Eq. (4.3). The expression $t_{T_R(\Delta)}$ is, of course, the same expression but with a tilde on all sets.

Let us now examine the quotient $t_\Delta / t_{T_R(\Delta)}$ again. Since $G = \widetilde{G}$, we find that the powers of $[G, G]$ and $[\widetilde{G}, \widetilde{G}]$ cancel each other. As already mentioned, so do the powers of the expressions $[G, C_k][G, C_{n-1-k}]$ and $[\widetilde{G}, \widetilde{C}_k][\widetilde{G}, \widetilde{C_{n-1-k}}]$ for every $(n+1)/2 \le k \le n-2$. Since after the cancelation of these expressions the numerator becomes g_Δ and the denominator becomes the expression obtained from it by putting a tilde on all the sets (which is known to be $g_{T_R(\Delta)}$), we conclude that $t_\Delta / t_{T_R(\Delta)} = g_\Delta / g_{T_R(\Delta)}$ and the theorem is proved. $\qquad\qquad\square$

4.3.3 The Description of h_Δ for Even n

We now turn to the case of even n. The ideas are the same as in the odd n case, but the details are a bit different, so we present this case as well. The pictorial description is now based on a matrix with n rows and $n + (n-2)/2$ columns, with similar labeling. The rows are again labeled by the sets C_j, $-1 \le j \le n-2$, in increasing order, and the columns are now labeled also by C_j, $-1 \le j \le n-2$, in increasing order, followed by C_j, $-1 \le j \le (n-6)/2$, in increasing order, or equivalently simply by C_j, $-1 \le j \le (3n-6)/2$, in increasing order, with $C_j = C_{j-n}$ for $j \ge n-1$. For example, for $n=8$ we have 8 rows and 11 columns, with the last 3 columns being labeled C_{-1}, C_0, and C_1 (which can be considered as C_7, C_8, and C_9). Again we consider all the diagonals of maximal length, now ending with the one running from $(C_{-1}, C_{(n-4)/2})$ to $(C_{n-2}, C_{(n-6)/2})$ (or equivalently $(C_{n-2}, C_{(3n-6)/2})$), and the expression we form is by taking, for each $0 \le k \le (n-2)/2$, all the elements on the kth diagonal, raise them to the power

$$n\frac{(n-2k)^2}{4} = n\left(\frac{n}{2} - k\right)^2 = n\left(\frac{n^2}{4} - k(n-k)\right),$$

and take the product of all of them. Also here the powers decrease with k, starting from $n(n^2/4)$ for $k=0$ and ending with n for $k=(n-2)/2$; they are all even integers (as they are divisible by n). This means that as a formula, we can write (again, allowing sets of the form C_j with $j \ge n-1$)

$$h_\Delta = \prod_{k=0}^{(n-2)/2} \prod_{i=-1}^{n-2} [C_i, C_{i+k}]^{n[n^2/4 - k(n-k)]}.$$

In order to see it in the pictorial description, we give here the matrix for $n=8$, where again the double vertical line marks the point to the right of which lie the columns labeled by "sets with two names".

128	72	32	8							
	128	72	32	8						
		128	72	32	8					
			128	72	32	8				
				128	72	32	8			
					128	72	32	8		
						128	72	32	8	
							128	72	32	8

Again we would like a formula which expresses h_Δ only with the usual sets C_j with $-1 \le j \le n-2$. As in the odd n case we can write the product over i corresponding to any $1 \le k \le (n-2)/2$ as

$$\prod_{j=-1}^{n-2-k} [C_j, C_{j+k}]^{n[n^2/4 - k(n-k)]} \prod_{j=-1}^{k-2} [C_j, C_{j+n-k}]^{n[n^2/4 - k(n-k)]},$$

and since it is clear that this power also has the property of being invariant under the operation of taking k to $n-k$, we look at the second product as the product corresponding to $n-k$. Since $1 \le k \le (n-2)/2$ we have $(n+2)/2 \le n-k \le n-1$, which gives

$$h_\Delta = \prod_{k=0}^{(n-2)/2} \prod_{j=-1}^{n-2-k} [C_j, C_{j+k}]^{n[n^2/4 - k(n-k)]} \prod_{k=(n+2)/2}^{n-1} \prod_{j=-1}^{n-2-k} [C_j, C_{j+k}]^{n[n^2/4 - k(n-k)]},$$

with only the one number $k = n/2$ missing. Since the corresponding power vanishes, we can add this value of k with no effect and obtain the formula

$$h_\Delta = \prod_{k=0}^{n-1} \prod_{j=-1}^{n-2-k} [C_j, C_{j+k}]^{n[n^2/4 - k(n-k)]}. \tag{4.8}$$

Also here we want to see, before proving the properties of this h_Δ, what the expression for h_Δ in Eq. (4.8) is, explicitly, for small values of even n. We start with $n = 2$, where k can be either 0 or 1. However, the power

$$n\left(\frac{n}{2} - k\right)^2 = 2(1-k)^2$$

vanishes for $k = 1$, leaving only $k = 0$ with the power 2. Indeed, in the pictorial description the number of columns is $n + (n-2)/2 = 2$ like the number $n = 2$ of rows, which means that there is only one long diagonal (corresponding to $k = 0$). Therefore the combination of Theorems 4.7 and 4.8 below contains the other main result of [EiF], which is also the result of [T1] and [T2], as the special case with $n = 2$. It is worth noting that in this case of $n = 2$, the denominator h_Δ coincides with g_Δ (both the reduced and unreduced). This happens since (as one can verify from Definition 2.16) the operator N in this case is trivial (since the φ_{P_0}-image of every relevant divisor is of order 2 in $J(X)$ and so is K_{P_0}).

As for $n = 4$, k can be 0, 1, 2 or 3, and the power

$$n\left(\frac{n}{2} - k\right)^2 = 4(2 - k)^2$$

vanishes for $k = 2$ and gives 16 for $k = 0$ and 4 for $k = 1$ and $k = 3$, the latter referring only to the expression $[C_{-1}, C_2]$. In the pictorial description the number of columns is $n + (n - 2)/2 = 5$, one more than the number $n = 4$ or rows, which gives two long diagonals, the zeroth for $k = 0$ and the first for $k = 1$ and $k = 3$. The example we have done with $n = 4$ is the one with $r = 1$ in Section 3.4, and we can see that Eq. (3.13) is the special case of Eq. (4.8) with $n = 4$ and $r = 1$. This means that Theorem 3.14 is the special case of Theorem 4.7 for $n = 4$ and $r = 1$, giving the Thomae formulae since $n = 4$ is small enough.

4.3.4 N-Invariance for Even n

As in the odd n case, we now prove

Theorem 4.7. *The expression* $\theta^{2n^2}[\Delta](0, \Pi)/h_\Delta$ *with* h_Δ *from Eq. (4.8) is invariant under the action of the operators N and T_R.*

Proof. The proof of the invariance under N is exactly the same as in the odd n case. By Eq. (1.5) in its form $\theta[N(\Delta)](0, \Pi) = \theta[\Delta](0, \Pi)$ we only have to show that the expression for h_Δ in Eq. (4.8) is invariant under N (or simply that $h_{N(\Delta)} = h_\Delta$), and Definition 2.16 gives the same description of the action of the operator N, leaving C_{-1} fixed and interchanges C_j with $0 \le j \le n - 2$ by C_{n-2-j} (it did not depend on the parity of n). Again the product with $0 \le j \le n - 2 - k$ (excluding $j = -1$) for some $0 \le k \le n - 1$ in Eq. (4.8) is also invariant under N by the same considerations, and the product $\prod_{k=0}^{n-1}[C_{-1}, C_{k-1}]^{n[n^2/4 - k(n-k)]}$ is invariant under N since the expression $[C_{-1}, C_{-1}]$ is fixed by N and N takes $[C_{-1}, C_{k-1}]$ with $k \ge 1$ (up to the usual sign) to $[C_{-1}, C_{n-1-k}]$. The power here is also invariant under the action of taking k to $n - k$ and so is the set of numbers from 1 to $n - 1$. Hence also here the expression for h_Δ in Eq. (4.8) is invariant under N, or $h_{N(\Delta)} = h_\Delta$.

As for the invariance under the operators T_R, the important ingredient is again the PMT, Proposition 4.4, and we again use the reduced expression for g_Δ, which is now Eq. (4.4) since n is even. We again use

$$\frac{\theta^{n^2}[\Delta](0, \Pi)}{h_\Delta} = \frac{\theta^{n^2}[\Delta](0, \Pi)}{g_\Delta} \Bigg/ \frac{h_\Delta}{g_\Delta}$$

and the (reduced) PMT from Proposition 4.4 to show that it suffices to show that for any divisor Δ (which determines the sets C_j, $-1 \le j \le n - 2$) and any branch point

$R \in C_0$ we have the equality $h_\Delta/g_\Delta = h_{T_R(\Delta)}/g_{T_R(\Delta)}$ or equivalently the equality $h_\Delta/h_{T_R(\Delta)} = g_\Delta/g_{T_R(\Delta)}$. We split h_Δ again as

$$h_\Delta = s_\Delta t_\Delta, \quad s_\Delta = \prod_{k=0}^{n-3} \prod_{j=1}^{n-2-k} [C_j, C_{j+k}]^{n[n^2/4-k(n-k)]},$$

and

$$t_\Delta = \prod_{k=0}^{n-2} [C_0, C_k]^{n[n^2/4-k(n-k)]} \prod_{k=-1}^{n-2} [C_{-1}, C_k]^{n[n^2/4-(k+1)(n-1-k)]}$$

(where in the elements with $j = -1$ we have replaced k by $k+1$), and we again want to show that $s_\Delta = s_{T_R(\Delta)}$. This is done in the same way as in the odd n case, as we still have $\widetilde{C}_j = C_{n-1-j}$ for all $1 \le j \le n-2$. Therefore in the quotient $h_\Delta/h_{T_R(\Delta)} = s_\Delta t_\Delta/s_{T_R(\Delta)} t_{T_R(\Delta)}$ we can again cancel s_Δ with $s_{T_R(\Delta)}$, leaving $t_\Delta/t_{T_R(\Delta)}$, and it remains to show that $t_\Delta/t_{T_R(\Delta)} = g_\Delta/g_{T_R(\Delta)}$.

We want to use again the fact that the set $G = C_{-1} \cup C_0$ equals $\widetilde{G} = \widetilde{C_{-1}} \cup \widetilde{C_0}$, and for this we want to write the expression for t_Δ with the set G. Using again the fact that

$$(k+1)(n-1-k) = k(n-k) + (n-1-2k)$$

we again split the products over k into 3 parts: one with $k = -1$ and $k = 0$, one with $1 \le k \le (n-2)/2$, and one with $n/2 \le k \le n-2$ (with no missing value here). In the part with $k = -1$ and $k = 0$ we find that $[C_{-1}, C_{-1}]$, $[C_0, C_0]$ and $[C_{-1}, C_0]$ now appear all to a power at least $n(n^2 - 4n + 4)/4$, and we again use Lemma 4.2 to see that $[C_{-1}, C_{-1}][C_0, C_0][C_{-1}, C_0] = [G, G]$ (up to sign). For $1 \le k \le (n-2)/2$ the power to which $[C_{-1}, C_k]$ appears is smaller than the power to which $[C_0, C_k]$ appears (again that minus sign), while for $n/2 \le k \le n-2$ the power to which $[C_0, C_k]$ appears is smaller than the power to which $[C_{-1}, C_k]$ appears, and we use again the fact that $[C_{-1}, C_k][C_0, C_k] = [G, C_k]$ (up to the usual sign). t_Δ now becomes

$$[G, G]^{n(n^2-4n+4)/4} ([C_{-1}, C_{-1}][C_0, C_0])^{n(n-1)}$$

$$\times \prod_{k=1}^{(n-2)/2} [G, C_k]^{n[n^2/4-(k+1)(n-1-k)]} [C_0, C_k]^{n(n-1-2k)}$$

$$\times \prod_{k=n/2}^{n-2} [G, C_k]^{n[n^2/4-k(n-k)]} [C_{-1}, C_k]^{n(2k-(n-1))}.$$

We again use the fact that for every $1 \le j \le n-2$, the product $[G, C_j][G, C_{n-1-j}]$ is the same as the expression obtained from it by putting a tilde on all the sets and the fact that for every $n/2 \le k \le n-2$ we have $1 \le n-1-k \le (n-2)/2$ and the power to which $[G, C_{n-1-k}]$ appears is n times the complement with respect to $n^2/4$ of

$$[(n-1-k)+1][n-1-(n-1-k)] = k(n-k)$$

to obtain the new expression

$$[G,G]^{n(n^2-4n+4)/4} \prod_{k=n/2}^{n-2} ([G,C_k][G,C_{n-1-k}])^{n[n^2/4-k(n-k)]}$$

$$\times ([C_{-1},C_{-1}][C_0,C_0])^{n(n-1)} \prod_{k=1}^{(n-2)/2} [C_0,C_k]^{n(n-1-2k)} \prod_{k=n/2}^{n-2} [C_{-1},C_k]^{n(2k-(n-1))}$$

for t_Δ. The second line here is now the reduced expression for g_Δ in Eq. (4.4). Again by putting a tilde on all the sets we get the expression $t_{T_R(\Delta)}$.

Examining now the quotient $t_\Delta/t_{T_R(\Delta)}$ we see that the powers of $[G,G]$ and $[\widetilde{G},\widetilde{G}]$ cancel each other again, and the same assertion holds also for the expressions $[G,C_k][G,C_{n-1-k}]$ and $[\widetilde{G},\widetilde{C}_k][\widetilde{G},\widetilde{C}_{n-1-k}]$ for $n/2 \leq k \leq n-2$. After these cancelations the numerator becomes g_Δ and the denominator becomes the expression obtained from it by putting a tilde on all the sets (which is $g_{T_R(\Delta)}$). We conclude that $t_\Delta/t_{T_R(\Delta)} = g_\Delta/g_{T_R(\Delta)}$, which proves the theorem. $\qquad\square$

4.4 Thomae Formulae for Nonsingular Z_n Curves

We are now ready to prove the Thomae formulae for a general nonsingular Z_n curve with equation

$$w^n = \prod_{i=0}^{nr-1} (z-\lambda_i).$$

We separate the case $r = 1$ from the case $r \geq 2$, starting with the latter as it is simpler. However, since Theorem 4.8 below shows that the Thomae formulae is based on the denominators h_Δ from Eq. (4.7) when n is odd and from Eq. (4.8) when n is even, we precede its proof with a remark about this dependence on the parity. We can actually remove this dependence at the price of obtaining a slightly weaker, and not completely reduced, result. This is clear once one notices that both equations depend on k in exactly the same manner, and the only difference is (except for the value of e) the fact that in Eq. (4.8) appears n^2 while in Eq. (4.7) appears $n^2 - 1$. The "parity merging" is now done by taking the 4th power of the denominator in Eq. (4.7) and the 8th power of the denominator in Eq. (4.8), and then multiplying the denominator in Eq. (4.7) by

$$\left[\bigcup_{j=-1}^{n-2} C_j, \bigcup_{j=-1}^{n-2} C_j \right]^{2n} = \prod_{0 \leq i < j \leq rn-1} (\lambda_i - \lambda_j)^{2n}.$$

This gives that the quotient $\theta^{16n^2}[\Delta](0,\Pi)/h_\Delta$ is invariant under the operators N and T_R (and hence is the Thomae formulae by Theorems 4.8 for $r \geq 2$ and 4.10 for $r = 1$) where

$$h_\Delta = \prod_{k=0}^{n-1} \prod_{j=-1}^{n-2-k} [C_j, C_{j+k}]^{2n(n-2k)^2}.$$

However, since we prefer our results to be the strongest possible and reduced, we shall stick with Eqs. (4.7) and (4.8), even though they depend on the parity of n.

4.4.1 The Case $r \geq 2$

As we saw in the example in Section 3.1, the Thomae formulae is the expression $\theta^{2en^2}[\Delta](0,\Pi)/h_\Delta$ which is invariant under the operators N and T_R, and this is because one can get from any divisor to any other divisor using a finite sequence of applications of the operators N and T_R. We now show that when $r \geq 2$ (and here is the place where it becomes important), this is the case for a general nonsingular Z_n curve, a statement expressed in the following

Theorem 4.8. *If $r \geq 2$, then the expression $\theta^{2en^2}[\Delta](0,\Pi)/h_\Delta$ with h_Δ from Eq. (4.7) when n is odd and from Eq. (4.8) when n is even is independent of the divisor Δ.*

Proof. By Theorems 4.6 and 4.7, this expression is invariant under the operators N and T_R. Therefore the proof will be complete once we show that one can get from any divisor to any other divisor using the operators N and T_R. Now, since by Theorems 2.6 and 2.9 all the non-special integral divisors of degree g supported on the branch points distinct from P_0 have the same form and every divisor of this form is non-special, it is clear that every divisor can be obtained from any given divisor by applying a permutation on the set of points $\{P_i\}_{i=1}^{nr-1}$, and every such permutation sends the given divisor to a divisor of the sort interesting us. Since the group S_{nr-1} of permutations is generated by transpositions, it will suffice to show that the action of any transposition on any divisor can be obtained by a finite sequence of operations of the operators N and T_R. This paragraph also shows that the set of divisors is a homogenous space for the group S_{nr-1}, which is not principal since for $r \geq 2$ some permutations can fix a given divisor—this is why a presentation like we did with $r = 1$ in Section 3.4 will not hold here.

For this let us denote by τ_{ij} the transposition that interchanges P_i and P_j, and let Δ be a divisor. We want to show that no matter in which sets C_k and C_l the points P_i and P_j lie, we can operate on Δ by a finite sequence of actions of N and T_R such that this action will give the action of τ_{ij}, i.e., will interchange the places of P_i and P_j and leave everything else fixed. It is clear that if P_i and P_j lie in the same set C_k, then the trivial action does that, hence we assume that they lie in different sets. Now, it suffices to assume that P_i lies in C_{-1}, since we claim that if we can interchange every point with a point in C_{-1}, then we can interchange any two points. The argument for this is as follows: If one point is in C_{-1} there is nothing to do. Otherwise for two points P_i and P_j in two sets C_k and C_l with k and l distinct and none of them being

-1, we can take a point P_h from C_{-1} and use the fact that $\tau_{ij} = \tau_{jh}\tau_{ij}\tau_{ih}$. We have replaced τ_{ij} with three transpositions, one of them being τ_{ij} itself, but now we see that each transposition involves an element that lies in the set C_{-1} of the divisor on which it acts: in the beginning we know it for P_h, after applying τ_{ih} it holds for P_i, and then after applying τ_{ij} it holds for P_j. Therefore it suffices to show that we can interchange a point from C_{-1} with any other point. We note that this is the place where we are using the assumption $r \geq 2$: if $r = 1$ the set C_{-1} is composed of P_0 alone, which cannot be replaced, so the existence of P_h assumes $r \geq 2$. This is the only place where we will use this, and Proposition 3.18 showed that this is indeed crucial; without it this claim is simply not true, and neither is the statement that one can get from every divisor to every other divisor using only N and T_R.

Hence assume that Δ is a given divisor (to which all the following sets will refer), that P_i is in C_{-1} and that P_j is in some other set C_k with $0 \leq k \leq n - 2$. We would like to interchange P_i and P_j, and leave everything else fixed. For this we examine the action of a few sequences of operators. We start by assuming that $R \in C_0$, and noticing that applying T_P for some $P \in C_{-1}$ distinct from P_0 to the divisor $T_R(\Delta)$, brings us back to Δ, but with R and P interchanged. In order to see this, we use the following observations. First, $\widetilde{C}_j = C_{n-1-j}$ for any $1 \leq j \leq n - 2$, and applying this twice leaves all these sets invariant. Second, every point of C_{-1} or C_0, which is not R, P or P_0, is taken by T_R to the other set (C_{-1} or C_0), and then the action of T_P takes it back to its original position. P_0 is fixed by every operator. Finally, since R is left fixed by T_R and is taken to C_{-1} by T_P, and P is taken to C_0 by T_R and is left fixed by T_P, we conclude that the action of $T_P T_R$ is as stated. This in particular solves the problem when $k = 0$, i.e., when we want to interchange P_i with a point P_j from C_0. Furthermore, since N interchanges C_0 and C_{n-2} we know that $NT_{P_i}T_{P_j}N$ interchanges P_i and P_j if P_j is in C_{n-2}. Indeed, N takes C_{n-2} to C_0 and fixes C_{-1}, which allows $T_{P_i}T_{P_j}$ to act as this transposition on $N(\Delta)$, and since N is an involution, applying it again would return everything to its place except for the interchanged P_i and P_j. This solves the problem for $k = n - 2$.

We now introduce another sequence of operators that is useful for our purposes. We know that $T_P T_R$ interchanges the places of a point R from C_0 and a point P from C_{-1} (or A), and since $|A| = r - 1$ and $|C_0| = r$ we can fix a point $S \in C_0$, and then apply $r - 1$ pairs of the sort $T_P T_R$ that will interchange all the $r - 1$ points from A and the $r - 1$ points of $C_0 \setminus S$. This will, of course, leave all the sets C_l with $1 \leq l \leq n - 2$ fixed, and now T_S will again interchange A and $C_0 \setminus S$, but will now in addition interchange C_l and C_{n-1-l} for every $1 \leq l \leq n - 2$. Let us denote this operator by W. Since we have seen that the action of W leaves C_{-1} and C_0 fixed and interchanges C_l and C_{n-1-l} for every $1 \leq l \leq n - 2$, we find that W is an involution and that it is independent of the choice of the point S we have chosen in order to define it. It is interesting to note that in the excluded case $r = 1$, this operator W is simply T_R with R being the unique point in the set C_0 (of cardinality $r = 1$), which was denoted simply by T in Section 3.4.

Let us return to the construction of the action of the transposition τ_{ij} where P_i is from C_{-1} and P_j is from some other set C_k (with k not being any of -1, 0 or $n - 2$ with which we already dealt). We now see that if $k = 1$, i.e., the point P_j is in C_1, then the sequence $WNT_{P_i}T_{P_j}NW$ does the desired interchange. Indeed, W takes C_1 (and in particular P_j) to C_{n-2} and fixes C_{-1}, $NT_{P_i}T_{P_j}N$ interchanges P_j with P_i, and then the involution W returns everything to its place except for the interchanged P_i and P_j. This solves the problem for $k = 1$. It is quite clear how to continue from here. Since N interchanges any set C_l, $0 \leq l \leq n-2$, with C_{n-2-l}, and W interchanges C_l, $1 \leq l \leq n-2$, with C_{n-1-l}, we see that if we write the sets C_l, , $-1 \leq l \leq n-2$, in the ordering

$$C_{-1}, C_0, C_{n-2}, C_1, C_{n-3}, C_2, C_{n-4}, \ldots \qquad (4.9)$$

(ending with $C_{(n-2)/2}$ when n is even and with $C_{(n-3)/2}, C_{(n-1)/2}$ when n is odd), then both the operators N and W (considering W as a single operator for this discussion) fix C_{-1} and interchange pairs of neighbors. Actually, W fixes also C_0, and if n is even, then N fixes the rightmost set $C_{(n-2)/2}$, while if n is odd, then W fixes the rightmost set $C_{(n-1)/2}$. However, for a generic set C_l (which is not C_{-1}, C_0 or the rightmost set), the action of N interchanges it with one of its "neighbors" in the ordering (4.9) and the action of W interchanges it with its other "neighbor" there. Therefore we can work inductively: Assume that P_j is in C_k, that the "left neighbor" of C_k in the ordering (4.9) is C_h for some h, and that we can interchange P_i from C_{-1} with a branch point from C_h. Then apply the operator V which interchanges C_k and C_h (V is either N or W, according to the specific situation of C_k and C_h), and since after that P_j will be moved to C_h (and P_i will remain fixed in C_{-1} as both N and W fix C_{-1}) we can then interchange P_i with P_j. Then apply the involution V again, which will return everything to its place except for the interchanged P_i and P_j, as desired. Since we already know that we can switch elements from C_{-1} with elements from C_0, this completes the proof of the Thomae formulae for nonsingular curves. Note that the way to replace elements from C_{-1} with elements from C_{n-2} or C_1 was obtained exactly using this induction step from the basic one with $k = 0$. $\square$

There is another way to formulate this proof, without permutations. This other formulation is based on the action of the operators on the divisors themselves, and not on operations on the branch points. It will also use this specific ordering (4.9) of the sets C_j, $-1 \leq j \leq n-2$, and will use a similar induction argument. Since we will give several proofs based on induction of this sort, let us introduce some notation which will make it more convenient. Let us write, for two numbers k and l between -1 and $n-2$, the symbol $k \prec l$ to denote the assertion "the set C_k appears to the left of the set C_l in the order (4.9)" (not including $k = l$, of course), and the symbol $k \preceq l$ to denote the assertion "either $k \prec l$ or $k = l$". We shall also use the reverse notation: $l \succ k$ is the same as $k \prec l$ and $l \succeq k$ is the same as $k \preceq l$. Thus the "left neighbor" of the set C_l in the ordering (4.9) is the set C_k with the "maximal" k (with respect to the ordering (4.9)) satisfying $k \prec l$ and the "right neighbor" of the set C_k in the ordering (4.9) is the set C_l with the "minimal" l (again with respect to the ordering (4.9)) satisfying $l \succ k$. However, we shall stick to the expressions "left neighbor" and "right neighbor" to state these assertions, and we may say that "l is a neighbor of k" to mean that "C_l is a neighbor of C_k". To show how we use this notation, we state that the induction step for the stage corresponding to k (or C_k) in the proof we now give for Theorem 4.8 is the following claim: Assume that

Δ and Ξ are two divisors that are distinct only in the sets C_j with $j \preceq k$. Assume that one can get, using the operators N and T_R, from any divisor to any other divisor which is distinct from it only in sets C_j with $j \prec k$ (not including C_k itself—this is the induction hypothesis). Then one can get from Δ to Ξ using N and T_R.

The beginning of this proof is the same: Using Theorems 4.6 and 4.7, we find that it suffices to show that one can get from any divisor to any other divisor using the operators N and T_R. If $k = -1$, then Δ and Ξ can be distinct only on the set C_{-1}, but since C_{-1} is the complement of the union of all the others, we must have $\Delta = \Xi$ then and there is nothing to prove. If $k = 0$, then Ξ is obtained from Δ by mixing C_{-1} and C_0, but with the cardinalities unchanged. This means that Ξ can be obtained from Δ using a finite number of interchanges of points from C_{-1} with points from C_0; we already saw that such an interchange of points P from C_{-1} and R from C_0 is obtained by the pair $T_P T_R$. This establishes the basis for the induction.

It turns out that in this proof the first induction step, for proving the claim for $k = n - 2$, is the trickiest one, and is in fact the one and only step which requires the assumption $r \geq 2$. We do assume in each induction step that $(C_k)_\Delta \neq (C_k)_\Xi$ (otherwise the induction hypothesis finishes immediately), but now for the first induction step we assume that $(C_k)_\Delta$ and $(C_k)_\Xi$ has nonempty intersection (this is the stage which assumes $r \geq 2$, as we want distinct sets of cardinality r with nonempty intersection, a situation that cannot happen if $r = 1$). Let U be the union $(C_{-1})_\Delta \cup (C_0)_\Delta \cup (C_{n-2})_\Delta$, which equals the union $(C_{-1})_\Xi \cup (C_0)_\Xi \cup (C_{n-2})_\Xi$ (as all the other sets C_j, $1 \leq j \leq n - 3$, or more suitably to here $j \succ n - 2$, are the same for Δ and Ξ), which is a set of cardinality $3r$. Now, since $(C_k)_\Delta$ and $(C_k)_\Xi$ has nonempty intersection we find that the cardinality of $U \setminus (C_{n-2})_\Delta \setminus (C_{n-2})_\Xi \setminus P_0$ is at least r. Indeed, we have taken out from a set of cardinality $3r$ one point and the union of two sets of cardinality r each, together $2r + 1$ points, but the nonempty intersection assures us that at least one point was counted twice in this calculation. This means that we have taken out at most $2r$ points, leaving at least r points. Therefore take a subset Y of $U \setminus (C_{n-2})_\Delta \setminus (C_{n-2})_\Xi \setminus P_0$ of cardinality r, and choose a divisor Γ with sets satisfying $(C_{n-2})_\Gamma = (C_{n-2})_\Delta$, $(C_0)_\Gamma = Y$, and $(C_j)_\Gamma = (C_j)_\Delta = (C_j)_\Xi$ for every j distinct from -1, 0 and $n - 2$ (or equivalently $j \succ n - 2$). Note that Γ is determined uniquely by our choice of Y, since it remains only to determine $(C_{-1})_\Gamma$ and it is determined as the complement of the union of the other sets, but this fact is not important to us. By the induction hypothesis we can get from Δ to Γ using the operators N and T_R, and therefore also to $N(\Gamma)$ by another single operation of N.

Now, by Definition 2.16 for N we find that $(C_{n-2})_{N(\Gamma)} = Y$, and every for j distinct from -1, 0 and $n - 2$ we have $(C_j)_{N(\Gamma)} = (C_{n-2-j})_\Gamma$, which equals also $(C_{n-2-j})_\Delta$ and $(C_{n-2-j})_\Xi$. In particular the elements of $(C_{n-2})_\Xi$ are contained in the union of $(C_0)_{N(\Gamma)}$ and $(C_{-1})_{N(\Gamma)}$, as they do not belong to any of the other sets. Now choose a divisor Υ satisfying $(C_{n-2})_\Upsilon = Y = (C_{n-2})_{N(\Gamma)}$, $(C_0)_\Upsilon = (C_{n-2})_\Xi$ and $(C_j)_\Upsilon = (C_j)_{N(\Gamma)}$ for j distinct from -1, 0 and $n - 2$. This Υ is again unique, but this will not serve us as well. Using the induction hypothesis again we find that we can get from $N(\Gamma)$ (and hence also from Δ) to Υ using the operators N and T_R, and another application of N takes us to $N(\Upsilon)$. Now, Definition 2.16 for N gives that

$(C_{n-2})_{N(\Upsilon)} = (C_0)_\Upsilon$ and hence equals $(C_{n-2})_\Xi$, and also that $(C_j)_{N(\Upsilon)} = (C_{n-2-j})_\Upsilon$ for all j distinct from -1, 0 and $n-2$. However, this equals $(C_{n-2-j})_{N(\Gamma)} = (C_j)_\Gamma$, hence $(C_j)_\Xi$ (and also $(C_j)_\Delta$, but this does not matter to us), so we can apply the induction hypothesis yet another time and find that we can get from $N(\Upsilon)$, and hence from Δ, to Ξ as desired.

All this, however, only proves the first induction step and under the assumption that $(C_{n-2})_\Delta$ and $(C_{n-2})_\Xi$ intersect nontrivially. However, we can now complete the first induction step easily: Assume that the intersection of $(C_{n-2})_\Delta$ and $(C_{n-2})_\Xi$ is empty. Then, since $r \geq 2$, we can take a subset Z of cardinality r of the set U (without P_0) which intersects both $(C_{n-2})_\Delta$ and $(C_{n-2})_\Xi$ (take one element from $(C_{n-2})_\Delta$ and one element from $(C_{n-2})_\Xi$, and complete to r elements), and then choose some divisor Σ such that $(C_{n-2})_\Sigma = Z$ and such that $(C_j)_\Sigma = (C_j)_\Delta = (C_j)_\Xi$ for every j distinct from -1, 0 and $n-2$. Then by what we have just proved one can get from Δ to Σ and from Σ to Ξ using the operators N and T_R (as $(C_{n-2})_\Delta$ and $(C_{n-2})_\Xi$ have nonempty intersection and $(C_{n-2})_\Sigma$ and $(C_{n-2})_\Xi$ have nonempty intersection), hence one can get from Δ to Ξ using the operators N and T_R and the first induction step is done. As for the other induction steps, they are done in a similar manner, but with less complications. Since at the stage where we deal with C_k with $k \neq -1, 0, n-2$ the set U (which is now the union $\bigcup_{j \prec k}(C_j)_\Delta$ and is the same as $\bigcup_{j \prec k}(C_j)_\Xi$) has now cardinality at least $4r$, we do not need the nonempty intersection condition in order to be able to take the needed set Y. Now we define Γ and Υ as above, but we notice that instead of N we must use the operator V which interchanges C_k with its left neighbor (V can still be N, but can also be the operator W defined in the first proof, and in the latter case W can be replaced by any operator T_R, as at this part of the proof we do not care about the internal partition between C_{-1} and C_0). If we denote that neighbor by C_h, then now the general induction step goes word for word as the first one with the nonempty intersection condition, except that every N is replaced by V, every C_{n-2} is replaced by C_k, every C_0 is replaced by C_h, every "j distinct from -1, 0 and $n-2$" is replaced by "$j \succ k$" (hence the equivalent $j \succ n-2$ may be better there), and the changes of the sets C_j, $j \succ k$ are done according to V instead of N. This induction clearly establishes another proof of Theorem 4.8.

The reason we make this point here is that the reader may at this stage already guess that permutations will not help us in the singular case, and thus the proof of this part for the singular case will be along the lines of the suggested second proof here. Second, both proofs are based (only at some initial, but critical stage) on the assumption that $r \geq 2$. As we already saw in Proposition 3.18, this condition is indeed necessary for this part (except for $n \leq 4$), but Theorems 4.6 and 4.7 hold for $r = 1$ as well. As already said, they also give Thomae formulae, but the proof is more complicated and needs the basepoint change, as we show at the end of this section.

4.4.2 Changing the Basepoint for $r \geq 2$

First we want to complete the case $r \geq 2$ by presenting the basepoint change and combining the results for the different basepoints. The Thomae formulae obtained from Theorems 4.6, 4.7 and 4.8, together with the nice expressions appearing in Eqs. (4.7) and (4.8), are indeed beautiful and symmetric, but as we saw in the examples, they include some dependence on the choice of the basepoint P_0. This means, in principal, that the value of the quotient $\theta[\varphi_{P_i}(\Delta) + K_{P_i}](0, \Pi)/h_\Delta$ (using full notation to emphasize the basepoint P_i) with h_Δ defined by Eqs. (4.7) and (4.8) (but recall that now C_{-1} is $A \cup P_i$ rather than $A \cup P_0$), which is independent of the divisor Δ (whose support now must not contain P_i rather than not contain P_0) by Theorems 4.6, 4.7 and 4.8 (which hold equally for any choice of basepoint P_i, not only P_0, with the necessary little changes), may depend on the basepoint P_i.

Since $r \geq 2$ we can use Corollary 1.13 to connect the values obtained by the different basepoints and obtain the symmetric Thomae formulae. Lemma 2.4 shows that the branch points satisfy the condition that $\varphi_{P_i}(P_j^n) = 0$ for every i and j. Since all the divisors we have are integral, of degree g, supported on the branch points (all of them—P_0 is no longer excluded), contain no branch point to the power n or higher, and contain some points to the power $n - 1$ (A is of cardinality $r - 1 > 0$ and is nonempty) we can apply Corollary 1.13. We begin with our divisors and basepoint P_0, fix another basepoint P_i with some $1 \leq i \leq nr - 1$, and we want to show that the quotients obtained using the basepoint P_i give the same constant as the quotients with the basepoint P_0. For this choose a divisor Δ (with basepoint P_0) such that P_i is in C_{-1} (or A—this is possible since $r \geq 2$), and then the divisor Γ_{P_i} from Corollary 1.13 is non-special by Theorem 2.6 and has the same sets C_j for $0 \leq j \leq n - 2$. As we saw in the example in Section 3.1, the set A is replaced by $A \setminus P_i \cup P_0 = C_{-1} \setminus P_i$ (which is not the same as A_Δ), but since for Γ_{P_i} the set C_{-1} is $A \cup P_i$ we do get the same C_{-1} as well. This means that Eqs. (4.7) and (4.8) will give the same expression for $h_{\Gamma_{P_i}}$ as for h_Δ. Since Corollary 1.13 shows that $\varphi_{P_i}(\Gamma_{P_i}) + K_{P_i} = \varphi_{P_0}(\Delta) + K_{P_0}$, we find that the constant quotient obtained from the basepoint P_i (represented by the divisor Γ_{P_i}) is the same as the one from P_0 (represented by the divisor Δ) as the representatives have the same numerator and denominator.

We now show, as we did in all the examples, that not only the constant quotients are the same for every basepoint, but so are the sets of characteristics, and thus so are the theta constants one obtains when going over all the divisors Theorems 2.6 and 2.9 give for any choice of basepoint. There are, as one can calculate using simple combinatorial considerations, exactly $(rn - 1)!/(r!)^{n-1}(r - 1)!$ divisors for any basepoint, and hence this is the number of the characteristics we obtain (for any basepoint, and from what we now show, for the full Thomae formulae as well) since we already saw that different non-special divisors give, with a fixed basepoint, different characteristics. We shall prove this, as usual, using Proposition 1.14, which is applicable by Lemma 2.5 and the fact that our divisors satisfy the assumptions of Proposition 1.14. For any characteristic $\varphi_{P_0}(\Delta) + K_{P_0}$ obtained from the basepoint P_0 and any $1 \leq i \leq rn - 1$, Proposition 1.14 gives that we have a divisor Γ_{P_i} such that $\varphi_{P_i}(\Gamma_{P_i}) + K_{P_i}$ gives the same characteristic, and since Γ_{P_i} is of the same form as Δ, Theorem 2.6 gives that Γ_{P_i} is non-special as well and with P_i not appearing in its support. The converse follows either by an argument of symmetry or by the fact that the maps $\varphi_{P_i}(\cdot) + K_{P_i}$ are one-to-one on the relevant sets of divisors and the fact that there is the same number of divisors Γ with the basepoint P_i as divisors Δ with the basepoint P_0. Therefore the Thomae formulae from Theorems 4.6 and 4.7 (using Theorem 4.8) with the rn distinct branch points P_i, $0 \leq i \leq rn - 1$, not only give the same constant, but involve the same theta constants when $r \geq 2$. This will also be true for $r = 1$, but with another proof, as we shall soon see.

As in the examples, we would like now to have new notation for the Thomae formulae from Theorems 4.6 and 4.7, notation that is independent of the choice of the basepoint. As always, for this we use the fact that if P_i appears in Δ to the power $n - 1$, then $P_0^{n-1}\Delta = P_i^{n-1}\Gamma_{P_i}$, and if we denote this divisor by Ξ (and hence $\Xi = \prod_{j=-1}^{n-2} C_j^{n-2-j}$, starting with $j = -1$) we can replace the notation $\theta[\Delta]$ with basepoint P_0 by $\theta[\Xi]$, which is the theta function with characteristic $\varphi_{P_t}(\Xi) + K_{P_t}$ for some $0 \leq t \leq nr - 1$. This does not depend on the choice of P_t by the second statement in Corollary 1.13. We also denote the denominators from Eqs. (4.7) and (4.8) by h_Ξ (this is again a more suitable correspondence since both Ξ and h_Ξ are expressed using C_{-1} and not A, as seen in the expressions for them), and this proves

Theorem 4.9. *If $r \geq 2$, then for every divisor $\Xi = \prod_{j=-1}^{n-2} C_j^{n-2-j}$ where the sets C_j, $-1 \leq j \leq n - 2$, are disjoint and contain r branch points each (hence form a partition of the set of nr branch points into n sets of cardinality r) define $\theta[\Xi]$ to be the theta function with characteristic $\varphi_{P_t}(\Xi) + K_{P_t}$ for some $0 \leq t \leq rn - 1$ (which is independent of the choice of P_t), and define h_Ξ to be the expression from Eq. (4.7) if n is odd and Eq. (4.8) if n is even. Then the quotient $\theta^{2en^2}[\Xi](0, \Pi)/h_\Xi$ is independent of the divisor Ξ.*

Theorem 4.9 is the most "symmetric" Thomae formulae for nonsingular Z_n curves with $r \geq 2$. As in the examples, unlike Theorems 4.6 and 4.7, in which every divisor gives a different characteristic, Theorem 4.9 yields $(rn)!/(r!)^n$ divisors, which split into $(rn-1)!/(r!)^{n-1}(r-1)!$ sets of size n that are the orbits of the action of the operator M defined after Proposition 1.14. These orbits give the $(rn-1)!/(r!)^{n-1}(r-1)!$ characteristics. It contains, of course, Theorem 3.3 from Section 3.1 as its special case with $n = 3$ and $r = 2$. We emphasize that the easy basepoint change using Corollary 1.13 also used the fact that $r \geq 2$.

4.4.3 The Case $r = 1$

We now turn to the case $r = 1$. We recall that in this case we have considered in Section 3.4 the divisors with basepoint P_0 (and equivalently the divisors with any basepoint) as a principal homogenous space of the group S_{n-1}, where each divisor Δ is of the form $\prod_{j=1}^{n-1} P_{i_j}^{n-1-j}$ with i_j, $1 \leq j \leq n-1$ being the numbers from 1 to $n-1$

in some order and $\sigma \in S_{n-1}$ takes this divisor Δ to $\prod_{j=1}^{n-1} P_{i_{\sigma^{-1}(j)}}^{n-1-j}$. Then the operator N and the operator T (which acts on the divisor Δ as the only possible operator T_R, with R the point in C_0, a set whose cardinality is $r = 1$) act simply like the elements

$$\begin{pmatrix} 1 & 2 & \dots & n-2 & n-1 \\ n-1 & n-2 & \dots & 2 & 1 \end{pmatrix} \text{ and } \begin{pmatrix} 1 & 2 & \dots & n-2 & n-1 \\ 1 & n & 1 & \dots & 3 & 2 \end{pmatrix} \text{ of } S_{n-1} \text{ respectively.}$$

We have defined the subgroup $G = \langle N, T \rangle$ of S_{n-1}, which, when $n = 3$, is of order 2 (hence all of $S_{n-1} = S_2$) and when $n \geq 4$ is of order $2(n-1)$ by Proposition 3.18, and this gave us the Thomae formulae in Theorem 3.16 for $n = 3$ and the division of the set of divisors into the $(n-2)!/2$ orbits of G on which the quotient from Eqs. (4.7) and (4.8) are constant by Theorems 4.6 and 4.7 respectively for $n \geq 4$ (in these theorems we have not assumed that $r \geq 2$). In particular we did (apart from the case $n = 3$) the cases $n = 4$ (where G was all of $S_{n-1} = S_3$ as $(n-2)!/2 = 1$ there and the corresponding Theorem 3.14 was already the full Thomae formulae) and $n = 5$ (which gave us Theorem 3.19, and the Thomae formulae in Theorem 3.20 required extra work). We remark again that the calculation of orbits does not hold for $n = 3$ (as Proposition 3.18 holds for $n \geq 4$ only and in any case $(n-2)!/2 = 1/2$ and is not integral for $n = 3$), but the Thomae formulae of Theorem 3.16 is still the special case with $n = 3$ here, as the proof itself does not use Proposition 3.18 or the assumption $n \geq 4$.

Now, when letting the basepoint vary, we obtain $n!$ quotients $((n-1)!$ for every choice of basepoint) that are divided into $n(n-2)!/2$ values, and these values are divided according to the choice of the basepoint. We want to show two things: first, that for $n \geq 4$ every basepoint gives the same $(n-2)!/2$ values, and second, that the basepoint change allows us to compare all these values and show that they are equal (the latter is trivial for $n = 4$ as we already saw, but the proof does not need the assumption $n \geq 5$ so we do not assume it—even the assumption $n \geq 4$ will not be used and hence it will not be made).

The tool for proving the two statements (in any order) is the basepoint change. As the statements allowing us to do the basepoint change (Corollary 1.13, Proposition 1.14, and also the one less relevant to this case, Proposition 1.15) all give that the more convenient way to look at this process and understand it is in the language of the divisors $\Xi = P_0^{n-1} \Delta$, we shall examine the situation in this language now. These divisors Ξ are of the form $\prod_{j=-1}^{n-2} C_j^{n-2-j}$ (as in the general nonsingular setting), and since here the cardinalities of all the sets must be $r = 1$, we can write each such Ξ as $P_0^{n-1} \prod_{j=1}^{n-1} P_{i_j}^{n-1-j}$ (as it comes from the basepoint P_0). If Ξ was obtained from a divisor with another basepoint P_k, then Ξ has the form $P_k^{n-1} \prod_{j=1}^{n-1} P_{i_j}^{n-1-j}$ where i_j, $1 \leq j \leq n-1$ are now all the numbers from 0 to $n-1$ except for k, and in general we can write each such Ξ as $\prod_{j=0}^{n-1} P_{i_j}^{n-1-j}$ (starting with $j = 0$). It is good to notice that for each such Ξ it is clear which is the basepoint to which the divisor yielding Ξ belonged—it has to be P_{i_0}, the unique point in the set C_{-1} (of cardinality $r = 1$). We emphasize again that this was not the case when $r \geq 2$.

4.4.4 Changing the Basepoint for $r = 1$

Let us now examine the action of basepoint change. As we saw in the examples in Section 3.4, even though Lemma 2.4 allows us to apply Corollary 1.13 we cannot really use it since the set A is empty for any of our divisors. However, Lemma 2.5 (together with the form of the divisors Δ from Theorems 2.6 and 2.9) gives us the other assumption in Proposition 1.14, and this will be the tool for basepoint change here. We recall from the remarks following Proposition 1.14 that every basepoint change is a combination of Corollary 1.13 with some power of M, and in this case as Corollary 1.13 does nothing we are left only with powers of M (which is known to be of order n). We recall that the operator M was defined only on the divisors $\varXi$, but since here we saw that we have a one-to-one correspondence between the divisors $\varXi$ and the union of divisors with all basepoints (once one knows the basepoint, which is P_{i_0} here, it is easy to find the divisor—simply divide $\varXi$ by the $(n-1)$th power of the basepoint), we can treat M and its powers as acting on the divisors Δ as well (but with the basepoint varying). We recall that the power of M needed to change the basepoint of the divisor Δ from P_0 to some point P_l is $k+1$, where P_l is in C_k (which is equivalent to saying that $l = i_{k-1}$), and that M acts as a "rotation" on the sets C_j, $-1 \leq j \leq n-2$ (as in the case $r \geq 2$). Now, Proposition 1.14 shows that $\varphi_{P_l}(\Gamma_{P_l}) + K_{P_l} = \varphi_{P_0}(\Delta) + K_{P_0}$, where Γ_{P_l} is obtained by the proper power of M (and is again non-special by Theorem 2.6), and we also know that as in the case $r \geq 2$, we have $h_{\Gamma_{P_l}} = h_\Delta$ for the denominators defined in Eqs. (4.7) and (4.8)—the proof there did not use the assumption that $r \geq 2$ and hence holds equally for $r = 1$. Therefore we have also in this case that the value of the quotient corresponding to Γ_{P_l} with the basepoint P_l equals the value of the quotient corresponding to Δ with the basepoint P_0.

As already mentioned, basepoint changes are more nicely formulated in the language of the divisors $\varXi$, and such a divisor here is $\prod_{j=0}^{n-1} P_{i_j}^{n-1-j}$. As we did in Section 3.4, we can treat the set of these divisors $\varXi$ as a principal homogenous space for the group S_n (but acting on the set of numbers between 0 and $n-1$ rather than between 1 and n). In this setting we look at S_{n-1}, one of whose orbits is the set of divisors $\varXi = P_0^{n-1}\Delta$, simply as the stabilizer of the index 0 (this is consistent with the fact that S_n acts on the set of numbers from 0 to $n-1$ and S_{n-1} acts on the set of numbers from 1 to $n-1$). As mentioned above, the set of $n!$ divisors $\varXi$ is the disjoint union of n sets of cardinality $(n-1)!$, each being the set of divisors with given basepoint. In the new formulation these are simply the orbits of S_{n-1}. It is clear that in this new setting we can look at N as the element $\begin{pmatrix} 0 & 1 & 2 & \dots & n-2 & n-1 \\ 0 & n-1 & n-2 & \dots & 2 & 1 \end{pmatrix}$ of $S_{n-1} \subseteq S_n$ and at T as the element $\begin{pmatrix} 0 & 1 & 2 & \dots & n-2 & n-1 \\ 0 & 1 & n-1 & \dots & 3 & 2 \end{pmatrix}$ of $S_{n-1} \subseteq S_n$ (which means that we now look at G as a subgroup of S_n). Since the explicit action of M is to take $\prod_{j=0}^{n-1} P_{i_j}^{n-1-j}$ to $\left(\prod_{j=1}^{n-1} P_{i_j}^{n-j}\right) P_{i_0}^0$, or more conveniently written to $\left(\prod_{j=0}^{n-2} P_{i_{j+1}}^{n-1-j}\right) P_{i_0}^0$ (replace the index j by $j-1$), we find that the action of M is the same as that of the element $\begin{pmatrix} 0 & 1 & 2 & \dots & n-2 & n-1 \\ n-1 & 0 & 1 & \dots & n-3 & n-2 \end{pmatrix}$ (note the σ^{-1} in the definition of the action!) of S_n. This demonstrates our old claim that applying the basepoint change operator M successively n times returns us to the initial divisor, as clearly this element is of order n. One should note that as a basepoint change operator, M takes Δ to the divisor $\Gamma_{P_{i_1}}$ with basepoint P_{i_1}, and it is important to understand that it is not the basepoint change operator from the point P_0 to some other fixed point P_l, but the basepoint to which it changes depends on the divisor on which it acts. This relates to the observation that the operation called M after Proposition 1.14 is, in this case, exactly this M, as is easily seen from the fact that applying it to Ξ increases the powers to which every point P_{i_j} with $1 \le j \le n-1$ (which is the same as every set C_j, $0 \le j \le n-2$) appears by 1 and decreases the power to which the point P_{i_0} (i.e., the set C_{-1}) appears by $n-1$. As mentioned above, the base change operator from P_0 to some other point P_l (which is the operation described in Proposition 1.14, and also in Proposition 1.15 for the singular case) acts also in this case as different powers of M on different divisors.

Let us get back to showing that all the divisors with basepoint P_0 give the same quotient, i.e., to proving the Thomae formulae. Theorems 4.6 and 4.7 give that the quotient is constant on every G-orbit, but since the action of M also preserves it, we can add to G more elements. We mean that if we change the basepoint to some other branch point, act with some element of G, and then change the basepoint back to P_0 then the divisor thus obtained gives us the same quotient as Δ, and thus the quotient is constant on the orbits of the group H generated by G and all these operations. Our task now is to prove that this group H is all of S_{n-1}. Therefore let $\Delta = \prod_{j=1}^{n-1} P_{i_j}^{n-1-j}$ be a divisor with basepoint P_0, and we examine the action of certain elements of H on Δ to get an idea of what H looks like.

At first one might think that H is not larger than G and nothing was added. This is because the operator M changes the basepoint of Δ to P_{i_1}, and if instead of Δ we apply M to $T(\Delta)$ (still changing to P_{i_1}) we obtain the image of the action of the element NTN of G on $M(\Delta)$, giving the relation $MTM^{-1} = NTN$. Similarly, if we want to change the basepoint of $N(\Delta)$ to P_{i_1} (hence applying now M^{-1} since P_{i_1} does not appear in $N(\Delta)$) then we obtain the image of the action of the element N of G on $M(\Delta)$, giving the relation $MNM = N$. One can verify these relations with the permutations, and calculate that both MTM^{-1} and NTN are the element $\begin{pmatrix} 0 & 1 & 2 & \dots & n-2 & n-1 \\ 0 & n-2 & n-3 & \dots & 1 & n-1 \end{pmatrix}$ in $S_{n-1} \subseteq S_n$ and that MNM gives the element $\begin{pmatrix} 0 & 1 & 2 & \dots & n-2 & n-1 \\ 0 & n-1 & n-2 & \dots & 2 & 1 \end{pmatrix}$ which equals again N in $S_{n-1} \subseteq S_n$. As N is an involution, these relations translate to the claim that the element NM in S_n is of order 2 and commutes with T, as we later show. However, let us apply M to the divisor Δ, and act with the element NT of G. The divisor $M(\Delta)$ is $\left(\prod_{j=2}^{n-1} P_{i_j}^{n-j}\right) P_0^0$ with basepoint P_{i_1}, and then (either by Definition 2.16 or using the permutations) we find that the image of this divisor by NT is $\left(\prod_{j=3}^{n-1} P_{i_j}^{n+1-j}\right) P_0^1 P_{i_2}^0$ (still with basepoint P_{i_1}). In order to return the basepoint of this divisor to P_0 we have to apply $M^{-?}$, and the resulting divisor is $P_{i_2}^{n-2} P_{i_1}^{n-3} \prod_{j=3}^{n-1} P_{i_j}^{n-1-j}$. This means that the action we just did is the action of the transposition τ_{12} in $S_{n-1} \subseteq S_n$—the reader can verify this by calculating the permutation $M^{-2}NTM$ in S_n (note that all this does hold also for $n = 3$, where M is of order 3, N is τ_{12} and T is trivial, and then $M^{-2}NTM = MNM = N$). This allows us to complete the proof of

Theorem 4.10. *The expression* $\theta^{2en^2}[\Delta](0,\Pi)/h_\Delta$ *with* h_Δ *from Eq.* (4.7) *when n is odd and from Eq.* (4.8) *when n is even is independent of the divisor* Δ *also in the case* $r = 1$, *giving the full Thomae formulae.*

Proof. As we already saw, this quotient is constant on the orbits of the subgroup H of S_{n-1}. Since the set of divisors Δ is a (principal, but this is unnecessary now) homogenous space of S_{n-1}, it suffices to show that H is all of S_{n-1}. Now, H contains G and (among other things) the transposition $M^{-2}NTM = \tau_{12}$. However, the subgroup G of H contains the element NT, which is the full cycle $\begin{pmatrix} 1 & 2\,3 \dots n-2\,n-1 \\ n-1 & 1\,1\,2 \dots n-3\,n-2 \end{pmatrix}$ in S_{n-1} as we saw in the proof of Proposition 3.18. The transposition τ_{12} of two elements adjacent in this full cycle is known to generate, together with the full cycle, all of S_{n-1}. Therefore $H = S_{n-1}$ and the theorem is proved.   □

An immediate corollary of all the discussion above is that the n constants we obtain from Theorem 4.10 with the n different basepoints are equal. We have already shown that the operator M leaves the value of the quotient invariant, and it is easy to see from the definition of M and from Proposition 1.14 that we can start with any divisor Δ with the basepoint P_0 and any other basepoint P_i, and then the divisor Γ_{P_i} is non-special (by Theorem 2.6) and the constant quotient obtained from the basepoint P_i (represented by the divisor Γ_{P_i}) is the same as the one from P_0 (represented by the divisor Δ). Moreover, the usual argument using Proposition 1.14 shows that we have the same $(n-1)!$ characteristics no matter which basepoint we choose. Now, we would like (as usual) to have a basepoint-independent notation for the Thomae formulae also in this case, and this will be (as always) based on the divisors $\Xi = P_0^{n-1}\Delta$. We note again, before going on, that in this particular setting of a nonsingular Z_n curve with $r = 1$ we have, in this language, that the sets of divisors with all the basepoints are united to be the set of divisors Ξ. This is unique to this setting (as one can see in the other situations).

Now, our ability to prove Theorem 4.10, which actually extends the validity of Theorem 4.8 to the case $r = 1$, suggests that the resulting theorem would be the extension of Theorem 4.9 to the case $r = 1$. This is indeed the case, as we now see, but it cannot be proved like Theorem 4.9 but rather like in the examples in Section 3.4. However, we already did all that we need to do for this proof, and all we need now is to give the notation again. For every divisor Ξ (which can be written either as $\prod_{j=-1}^{n-2} C_j^{n-2-j}$ or as $\prod_{j=0}^{n-1} P_{i_j}^{n-1-j}$ as we already saw) we can define $\theta[\Xi]$ (which replaces the notation $\theta[\Delta]$ with basepoint P_0) to be the theta function with characteristic $\varphi_{P_t}(\Xi) + K_{P_t}$ for some $0 \le t \le r-1$, and this does not depend on the choice of P_t by the second statement in Corollary 1.13. Once again we denote the denominators from Eqs. (4.7) and (4.8) by h_Ξ (which is more suitable again by the usual considerations), and since we know that the operator M leaves the quotient invariant (this is what made the constants with different basepoints equal) we have

Theorem 4.11. *For any divisor* $\Xi = \prod_{j=-1}^{n-2} C_j^{n-2-j}$ *with* C_j, $-1 \leq j \leq n-2$, *disjoint sets of cardinality 1 (hence form a partition of the set of n branch points into n sets of cardinality 1) define* $\theta[\Xi]$ *to be the theta function with characteristic* $\varphi_{P_t}(\Xi) + K_{P_t}$ *for some* $0 \leq t \leq rn - 1$ *(which is independent of the choice of* P_t), *and define* h_Ξ *to be the expression from Eq.* (4.7) *if n is odd and Eq.* (4.8) *if n is even. Then the quotient* $\theta^{2en^2}[\Xi](0,\Pi)/h_\Xi$ *is independent of* Ξ.

Theorem 4.11 is the "symmetric" way to express the Thomae formulae we obtained for a general nonsingular Z_n curve with $r = 1$, and knowing it would give as a corollary both Theorem 4.10 and the fact that the different basepoints give the same constant. We state this here as one can prove Theorem 4.11 directly from Theorems 4.6 and 4.7, with a proof similar to the way we proved these two conclusions but even a bit shorter. As we have defined N and T as elements in S_n rather than S_{n-1}, we can let them act on the divisor $\Xi = \prod_{j=0}^{n-1} P_{i_j}^{n-1-j}$ to give $N(\Xi) = P_{i_0}^{n-1} \prod_{j=1}^{n-1} P_{i_j}^{j-1}$ and $T(\Xi) = P_{i_0}^{n-1} P_{i_1}^{n-2} \prod_{j=2}^{n-1} P_{i_j}^{j-2}$ (which is the same as defining, for Ξ written as $\prod_{j=-1}^{n-2} C_j^{n-2-j}$, $N(\Xi)$ to be $C_{-1}^{n-1} \prod_{j=0}^{n-2} C_j^{j}$ and $T(\Xi)$ to be $C_{-1}^{n-1} C_0^{n-2} \prod_{j=1}^{n-2} C_j^{j-1}$— this explains the remark about the operator W from the proof of Theorem 4.8 being simply T if $r = 1$). Then since Theorems 4.6 and 4.7 give the invariance of the quotient $\theta^{2en^2}[\Xi](0,\Pi)/h_\Xi$ under N and T and we also have invariance under M, it suffices to prove that M, N and T generate all of S_n (as the set of divisors Ξ is a principal homogenous space for S_n). Verifying the relations $MTM^{-1} = NTN$ and $MNM = N$ is the same as verifying that NM is of order 2 and commutes with T, and calculating it gives the element $\begin{pmatrix} 0 & 1 & 2 & \dots & n-2 & n-1 \\ 1 & 0 & n-1 & \dots & 3 & 2 \end{pmatrix}$. It is clearly of order 2, and multiplying it by T from either side gives that both TNM and NMT are the transposition τ_{01}. Now, since we also have the full cycle M in the group and 0 and 1 are adjacent in this cycle, we know that M and τ_{01} generate all of S_n. This proves Theorem 4.11 directly, and consequently Theorem 4.10 and the equality of the constants for all the different basepoints. However, even though this proof is more "symmetric" and intrinsic to the situation (S_n is the "right" group to look at here, and not the subgroup S_{n-1}), we gave the proof as we did to maintain the logical line of thought we have in all the other cases, of first proving everything with the basepoint fixed and only then combining the results and giving the new notation. We only mention that as we remarked at the end of Section 3.4 we do not need all of G for generation in either proof and the element NT suffices—in the first proof we used NT and its "conjugate" $M^{-2}NTM$, and in the second proof we used M and TNM, where the last element is the product of M and the inverse TN of NT.

We see that Theorem 4.11 is indeed the extension of Theorem 4.9 for the case $r = 1$, which shows that the same formula holds for $r \geq 2$ and for $r = 1$. As we have seen, we had to split this result into two different theorems as their proofs were too different. One might ask whether we could have combined Theorems 4.9 and 4.11 into one theorem, with proof as we did for the latter, as we know that the operator

M does exist and is defined the same for $r \geq 2$ as for $r = 1$. While the answer is yes we do not feel that anything is gained from this approach and it makes things more complicated rather than easier, so we do not present this proof here.

Theorem 4.11 contains Theorems 3.15, 3.17 and 3.21 as the special cases of $n = 4$ (as $(n-2)!/2 = 1$ there), $n = 3$ (as we said) and $n = 5$ respectively. We note that the fact that in the cases $n = 4$ and $n = 3$ we obtain the Thomae formulae without the basepoint change process does not matter for the final result. Finally, since Theorem 4.11 extends the validity of Theorem 4.9 to the case $r = 1$ as well, we can say that (ignoring the fact that the proofs were different) we have the same result for every nonsingular Z_n curve, regardless of the value of r, which is a very beautiful and symmetric result in view of the fact that the proofs were indeed different for the cases $r \geq 2$ and $r = 1$.

Chapter 5

Thomae Formulae for Singular Z_n Curves

In this chapter we shall present a proof of the Thomae formula for the singular Z_n curve associated to the equation

$$w^n = \prod_{i=0}^{m-1}(z - \lambda_i) \prod_{i=1}^{m}(z - \mu_i)^{n-1},$$

where $\lambda_i \neq \lambda_j$ for $i \neq j$, $\mu_i \neq \mu_j$ for $i \neq j$, and $\lambda_i \neq \mu_j$ for any i and j. Also here, with the examples of Chapter 3 as our guide we can continue along with general statements, and distinguish between even and odd values of n at the necessary stage. We shall of course assume that $m \geq 2$, in order for the genus $g = (n-1)(m-1)$ to be positive.

We shall continue to make use of the useful Definition 4.1, which of course extends its validity to this case as well. We note that one should be careful with the expression $[Y, Y]$ in Definition 4.1 (as it depends on some order fixed on the branch points), but as we continue to work only with expressions raised to even powers (multiples of en), we can ignore all the signs and the choice of this order changes nothing. All the remarks following Definition 4.1 hold equally here, and the same applies for Lemma 4.2 and Corollary 4.3, which continue to give exact equalities in all our applications.

5.1 The Poor Man's Thomae Formulae

As in the nonsingular case, we start by constructing the PMT for a general singular Z_n curve of this form. Once again, as in the examples in Chapter 3, the way to obtain the PMT involving $\theta[\Delta](0, \Pi)$ for a non-special integral divisor $\Delta = A^{n-1}B^0 \prod_{j=0}^{n-2} C_j^{n-2-j} D_j^{j+1}$ of degree g with support in the branch points distinct from P_0 passes through the quotient

H.M. Farkas and S. Zemel, *Generalizations of Thomae's Formula for Z_n Curves*,
Developments in Mathematics 21, DOI 10.1007/978-1-4419-7847-9_5,
© Springer Science+Business Media, LLC 2011

$$f(P) = \frac{\theta[\Delta]^{en^2}(\varphi_{P_0}(P), \Pi)}{\theta[\Xi]^{en^2}(\varphi_{P_0}(P), \Pi)},$$

for some other such divisor Ξ. As before, the considerations of Section 2.6 show that the "best" divisors to take in place of Ξ are $T_R(\Delta)$ for some branch point $R \in C_0$, and here also $T_S(\Delta)$ for some branch point $S \in B$. In these cases Proposition 2.21 shows that this function is

$$c \prod_Q (z - z(Q))^{en[v_Q(N(\Delta)) - v_Q(N(T_R(\Delta)))]}$$

in the case $\Xi = T_R(\Delta)$ and

$$d \prod_Q (z - z(Q))^{en[v_Q(N(\Delta)) - v_Q(N(T_S(\Delta)))]}$$

in the case $\Xi = T_S(\Delta)$ where c and d are nonzero complex constants. It turns out that the way to obtain the PMT is to deal with the operators T_R alone (which resembles the way to obtain the PMT in the nonsingular case), then to deal with the operators T_S alone (which is quite a similar process, but with slightly different results), and then merge them together.

5.1.1 First Identities Between Theta Constants Based on the Branch Point R

Let us thus start with $\Xi = T_R(\Delta)$. Calculating $N(\Delta)$, $T_R(\Delta)$ and $N(T_R(\Delta))$ using Definition 2.18 shows that

$$N(\Delta) = A^{n-1} \prod_{j=0}^{n-2} C_j^j D_j^{n-1-j}, \quad T_R(\Delta) = (C_0 \setminus R)^{n-1}(A \cup R)^{n-2} B^1 \prod_{j=1}^{n-2} C_j^{j-1} D_j^{n-j},$$

and

$$N(T_R(\Delta)) = (C_0 \setminus R)^{n-1} B^{n-1} \prod_{j=1}^{n-2} C_j^{n-1-j} D_j^j.$$

Hence when we compare the powers to which a branch point from each set appears in $N(\Delta)$ and $N(T_R(\Delta))$ we find that the explicit expressions for the functions in Proposition 2.21 are as follows. For odd n it is

$$c \frac{[z,A]^{2n(n-1)} \left(\prod_{j=(n+1)/2}^{n-2} [z,C_j]^{2n(2j-(n-1))} \right) [z,D_0]^{2n(n-1)} \prod_{j=1}^{(n-3)/2} [z,D_j]^{2n(n-1-2j)}}{[z,C_0 \setminus R]^{2n(n-1)} \left(\prod_{j=1}^{(n-3)/2} [z,C_j]^{2n(n-1-2j)} \right) [z,B]^{2n(n-1)} \prod_{j=(n+1)/2}^{n-2} [z,D_j]^{2n(2j-(n-1))}}$$

(with the powers of $[z, C_{(n-1)/2}]$ and $[z, D_{(n-1)/2}]$ canceling), and for even n it is

$$
c\frac{[z,A]^{n(n-1)}\left(\prod_{j=n/2}^{n-2}[z,C_j]^{n(2j-(n-1))}\right)[z,D_0]^{n(n-1)}\prod_{j=1}^{(n-2)/2}[z,D_j]^{n(n-1-2j)}}{[z,C_0\setminus R]^{n(n-1)}\left(\prod_{j=1}^{(n-2)/2}[z,C_j]^{n(n-1-2j)}\right)[z,B]^{n(n-1)}\prod_{j=n/2}^{n-2}[z,D_j]^{n(2j-(n-1))}}
$$

(with all the sets now appearing). Note that when we take from the expression concerning $T_R(\Delta)$ only the parts with the sets A and C_j, $0 \le j \le n-2$, we obtain exactly the corresponding terms that appear in the nonsingular case. This will continue to be the case, and as already mentioned, the reason for this is that the nonsingular case and this singular case are special cases of one larger family of Z_n curves. For all of the Z_n curves in this family similar Thomae formulae hold, and with a similar proof as explained in Appendix A.

As in the nonsingular case, the only difference between the even and odd n here is (except for the value of e) "what happens in the middle sets". The explicit description is just as in the nonsingular case, where every statement asserted for some C_j extends equally to the corresponding D_j. Again we prefer to work independently of the parity of n, so that here the form we obtain for the function from Proposition 2.21 only by comparing the numbers $v_Q(N(\Delta))$ and $v_Q(N(T_R(\Delta)))$ for any such Q is

$$
c([z,A][z,D_0])^{en(n-1)}([z,C_0\setminus R][z,B])^{-en(n-1)}\prod_{j=1}^{n-2}[z,C_j]^{en(2j-(n-1))}[z,D_j]^{en(n-1-2j)},
$$

independently of the parity of n. This is obtained since we still have only differences between powers that are either enj or $en(n-1-j)$ (except for the initial part, which contains expressions appearing only in the numerator or only in the denominator and to the power $en(n-1)$). It will be more convenient to write this function $f(P)$ in the form

$$
\frac{\theta[\Delta]^{en^2}(\varphi_{P_0}(P),\Pi)}{\theta[T_R(\Delta)]^{en^2}(\varphi_{P_0}(P),\Pi)}=c\frac{([z,A][z,D_0])^{en(n-1)}\prod_{j=1}^{n-2}[z,C_j]^{enj}[z,D_j]^{en(n-1-j)}}{([z,C_0\setminus R][z,B])^{en(n-1)}\prod_{j=1}^{n-2}[z,C_j]^{en(n-1-j)}[z,D_j]^{enj}},
$$

which is not "reduced" like the expressions for odd and even n, but is independent of the parity of n. As in the nonsingular case, this expression is based on the explicit form of $N(\Delta)$ and $N(T_R(\Delta))$.

As we always do, we substitute $P=P_0$ to find the constant c, from which we obtain

$$
\frac{\theta[\Delta]^{en^2}(0,\Pi)}{\theta[T_R(\Delta)]^{en^2}(0,\Pi)}=c\frac{([P_0,A][P_0,D_0])^{en(n-1)}\prod_{j=1}^{n-2}[P_0,C_j]^{enj}[P_0,D_j]^{en(n-1-j)}}{([P_0,C_0\setminus R][P_0,B])^{en(n-1)}\prod_{j=1}^{n-2}[P_0,C_j]^{en(n-1-j)}[P_0,D_j]^{enj}},
$$

and thus

$$f(P) = \frac{\theta[\Delta]^{en^2}(0,\Pi)}{\theta[T_R(\Delta)]^{en^2}(0,\Pi)} \times \frac{([P_0,C_0\setminus R][P_0,B])^{en(n-1)}\prod_{j=1}^{n-2}[P_0,C_j]^{en(n-1-j)}[P_0,D_j]^{enj}}{([P_0,A][P_0,D_0])^{en(n-1)}\prod_{j=1}^{n-2}[P_0,C_j]^{enj}[P_0,D_j]^{en(n-1-j)}}$$

$$\times \frac{([z,A][z,D_0])^{en(n-1)}\prod_{j=1}^{n-2}[z,C_j]^{enj}[z,D_j]^{en(n-1-j)}}{([z,C_0\setminus R][z,B])^{en(n-1)}\prod_{j=1}^{n-2}[z,C_j]^{en(n-1-j)}[z,D_j]^{enj}}.$$

We now substitute $P = R$, which gives us, on the one hand,

$$f(R) = \frac{\theta[\Delta]^{en^2}(0,\Pi)}{\theta[T_R(\Delta)]^{en^2}(0,\Pi)} \times \frac{([P_0,C_0\setminus R][P_0,B])^{en(n-1)}\prod_{j=1}^{n-2}[P_0,C_j]^{en(n-1-j)}[P_0,D_j]^{enj}}{([P_0,A][P_0,D_0])^{en(n-1)}\prod_{j=1}^{n-2}[P_0,C_j]^{enj}[P_0,D_j]^{en(n-1-j)}}$$

$$\times \frac{([R,A][R,D_0])^{en(n-1)}\prod_{j=1}^{n-2}[R,C_j]^{enj}[R,D_j]^{en(n-1-j)}}{([R,C_0\setminus R][R,B])^{en(n-1)}\prod_{j=1}^{n-2}[R,C_j]^{en(n-1-j)}[R,D_j]^{enj}},$$

while on the other hand, Eq. (2.3) gives us

$$f(R) = \frac{\theta[\Delta]^{en^2}(\varphi_{P_0}(R),\Pi)}{\theta[T_R(\Delta)]^{en^2}(\varphi_{P_0}(R),\Pi)} = \frac{\theta[T_R(\Delta)]^{en^2}(0,\Pi)}{\theta[\Delta]^{en^2}(0,\Pi)}.$$

We therefore combine the two equations and obtain

$$\frac{\theta[T_R(\Delta)]^{2en^2}(0,\Pi)}{\theta[\Delta]^{2en^2}(0,\Pi)} = \frac{([P_0,C_0\setminus R][P_0,B])^{en(n-1)}\prod_{j=1}^{n-2}[P_0,C_j]^{en(n-1-j)}[P_0,D_j]^{enj}}{([P_0,A][P_0,D_0])^{en(n-1)}\prod_{j=1}^{n-2}[P_0,C_j]^{enj}[P_0,D_j]^{en(n-1-j)}}$$

$$\times \frac{([R,A][R,D_0])^{en(n-1)}\prod_{j=1}^{n-2}[R,C_j]^{enj}[R,D_j]^{en(n-1-j)}}{([R,C_0\setminus R][R,B])^{en(n-1)}\prod_{j=1}^{n-2}[R,C_j]^{en(n-1-j)}[R,D_j]^{enj}},$$

an equality which is equivalent to

$$\frac{\theta[\Delta]^{2en^2}(0,\Pi)}{[P_0,A]^{en(n-1)}[P_0,D_0]^{en(n-1)}[R,C_0\setminus R]^{en(n-1)}[R,B]^{en(n-1)}}$$

$$\times \frac{1}{\prod_{j=1}^{n-2}[P_0,C_j]^{enj}[R,D_j]^{enj}[P_0,D_j]^{en(n-1-j)}[R,C_j]^{en(n-1-j)}}$$

$$= \frac{\theta[T_R(\Delta)]^{2en^2}(0,\Pi)}{[P_0,C_0\setminus R]^{en(n-1)}[P_0,B]^{en(n-1)}[R,A]^{en(n-1)}[R,D_0]^{en(n-1)}}$$

$$\times \frac{1}{\prod_{j=1}^{n-2}[P_0,C_j]^{en(n-1-j)}[R,D_j]^{en(n-1-j)}[P_0,D_j]^{enj}[R,C_j]^{enj}}. \tag{5.1}$$

Also here we see that Eq. (5.1), which gives the first relations between the theta constants on X, is proved by a calculation once we have defined the operators T_R and N with the correct properties. In order to obtain the PMT we shall have to symmetrize the denominators in Eq. (5.1) to eliminate the dependence on the point R, and later combine this with the results obtained using the operators T_S. Let us finish the part concerning the operators T_R before turning to the action of the operators T_S.

5.1.2 Symmetrization over R

As we did in the nonsingular case, we denote the sets corresponding to $T_R(\Delta)$ by adding a tilde to the usual notation in order to show that the relation that Eq. (5.1) gives for a divisor Δ and a branch point $R \in C_0$ is exactly the same relation we would have obtained if we had started with the divisor $T_R(\Delta)$ and the branch point R. As in the nonsingular case, we shall keep track of our equations and make sure that this property is preserved at every stage. Now we have from Definition 2.18 that

$$T_R(\Delta) = (C_0 \setminus R)^{n-1} D_0^0 (A \cup R)^{n-2} B^1 \prod_{j=1}^{n-2} C_j^{j-1} D_j^{n-j},$$

and hence $\widetilde{A} = C_0 \setminus R$, $\widetilde{C_0} = A \cup R$, and $\widetilde{C_j} = C_{n-1-j}$ for $1 \leq j \leq n-2$ as before, and also $\widetilde{B} = D_0$, $\widetilde{D_0} = B$, and $\widetilde{D_j} = D_{n-1-j}$ for $1 \leq j \leq n-2$. Hence if we write the denominator under $\theta[T_R(\Delta)](0,\Pi)$ in Eq. (5.1) using the sets with the tilde we get

$$([P_0,\widetilde{A}][P_0,\widetilde{D_0}][R,\widetilde{C_0} \setminus R][R,\widetilde{B}])^{en(n-1)} \prod_{j=1}^{n-2} ([P_0,\widetilde{C_j}][R,\widetilde{D_j}])^{enj}([P_0,\widetilde{D_j}][R,\widetilde{C_j}])^{en(n-1-j)}$$

(changing the multiplication index j to $n-1-j$)—and this is exactly the denominator appearing under $\theta[\Delta](0,\Pi)$ in Eq. (5.1) but with a tilde on all the sets. Therefore we see that starting with $T_R(\Delta)$ and R instead of Δ and R gives exactly the same Eq. (5.1), a property which we know we would indeed like our expressions to have.

We now want to replace the denominators appearing in Eq. (5.1) by expressions which have the property that the denominator appearing under the theta constant corresponding to a divisor Δ in it must be constructed from Δ alone. Again, Eq. (5.1) does not have this property, as the denominator appearing under $\theta[\Delta](0,\Pi)$ in it depends strongly on the branch point R chosen from C_0. The way to eliminate this dependence on R, which will also allow us to merge Eqs. (5.1) for all the different points R into one large equation that arises from $\theta[\Delta](0,\Pi)$ for our divisor Δ, is by symmetrizing it over R. This symmetrization is carried out exactly as we did in the nonsingular case, but now we have to symmetrize the expressions with the sets B and D_j, $0 \leq j \leq n-2$, as well. Again, we must preserve the property that the equation must be the same no matter from which divisor appearing in it we begin, a property which our current Eq. (5.1) does have.

In order to do so, we examine the expressions depending on R in this denominator. These are $[R, C_0 \setminus R]^{en(n-1)} = [R, C_0]^{en(n-1)}$, $[R, B]^{en(n-1)}$, and $[R, C_j]^{en(n-1-j)}$ and $[R, D_j]^{enj}$ for $1 \leq j \leq n-2$. In order to make this denominator independent of R we see that we have to multiply these expressions (i.e., divide Eq. (5.1)) by the following expressions. We correct the first one by $[C_0 \setminus R, C_0 \setminus R]$, a correction which by Lemma 4.2 changes it to $[C_0, C_0]$. The other expressions are of the form $[R, Y]$ with $Y \cap C_0 = \emptyset$, and will be corrected by $[C_0 \setminus R, Y]$ with the corresponding powers to yield simply $[C_0, Y]$ (which is also independent of R). The sign appearing in all these actions does not matter, since all these expressions are taken to an even power. Altogether our first try is to divide Eq. (5.1) by

$$([C_0 \setminus R, C_0 \setminus R][C_0 \setminus R, B])^{en(n-1)} \prod_{j=1}^{n-2} [C_0 \setminus R, C_j]^{en(n-1-j)} [C_0 \setminus R, D_j]^{enj}.$$

However, in the notation with sets with the tilde we find that the expression by which we divided Eq. (5.1) is

$$([\widetilde{A}, \widetilde{A}][\widetilde{A}, \widetilde{D_0}])^{en(n-1)} \prod_{j=1}^{n-2} [\widetilde{A}, \widetilde{C_j}]^{enj} [\widetilde{A}, \widetilde{D_j}]^{en(n-1-j)}$$

(with the multiplication index j replaced by $n - 1 - j$). Since we want to preserve the property that starting with $T_R(\Delta)$ and R gives us the same equation as starting with Δ and R, we see that we must divide Eq. (5.1) also by

$$([A,A][A,D_0])^{en(n-1)} \prod_{j=1}^{n-2} [A, C_j]^{enj} [A, D_j]^{en(n-1-j)},$$

which in the notation of the sets with the tilde equals

$$([\widetilde{C_0} \setminus R, \widetilde{C_0} \setminus R][\widetilde{C_0} \setminus R, \widetilde{B}])^{en(n-1)} \prod_{j=1}^{n-2} [\widetilde{C_0} \setminus R, \widetilde{C_j}]^{en(n-1-j)} [\widetilde{C_0} \setminus R, \widetilde{D_j}]^{enj}$$

(replacing j with $n - 1 - j$ again). As the product of these two expressions looks the same in the notation with the tilde as in the notation without it (since also in this case the operator T_R is an involution), this product is a "good" expression to divide Eq. (5.1) by. Let us now examine the result of this division.

Again let us denote the set $A \cup P_0$ by C_{-1}. Then we see that apart from the merging of the expressions involving R into expressions involving C_0 for which we have done these manipulations we have further merging. The new expression $[A,A]$ combines, by Lemma 4.2, with $[P_0, A]$ to $[C_{-1}, C_{-1}]$ (as $A = C_{-1} \setminus P_0$), and all the other expressions are of the form $[A, Y]$ with $Y \cap C_{-1} = \emptyset$ and combine with $[P_0, Y]$ to give $[C_{-1}, Y]$. The reader can verify that every pair of expressions asserted to be merged here indeed appears to the same power and that power is even, so that the merging indeed occurs as stated and without any sign problems. Now, all this merging happens in the denominator appearing under $\theta[\Delta](0, \Pi)$, and it is clear that under $\theta[T_R(\Delta)](0, \Pi)$ the same thing happens but with a tilde on all the sets (again $\widetilde{C_{-1}}$ is $\widetilde{A} \cup P_0 = C_0 \setminus R \cup P_0$). Therefore this action preserves the property that starting with $T_R(\Delta)$ and R gives us the same equation as starting with Δ and R, and we obtain the formula

$$\frac{\theta[\Delta]^{2en^2}(0,\Pi)}{[C_{-1},C_{-1}]^{en(n-1)}[C_{-1},D_0]^{en(n-1)}[C_0,C_0]^{en(n-1)}[C_0,B]^{en(n-1)}}$$

$$\times \frac{1}{\prod_{j=1}^{n-2}[C_{-1},C_j]^{enj}[C_0,D_j]^{enj}[C_{-1},D_j]^{en(n-1-j)}[C_0,C_j]^{en(n-1-j)}}$$

$$= \frac{\theta[T_R(\Delta)]^{2en^2}(0,\Pi)}{[\widetilde{C_{-1}},\widetilde{C_{-1}}]^{en(n-1)}[\widetilde{C_{-1}},\widetilde{D_0}]^{en(n-1)}[\widetilde{C_0},\widetilde{C_0}]^{en(n-1)}[\widetilde{C_0},\widetilde{B}]^{en(n-1)}}$$

$$\times \frac{1}{\prod_{j=1}^{n-2}[\widetilde{C_{-1}},\widetilde{C_j}]^{enj}[\widetilde{C_0},\widetilde{D_j}]^{enj}[\widetilde{C_{-1}},\widetilde{D_j}]^{en(n-1-j)}[\widetilde{C_0},\widetilde{C_j}]^{en(n-1-j)}}. \tag{5.2}$$

We describe the meaning of Eq. (5.2) in the following

Proposition 5.1. *For a divisor* $\Delta = A^{n-1}B^0\prod_{j=0}^{n-2}C_j^{n-2-j}D_j^{j+1}$, *define*

$$p_\Delta = ([C_{-1},C_{-1}][C_{-1},D_0][C_0,C_0][C_0,B])^{en(n-1)}$$

$$\times \prod_{j=1}^{n-2}([C_{-1},C_j][C_0,D_j])^{enj}([C_{-1},D_j][C_0,C_j])^{en(n-1-j)}$$

where $C_{-1} = A \cup P_0$. *Then the quotient* $\theta^{2en^2}[\Delta](0,\Pi)/p_\Delta$ *is invariant under all the operators* T_R.

Proof. The way of constructing p_Δ from Δ, the fact that the construction depends on Δ alone, and the fact that Eq. (5.2) is formulated in this notation as

$$\frac{\theta[\Delta](0,\Pi)}{p_\Delta} = \frac{\theta[T_R(\Delta)](0,\Pi)}{p_{T_R(\Delta)}}$$

are all clear. Therefore, we see that an application of any permissible T_R on the quotient based on any divisor Δ does not change the quotient, whence the proposition. $\square$

As in the nonsingular case, Proposition 5.1 shows that instead of looking at Eq. (5.2) as an equality between two terms, we can merge it into an equality between the quotients based on any divisor to which one can get from Δ using successive applications of the operators T_R. This is understood by considerations identical to those of the nonsingular case, except for the remark that starting from a given divisor Δ, after the first merging, we no longer have an equality between a fixed number of terms ($r+1$ in the nonsingular case), but here this number depends on Δ (and equals $|C_0|+1$). Therefore, for a given Δ there may not be many divisors in this equality, since $|C_0|$ may be small and even empty, the latter case leaving Δ not connected to any other divisor via some operator T_R. Nonetheless, continuing with iterated applications, the number of divisors can increase significantly. This is not always the case, since we already saw in Section 3.2 that also the final number of divisors appearing in the equality corresponding to Δ may still be very small for some divisors, but it can be larger for others. In any case, in comparison with the nonsingular case, it does depend on Δ.

Once again we would like to remark that for a given minimal set of divisors which is closed under the action of the operators T_R (on which the quotient from Proposition 5.1 is constant by the proposition) the quotients are again not reduced. We just state here that the set $G = C_{-1} \cup C_0$ again satisfies $\widetilde{G} = G$, and the expressions $[G, C_j]^{en \min\{j, n-1-j\}}$ for $1 \le j \le n-2$, which appeared as common denominators in the nonsingular case (and are again the same as the expressions $[\widetilde{G}, \widetilde{C}_j]^{en \min\{j, n-1-j\}}$ for $1 \le j \le n-2$ when replacing j by $n-1-j$), are common denominators also here. However, in our case also $[G, D_j]^{en \min\{j, n-1-j\}}$ with $1 \le j \le n-2$ (which are also the same as $[\widetilde{G}, \widetilde{D}_j]^{en \min\{j, n-1-j\}}$ for $1 \le j \le n-2$ when replacing j by $n-1-j$) appear as common denominators. Once again, the fact that applications of the operators T_R mixes C_{-1} and C_0 more and more shows that there is no further common factor. As before, these common expressions appear since when we merged the expressions for the function $f(P)$ as a rational function of z for odd n and even n we used unreduced quotients. However, we would like to keep the unreduced Eq. (5.2) in Proposition 5.1 since we want to merge it with Eq. (5.4) in Proposition 5.2 below and we do not want to have to deal with the odd n and even n cases separately until we really have to.

5.1.3 First Identities Between Theta Constants Based on the Branch Point S

We now take $\Xi = T_S(\Delta)$ in the process. Here we obtain from Definition 2.18 that

$$N(\Delta) = A^{n-1}B^0 \prod_{j=0}^{n-2} C_j^j D_j^{n-1-j}, \quad T_S(\Delta) = C_0^{n-1}(D_0 \cup S)^0 A^{n-2}(B \setminus S)^1 \prod_{j=1}^{n-2} C_j^{j-1} D_j^{n-j},$$

and

$$N(T_S(\Delta)) = C_0^{n-1}(D_0 \cup S)^0 (B \setminus S)^{n-1} \prod_{j=1}^{n-2} C_j^{n-1-j} D_j^j;$$

hence the explicit expressions for the functions in Proposition 2.21 are the following. For odd n it is

$$d\frac{[z,A]^{2n(n-1)} \left(\prod_{j=(n+1)/2}^{n-2} [z,C_j]^{2n(2j-(n-1))} \right) [z,D_0]^{2n(n-1)} \prod_{j=1}^{(n-3)/2} [z,D_j]^{2n(n-1-2j)}}{[z,C_0]^{2n(n-1)} \left(\prod_{j=1}^{(n-3)/2} [z,C_j]^{2n(n-1-2j)} \right) [z,B \setminus S]^{2n(n-1)} \prod_{j=(n+1)/2}^{n-2} [z,D_j]^{2n(2j-(n-1))}},$$

and for even n it is

$$d\,\frac{[z,A]^{n(n-1)}\big(\prod_{j=n/2}^{n-2}[z,C_j]^{n(2j-(n-1))}\big)[z,D_0]^{n(n-1)}\prod_{j=1}^{(n-2)/2}[z,D_j]^{n(n-1-2j)}}{[z,C_0]^{n(n-1)}\big(\prod_{j=1}^{(n-2)/2}[z,C_j]^{n(n-1-2j)}\big)[z,B\setminus S]^{n(n-1)}\prod_{j=n/2}^{n-2}[z,D_j]^{n(2j-(n-1))}}$$

(with the same remark about sets appearing or canceling). We note that here the terms with the sets A and C_j, $0\le j\le n-2$, are no longer the corresponding terms for the nonsingular case, as the operators T_S do not exist in the nonsingular case, and we can only compare the actions of the operators T_R (and N).

The same remark about the difference between the even and odd n cases we had with the divisor $T_R(\Delta)$ also applies here, and the expression for the function from Proposition 2.21 using the comparison of the numbers $v_Q(N(\Delta))$ and $v_Q(N(T_S(\Delta)))$ for any Q is just

$$d([z,A][z,D_0])^{en(n-1)}([z,C_0][z,B\setminus S])^{-en(n-1)}\prod_{j=1}^{n-2}[z,C_j]^{en(2j-(n-1))}[z,D_j]^{en(n-1-2j)},$$

again a form for $f(P)$ which is independent of the parity of n. The more convenient form for $f(P)$ that we use here is

$$\frac{\theta[\Delta]^{en^2}(\varphi_{P_0}(P),\Pi)}{\theta[T_S(\Delta)]^{en^2}(\varphi_{P_0}(P),\Pi)}=d\,\frac{([z,A][z,D_0])^{en(n-1)}\prod_{j=1}^{n-2}[z,C_j]^{enj}[z,D_j]^{en(n-1-j)}}{([z,C_0][z,B\setminus S])^{en(n-1)}\prod_{j=1}^{n-2}[z,C_j]^{en(n-1-j)}[z,D_j]^{enj}},$$

which is again not "reduced", independent of the parity of n, and based directly on $N(\Delta)$ and $N(T_S(\Delta))$.

As usual, the constant d is found by substituting $P=P_0$, which yields

$$\frac{\theta[\Delta]^{en^2}(0,\Pi)}{\theta[T_S(\Delta)]^{en^2}(0,\Pi)}=d\,\frac{([P_0,A][P_0,D_0])^{en(n-1)}\prod_{j=1}^{n-2}[P_0,C_j]^{enj}[P_0,D_j]^{en(n-1-j)}}{([P_0,C_0][P_0,B\setminus S])^{en(n-1)}\prod_{j=1}^{n-2}[P_0,C_j]^{en(n-1-j)}[P_0,D_j]^{enj}}$$

and thus

$$f(P)=\frac{\theta[\Delta]^{en^2}(0,\Pi)}{\theta[T_S(\Delta)]^{en^2}(0,\Pi)}\times\frac{([P_0,C_0][P_0,B\setminus S])^{en(n-1)}\prod_{j=1}^{n-2}[P_0,C_j]^{en(n-1-j)}[P_0,D_j]^{enj}}{([P_0,A][P_0,D_0])^{en(n-1)}\prod_{j=1}^{n-2}[P_0,C_j]^{enj}[P_0,D_j]^{en(n-1-j)}}$$
$$\times\frac{([z,A][z,D_0])^{en(n-1)}\prod_{j=1}^{n-2}[z,C_j]^{enj}[z,D_j]^{en(n-1-j)}}{([z,C_0][z,B\setminus S])^{en(n-1)}\prod_{j=1}^{n-2}[z,C_j]^{en(n-1-j)}[z,D_j]^{enj}}.$$

Next we substitute $P=S$, which, on the one hand, gives us

$$f(S)=\frac{\theta[\Delta]^{en^2}(0,\Pi)}{\theta[T_S(\Delta)]^{en^2}(0,\Pi)}\times\frac{([P_0,C_0][P_0,B\setminus S])^{en(n-1)}\prod_{j=1}^{n-2}[P_0,C_j]^{en(n-1-j)}[P_0,D_j]^{enj}}{([P_0,A][P_0,D_0])^{en(n-1)}\prod_{j=1}^{n-2}[P_0,C_j]^{enj}[P_0,D_j]^{en(n-1-j)}}$$
$$\times\frac{([S,A][S,D_0])^{en(n-1)}\prod_{j=1}^{n-2}[S,C_j]^{enj}[S,D_j]^{en(n-1-j)}}{([S,C_0][S,B\setminus S])^{en(n-1)}\prod_{j=1}^{n-2}[S,C_j]^{en(n-1-j)}[S,D_j]^{enj}}.$$

and on the other hand, we have from Eq. (2.4) the equality

$$f(S) = \frac{\theta[\Delta]^{en^2}(\varphi_{P_0}(S),\Pi)}{\theta[T_S(\Delta)]^{en^2}(\varphi_{P_0}(S),\Pi)} = \frac{\theta[T_S(\Delta)]^{en^2}(0,\Pi)}{\theta[\Delta]^{en^2}(0,\Pi)}.$$

Combining the two equations we obtain

$$\frac{\theta[T_S(\Delta)]^{2en^2}(0,\Pi)}{\theta[\Delta]^{2en^2}(0,\Pi)} = \frac{([P_0,C_0][P_0,B\setminus S])^{en(n-1)} \prod_{j=1}^{n-2}[P_0,C_j]^{en(n-1-j)}[P_0,D_j]^{enj}}{([P_0,A][P_0,D_0])^{en(n-1)} \prod_{j=1}^{n-2}[P_0,C_j]^{enj}[P_0,D_j]^{en(n-1-j)}}$$

$$\times \frac{([S,A][S,D_0])^{en(n-1)} \prod_{j=1}^{n-2}[S,C_j]^{enj}[S,D_j]^{en(n-1-j)}}{([S,C_0][S,B\setminus S])^{en(n-1)} \prod_{j=1}^{n-2}[S,C_j]^{en(n-1-j)}[S,D_j]^{enj}},$$

or equivalently

$$\frac{\theta[\Delta]^{2en^2}(0,\Pi)}{[P_0,A]^{en(n-1)}[P_0,D_0]^{en(n-1)}[S,C_0]^{en(n-1)}[S,B\setminus S]^{en(n-1)}}$$

$$\times \frac{1}{\prod_{j=1}^{n-2}[P_0,C_j]^{enj}[S,D_j]^{enj}[P_0,D_j]^{en(n-1-j)}[S,C_j]^{en(n-1-j)}}$$

$$= \frac{\theta[T_S(\Delta)]^{2en^2}(0,\Pi)}{[P_0,C_0]^{en(n-1)}[P_0,B\setminus S]^{en(n-1)}[S,A]^{en(n-1)}[S,D_0]^{en(n-1)}}$$

$$\times \frac{1}{\prod_{j=1}^{n-2}[P_0,C_j]^{en(n-1-j)}[S,D_j]^{en(n-1-j)}[P_0,D_j]^{enj}[S,C_j]^{enj}}. \tag{5.3}$$

As before, the basic relation between the theta constants on X expressed in Eq. (5.3) is just a calculation away from the correct definition of the operators T_S and N. We now symmetrize the denominators in Eq. (5.3) to eliminate the dependence on the point S, and then combine the result with Eq. (5.2) and Proposition 5.1 to obtain the PMT.

5.1.4 Symmetrization over S

In order to distinguish between the operators T_R and T_S, we denote the sets corresponding to $T_S(\Delta)$ by adding a hat to the usual notation. As with $T_R(\Delta)$, we use this in order to show that the relation which Eq. (5.3) gives for a divisor Δ and a branch point $S \in B$ is exactly the same relation we would have obtained if we had started with the divisor $T_S(\Delta)$ and the branch point S. We shall be careful to preserve this property as well throughout the process of our work. Definition 2.18 gives us

$$T_S(\Delta) = C_0^{n-1}(D_0 \cup S)^0 A^{n-2}(B\setminus S)^1 \prod_{j=1}^{n-2} C_j^{j-1} D_j^{n-i},$$

and hence we have $\widehat{A} = C_0$, $\widehat{C_0} = A$ and $\widehat{C_j} = C_{n-1-j}$ for $1 \leq j \leq n-2$, and we also have $\widehat{B} = D_0 \cup S$, $\widehat{D_0} = B \setminus S$, and $\widehat{D_j} = D_{n-1-j}$ for $1 \leq j \leq n-2$. Using the sets with the hat the denominator under $\theta[T_S(\Delta)](0, \Pi)$ in Eq. (5.3) becomes

$$([P_0, \widehat{A}][P_0, \widehat{D_0}][S, \widehat{C_0}][S, \widehat{B \setminus S}])^{en(n-1)} \prod_{j=1}^{n-2} ([P_0, \widehat{C_j}][S, \widehat{D_j}])^{enj} ([S, \widehat{C_j}][P_0, \widehat{D_j}])^{en(n-1-j)},$$

(with the same multiplication index change), which is precisely the denominator appearing under $\theta[\Delta](0, \Pi)$ in Eq. (5.3) but with a hat on all the sets. This means that we do begin with expressions possessing the desired property that starting with $T_S(\Delta)$ and S instead of Δ and S gives exactly the same Eq. (5.3).

As with the operators T_R, we find that the denominators appearing in Eq. (5.3) depend on the point S, a dependence that we now want to remove. This is done by a symmetrization which is very similar to the one we did for Eq. (5.1) in order to obtain Eq. (5.2). We shall again have to be careful to preserve the property that the equation must be the same no matter from which divisor appearing in it we begin. We have noted that Eq. (5.3) does have this property.

The expressions depending on S in the denominator in question are the expressions $[S, C_0]^{en(n-1)}$, $[S, B \setminus S]^{en(n-1)} = [S, B]^{en(n-1)}$, $[S, C_j]^{en(n-1-j)}$ and $[S, D_j]^{enj}$ for $1 \leq j \leq n-2$. We correct the second one by $[B \setminus S, B \setminus S]$, and since the other expressions are of the form $[S, Y]$ with $Y \cap B = \emptyset$ the corresponding corrections will be $[B \setminus S, Y]$. Since all the powers are even we find that the correction to the second expression yields, by Lemma 4.2, $[B, B]$, and the other expressions become $[B, Y]$ for the corresponding set Y. All these expressions are indeed independent of S, and are obtained after we divide Eq. (5.3) by

$$([B \setminus S, B \setminus S][B \setminus S, C_0])^{en(n-1)} \prod_{j=1}^{n-2} [B \setminus S, C_j]^{en(n-1-j)} [B \setminus S, D_j]^{enj}.$$

In the notation with the hat we find that the expression by which we divided Eq. (5.3) is

$$[\widehat{D_0}, \widehat{D_0}]^{en(n-1)} [\widehat{D_0}, \widehat{A}]^{en(n-1)} \prod_{j=1}^{n-2} [\widehat{D_0}, \widehat{C_j}]^{enj} [\widehat{D_0}, \widehat{D_j}]^{en(n-1-j)}$$

(with j replaced by $n - 1 - j$ again), so that in order to preserve the property that starting with $T_S(\Delta)$ and S gives us the same equation as starting with Δ and S we must divide Eq. (5.3) also by

$$([D_0, D_0][D_0, A])^{en(n-1)} \prod_{j=1}^{n-2} [D_0, C_j]^{enj} [D_0, D_j]^{en(n-1-j)}.$$

This expression, when written in the notation of the sets with the hat, is

$$[\widehat{B} \setminus S, \widehat{B} \setminus S]^{en(n-1)} [\widehat{B} \setminus S, \widehat{C_0}]^{en(n-1)} \prod_{j=1}^{n-2} [\widehat{B} \setminus S, \widehat{C_j}]^{en(n-1-j)} [\widehat{B} \setminus S, \widehat{D_j}]^{enj}$$

(with the usual index interchange), and the product of these two expressions is a "good" expression to divide Eq. (5.3) by since it looks the same in the notation with the hat as in the notation without it (a fact which is related to T_S being an involution as well).

In order to examine the further merging we have here let us denote the set $D_0 \cup P_0$ by H. Using this notation we see that by Lemma 4.2 the expression $[D_0, D_0]$ merges with $[P_0, D_0]$ to give $[H, H]$ (since $D_0 = H \setminus P_0$), and since the other expressions containing D_0 are of the form $[D_0, Y]$ with $Y \cap D_0 = \emptyset$ and merge with the corresponding expressions $[P_0, Y]$ to give $[H, Y]$. Once again the reader should verify that the powers to which the two expressions in each merging pair appear are equal and even, so the merging does work as asserted. This completes the discussion about the merging occurring in the denominator appearing under $\theta[\Delta](0, \Pi)$, and under $\theta[T_S(\Delta)](0, \Pi)$ the same thing happens with a hat on all the sets (with $\widehat{H}$ being $\widehat{D_0} \cup P_0 = B \setminus S \cup P_0$). Thus the property that starting with $T_S(\Delta)$ and S gives us the same equation as starting with Δ and S is preserved, and the formula this process yields is

$$\frac{\theta[\Delta]^{2en^2}(0, \Pi)}{[H, A]^{en(n-1)}[H, H]^{en(n-1)}[B, C_0]^{en(n-1)}[B, B]^{en(n-1)}}$$

$$\times \frac{1}{\prod_{j=1}^{n-2}[H, C_j]^{enj}[B, D_j]^{enj}[H, D_j]^{en(n-1-j)}[B, C_j]^{en(n-1-j)}}$$

$$= \frac{\theta[T_S(\Delta)]^{2en^2}(0, \Pi)}{[\widehat{H}, \widehat{A}]^{en(n-1)}[\widehat{H}, \widehat{H}]^{en(n-1)}[\widehat{B}, \widehat{C_0}]^{en(n-1)}[\widehat{B}, \widehat{B}]^{en(n-1)}}$$

$$\times \frac{1}{\prod_{j=1}^{n-2}[\widehat{H}, \widehat{C_j}]^{enj}[\widehat{B}, \widehat{D_j}]^{enj}[\widehat{H}, \widehat{D_j}]^{en(n-1-j)}[\widehat{B}, \widehat{C_j}]^{en(n-1-j)}}. \qquad (5.4)$$

What Eq. (5.4) means is

Proposition 5.2. *For a divisor* $\Delta = A^{n-1}B^0 \prod_{j=0}^{n-2} C_j^{n-2-j} D_j^{j+1}$, *define*

$$q_\Delta = ([H, A][H, H][B, C_0][B, B])^{en(n-1)}$$

$$\times \prod_{j=1}^{n-2} ([H, C_j][B, D_j])^{enj}([H, D_j][B, C_j])^{en(n-1-j)},$$

where $H = D_0 \cup P_0$. *Then the quotient* $\theta^{2en^2}[\Delta](0, \Pi)/q_\Delta$ *is invariant under all the operators* T_S.

Proof. As in the proof of Proposition 5.1, the necessary properties of the construction of q_Δ from Δ are clear, and so is the fact that the formulation of Eq. (5.4) in this notation is

$$\frac{\theta[\Delta](0, \Pi)}{q_\Delta} = \frac{\theta[T_S(\Delta)](0, \Pi)}{q_{T_S(\Delta)}}.$$

Hence we obtain that for any divisor Δ, applying any permissible T_S does not change the quotient, which proves the proposition. $\qquad\square$

The same remark we had after Proposition 5.1 about the dependence of the number of divisors connected to a given divisor Δ holds equally here (but with every T_R replaced by T_S). The only difference is that now the set whose cardinality matters for this is B, and this makes a difference since contrary to C_0, the set B has cardinality equal to that of C_{-1}, and is thus never empty. However, we have seen in Section 3.2 that this does not prevent the divisors from being isolated when we talk about connections using the operators T_S. Once again, the number of divisors to which one can get from a given divisor Δ depends on Δ.

Also here we make the same remark about reduction. Since the common denominator here is not the same as with the operators T_R, we state what it is. Denote the set $H \cup B$ by J. Then since $\widehat{H} = B \setminus S \cup P_0$ and $\widehat{B} = D_0 \cup S$ we find that their union (which is of course $\widehat{J}$) is $B \cup D_0 \cup P_0 = B \cup H$ and equals J. Therefore the expressions $[J, C_j]^{en\min\{j, n-1-j\}}$ for $1 \le j \le n-2$ (which are the same as $[\widehat{J}, \widehat{C_j}]^{en\min\{j, n-1-j\}}$ for $1 \le j \le n-2$ when replacing j by $n-1-j$) are all common denominators, and so are the expressions $[J, D_j]^{en\min\{j, n-1-j\}}$ for $1 \le j \le n-2$ (which are the same as $[\widehat{J}, \widehat{D_j}]^{en\min\{j, n-1-j\}}$ for $1 \le j \le n-2$ when replacing j by $n-1-j$). Since applying T_S iteratively mixes B and $D_0 \subseteq H$ more and more, these are all the common factors we have here. The reason for these common expressions is as usual the fact that we have started with unreduced quotients for expressing the function $f(P)$. We continue to work with the unreduced Eq. (5.4) in Proposition 5.2 since we can merge it with Eq. (5.2) in Proposition 5.1 above regardless of the parity of n.

5.1.5 The Poor Man's Thomae

We now combine, for a given divisor $\Delta = A^{n-1}B^0 \prod_{j=0}^{n-2} C_j^{n-2-j} D_j^{j+1}$, Eqs. (5.2) and (5.4) from Propositions 5.1 and 5.2 into one bigger equality. This will be done by multiplying the expression p_Δ (i.e., dividing Eq. (5.2)) by some expression which is invariant under all the operators T_R and multiplying the expression q_Δ (i.e., dividing Eq. (5.4)) by some expression which is invariant under all the operators T_S such that the resulting denominators will be the same one. For this we notice that except for the common expressions $[C_0, B]^{en(n-1)}$ and $[A, D_0]^{en(n-1)}$ (the latter being part of $[C_{-1}, D_0]^{en(n-1)}$ in Eq. (5.2) and part of $[H, A]^{en(n-1)}$ in Eq. (5.2)), the sets B and D_0 appear only in q_Δ (recall that $H = D_0 \cup P_0$) and the sets A and C_0 appear only in p_Δ (recall that $C_{-1} = A \cup P_0$). So let us deal with these missing expressions and see how to correct them.

The set B appears in q_Δ in the expression $[B, B]^{en(n-1)}$ and in the expressions $[B, C_j]^{en(n-1-j)}$ and $[B, D_j]^{enj}$ for $1 \le j \le n-2$. Therefore in order to obtain the common expression we shall need to divide Eq. (5.2) by the product of these expressions. For this we have to write them also in the language of the sets with the tilde, which gives the expressions $[\widetilde{D_0},\widetilde{D_0}]^{en(n-1)}$, $[\widetilde{D_0},\widetilde{C_{n-1-j}}]^{en(n-1-j)}$ and $[\widetilde{D_0},\widetilde{D_{n-1-j}}]^{enj}$ for $1 \leq j \leq n-2$. This means that in order for the expression by which we divide Eq. (5.2) to be invariant under the operators T_R we have to divide it also by $[D_0,D_0]^{en(n-1)}$ and by $[D_0,C_j]^{enj}$ and $[D_0,D_j]^{en(n-1-j)}$ for $1 \leq j \leq n-2$ (with the usual replacement of j by $n-1-j$). This is very good for our purposes, since these are exactly the expressions in q_Δ in which D_0 appears (they are contained in $[H,H]^{en(n-1)}$ and in $[H,C_j]^{enj}$ and $[H,D_j]^{en(n-1-j)}$ for $1 \leq j \leq n-2$ respectively). In the language of the sets with the tilde these are the expressions $[\widetilde{B},\widetilde{B}]^{en(n-1)}$, $[\widetilde{B},\widetilde{C_{n-1-j}}]^{enj}$ and $[\widetilde{B},\widetilde{D_{n-1-j}}]^{en(n-1-j)}$ for $1 \leq j \leq n-2$, which means that altogether we have shown that the quotient composed of $\theta^{2en^2}[\Delta](0,\Pi)$ divided by

$$p_\Delta([B,B][D_0,D_0])^{en(n-1)} \prod_{j=1}^{n-2}([B,C_j][D_0,D_j])^{en(n-1-j)}([D_0,C_j][B,D_j])^{enj}$$

$$= ([C_{-1},C_{-1}][C_{-1},D_0][D_0,D_0][C_0,C_0][C_0,B][B,B])^{en(n-1)}$$

$$\times \prod_{j=1}^{n-2}([C_{-1},C_j][D_0,C_j][C_0,D_j][B,D_j])^{enj}([C_{-1},D_j][D_0,D_j][C_0,C_j][B,C_j])^{en(n-1-j)}$$

is invariant under the operators T_R.

Similarly, the set A appears in p_Δ in the expression $[C_{-1},C_{-1}]^{en(n-1)}$ and in the expressions $[C_{-1},C_j]^{enj}$ and $[C_{-1},D_j]^{en(n-1-j)}$ for $1 \leq j \leq n-2$. We thus want to divide Eq. (5.4) by the expression $[A,A]^{en(n-1)}$ and the product of the expressions $[A,C_j]^{enj}$ and $[A,D_j]^{en(n-1-j)}$ for $1 \leq j \leq n-2$. Since in the language of the sets with the hat these are the expression $[\widehat{C_0},\widehat{C_0}]^{en(n-1)}$ and the expressions $[\widehat{C_0},\widehat{C_{n-1-j}}]^{enj}$ and $[\widehat{C_0},\widehat{D_{n-1-j}}]^{en(n-1-j)}$ for $1 \leq j \leq n-2$, we see that in order to divide Eq. (5.4) by an expression that is invariant under the operators T_S, we shall have to divide it also by $[C_0,C_0]^{en(n-1)}$ and by $[C_0,C_j]^{en(n-1-j)}$ and $[C_0,D_j]^{enj}$ for $1 \leq j \leq n-2$ (j changed to $n-1-j$ again). Since these are exactly the expressions in p_Δ in which C_0 appears, this operation serves our purposes very well. In the language of the sets with the hat the new expressions are $[\widehat{A},\widehat{A}]^{en(n-1)}$ and the expressions $[\widehat{A},\widehat{C_{n-1-j}}]^{en(n-1-j)}$ and $[\widehat{A},\widehat{D_{n-1-j}}]^{enj}$ for $1 \leq j \leq n-2$, and we conclude that the quotient composed of $\theta^{2en^2}[\Delta](0,\Pi)$ divided by

$$q_\Delta([A,A][C_0,C_0])^{en(n-1)} \prod_{j=1}^{n-2}([A,C_j][C_0,D_j])^{enj}([C_0,C_j][A,D_j])^{en(n-1-j)}$$

$$= ([H,H][H,A][A,A][B,B][B,C_0][C_0,C_0])^{en(n-1)}$$

$$\times \prod_{j=1}^{n-2}([H,C_j][A,C_j][B,D_j][C_0,D_j])^{enj}([H,D_j][A,D_j][B,C_j][C_0,C_j])^{en(n-1-j)}$$

is invariant under the operators T_S.

Let us now compare these last two denominators. We claim that they are the same. It is clear that all the expressions containing B and C_0 indeed appear in both of them to the same powers, which leaves the expressions containing A, D_0, H and C_{-1}. Our first observation here is that $H \cup A = C_{-1} \cup D_0$ (as they both equal $A \cup D_0 \cup P_0$). This would give us that for every $1 \le j \le n-2$ we have (up to the usual sign) the equalities

$$[C_{-1},C_j][D_0,C_j] = [C_{-1} \cup D_0, C_j] = [H \cup A, C_j] = [H,C_j][A,C_j]$$

and

$$[C_{-1},D_j][D_0,D_j] = [C_{-1} \cup D_0, D_j] = [H \cup A, D_j] = [H,D_j][A,D_j].$$

Since the powers enj and $en(n-1-j)$ are even, we can compare these parts as well. The remaining expressions, which are the $en(n-1)$th powers of the expressions $[C_{-1},C_{-1}][C_{-1},D_0][D_0,D_0]$ and of $[H,H][H,A][H,A]$, are now equal by our observation and by Corollary 4.3 (the power $en(n-1)$ is also even). Altogether the two expressions are indeed equal.

We shall denote this common expression by g_Δ, and we would like to obtain a "shorter" form of it. For this we notice that the sets B and C_0 come only in expressions containing them to the same power and so do C_{-1} and D_0 in the first expression for g_Δ, and H and A in the second expression for g_Δ. Let us thus denote $H \cup A = C_{-1} \cup D_0$ by E and $C_0 \cup B$ by F. Thus

$$\widetilde{E} = \widetilde{H} \cup \widetilde{A} = \widetilde{C_{-1}} \cup \widetilde{D_0} = C_0 \setminus R \cup P_0 \cup B = F \cup P_0 \setminus R$$

and

$$\widetilde{F} = \widetilde{C_0} \cup \widetilde{B} = A \cup R \cup D_0 = E \setminus P_0 \cup R.$$

Furthermore,

$$\widehat{E} = \widehat{H} \cup \widehat{A} = \widehat{C_{-1}} \cup \widehat{D_0} = C_0 \cup P_0 \cup B \setminus S = F \cup P_0 \setminus S$$

and

$$\widehat{F} = \widehat{C_0} \cup \widehat{B} = A \cup D_0 \cup S = E \setminus P_0 \cup S.$$

This means that in this form we can merge Eqs. (5.2) and (5.4) into the one larger equation

$$\frac{\theta[\Delta]^{2en^2}(0,\Pi)}{([E,E][F,F])^{en(n-1)} \prod_{j=1}^{n-2}([E,C_j][F,D_j])^{enj}([E,D_j][F,C_j])^{en(n-1-j)}}$$

$$= \frac{\theta[T_R(\Delta)]^{2en^2}(0,\Pi)}{([\widetilde{E},\widetilde{E}][\widetilde{F},\widetilde{F}])^{en(n-1)} \prod_{j=1}^{n-2}([\widetilde{E},\widetilde{C_j}][\widetilde{F},\widetilde{D_j}])^{enj}([\widetilde{E},\widetilde{D_j}][\widetilde{F},\widetilde{C_j}])^{en(n-1-j)}}$$

$$= \frac{\theta[T_S(\Delta)]^{2en^2}(0,\Pi)}{([\widehat{E},\widehat{E}][\widehat{F},\widehat{F}])^{en(n-1)} \prod_{j=1}^{n-2}([\widehat{E},\widehat{C_j}][\widehat{F},\widehat{D_j}])^{enj}([\widehat{E},\widehat{D_j}][\widehat{F},\widehat{C_j}])^{en(n-1-j)}}, \quad (5.5)$$

where we have also used Lemma 4.2. This gives us the PMT, which will be concluded in Proposition 5.3 below.

Before we move on we remark that the appearance of the sets E and F in Eq. (5.5), which by Proposition 5.3 is the PMT for this type of singular Z_n curves, is quite natural when we examine the case $n = 2$. As is easily seen from the equation defining the Z_n curve, a singular Z_2 curve of this type with a given value of m is actually a nonsingular Z_2 curve with $r = m$. When comparing the two ways of looking at this curve we see that a divisor $\Delta = A^1 B^0 C_0^0 D_0^1$ can be written as $(E \setminus P_0)^1 F^0$ in this language, with these sets E and F. The permissible operators are T_R with $R \in C_0$ and T_S with $S \in B$, and here it is just the same as taking points from F, which is the "new C_0". Since adding P_0 to the "new A" (which is $E \setminus P_0$) we clearly get E again (which is the "new C_{-1}"), this is indeed the expected PMT. Returning to work with general n, we have chosen, for later convenience, to reconvert the expressions containing E and F to expressions involving C_{-1} and D_0, and to use the first form of g_Δ. This is because we prefer to work with the sets C_{-1} and D_0, rather than A and the ad hoc set H. We also note that the expression for g_Δ defined in Proposition 5.3 below becomes much more symmetric in the sets C_j and D_j, $-1 \leq j \leq n-2$ after we denote B by D_{-1}. Therefore we do this replacement from now on.

Now we are ready to give the PMT.

Proposition 5.3. *For a divisor* $\Delta = A^{n-1} B^0 \prod_{j=0}^{n-2} C_j^{n-2-j} D_j^{j+1}$, *define*

$$g_\Delta = ([C_{-1},C_{-1}][C_{-1},D_0][D_0,D_0][C_0,C_0][C_0,D_{-1}][D_{-1},D_{-1}])^{en(n-1)}$$

$$\times \prod_{j=1}^{n-2}([C_{-1},C_j][D_0,C_j][C_0,D_j][D_{-1},D_j])^{enj}$$

$$\times \prod_{j=1}^{n-2}([C_{-1},D_j][D_0,D_j][C_0,C_j][D_{-1},C_j])^{en(n-1-j)},$$

where $C_{-1} = A \cup P_0$ *and* $D_{-1} = B$. *Then the quotient* $\theta^{2en^2}[\Delta](0,\Pi)/g_\Delta$ *is invariant under all the operators* T_R *and all the operators* T_S *and is the PMT for our singular* Z_n *curve* X.

Proof. It is clear how to construct g_Δ from Δ and that the construction depends only on Δ, which is a necessary property of the PMT. We now write the quotient in question as

$$\frac{\theta^{2en^2}[\Delta](0,\Pi)}{g_\Delta} = \frac{\theta^{2en^2}[\Delta](0,\Pi)}{p_\Delta} \bigg/ \frac{g_\Delta}{p_\Delta} = \frac{\theta^{2en^2}[\Delta](0,\Pi)}{q_\Delta} \bigg/ \frac{g_\Delta}{q_\Delta}.$$

Now, since Proposition 5.1 shows us that $\theta[\Delta](0,\Pi)/p_\Delta$ is T_R-invariant and we know that the expression

$$\frac{g_\Delta}{p_\Delta} = ([B,B][D_0,D_0])^{en(n-1)} \prod_{j=1}^{n-2}([B,C_j][D_0,D_j])^{en(n-1-j)}([D_0,C_j][B,D_j])^{enj}$$

is also invariant under the operators T_R, we deduce the T_R-invariance of the quotient $\theta[\Delta](0,\Pi)/g_\Delta$. Moreover, Proposition 5.2 gives that $\theta[\Delta](0,\Pi)/q_\Delta$ is invariant under the operators T_S, and as the expression

$$\frac{g_\Delta}{q_\Delta} = ([A,A][C_0,C_0])^{en(n-1)} \prod_{j=1}^{n-2} ([A,C_j][C_0,D_j])^{enj} ([C_0,C_j][A,D_j])^{en(n-1-j)}$$

is also T_S-invariant, we conclude that the quotient $\theta[\Delta](0,\Pi)/g_\Delta$ is invariant also under the operators T_S. This proves the proposition. Alternatively, it follows directly from Eq. (5.5) by replacing E with $C_{-1} \cup D_0$ and F with $C_0 \cup D_{-1}$, and splitting the corresponding expressions.   □

5.1.6 Reduced Formulae

As in the nonsingular case, the expression for g_Δ in Proposition 5.3 is not reduced. We recall that the set $G = C_{-1} \cup C_0$ satisfies $\widetilde{G} = G$ and that the set $J = H \cup B$ satisfies $\widehat{J} = J$. Since P_0 is fixed this means that the set $I = J \setminus P_0 = D_{-1} \cup D_0$ also satisfies $\widehat{I} = I$. The equalities

$$\widehat{C_0} = A, \quad \widehat{C_{-1}} = \widehat{A} \cup P_0 = C_0 \cup P_0, \quad \widetilde{D_{-1}} = \widetilde{B} = D_0, \quad \widetilde{D_0} = B = D_{-1}$$

imply that $\widehat{G} = G$ and $\widetilde{I} = I$ as well. This means that for a given divisor Δ, the sets G and I are the same sets of points in all the divisors participating in Proposition 5.3 arising from Δ. This is thus true for the set $K = G \cup I$ as well, which also equals $E \cup F$, and the equality $\widetilde{K} = \widehat{K} = K$ can thus be obtained also from the expressions for $\widetilde{E}$, $\widetilde{F}$, $\widehat{E}$ and $\widehat{F}$. Therefore when we examine the PMT in Eq. (5.5) we see that for a given divisor Δ the expressions

$$[K,C_j]^{en\min\{j,n-1-j\}} = [\widetilde{K},\widetilde{C_{n-1-j}}]^{en\min\{j,n-1-j\}} = [\widehat{K},\widehat{C_{n-1-j}}]^{en\min\{j,n-1-j\}}$$

and the expressions

$$[K,D_j]^{en\min\{j,n-1-j\}} = [\widetilde{K},\widetilde{D_{n-1-j}}]^{en\min\{j,n-1-j\}} = [\widehat{K},\widehat{D_{n-1-j}}]^{en\min\{j,n-1-j\}}$$

for $1 \le j \le n-2$ appear in the products in all three terms of Eq. (5.5). As in the nonsingular case (where the common factors $[G,C_j]^{en\min\{j,n-1-j\}}$ there are exactly the C-sets part of these expressions), these are the only common expressions which appear under every theta constant in this set of divisors. Since when j runs from 1 to $n-2$, so does $n-1-j$, and since the powers are invariant under $j \mapsto n-1-j$, we see that this common denominator is simply the product of

$$\prod_{j=1}^{n-2} [K,C_j]^{en\min\{j,n-1-j\}} = \prod_{j=1}^{n-2} [\widetilde{K},\widetilde{C}_j]^{en\min\{j,n-1-j\}} = \prod_{j=1}^{n-2} [\widehat{K},\widehat{C}_j]^{en\min\{j,n-1-j\}}$$

and

$$\prod_{j=1}^{n-2} [K,D_j]^{en\min\{j,n-1-j\}} = \prod_{j=1}^{n-2} [\widetilde{K},\widetilde{D}_j]^{en\min\{j,n-1-j\}} = \prod_{j=1}^{n-2} [\widehat{K},\widehat{D}_j]^{en\min\{j,n-1-j\}},$$

an expression which is indeed clearly invariant under all the operators T_R and T_S.

Again, these common expressions appear since we used unreduced expressions throughout the process. Let us now indicate what happens if one does start with the reduced expressions (and thus separating into cases according to the parity of n from the start). We leave as an exercise for the reader to verify the details, but we state the results. In the reduced version of Eq. (5.1) we obtain, when n is odd, an equality where the denominator under $\theta[\Delta]^{4n^2}(0,\Pi)$ is

$$([P_0,A][P_0,D_0][R,C_0 \setminus R][R,B])^{2n(n-1)}$$

$$\times \prod_{j=(n+1)/2}^{n-2} ([P_0,C_j][R,D_j])^{2n(2j-(n-1))} \prod_{j=1}^{(n-3)/2} ([P_0,D_j][R,C_j])^{2n(n-1-2j)},$$

and the denominator under $\theta[T_R(\Delta)]^{4n^2}(0,\Pi)$ is

$$([P_0,C_0 \setminus R][P_0,B][R,A][R,D_0])^{2n(n-1)}$$

$$\times \prod_{j=1}^{(n-3)/2} ([P_0,C_j][R,D_j])^{2n(n-1-2j)} \prod_{j=(n+1)/2}^{n-2} ([P_0,D_j][R,C_j])^{2n(2j-(n-1))}.$$

When n is even, we obtain an equality where the denominator under $\theta[\Delta]^{2n^2}(0,\Pi)$ is

$$([P_0,A][P_0,D_0][R,C_0 \setminus R][R,B])^{n(n-1)}$$

$$\times \prod_{j=n/2}^{n-2} ([P_0,C_j][R,D_j])^{n(2j-(n-1))} \prod_{j=1}^{(n-2)/2} ([P_0,D_j][R,C_j])^{n(n-1-2j)},$$

and the denominator under $\theta[T_R(\Delta)]^{2n^2}(0,\Pi)$ is

$$([P_0,C_0 \setminus R][P_0,B][R,A][R,D_0])^{n(n-1)}$$

$$\times \prod_{j=1}^{(n-2)/2} ([P_0,C_j][R,D_j])^{n(n-1-2j)} \prod_{j=n/2}^{n-2} ([P_0,D_j][R,C_j])^{n(2j-(n-1))}.$$

Before we turn to the symmetrization and the reduced expression for p_Δ, let us introduce the reduced version of Eq. (5.3). When n is odd, we obtain an equality where the denominator under $\theta[\Delta]^{4n^2}(0,\Pi)$ is

$$([P_0,A][P_0,D_0][S,C_0][S,B\setminus S])^{2n(n-1)}$$

$$\times \prod_{j=(n+1)/2}^{n-2} ([P_0,C_j][S,D_j])^{2n(2j-(n-1))} \prod_{j=1}^{(n-3)/2} ([P_0,D_j][S,C_j])^{2n(n-1-2j)},$$

and the denominator under $\theta[T_S(\Delta)]^{4n^2}(0,\Pi)$ is

$$([P_0,C_0][P_0,B\setminus S][S,A][S,D_0])^{2n(n-1)}$$

$$\times \prod_{j=1}^{(n-3)/2} ([P_0,C_j][S,D_j])^{2n(n-1-2j)} \prod_{j=(n+1)/2}^{n-2} ([P_0,D_j][S,C_j])^{2n(2j-(n-1))}.$$

When n is even, we get an equality where the denominator under $\theta[\Delta]^{2n^2}(0,\Pi)$ is

$$([P_0,A][P_0,D_0][S,C_0][S,B\setminus S])^{n(n-1)}$$

$$\times \prod_{j=n/2}^{n-2} ([P_0,C_j][S,D_j])^{n(2j-(n-1))} \prod_{j=1}^{(n-2)/2} ([P_0,D_j][S,C_j])^{n(n-1-2j)},$$

and the denominator under $\theta[T_S(\Delta)]^{2n^2}(0,\Pi)$ is

$$([P_0,C_0][P_0,B\setminus S][S,A][S,D_0])^{n(n-1)}$$

$$\times \prod_{j=1}^{(n-2)/2} ([P_0,C_j][S,D_j])^{n(n-1-2j)} \prod_{j=n/2}^{n-2} ([P_0,D_j][S,C_j])^{n(2j-(n-1))}.$$

One can now observe that the difference between the reduced formulae and the unreduced ones is as described in Chapter 4. Explicitly, the expressions which appeared to the power $en(n-1)$ remain untouched, and the expressions whose power depend on j are changed only according to their power and to the parity of n just as they are changed in the nonsingular case. Every expression which appeared to the power enj now appears to the power $en(2j-(n-1))$ and is taken in the case of odd n only for $(n+1)/2 \le j \le n-2$ and in the case of even n only for $n/2 \le j \le n-2$. Every expression which appeared to the power $en(n-1-j)$ now appears to the power $en(n-1-2j)$ and is taken only for $1 \le j \le (n-3)/2$ if n is odd and for $1 \le j \le (n-3)/2$ if n is even. The remark about the value $j = (n-1)/2$ for odd n in the nonsingular case holds equally here. Following this description one sees that the symmetrization of the reduced Eq. (5.1) over R replaces Proposition 5.1 and Eq. (5.2) by similar equations but with the reduced form of p_Δ, which is obtained from p_Δ by the same "reduction mechanism". Similarly, after the symmetrization of the reduced Eq. (5.3) over S one gets an analog of Proposition 5.2 and Eq. (5.4) but with the reduced q_Δ (with similar reduction). Finally, in the combination of the T_R-invariant part and the T_S-invariant part we can reprove Proposition 5.3 with the reduced form of Eq. (5.5), where g_Δ is

$$([E,E][F,F])^{2n(n-1)} \prod_{j=(n+1)/2}^{n-2} ([E,C_j][F,D_j])^{2n(2j-(n-1))} \prod_{j=1}^{(n-3)/2} ([E,D_j][F,C_j])^{2n(n-1-2j)}$$

$$= ([C_{-1},C_{-1}][C_{-1},D_0][D_0,D_0][C_0,C_0][C_0,D_{-1}][D_{-1},D_{-1}])^{2n(n-1)}$$

$$\times \prod_{j=(n+1)/2}^{n-2} ([C_{-1},C_j][D_0,C_j][C_0,D_j][D_{-1},D_j])^{2n(2j-(n-1))}$$

$$\times \prod_{j=1}^{(n-3)/2} ([C_{-1},D_j][D_0,D_j][C_0,C_j][D_{-1},C_j])^{2n(n-1-2j)} \tag{5.6}$$

when n is odd, and

$$([E,E][F,F])^{n(n-1)} \prod_{j=n/2}^{n-2} ([E,C_j][F,D_j])^{n(2j-(n-1))} \prod_{j=1}^{(n-2)/2} ([E,D_j][F,C_j])^{n(2j-(n-1))}$$

$$= ([C_{-1},C_{-1}][C_{-1},D_0][D_0,D_0][C_0,C_0][C_0,D_{-1}][D_{-1},D_{-1}])^{n(n-1)}$$

$$\times \prod_{j=n/2}^{n-2} ([C_{-1},C_j][D_0,C_j][C_0,D_j][D_{-1},D_j])^{n(2j-(n-1))}$$

$$\times \prod_{j=1}^{(n-2)/2} ([C_{-1},D_j][D_0,D_j][C_0,C_j][D_{-1},C_j])^{n(2j-(n-1))} \tag{5.7}$$

when n is even. As in the nonsingular case, one can obtain at every stage the reduced equations from the unreduced ones by by canceling the common factor appearing in every stage, as the reader can easily verify. The Thomae formulae can now be obtained either from these reduced equations or just from Proposition 5.3 and Eq. (5.5).

We now want to improve the PMT of Proposition 5.3 in the same way as before: replacing the denominator g_Δ by some other denominator h_Δ which will be constructed from Δ alone and invariant under N, while preserving the property that the quotient $\theta[\Delta](0,\Pi)/h_\Delta$ is invariant under the operators T_R and T_S. As usual, the N-invariance of the denominator h_Δ translates, via Eq. (1.5) in the form $\theta[N(\Delta)](0,\Pi) = \theta[\Delta](0,\Pi)$ to the proper power, to N-invariance of the quotient $\theta[\Delta](0,\Pi)/h_\Delta$. As in the nonsingular case, before we prove the general case we do the example of $n=5$, which will hopefully illuminate the process of the general case (again $n=3$ is too small).

5.2 Example with $n=5$ and General m

The PMT in Proposition 5.3 and Eq. (5.5) states, when $n=5$ (and hence $e=2$) and arbitrary $m \geq 2$, that the expression $\theta^{100}[\Delta](0,\Pi)/g_\Delta$ with

$$\begin{aligned}
g_\Delta = &([C_{-1},C_{-1}][C_{-1},D_0][D_0,D_0][C_0,C_0][C_0,D_{-1}][D_{-1},D_{-1}])^{40} \qquad (5.8)\\
&\times ([C_{-1},C_1][D_0,C_1][C_0,D_1][D_{-1},D_1])^{10}([C_{-1},D_1][D_0,D_1][C_0,C_1][D_{-1},C_1])^{30}\\
&\times ([C_{-1},C_2][D_0,C_2][C_0,D_2][D_{-1},D_2])^{20}([C_{-1},D_2][D_0,D_2][C_0,C_2][D_{-1},C_2])^{20}\\
&\times ([C_{-1},C_3][D_0,C_3][C_0,D_3][D_{-1},D_3])^{30}([C_{-1},D_3][D_0,D_3][C_0,C_3][D_{-1},C_3])^{10}
\end{aligned}$$

(the divisor Δ is $A^4 B^0 C_0^3 D_0^1 C_1^2 D_1^2 C_2^1 D_2^3 C_3^0 D_3^4$ when $n = 5$) is invariant under the operators T_R and T_S. When one works with the reduced PMT in Eq. (5.6) one finds that the quotient with

$$\begin{aligned}
g_\Delta = &([C_{-1},C_{-1}][C_{-1},D_0][D_0,D_0][C_0,C_0][C_0,D_{-1}][D_{-1},D_{-1}])^{40}\\
&\times ([C_{-1},C_3][D_0,C_3][C_0,D_3][D_{-1},D_3])^{20}([C_{-1},D_1][D_0,D_1][C_0,C_1][D_{-1},C_1])^{20}
\end{aligned}$$

has the same properties. We shall, however, continue to work with the unreduced formula. The sets with the tilde, corresponding to $T_R(\Delta)$, are

$$\widetilde{A} = C_0 \setminus R, \quad \widetilde{C_{-1}} = \widetilde{A} \cup P_0 = C_0 \setminus R \cup P_0, \quad \widetilde{C_0} = A \cup R,$$

$$\widetilde{C_1} = C_3, \quad \widetilde{C_2} = C_2, \quad \text{and} \quad \widetilde{C_3} = C_1,$$

and also

$$\widetilde{D_{-1}} = \widetilde{B} = D_0, \quad \widetilde{D_0} = B = D_{-1}, \quad \widetilde{D_1} = D_3, \quad \widetilde{D_2} = D_2, \quad \text{and} \quad \widetilde{D_3} = D_1.$$

Similarly, sets with the hat that correspond to $T_S(\Delta)$ are

$$\widehat{A} = C_0, \quad \widehat{C_{-1}} = \widehat{A} \cup P_0 = C_0 \cup P_0, \quad \widehat{C_0} = A, \quad \widehat{C_1} = C_3, \quad \widehat{C_2} = C_2, \quad \text{and} \quad \widehat{C_3} = C_1,$$

and also

$$\widehat{D_{-1}} = \widehat{B} = D_0 \cup S, \quad \widehat{D_0} = B \setminus S = D_{-1} \setminus S, \quad \widehat{D_1} = D_3, \quad \widehat{D_2} = D_2, \quad \text{and} \quad \widehat{D_3} = D_1.$$

We want to make the denominator invariant under the operator N. Here the construction will be similar to the nonsingular case, but with additional parts which deal with the sets D_j, $-1 \le j \le 3$.

The divisor $N(\Delta) = A^{n-1} B^0 \prod_{j=0}^{n-2} C_j^j D_j^{n-1-j}$, from Definition 2.18, becomes just $A^4 B^0 C_3^3 D_3^1 C_2^2 D_2^2 C_1^1 D_1^3 C_0^0 D_0^4$ when $n = 5$, so that the denominator $g_{N(\Delta)}$ appearing under $\theta^{100}[N(\Delta)](0,\Pi)$ in the corresponding PMT is obtained from g_Δ by replacing C_0 by C_3, D_0 by D_3, C_1 by C_2, D_1 by D_2, C_2 by C_1, D_2 by D_1, C_3 by C_0, and D_3 by D_0 (and fixing every C_{-1} and every $B = D_{-1}$). Hence we have

$$\begin{aligned}
g_{N(\Delta)} = &([C_{-1},C_{-1}][C_{-1},D_3][D_3,D_3][C_3,C_3][C_3,D_{-1}][D_{-1},D_{-1}])^{40}\\
&\times ([C_{-1},C_2][D_3,C_2][C_3,D_2][D_{-1},D_2])^{10}([C_{-1},D_2][D_3,D_2][C_3,C_2][D_{-1},C_2])^{30}\\
&\times ([C_{-1},C_1][D_3,C_1][C_3,D_1][D_{-1},D_1])^{20}([C_{-1},D_1][D_3,D_1][C_3,C_1][D_{-1},C_1])^{20}\\
&\times ([C_{-1},C_0][D_3,C_0][C_3,D_0][D_{-1},D_0])^{30}([C_{-1},D_0][D_3,D_0][C_3,C_0][D_{-1},C_0])^{10}.
\end{aligned}$$

As in the nonsingular case, we shall multiply the denominator g_Δ from Eq. (5.9) by expressions which appear in $g_{N(\Delta)}$ to powers larger than those to which they appear in g_Δ, and once again the order of doing these corrections will be important. In the beginning there are 4 types of expressions: $[C_i, C_j]$, $[D_i, D_j]$, $[C_i, D_j]$ and $[D_i, C_j]$ with i being -1 or 0 and $j \geq 1$ (hence the two last types are distinct), and later i and j can attain other values (so that the last two types may then coincide). In the process presented here we shall do one stage at a time for every type, although the corrections of the different types are independent of one another. Of course, the first two types behave exactly the same and the last two types behave exactly the same, but the two pairs exhibit different behaviors. The reader can always compare what we do here with Section 4.2 and see that what we do there is exactly the corrections of the first type here. This is what will happen also in the general case.

5.2.1 Correcting the Expressions Involving C_{-1} and D_{-1}

The first expressions we deal with are the pair of the expressions $[C_{-1}, C_1]^{10}$ and $[D_{-1}, D_1]^{10}$ and the pair $[C_{-1}, D_3]^{30}$ and $[D_{-1}, C_3]^{30}$. All these expressions appear in both g_Δ and $g_{N(\Delta)}$ but to different powers. In the notation of the sets with the tilde we have that these expressions are $[\widetilde{C_0 \setminus R \cup P_0}, \widetilde{C_3}]^{10}$, $[\widetilde{D_0}, \widetilde{D_3}]^{10}$, $[\widetilde{C_0 \setminus R \cup P_0}, \widetilde{D_1}]^{30}$ and $[\widetilde{D_0}, \widetilde{C_1}]^{30}$, and in the notation of the sets with the hat they are $[\widehat{C_0 \cup P_0}, \widehat{C_3}]^{10}$, $[\widehat{D_0 \cup S}, \widehat{D_3}]^{10}$, $[\widehat{C_0 \cup P_0}, \widehat{D_1}]^{30}$ and $[\widehat{D_0 \cup S}, \widehat{C_1}]^{30}$. As in the nonsingular case, in order to keep the invariance of the whole expression under the operators T_R and T_S it does not suffice to divide Eq. (5.9) also by these expressions with the tilde or hat omitted, and we must divide Eq. (5.9) by the invariant expressions

$$[C_{-1}, C_1]^{10}[C_0, C_1]^{10}[C_{-1}, C_3]^{10}[C_0, C_3]^{10},$$

$$[D_{-1}, D_1]^{10}[D_0, D_1]^{10}[D_{-1}, D_3]^{10}[D_0, D_3]^{10},$$

$$[C_{-1}, D_3]^{30}[C_0, D_3]^{30}[C_{-1}, D_1]^{30}[C_0, D_1]^{30},$$

and

$$[D_{-1}, C_3]^{30}[D_0, C_3]^{30}[D_{-1}, C_1]^{30}[D_0, C_1]^{30}.$$

These expressions are indeed invariant under the operators T_R and T_S, i.e., are equal to the expressions obtained from them by putting a tilde or a hat on all the sets, since they equal the 10th or 30th powers of $[G, C_1][G, C_3]$, $[I, D_1][I, D_3]$, $[G, D_3][G, D_1]$ and $[I, C_3][I, C_1]$ respectively. We have the equalities

$$[\widetilde{G}, \widetilde{C_1}] = [\widehat{G}, \widehat{C_1}] = [G, C_3], \quad [\widetilde{G}, \widetilde{C_3}] = [\widehat{G}, \widehat{C_3}] = [G, C_1]$$

for the first expression,

$$[\widetilde{I}, \widetilde{D_1}] = [\widehat{I}, \widehat{D_1}] = [I, D_3], \quad [\widetilde{I}, \widetilde{D_3}] = [\widehat{I}, \widehat{D_3}] = [I, D_1]$$

for the second one,

$$[\widetilde{G},\widetilde{D_3}] = [\widehat{G},\widehat{D_3}] = [G,D_1], \quad [\widetilde{G},\widetilde{D_1}] = [\widehat{G},\widehat{D_1}] = [G,D_3]$$

for the third, and

$$[\widetilde{I},\widetilde{C_3}] = [\widehat{I},\widehat{C_3}] = [I,C_1], \quad [\widetilde{I},\widetilde{C_1}] = [\widehat{I},\widehat{C_1}] = [I,C_3]$$

for the fourth (as we already saw that $\widetilde{G} = \widehat{G} = G$ and $\widetilde{I} = \widehat{I} = I$), all up to sign but the powers 10 and 30 are even. Doing so yields the new denominator

$$([C_{-1},C_{-1}][C_{-1},D_0][D_0,D_0][C_0,C_0][C_0,D_{-1}][D_{-1},D_{-1}])^{40}$$
$$\times ([C_{-1},C_1][D_{-1},D_1])^{20}([D_0,C_1][C_0,D_1])^{40}([C_{-1},D_1][D_{-1},C_1])^{60}([D_0,D_1][C_0,C_1])^{40}$$
$$\times ([C_{-1},C_2][D_0,C_2][C_0,D_2][D_{-1},D_2])^{20}([C_{-1},D_2][D_0,D_2][C_0,C_2][D_{-1},C_2])^{20}$$
$$\times ([C_{-1},C_3][D_{-1},D_3])^{40}([D_0,C_3][C_0,D_3])^{60}([C_{-1},D_3][D_{-1},C_3])^{40}([D_0,D_3][C_0,C_3])^{20},$$

and since we see that the expressions by which we multiplied $g_{N(\Delta)}$ are

$$[C_{-1},C_2]^{10}[C_3,C_2]^{10}[C_{-1},C_0]^{10}[C_3,C_0]^{10},$$

$$[D_{-1},D_2]^{10}[D_3,D_2]^{10}[D_{-1},D_0]^{10}[D_3,D_0]^{10},$$
$$[C_{-1},D_0]^{30}[C_3,D_0]^{30}[C_{-1},D_2]^{30}[C_3,D_2]^{30},$$

and

$$[D_{-1},C_0]^{30}[D_3,C_0]^{30}[D_{-1},C_2]^{30}[D_3,C_2]^{30},$$

we find that it becomes the expression

$$([C_{-1},C_{-1}][C_{-1},D_3][D_3,D_3][C_3,C_3][C_3,D_{-1}][D_{-1},D_{-1}])^{40}$$
$$\times ([C_{-1},C_2][D_{-1},D_2])^{20}([D_3,C_2][C_3,D_2])^{40}([C_{-1},D_2][D_{-1},C_2])^{60}([D_3,D_2][C_3,C_2])^{40}$$
$$\times ([C_{-1},C_1][D_3,C_1][C_3,D_1][D_{-1},D_1])^{20}([C_{-1},D_1][D_3,D_1][C_3,C_1][D_{-1},C_1])^{20}$$
$$\times ([C_{-1},C_0][D_{-1},D_0])^{40}([D_3,C_0][C_3,D_0])^{60}([C_{-1},D_0][D_{-1},C_0])^{40}([D_3,D_0][C_3,C_0])^{20}.$$

The next part we want to fix is the part containing the expressions $[C_{-1},C_0]^{40}$ and $[D_{-1},D_0]^{40}$, together with $[C_{-1},D_2]^{40}$ and $[D_{-1},C_2]^{40}$. The first two expressions appear in $g_{N(\Delta)}$ but not in g_Δ, and the other two appear in both of them but to different powers. As for the order, note that we should indeed do these corrections after the previous ones as the last corrections changed the first two powers from 30 to 40 and the other two powers from 10 to 40. These expressions are, in the notation of the sets with the tilde, $[\widetilde{C_0 \setminus R \cup P_0},\widetilde{A \cup R}]^{40}$, $[\widetilde{D_0},\widetilde{D_{-1}}]^{40}$, $[\widetilde{C_0 \setminus R \cup P_0},\widetilde{D_2}]^{40}$ and $[\widetilde{D_0},\widetilde{C_2}]^{40}$ respectively, and in the notation of the sets with the hat we see that they are $[\widehat{C_0 \cup P_0},\widehat{A}]^{40}$, $[\widehat{D_0 \cup S},\widehat{D_{-1} \setminus S}]^{40}$, $[\widehat{C_0 \cup P_0},\widehat{D_2}]^{40}$ and $[\widehat{D_0 \cup S},\widehat{C_2}]^{40}$ respectively. Again this means that we shall have to divide Eq. (5.9) by these expressions without the tilde or hat, but as before this will not suffice and we must divide Eq. (5.9) by the invariant expressions

$$[C_{-1}, C_{-1}]^{40}[C_0, C_0]^{40}[C_{-1}, C_0]^{40}, \quad [D_{-1}, D_{-1}]^{40}[D_0, D_0]^{40}[D_{-1}, D_0]^{40},$$

$$[C_{-1}, D_2]^{40}[C_0, D_2]^{40}, \quad \text{and} \quad [D_{-1}, C_2]^{40}[D_0, C_2]^{40}.$$

These expressions are indeed invariant under the operators T_R and T_S since they equal the 40th powers of the expressions $[G, G]$, $[I, I]$, $[G, D_2]$ and $[I, C_2]$ respectively and we have the corresponding equalities

$$[\widetilde{G}, \widetilde{G}] = [\widehat{G}, \widehat{G}] = [G, G], \quad [\widetilde{I}, \widetilde{I}] = [\widehat{I}, \widehat{I}] = [I, I],$$

$$[\widetilde{G}, \widetilde{C_2}] = [\widehat{G}, \widehat{C_2}] = [G, C_2], \quad \text{and} \quad [\widetilde{I}, \widetilde{D_2}] = [\widehat{I}, \widehat{D_2}] = [I, D_2]$$

(where we recall again that $\widetilde{G} = \widehat{G} = G$ and $\widetilde{I} = \widehat{I} = I$), again up to sign but the power 40 is even (we have used Corollary 4.3 for the first two equalities). After this action we obtain

$$([C_{-1}, C_{-1}][D_0, D_0][C_0, C_0][D_{-1}, D_{-1}])^{80}([C_{-1}, D_0][C_{-1}, C_0][C_0, D_{-1}][D_{-1}, D_0])^{40}$$
$$\times ([C_{-1}, C_1][D_{-1}, D_1])^{20}([D_0, C_1][C_0, D_1])^{40}([C_{-1}, D_1][D_{-1}, C_1])^{60}([D_0, D_1][C_0, C_1])^{40}$$
$$\times ([C_{-1}, C_2][D_{-1}, D_2])^{20}([D_0, C_2][C_0, D_2])^{60}([C_{-1}, D_2][D_{-1}, C_2])^{60}([D_0, D_2][C_0, C_2])^{20}$$
$$\times ([C_{-1}, C_3][D_{-1}, D_3])^{40}([D_0, C_3][C_0, D_3])^{60}([C_{-1}, D_3][D_{-1}, C_3])^{40}([D_0, D_3][C_0, C_3])^{20}.$$

The expressions by which we multiplied $g_{N(\Delta)}$ in this process are

$$[C_{-1}, C_{-1}]^{40}[C_3, C_3]^{40}[C_{-1}, C_3]^{40}, \quad [D_{-1}, D_{-1}]^{40}[D_3, D_3]^{40}[D_{-1}, D_3]^{40},$$

$$[C_{-1}, D_1]^{40}[C_3, D_1]^{40}, \quad \text{and} \quad [D_{-1}, C_1]^{40}[D_3, C_1]^{40},$$

which yields

$$([C_{-1}, C_{-1}][D_3, D_3][C_3, C_3][D_{-1}, D_{-1}])^{80}([C_{-1}, D_3][C_{-1}, C_3][C_3, D_{-1}][D_{-1}, D_3])^{40}$$
$$\times ([C_{-1}, C_2][D_{-1}, D_2])^{20}([D_3, C_2][C_3, D_2])^{40}([C_{-1}, D_2][D_{-1}, C_2])^{60}([D_3, D_2][C_3, C_2])^{40}$$
$$\times ([C_{-1}, C_1][D_{-1}, D_1])^{20}([D_3, C_1][C_3, D_1])^{60}([C_{-1}, D_1][D_{-1}, C_1])^{60}([D_3, D_1][C_3, C_1])^{20}$$
$$\times ([C_{-1}, C_0][D_{-1}, D_0])^{40}([D_3, C_0][C_3, D_0])^{60}([C_{-1}, D_0][D_{-1}, C_0])^{40}([D_3, D_0][C_3, C_0])^{20}.$$

Comparing the expressions involving the sets C_{-1} and D_{-1} in the last two expressions shows that all of them appear in both of them to the same powers, so that we are done with the corrections concerning them.

5.2.2 Correcting the Expressions Not Involving C_{-1} and D_{-1}

We now turn to correct the 5 pairs of expressions $[C_3, C_3]^{80}$ and $[D_3, D_3]^{80}$, $[C_3, C_2]^{40}$ and $[D_3, D_2]^{40}$, $[C_3, C_1]^{20}$ and $[D_3, D_1]^{20}$, $[C_3, D_2]^{40}$ and $[D_3, C_2]^{40}$, and $[C_3, D_1]^{60}$ and $[D_3, C_1]^{60}$, all of which appear in $g_{N(\Delta)}$ but not in g_Δ. Here as well the powers 80 and 40 in the first expressions were originally 40 and 30 and the powers 40 and

60 in the last expressions were originally 10 and 20, so these corrections should indeed be done after the two others. In the notation with the tilde or hat we see that these expressions are

$$[\widetilde{C_1, C_1}]^{80} = [\widehat{C_1, C_1}]^{80}, \quad [\widetilde{D_1, D_1}]^{80} = [\widehat{D_1, D_1}]^{80}, \quad [\widetilde{C_1, C_2}]^{40} = [\widehat{C_1, C_2}]^{40},$$

$$[\widetilde{D_1, D_2}]^{40} = [\widehat{C_1, D_2}]^{40}, \quad [\widetilde{C_1, C_3}]^{20} = [\widehat{C_1, C_3}]^{20}, \quad [\widetilde{D_1, D_3}]^{20} = [\widehat{D_1, D_3}]^{20},$$

$$[\widetilde{C_1, D_2}]^{40} = [\widehat{C_1, D_2}]^{40}, \quad [\widetilde{D_1, C_2}]^{40} = [\widehat{D_1, C_2}]^{40}, \quad [\widetilde{C_1, D_3}]^{60} = [\widehat{C_1, D_3}]^{60},$$

and

$$[\widetilde{D_1, C_3}]^{60} = [\widehat{D_1, C_3}]^{60}.$$

We thus need to divide Eq. (5.9) also by $[C_1, C_1]^{80}$, $[D_1, D_1]^{80}$, $[C_1, C_2]^{40}$, $[D_1, D_2]^{40}$, $[C_1, D_2]^{40}$ and $[D_1, C_2]^{40}$. Therefore the correct expression by which we need to divide Eq. (5.9) now is

$$[C_3, C_3]^{80} [D_3, D_3]^{80} [C_3, C_2]^{40} [D_3, D_2]^{40} [C_3, C_1]^{20} [D_3, D_1]^{20} [C_1, C_1]^{80} [D_1, D_1]^{80}$$
$$\times [C_1, C_2]^{40} [D_1, D_2]^{40} [C_3, D_2]^{40} [D_3, C_2]^{40} [C_3, D_1]^{60} [D_3, C_1]^{60} [C_1, D_2]^{40} [D_1, C_2]^{40},$$

which is invariant under the operators T_R and T_S since when no set with index -1 or 0 is involved putting a tilde or a hat simply interchanges the indices 1 and 3 and one easily sees that this interchange leaves the last expression invariant (with no sign problems since the powers 20, 40, 60 and 80 are even). In order to be able to continue with the new denominator we shall reorder the expressions in it according to their type, and this way we see that this denominator is

$$[C_{-1}, C_{-1}]^{80} [C_{-1}, C_0]^{40} [C_0, C_0]^{80} [C_{-1}, C_1]^{20} [C_0, C_1]^{40} [C_{-1}, C_2]^{20} [C_0, C_2]^{20}$$
$$\times [C_{-1}, C_3]^{40} [C_0, C_3]^{20} [C_3, C_3]^{80} [C_3, C_2]^{40} [C_3, C_1]^{20} [C_1, C_1]^{80} [C_1, C_2]^{40}$$
$$\times [D_{-1}, D_{-1}]^{80} [D_{-1}, D_0]^{40} [D_0, D_0]^{80} [D_{-1}, D_1]^{20} [D_0, D_1]^{40} [D_{-1}, D_2]^{20} [D_0, D_2]^{20}$$
$$\times [D_{-1}, D_3]^{40} [D_0, D_3]^{20} [D_3, D_3]^{80} [D_3, D_2]^{40} [D_3, D_1]^{20} [D_1, D_1]^{80} [D_1, D_2]^{40}$$
$$\times [C_{-1}, D_0]^{40} [C_{-1}, D_1]^{60} [C_0, D_1]^{40} [C_{-1}, D_2]^{60} [C_0, D_2]^{60} [C_{-1}, D_3]^{40} [C_0, D_3]^{60}$$
$$\times [C_3, D_2]^{40} [D_3, C_2]^{40} [C_3, D_1]^{60} [D_3, C_1]^{60} [C_1, D_2]^{40} [D_1, C_2]^{40}$$
$$\times [D_{-1}, C_0]^{40} [D_{-1}, C_1]^{60} [D_0, C_1]^{40} [D_{-1}, C_2]^{60} [D_0, C_2]^{60} [D_{-1}, C_3]^{40} [D_0, C_3]^{60}.$$

This means that $g_{N(\Delta)}$ is being multiplied by

$$[C_0, C_0]^{80} [D_0, D_0]^{80} [C_0, C_1]^{40} [D_0, D_1]^{40} [C_0, C_2]^{20} [D_0, D_2]^{20} [C_2, C_2]^{80} [D_2, D_2]^{80}$$
$$\times [C_2, C_1]^{40} [D_2, D_1]^{40} [C_0, D_1]^{40} [D_0, C_1]^{40} [C_0, D_2]^{60} [D_0, C_2]^{60} [C_2, D_1]^{40} [D_2, C_1]^{40},$$

which gives the denominator

$[C_{-1},C_{-1}]^{80}[C_{-1},C_3]^{40}[C_3,C_3]^{80}[C_{-1},C_2]^{20}[C_3,C_2]^{40}[C_{-1},C_1]^{20}[C_3,C_1]^{20}$

$\times[C_{-1},C_0]^{40}[C_3,C_0]^{20}[C_0,C_0]^{80}[C_0,C_1]^{40}[C_0,C_2]^{20}[C_2,C_2]^{80}[C_2,C_1]^{40}$

$\times[D_{-1},D_{-1}]^{80}[D_{-1},D_3]^{40}[D_3,D_3]^{80}[D_{-1},D_2]^{20}[D_3,D_2]^{40}[D_{-1},D_1]^{20}[D_3,D_1]^{20}$

$\times[D_{-1},D_0]^{40}[D_3,D_0]^{20}[D_0,D_0]^{80}[D_0,D_1]^{40}[D_0,D_2]^{20}[D_2,D_2]^{80}[D_2,D_1]^{40}$

$\times[C_{-1},D_3]^{40}[C_{-1},D_2]^{60}[C_3,D_2]^{40}[C_{-1},D_1]^{60}[C_3,D_1]^{60}[C_{-1},D_0]^{40}[C_3,D_0]^{60}$

$\times[C_0,D_1]^{40}[D_0,C_1]^{40}[C_0,D_2]^{60}[D_0,C_2]^{60}[C_2,D_1]^{40}[D_2,C_1]^{40}$

$\times[D_{-1},C_3]^{40}[D_{-1},C_2]^{60}[D_3,C_2]^{40}[D_{-1},C_1]^{60}[D_3,C_1]^{60}[D_{-1},C_0]^{40}[D_3,C_0]^{60}.$

Our work is almost done, since the careful reader can verify that the only difference is that the expression corresponding to Δ contains $[C_1,C_1]^{80}$ and $[D_1,D_1]^{80}$, and the one corresponding to $N(\Delta)$ contains $[C_2,C_2]^{80}$ and $[D_2,D_2]^{80}$. Hence the last division of Eq. (5.9) we do is by the expressions $[C_2,C_2]^{80}$ and $[D_2,D_2]^{80}$, and since these expressions also equal $[\widetilde{C_2,C_2}]^{80}$ and $[\widetilde{D_2,D_2}]^{80}$ and also $[\widehat{C_2,C_2}]^{80}$ and $[\widehat{D_2,D_2}]^{80}$, these expressions are invariant under the operators T_R and T_S and are good to work with. This operation gives the denominator (which we shall denote by h_Δ for reasons which will soon be clear)

$[C_{-1},C_{-1}]^{80}[C_{-1},C_0]^{40}[C_0,C_0]^{80}[C_{-1},C_1]^{20}[C_0,C_1]^{40}[C_{-1},C_2]^{20}[C_0,C_2]^{20}$

$\times[C_{-1},C_3]^{40}[C_0,C_3]^{20}[C_3,C_3]^{80}[C_3,C_2]^{40}[C_3,C_1]^{20}[C_1,C_1]^{80}[C_1,C_2]^{40}[C_2,C_2]^{80}$

$\times[D_{-1},D_{-1}]^{80}[D_{-1},D_0]^{40}[D_0,D_0]^{80}[D_{-1},D_1]^{20}[D_0,D_1]^{40}[D_{-1},D_2]^{20}[D_0,D_2]^{20}$

$\times[D_{-1},D_3]^{40}[D_0,D_3]^{20}[D_3,D_3]^{80}[D_3,D_2]^{40}[D_3,D_1]^{20}[D_1,D_1]^{80}[D_1,D_2]^{40}[D_2,D_2]^{80}$

$\times[C_{-1},D_0]^{40}[C_{-1},D_1]^{60}[C_0,D_1]^{40}[C_{-1},D_2]^{60}[C_0,D_2]^{60}[C_{-1},D_3]^{40}[C_0,D_3]^{60}$

$\times[C_3,D_2]^{40}[D_3,C_2]^{40}[C_3,D_1]^{60}[D_3,C_1]^{60}[C_1,D_2]^{40}[D_1,C_2]^{40}$

$\times[D_{-1},C_0]^{40}[D_{-1},C_1]^{60}[D_0,C_1]^{40}[D_{-1},C_2]^{60}[D_0,C_2]^{60}[D_{-1},C_3]^{40}[D_0,C_3]^{60},$

and since the denominator for $N(\Delta)$ is multiplied by $[C_1,C_1]^{80}$ and $[D_1,D_1]^{80}$ in this operation, we obtain that $h_{N(\Delta)}$ is

$[C_{-1},C_{-1}]^{80}[C_{-1},C_3]^{40}[C_3,C_3]^{80}[C_{-1},C_2]^{20}[C_3,C_2]^{40}[C_{-1},C_1]^{20}[C_3,C_1]^{20}$

$\times[C_{-1},C_0]^{40}[C_3,C_0]^{20}[C_0,C_0]^{80}[C_0,C_1]^{40}[C_0,C_2]^{20}[C_2,C_2]^{80}[C_2,C_1]^{40}[C_1,C_1]^{80}$

$\times[D_{-1},D_{-1}]^{80}[D_{-1},D_3]^{40}[D_3,D_3]^{80}[D_{-1},D_2]^{20}[D_3,D_2]^{40}[D_{-1},D_1]^{20}[D_3,D_1]^{20}$

$\times[D_{-1},D_0]^{40}[D_3,D_0]^{20}[D_0,D_0]^{80}[D_0,D_1]^{40}[D_0,D_2]^{20}[D_2,D_2]^{80}[D_2,D_1]^{40}[D_1,D_1]^{80}$

$\times[C_{-1},D_3]^{40}[C_{-1},D_2]^{60}[C_3,D_2]^{40}[C_{-1},D_1]^{60}[C_3,D_1]^{60}[C_{-1},D_0]^{40}[C_3,D_0]^{60}$

$\times[C_0,D_1]^{40}[D_0,C_1]^{40}[C_0,D_2]^{60}[D_0,C_2]^{60}[C_2,D_1]^{40}[D_2,C_1]^{40}$

$\times[D_{-1},C_3]^{40}[D_{-1},C_2]^{60}[D_3,C_2]^{40}[D_{-1},C_1]^{60}[D_3,C_1]^{60}[D_{-1},C_0]^{40}[D_3,C_0]^{60},$

which equals h_Δ. This is why we have denoted the previous expression by h_Δ at this point. The fact that throughout the process we preserved the invariance under the operators T_R and T_S assures us that $\theta^{100}[\Delta](0,\Pi)/h_\Delta$ with this h_Δ is still invariant under the operators T_R and T_S. The the fact that $h_{N(\Delta)}=h_\Delta$ shows, together with Eq.

(1.5) in the form $\theta[N(\Delta)](0,\Pi) = \theta[\Delta](0,\Pi)$ to the proper power, that this quotient is invariant also under N.

We now remark that also here the last two stages can be merged into one, in which we divide Eq. (5.9) by

$$[C_3,C_3]^{80}[D_3,D_3]^{80}[C_3,C_2]^{40}[D_3,D_2]^{40}[C_3,C_1]^{20}[D_3,D_1]^{20}[C_1,C_1]^{80}[D_1,D_1]^{80}$$
$$\times [C_1,C_2]^{40}[D_1,D_2]^{40}[C_2,C_2]^{80}[D_2,D_2]^{80}[C_3,D_2]^{40}[D_3,C_2]^{40}[C_3,D_1]^{60}[D_3,C_1]^{60}$$
$$\times [C_1,D_2]^{40}[D_1,C_2]^{40}.$$

We give the following description for this expression. Take the product of all the expressions $[C_j,C_j]^{80}$ and $[D_j,D_j]^{80}$ with $1 \leq j \leq 3$, the product of all the expressions $[C_j,C_{j+1}]^{40}$ and $[D_j,D_{j+1}]^{40}$ with $1 \leq j \leq 2$, and the expressions $[C_j,C_{j+2}]^{20}$ and $[D_j,D_{j+2}]^{20}$ for $j = 1$, and multiply it by the product of all the expressions $[C_j,D_{j+1}]^{40}$ and $[D_j,C_{j+1}]^{40}$ with $1 \leq j \leq 2$, and the expressions $[C_j,D_{j+2}]^{20}$ and $[D_j,C_{j+2}]^{60}$ for $j = 1$. This is the same as the product of all the expressions $[C_j,C_i]$, $[D_j,D_i]$, $[C_j,D_i]$, and $[D_j,C_i]$, with $i \geq j$, which do not involve the indices 0 and -1, and to powers depending only on the type and on $i - j$ (where the last two expressions do not appear, or equivalently appear to the power 0, if $i - j = 0$). Similarly, for $N(\Delta)$ the expression by which we divided was

$$[C_0,C_0]^{80}[D_0,D_0]^{80}[C_0,C_1]^{40}[D_0,D_1]^{40}[C_0,C_2]^{20}[D_0,D_2]^{20}[C_2,C_2]^{80}[D_2,D_2]^{80}$$
$$\times [C_2,C_1]^{40}[D_2,D_1]^{40}[C_1,C_1]^{80}[D_1,D_1]^{80}[C_0,D_1]^{40}[D_0,C_1]^{40}[C_0,D_2]^{60}[D_0,C_2]^{60}$$
$$\times [C_2,D_1]^{40}[D_2,C_1]^{40}.$$

The expression here is obtained by taking the product of all the expressions $[C_j,C_j]^{80}$ and $[D_j,D_j]^{80}$ with $0 \leq j \leq 2$, the product of all the expressions $[C_j,C_{j+1}]^{40}$ and $[D_j,D_{j+1}]^{40}$ with $0 \leq j \leq 1$, and the expressions $[C_j,C_{j+2}]^{20}$ and $[D_j,D_{j+2}]^{20}$ for $j = 0$, and multiplying it by the product of all the expressions $[C_j,D_{j+1}]^{40}$ and $[D_j,C_{j+1}]^{40}$ with $0 \leq j \leq 1$, and the expressions $[C_j,D_{j+2}]^{20}$ and $[D_j,C_{j+2}]^{60}$ for $j = 0$. This give the product of all the expressions $[C_j,C_i]$, $[D_j,D_i]$, $[C_j,D_i]$, and $[D_j,C_i]$, with $i \geq j$, which do not involve the indices 3 and -1, and to powers depending only on the type and on $i - j$ (again with power 0 for the last two expressions when $i - j = 0$). Once again partial expressions will not work, so that in the general case it is again possible to merge all these stages to one. For details see Appendix A.

5.2.3 Reduction and the Thomae Formulae for $n = 5$

As in the nonsingular case, we do not take this expression for h_Δ to be the denominator appearing in our Thomae formulae (even though it is mathematically correct), since it is not reduced. As one can verify, every expression $[C_i,C_j]$ and every expression $[D_i,D_j]$ appears in it at least to the 20th power. Thus the expressions

$$\left[\bigcup_{j=-1}^{3} C_j, \bigcup_{j=-1}^{3} C_j \right]^{20} = \prod_{0 \le i < j \le m-1} (\lambda_i - \lambda_j)^{20}$$

and

$$\left[\bigcup_{j=-1}^{3} D_j, \bigcup_{j=-1}^{3} D_j \right]^{20} = \prod_{1 \le i < j \le m} (\mu_i - \mu_j)^{20}$$

(which are of course invariant under every operation) are "common divisors" that appear in h_Δ for every Δ. In order to cancel them we multiply Eq. (5.9) by these expressions, and obtain that $\theta^{100}[\Delta](0,\Pi)/h_\Delta$ with h_Δ being

$$
\begin{aligned}
& [C_{-1},C_{-1}]^{60}[C_{-1},C_0]^{20}[C_0,C_0]^{60}[C_0,C_1]^{20}[C_{-1},C_3]^{20}[C_3,C_3]^{60}[C_3,C_2]^{20} \\
& \times [C_1,C_1]^{60}[C_1,C_2]^{20}[C_2,C_2]^{60}[D_1,D_1]^{60}[D_1,D_2]^{20}[D_2,D_2]^{60} \\
& \times [D_{-1},D_{-1}]^{60}[D_{-1},D_0]^{20}[D_0,D_0]^{60}[D_0,D_1]^{20}[D_{-1},D_3]^{20}[D_3,D_3]^{60}[D_3,D_2]^{20} \\
& \times [C_{-1},D_0]^{40}[C_{-1},D_1]^{60}[C_0,D_1]^{40}[C_{-1},D_2]^{60}[C_0,D_2]^{60}[C_{-1},D_3]^{40}[C_0,D_3]^{60} \\
& \times [C_3,D_2]^{40}[D_3,C_2]^{40}[C_3,D_1]^{60}[D_3,C_1]^{60}[C_1,D_2]^{40}[D_1,C_2]^{40} \qquad\qquad (5.9) \\
& \times [D_{-1},C_0]^{40}[D_{-1},C_1]^{60}[D_0,C_1]^{40}[D_{-1},C_2]^{60}[D_0,C_2]^{60}[D_{-1},C_3]^{40}[D_0,C_3]^{60}
\end{aligned}
$$

is invariant under the operators T_R, T_S and N. This is the main argument in proving

Theorem 5.4. *The expression $\theta^{100}[\Delta](0,\Pi)/h_\Delta$ with h_Δ from Eq. (5.10) is independent of the divisor Δ, and is the Thomae formulae for the nonsingular Z_5 curve X.*

In order to complete the proof of Theorem 5.4 we have to show that one can get from any divisor to any other divisor using the successive operations of T_R, T_S and N. We leave this part of the proof for the treatment of the general case in Section 5.4. We just note that we can describe h_Δ of Eq. (5.10) as the product of

$$\prod_{j=-1}^{3} [C_j,C_j]^{60}[D_j,D_j]^{60}, \qquad \prod_{j=-1}^{2} [C_j,C_{j+1}]^{20}[D_j,D_{j+1}]^{20},$$

and $[C_j,C_{j+4}]^{20}[D_j,D_{j+4}]^{20}$ with $j=-1$ multiplied by the product of

$$\prod_{j=-1}^{2} [C_j,D_{j+1}]^{40}[D_j,C_{j-1}]^{40}, \qquad \prod_{j=-1}^{1} [C_j,D_{j+2}]^{60}[D_j,C_{j+2}]^{60},$$

$$\prod_{j=-1}^{0} [C_j,D_{j+3}]^{60}[D_j,C_{j+3}]^{60},$$

and finally the expression $[C_j,D_{j+4}]^{40}[D_j,C_{j+4}]^{40}$ with $j=-1$.

Also here, the reason we have to do this reduction is the fact that Eq. (5.9) with which we have started comes from the unreduced PMT in Eq. (5.5) and not from the reduced one in Eq. (5.6). Starting with the reduced PMT, we see from Eq. (5.6) with $n = 5$ that $\theta[\Delta]^{100}(0,\Pi)/g_\Delta$ with

$$g_\Delta = ([C_{-1},C_{-1}][C_{-1},D_0][D_0,D_0][C_0,C_0][C_0,D_{-1}][D_{-1},D_{-1}])^{40}$$
$$\times ([C_{-1},C_3][D_0,C_3][C_0,D_3][D_{-1},D_3])^{20}([C_{-1},D_1][D_0,D_1][C_0,C_1][D_{-1},C_1])^{20}$$

is invariant under the operators T_R and T_S. The reader can check as an exercise that doing the following stages leads us from this formula to the reduced Thomae formulae directly. The first stage is dividing by

$$[C_{-1},D_3]^{40}[C_0,D_3]^{40}[C_{-1},D_1]^{40}[C_0,D_1]^{40}[D_{-1},C_3]^{40}[D_0,C_3]^{40}[D_{-1},C_1]^{40}[D_0,C_1]^{40}$$

(coming from $[C_{-1},D_3]^{40}$ and $[D_{-1},C_3]^{40}$). In the second one divide by

$$[C_{-1},C_{-1}]^{20}[C_0,C_0]^{20}[C_{-1},C_0]^{20}[D_{-1},D_{-1}]^{20}[D_0,D_0]^{20}[D_{-1},D_0]^{20}$$

and

$$[C_{-1},D_2]^{60}[C_0,D_2]^{60}D_{-1},C_2]^{60}[D_0,C_2]^{60}$$

(coming from $[C_{-1},C_0]^{20}$, $[D_{-1},D_0]^{20}$, $[C_{-1},D_2]^{60}$ and $[D_{-1},C_2]^{60}$). Then divide by

$$[C_3,C_3]^{60}[D_3,D_3]^{60}[C_3,C_2]^{20}[D_3,D_2]^{20}[C_1,C_1]^{60}[D_1,D_1]^{60}[C_1,C_2]^{20}[D_1,D_2]^{20}$$
$$\times[C_3,D_2]^{40}[D_3,C_2]^{40}[C_3,D_1]^{60}[D_3,C_1]^{60}[C_1,D_2]^{40}[D_1,C_2]^{40}$$

(which comes from the expressions $[C_3,C_3]^{60}$, $[D_3,D_3]^{60}$, $[C_3,C_2]^{20}$, $[D_3,D_2]^{20}$, $[C_3,D_2]^{40}$, $[D_3,C_2]^{40}$, $[C_3,D_1]^{60}$ and $[D_3,C_1]^{60}$), and finally by the expressions $[C_2,C_2]^{60}$ and $[D_2,D_2]^{60}$. After doing so one obtains Eq. (5.10) and Theorem 5.4 as stated, without the reduction. Again this process is simpler, but we preferred to show the longer process in order to demonstrate better the general process in the singular case, and to give the reader the possibility to compare the singular and non-singular cases. Clearly here as well the stages should be done in the proper order also when starting with the reduced PMT in Eqs. (5.6) and (5.7) and not only with the unreduced one in Eq. (5.5). This is done in detail in Appendix A.

5.3 Invariance also under N

In this section we obtain the Thomae formulae for a general singular Z_n curve, but as in the nonsingular case the proof that what we get is indeed the full Thomae formulae will have to wait until Section 5.4. This is again done by dividing the PMT from Proposition 5.3 and Eq. (5.5) (or the reduced Eqs. (5.6) and (5.7)) by some invariant expressions as we did in the example of $n = 5$ in Section 5.2, so that the obtained denominators h_Δ will still give quotients invariant under the operators T_R and T_S, but now also under N. The exact form of the corrections, which again depend on the parity of n, appears in detail in Appendix A. Here we shall start directly from

the description of the denominators h_Δ in both cases of odd and even n, and prove
that the denominators we are writing have all the desired properties. The remark
about "merging" the two cases of odd and even n into one holds equally here and
will be explained later.

5.3.1 The Description of h_Δ for Odd n

We begin with the case of odd n. As we did in the nonsingular case, here also we
can give a pictorial description of h_Δ which is based on a matrix. This matrix now
must contain both the sets C_j, $-1 \le j \le n-2$, and the sets D_j, $-1 \le j \le n-2$,
as labels, so one might expect that for a given value of n the matrix would contain
twice as many rows and twice as many columns as the matrix from the nonsingular
case. This is, however, not accurate, and our matrix now has indeed $2n$ rows, but
$2n + (n-1)$, rather than $2n + (n-3)$, columns. The rows are labeled by the sets
C_j and D_j, $-1 \le j \le n-2$, in increasing order (this means increasing order of j,
where in the comparison of C_j and D_j with the same index j we always let the set
C_j come right before the set D_j). The columns are labeled similarly by the sets C_j
and D_j with $-1 \le j \le n-2$, in increasing order, and, as expected, these labels are
followed by C_j and D_j, $-1 \le j \le (n-5)/2$, in increasing order. It may again be
more convenient, at least in the beginning, to label the columns by the sets C_j and
D_j, $-1 \le j \le (3n-5)/2$, in increasing order and define $C_j = C_{j-n}$ and $D_j = D_{j-n}$
for $j \ge n-1$. For example, for $n=5$ we have 10 rows and 14 columns, with the
last 4 columns being labeled C_{-1}, D_{-1}, C_0, and D_0 (which can be considered as the
labels C_4, D_4, C_5, and D_5). The reason for the extra pair of columns will soon be
clear. We note that the rows and the columns are gathered in pairs, and thus we shall
look at the entries of this matrix as if they are arranged in squares of 2×2, and not
each entry on its own.

Once again we consider the diagonals of maximal length in this matrix, but now
we look at diagonals of squares of 2×2 of entries rather than diagonals of the entries
themselves. We now have diagonals labeled by $0 \le k \le (n-1)/2$ (rather than up
to $(n-3)/2$), and in each diagonal appear the four entries (C_i, C_{i+k}), (C_i, D_{i+k}),
(D_i, C_{i+k}) and (D_i, D_{i+k}) with i running from -1 to $n-2$ (in the notation with the
sets C_j and D_j with $j \ge n-1$). Now, for each such k, we take all the squares on the
kth diagonal, raise their (C_i, C_{i+k}) and (D_i, D_{i+k}) elements to the power

$$2n\frac{(n-2k)^2 - 1}{4} = 2n\left(\frac{n-1}{2} - k\right)\left(\frac{n+1}{2} - k\right) = 2n\left(\frac{n^2-1}{4} - k(n-k)\right),$$

raise their (C_i, D_{i+k}) and (D_i, C_{i+k}) elements to the power $2nk(n-k)$ (the complement of the previous power with respect to $2n(n^2-1)/4$), and take the product
of all these expressions. It will be convenient to refer, in every such 2×2 square,
to the entries (C_i, C_{i+k}) and (D_i, D_{i+k}) as the *main diagonal* entries of this square,
and to the entries (C_i, D_{i+k}) and (D_i, C_{i+k}) as the *complementary diagonal* entries.

Now, the powers of the main diagonals decrease with k, starting from $2n(n^2-1)/4$ for $k=0$ and ending with 0 for $k=(n-1)/2$, while the powers of the complementary diagonals increase with k, starting from 0 for $k=0$ and ending with $2n(n^2-1)/4$ for $k=(n-1)/2$. All of these are even integers (as they are divisible by $2n$). Note that taking only the (C_i, C_{i+k}) part of each square gives us the pictorial description of the nonsingular h_Δ, which explains the reason for us needing also the $[(n-1)/2]$th diagonal—the (C_i, C_{i+k}) part vanishes there and is unnecessary, but here the (C_i, D_{i+k}) and (D_i, C_{i+k}) parts also appear and they do not vanish for $k=(n-1)/2$. All this means that as a formula we have (with the sets of the form C_j and D_j with $j \geq n-1$ allowed)

$$h_\Delta = \prod_{k=0}^{(n-1)/2} \prod_{i=-1}^{n-2} ([C_i, C_{i+k}][D_i, D_{i+k}])^{2n[(n^2-1)/4-k(n-k)]} ([C_i, D_{i+k}][D_i, C_{i+k}])^{2nk(n-k)}.$$

For visualization of this description we give here the matrix for $n=5$, where as usual the double vertical line marks the point to the right of which lie the columns labeled by "sets with two names".

60 0	20 40	0 60						
0 60	40 20	60 0						
	60 0	20 40	0 60					
	0 60	40 20	60 0					
		60 0	20 40	0 60				
		0 60	40 20	60 0				
			60 0	20 40	0 60			
			0 60	40 20	60 0			
				60 0	20 40	0 60		
				0 60	40 20	60 0		

We want to write this h_Δ using only the usual sets C_j with $-1 \leq j \leq n-2$ and D_j with $-1 \leq j \leq n-2$. For this we let Y and Z be either C or D each (independently), and we note that for a fixed $1 \leq k \leq (n-1)/2$ (again for $k=0$ we do not have to do anything), and for any $i \geq n-1-k$, the expression $[Y_i, Z_{i+k}]$ is actually $[Y_i, Z_{i+k-n}]$. Up to sign, this expression is just $[Z_j, Y_{j+n-k}]$ for $j = i+k-n$ (and this j satisfies $-1 \leq j \leq k-2$); since replacing Y and Z leaves the power invariant (this is the statement that $[C_j, D_l]$ and $[D_j, C_l]$ appear to the same power in h_Δ for every relevant j and l), we find that for our k the product over i is

$$\prod_{j=-1}^{n-2-k} ([C_j, C_{j+k}][D_j, D_{j+k}])^{2n[(n^2-1)/4-k(n-k)]} ([C_j, D_{j+k}][D_j, C_{j+k}])^{2nk(n-k)}$$

$$\times \prod_{j=-1}^{k-2} ([C_j, C_{j+n-k}][D_j, D_{j+n-k}])^{2n[(n^2-1)/4-k(n-k)]} ([C_j, D_{j+n-k}][D_j, C_{j+n-k}])^{2nk(n-k)}$$

(again we want only j as our multiplication index). Since we still have invariance of all these powers under $k \mapsto n - k$, the second product is the first one corresponding to $n - k$, which satisfies $(n+1)/2 \leq n - k \leq n - 1$ since k satisfies $1 \leq k \leq (n-1)/2$. Hence we write

$$h_\Delta = \prod_{k=0}^{n-1} \prod_{j=-1}^{n-2-k} ([C_j, C_{j+k}][D_j, D_{j+k}])^{2n[(n^2-1)/4-k(n-k)]} ([C_j, D_{j+k}][D_j, C_{j+k}])^{2nk(n-k)}$$

(5.10)

where the parts with $(n+1)/2 \leq k \leq n - 1$ came as before from the $n - k$ parts for $1 \leq k \leq (n-1)/2$. We remark that here the values $k = (n-1)/2$ and $k = (n+1)/2$ appear from the start and have contributions, so that we do not have to artificially add them in order to have the compact formula in Eq. (5.10).

We again want to get a taste of how this h_Δ from Eq. (5.10) looks like in examples of small odd n before proving its properties. Let us begin with $n = 3$, where $2n = 6$ and k can be either 0, 1 or 2. We recall from the nonsingular case with $n = 3$ that the power

$$2n \left(\frac{n+1}{2} - k \right) \left(\frac{n-1}{2} - k \right) = 6(2-k)(1-k)$$

vanishes for $k = 1$ and $k = 2$ and leaves $k = 0$ with the power 12. Its complement with respect to $2n(n^2 - 1)/4 = 12$ is 12 for $k = 1$ and $k = 2$ and vanishes for $k = 0$. When we want to see it in the pictorial description, we find that the number of columns is $2n + (n - 1) = 8$, 2 more than the number $2n = 6$ of rows. Hence there are two long diagonals of squares, the zeroth (which contain only elements from main diagonals of the squares) and the first (which is the last and thus contains only elements from complementary diagonals of the squares). Note that in the case $m = 3$ this gives exactly Eq. (3.7) from Section 3.2, which means that the combined result of Theorems 5.5 and 5.7 below contains Theorem 3.6 from Section 3.2 as the special case with $n = 3$ and $m = 3$.

We now look at $n = 5$, where we again have $2n = 10$ and k varies from 0 to 4. We have seen in the nonsingular case that the power

$$2n \left(\frac{n+1}{2} - k \right) \left(\frac{n-1}{2} - k \right) = 10(3-k)(2-k)$$

vanishes for $k = 2$ and $k = 3$ and gives 60 for $k = 0$ and 20 for $k = 1$ and $k = 4$. The complement of this power with respect to $2n(n^2 - 1)/4 = 60$ is 60 for $k = 2$ and $k = 3$, 40 for $k = 1$ and $k = 4$, and vanishes for $k = 0$. Visualizing this in the pictorial description, one sees that the number of columns is $2n + (n - 1) = 14$, 4 more than the number $2n = 10$ of rows. This gives three long diagonals of squares, the zeroth (containing only elements from main diagonals of the squares) for $k = 0$, the first for $k = 1$ and $k = 4$, and the second (containing only elements from complementary diagonals of the squares, being the last diagonal) for $k = 2$ and $k = 3$. From this one can verify that Eq. (5.10) gives Eq. (5.10) as its special case with $n = 5$, and therefore Theorem 5.5 below contains Theorem 5.4 as the special case with $n = 5$, so that proving Theorem 5.7 completes the proof that Theorem 4.4 gives the Thomae formulae for general singular Z_5 curves of this form.

Another special case we have, but with small m rather than small n, is if we take the square of Theorem 3.13 (and hence weaken it a bit). To see this we notice that for the divisor $P_1^k Q_i^{n-1-k}$ with $0 \leq k \leq n-2$, we have $C_{-1} = \{P_0\}$, $D_{-1} = \{Q_j\}$, $C_{n-2-k} = \{P_1\}$, $D_{n-2-k} = \{Q_i\}$, and the other sets empty (with i and j being 1 and 2 and distinct), and for the divisor P_1^{n-1}, we have $C_{-1} = \{P_0\}$, $D_{-1} = \{Q_1, Q_2\}$, and the other sets empty. Then the fact that the square of Theorem 3.13 is the special case with $m = 2$ of the general result composed of Theorems 5.5 and 5.7 is clear once we substitute λ instead of μ_1 and erase all the expressions involving μ_2 (which is now ∞). As mentioned in Section 3.3, Theorem 3.13 there itself is a bit stronger than the special case of the general result.

5.3.2 N-Invariance for Odd n

Now we prove

Theorem 5.5. *The expression* $\theta^{4n^2}[\Delta](0, \Pi)/h_\Delta$ *with* h_Δ *from Eq. (5.10) is invariant under the action of the operators* N, T_R *and* T_S.

Proof. Also here N-invariance is equivalent to the assertion that $h_{N(\Delta)} = h_\Delta$ by Eq. (1.5), and hence it is simpler and we start with it. Definition 2.18 gives that the operator N fixes C_{-1} and D_{-1} and replaces C_j and D_j with $0 \leq j \leq n-2$ by C_{n-2-j} and D_{n-2-j} respectively, so that again we deal with the expressions involving C_{-1} and D_{-1} separately from the others. Starting with the expressions not involving C_{-1} and D_{-1} we see that, for a fixed $0 \leq k \leq n-1$, the product over $0 \leq j \leq n-2-k$ (without $j = -1$) in Eq. (5.10) is over expressions $[Y_j, Z_{j+k}]$ with Y and Z again C and D and independent, and the powers to which they appear depends only on k and on whether $Y = Z$ (as a type of sets C or D) or not. Now, N takes $[Y_j, Z_{j+k}]$ to $[Y_{n-2-j}, Z_{n-2-j-k}]$, which equals $[Z_i, Y_{i+k}]$ for $i = n-2-j-k$ up to the usual sign, and when j runs through the numbers from 0 to $n-2-k$, so does i (again in decreasing order, a fact which we can ignore). Since the power depends only on k and on whether $Y = Z$ as type of sets, this shows that the action of N on this product leaves it invariant for each $0 \leq k \leq n-1$ (again for $k = n-1$ this product is empty so one can equally say $0 \leq k \leq n-2$).

The expressions containing C_{-1} and D_{-1} are

$$\prod_{k=0}^{n-1} ([C_{-1}, C_{k-1}][D_{-1}, D_{k-1}])^{2n[(n^2-1)/4 - k(n-k)]}$$

and

$$\prod_{k=0}^{n-1} ([C_{-1}, D_{k-1}][D_{-1}, C_{k-1}])^{2n\lambda(n-\lambda)},$$

and we want to show that they are invariant under N. For $[Y_{-1}, Z_{-1}]$ this is trivial, which deals with the expressions with $k = 0$. For another value of k we find that N takes $[Y_{-1}, Z_{k-1}]$ to $[Y_{-1}, Z_{n-1-k}]$, which we write as $[Y_{-1}, Z_{(n-k)-1}]$. Therefore the action of N on the products with $1 \leq k \leq n-1$ is just taking k to $n-k$, an action under which the powers are invariant, and since when k runs from 1 to $n-1$ so does $n-k$ (still in decreasing order, which still does not matter) this product is invariant under N. Therefore the expression for h_Δ in Eq. (5.10) is invariant under N, or equivalently $h_{N(\Delta)} = h_\Delta$.

As before, the PMT in Proposition 5.3 will be the key ingredient for proving of the invariance under the operators T_R and T_S, and we again choose to use the reduced expression for g_Δ from Eq. (5.6) (n is odd). We do the usual trick of writing

$$\frac{\theta^{2n^2}[\Delta](0, \Pi)}{h_\Delta} = \frac{\theta^{2n^2}[\Delta](0, \Pi)}{g_\Delta} \bigg/ \frac{h_\Delta}{g_\Delta},$$

and the (reduced) PMT in Proposition 5.3 reduces what we have to show to invariance of the denominator here under the operators T_R and T_S (since it gives invariance of the numerator). Since no theta constants appear there, this is again just a verification of explicit expressions in the parameters λ_i, $0 \leq i \leq m-1$, and μ_i, $1 \leq i \leq m$. We start from a fixed divisor Δ (and thus the sets C_j and D_j, $-1 \leq j \leq n-2$ are fixed), and choose arbitrary branch points $R \in C_0$ and $S \in B = D_{-1}$. Then we write the expressions corresponding to $T_R(\Delta)$ using the sets with the tilde and the expressions corresponding to $T_S(\Delta)$ using the sets with the hat in order to make the comparison easier, and what we need to show is that $h_\Delta/g_\Delta = h_{T_R(\Delta)}/g_{T_R(\Delta)} = h_{T_S(\Delta)}/g_{T_S(\Delta)}$. This is equivalent to $h_\Delta/h_{T_R(\Delta)} = g_\Delta/g_{T_R(\Delta)}$ and $h_\Delta/h_{T_S(\Delta)} = g_\Delta/g_{T_S(\Delta)}$, and these equalities are what we in fact prove.

It turns out that it is again useful to split the expressions involving C_{-1}, D_{-1}, C_0, or D_0 from the others. Hence we write expression for h_Δ from Eq. (5.10) as $h_\Delta = s_\Delta t_\Delta$, with

$$s_\Delta = \prod_{k=0}^{n-3} \prod_{j=1}^{n-2-k} ([C_j, C_{j+k}][D_j, D_{j+k}])^{2n[(n^2-1)/4 - k(n-k)]} ([C_j, D_{j+k}][D_j, C_{j+k}])^{2nk(n-k)}$$

and

$$t_\Delta = \prod_{k=0}^{n-2} ([C_0, C_k][D_0, D_k])^{2n[(n^2-1)/4 - k(n-k)]} ([C_0, D_k][D_0, C_k])^{2nk(n-k)}$$

$$\times \prod_{k=-1}^{n-2} ([C_{-1}, C_k][D_{-1}, D_k])^{2n[(n^2-1)/4 - (k+1)(n-1-k)]} ([C_{-1}, D_k][D_{-1}, C_k])^{2n(k+1)(n-1-k)}$$

with the usual replacement of k by $k+1$ in the product for $j = -1$. We first deal with s_Δ, where we claim that $s_\Delta = s_{T_R(\Delta)} = s_{T_S(\Delta)}$. We again interpret the product corresponding to any $0 \leq k \leq n-3$ in the expression for s_Δ as the product of expressions of the form $[Y_j, Z_{j+k}]$ with Y and Z as usual and taken to powers depending only on k and on whether $Y = Z$ (in the usual meaning). We recall that in this notation we have

$\widetilde{Y}_j = \widehat{Y}_j = Y_{n-1-j}$ and $\widetilde{Z}_j = \widehat{Z}_j = Z_{n-1-j}$ for $1 \leq j \leq n-2$, so that both $[\widetilde{Y}_j, \widetilde{Z_{j+k}}]$ and $[\widehat{Y}_j, \widehat{Z_{j+k}}]$ equal $[Y_{n-1-j}, Z_{n-1-j-k}]$ for any $1 \leq j \leq n-2-k$. Up to sign this equals $[Z_i, Y_{i+k}]$ for $i = n-1-j-k$, which runs through the numbers from 1 to $n-2-k$ (in the usual decreasing order) when j does. This proves the invariance of the expression s_Δ under T_R and T_S, so that in the quotients $h_\Delta/h_{T_R(\Delta)} = s_\Delta t_\Delta/s_{T_R(\Delta)}t_{T_R(\Delta)}$ and $h_\Delta/h_{T_S(\Delta)} = s_\Delta t_\Delta/s_{T_S(\Delta)}t_{T_S(\Delta)}$ we can cancel s_Δ with $s_{T_R(\Delta)}$ or with $s_{T_S(\Delta)}$, leaving $t_\Delta/t_{T_R(\Delta)}$ and $t_\Delta/t_{T_S(\Delta)}$ respectively. We are thus reduced to showing that $t_\Delta/t_{T_R(\Delta)} = g_\Delta/g_{T_R(\Delta)}$ and $t_\Delta/t_{T_S(\Delta)} = g_\Delta/g_{T_S(\Delta)}$.

We now want apply the fact that the sets $G = C_{-1} \cup C_0$ and $I = D_{-1} \cup D_0$ are left invariant when putting a tilde or a hat on them. For this we gather as many expressions from t_Δ as we can into expressions involving these unions, which will allow us to do further cancelations in the quotients $t_\Delta/t_{T_R(\Delta)}$ and $t_\Delta/t_{T_S(\Delta)}$. For this we want to compare the powers to which expressions of the form $[Y_0, Z_k]$ and $[Y_{-1}, Z_k]$ (where Y and Z have the usual meaning) appear in t_Δ, which will be done again using the equality

$$(k+1)(n-1-k) = k(n-k) + (n-1-2k).$$

In order to express the results we split the products on k into 4 parts: one with $k = -1$ and $k = 0$, one with $1 \leq k \leq (n-3)/2$, one with $(n+1)/2 \leq k \leq n-2$, and one with the intermediate value $k = (n-1)/2$ (we remark again that unlike the nonsingular case, this value does make a contribution now). Dealing with the part with $k = -1$ and $k = 0$ we find that the expressions $[Y_{-1}, Z_{-1}]$ and $[Y_0, Z_0]$ with $Y \neq Z$ do not appear at all while the expressions $[Y_{-1}, Z_{-1}]$, $[Y_0, Z_0]$ and $[Y_{-1}, Z_0]$ with $Y = Z$ all appear at least to the power $2n(n^2 - 4n + 3)/4$. Since Lemma 4.2 give us (up to sign) $[C_{-1}, C_{-1}][C_0, C_0][C_{-1}, C_0] = [G, G]$ and $[D_{-1}, D_{-1}][D_0, D_0][D_{-1}, D_0] = [I, I]$, this is the merging we do for the $Y = Z$ part. Now, for $1 \leq k \leq (n-3)/2$, the power to which $[Y_{-1}, Z_k]$ with $Y = Z$ appear is smaller than the power to which $[Y_0, Z_k]$ appear, and the power to which $[Y_0, Z_k]$ with $Y \neq Z$ appear is smaller than the power to which $[Y_{-1}, Z_k]$ appear. On the other hand, for $(n+1)/2 \leq k \leq n-2$, the power to which $[Y_0, Z_k]$ with $Y = Z$ appear is smaller than the power to which $[Y_{-1}, Z_k]$ appear, and the power to which $[Y_{-1}, Z_k]$ with $Y \neq Z$ appear is smaller than the power to which $[Y_0, Z_k]$ appear. For the remaining value $k = (n-1)/2$, both numbers $k(n-k)$ and $(k+1)(n-1-k)$ equal $(n^2-1)/4$, meaning that $[Y_0, Z_k]$ and $[Y_{-1}, Z_k]$ with $Y = Z$ do not appear at all, and $[Y_0, Z_k]$ and $[Y_{-1}, Z_k]$ with $Y \neq Z$ appear to the power $2n(n^2-1)/4$. In any case, we can use the four equalities

$$[C_{-1}, C_k][C_0, C_k] = [G, C_k], \quad [D_{-1}, D_k][D_0, D_k] = [I, D_k],$$

$$[C_{-1}, D_k][C_0, D_k] = [G, D_k], \quad \text{and} \quad [D_{-1}, C_k][D_0, C_k] = [I, C_k]$$

(up to the usual sign). All this means that we can write the expression for t_Δ as

$$([C_{-1},C_{-1}][D_{-1},D_{-1}][C_0,C_0][D_0,D_0][C_{-1},D_0][D_{-1},C_0])^{2n(n-1)}$$

$$\times \prod_{k=1}^{(n-3)/2} ([G,C_k][I,D_k])^{2n[(n^2-1)/4-(k+1)(n-1-k)]} ([C_0,C_k][D_0,D_k])^{2n(n-1-2k)}$$

$$\times \prod_{k=1}^{(n-3)/2} ([G,D_k][I,C_k])^{2nk(n-k)} ([C_{-1},D_k][D_{-1},C_k])^{2n(n-1-2k)}$$

$$\times ([G,G][I,I])^{2n(n^2-4n+3)/4} ([G,D_{(n-1)/2}][I,C_{(n-1)/2}])^{2n(n^2-1)/4}$$

$$\times \prod_{k=(n+1)/2}^{n-2} ([G,C_k][I,D_k])^{2n[(n^2-1)/4-k(n-k)]} ([C_{-1},C_k][D_{-1},D_k])^{2n(2k-(n-1))}$$

$$\times \prod_{k=(n+1)/2}^{n-2} ([G,D_k][I,C_k])^{2n(k+1)(n-1-k)} ([C_0,D_k][D_0,C_k])^{2n(2k-(n-1))}$$

(with the expressions for the value $k=(n-1)/2$ appearing, together with the G and I parts from the part with $k=-1$ and $k=0$, in the fourth row here).

Apart from the equalities $\widetilde{G}=\widehat{G}=G$ and $\widetilde{I}=\widehat{I}=I$, in order to obtain expressions invariant under T_R and T_S we also recall that putting a tilde or a hat interchanges any index $1 \leq j \leq n-2$ with $n-1-j$. Therefore putting a tilde or a hat on all the sets in any of the expressions

$$[G,C_j][G,C_{n-1-j}], \quad [I,D_j][I,D_{n-1-j}],$$

$$[G,D_j][G,D_{n-1-j}], \quad \text{and} \quad [I,C_j][I,C_{n-1-j}],$$

for such j, leaves that expression invariant. Moreover, in our case of odd n the value $j=(n-1)/2$ is exceptional since $\widetilde{C}_j=\widehat{C}_j=C_j$ and $\widetilde{D}_j=\widehat{D}_j=D_j$ there, meaning that $[G,D_j]$ and $[I,C_j]$ (and also $[G,C_j]$ and $[I,D_j]$, but they do not appear here) are the same as the expressions obtained from them by putting a tilde or a hat on all the sets. Hence we want to gather as many of these expression as possible in the expression for t_Δ above, since these expressions will be easier to cancel in the quotients $t_\Delta/t_{T_R(\Delta)}$ and $t_\Delta/t_{T_S(\Delta)}$. Now, for any $(n+1)/2 \leq k \leq n-2$ we have $1 \leq n-1-k \leq (n-3)/2$, and the power to which the expressions $[G,C_{n-1-k}]$ and $[I,D_{n-1-k}]$ appear is $2n$ times the complement with respect to $(n^2-1)/4$ of

$$[(n-1-k)+1][n-1-(n-1-k)]=k(n-k).$$

On the other hand, for $1 \leq k \leq (n-3)/2$ we have $(n+1)/2 \leq n-1-k \leq n-2$ and the power to which $[G,D_{n-1-k}]$ and $[I,C_{n-1-k}]$ appear is $2n$ times

$$[(n-1-k)+1][n-1-(n-1-k)]=k(n-k).$$

Therefore we can write the last expression for t_Δ (after some reordering) as

$$\prod_{k=(n+1)/2}^{n-2} ([G,C_k][G,C_{n-1-k}][I,D_k][I,D_{n-1-k}])^{2n[(n^2-1)/4-k(n-k)]}$$

$$\times \prod_{k=1}^{(n-3)/2} ([G,D_k][G,D_{n-1-k}][I,C_k][I,C_{n-1-k}])^{2nk(n-k)}$$

$$\times ([G,G][I,I])^{2n(n^2-4n+3)/4}([G,D_{(n-1)/2}][I,C_{(n-1)/2}])^{2n(n^2-1)/4}$$

$$\times ([C_{-1},C_{-1}][D_{-1},D_{-1}][C_0,C_0][D_0,D_0][C_{-1},D_0][D_{-1},C_0])^{2n(n-1)}$$

$$\times \prod_{k=1}^{(n-3)/2} ([C_0,C_k][D_0,D_k])^{2n(n-1-2k)}([C_{-1},D_k][D_{-1},C_k])^{2n(n-1-2k)}$$

$$\times \prod_{k=(n+1)/2}^{n-2} ([C_{-1},C_k][D_{-1},D_k])^{2n(2k-(n-1))}([C_0,D_k][D_0,C_k])^{2n(2k-(n-1))}.$$

We now see that the last three lines are exactly the reduced expression for g_Δ in Eq. (5.6). Clearly, the expressions $t_{T_R(\Delta)}$ and $t_{T_S(\Delta)}$ are the expressions obtained from this by putting a tilde or a hat respectively on all the sets.

We now re-examine the quotients $t_\Delta/t_{T_R(\Delta)}$ and $t_\Delta/t_{T_S(\Delta)}$. In the numerator of each of these two quotients, we can see the expressions $[G,G]$, $[I,I]$, $[G,D_{(n-1)/2}]$, and $[I,C_{(n-1)/2}]$. In addition we find the pairs of expressions $[G,C_k][G,C_{n-1-k}]$ and $[I,D_k][I,D_{n-1-k}]$ for every $(n+1)/2 \le k \le n-2$, and also the pairs of expressions $[G,D_k][G,D_{n-1-k}]$ and $[I,C_k][I,C_{n-1-k}]$ for every $1 \le k \le (n-3)/2$. By all that was said above we see that these expressions cancel with the corresponding expressions with a tilde on all the sets in the denominator of the first quotient and with the corresponding expressions with a hat on all the sets in the denominator of the second quotient. After these cancelations the numerators become just g_Δ and the denominators become the expressions obtained from it by putting a tilde or a hat on all the sets (which are evidently $g_{T_R(\Delta)}$ or $g_{T_S(\Delta)}$ respectively). Hence we conclude that $t_\Delta/t_{T_R(\Delta)} = g_\Delta/g_{T_R(\Delta)}$ and $t_\Delta/t_{T_S(\Delta)} = g_\Delta/g_{T_S(\Delta)}$ and the theorem is proved. $\square$

5.3.3 The Description of h_Δ for Even n

We now deal with the case of even n. The usual changes one has to do from the odd n case appear also here, but now we have an additional twist. Therefore we present this case as well with all the details. The matrix in the pictorial description here has $2n$ rows and $2n+n$ columns. As usual, the rows are labeled by the sets C_j and D_j, $-1 \le j \le n-2$, in increasing order (with every set C_j preceding the corresponding D_j), and the columns are labeled also by the sets C_j and D_j, $-1 \le j \le n-2$, in increasing order, followed by C_j and D_j, $-1 \le j \le (n-4)/2$, in increasing order. As before, we can label the columns simply by C_j and D_j, $-1 < j < (3n-4)/2$, in increasing order, with the usual definition of $C_j = C_{j-n}$ and $D_j = D_{j-n}$ for $j \ge n-1$.

In fact we shall need only $2n + (n-1)$ columns, but we leave the last one as well since we prefer to keep working with 2×2 squares. For example, for $n = 4$ we have 8 rows and 12 columns, with the last 4 columns being labeled C_{-1}, D_{-1}, C_0, and D_0 (which can be considered as C_3, D_3, C_4, and D_4). Again we consider diagonals of maximal length, where we still look at diagonals of squares of 2×2 of entries rather than diagonals of the entries themselves. We have our diagonals labeled by $0 \leq k \leq n/2$ (rather than $0 \leq k \leq (n-2)/2$), but the last diagonal with $k = n/2$ will now behave differently. For every $0 \leq k \leq (n-2)/2$, we raise the main diagonal elements of every 2×2 square in it to the power

$$n\frac{(n-2k)^2}{4} = n\left(\frac{n}{2} - k\right)^2 = n\left(\frac{n^2}{4} - k(n-k)\right),$$

and the complementary diagonal elements of every 2×2 square in it to the power $nk(n-k)$ (the complement with respect to $n(n^2/4)$). However, for the diagonal corresponding to $k = n/2$, we still take the main diagonal elements of every 2×2 square in it to the power $n(n/2 - k)^2 = 0$ (and thus change nothing), but in the complementary diagonal we take only the lower left entry $(D_i, C_{i+n/2})$ to the power $nk(n-k) = n(n^2/4)$. The expression we form is, as usual, the product of all of these expressions. Again the powers of the main diagonals decrease with k, starting from $n(n^2/4)$ for $k = 0$ and ending with 0 for $k = n/2$, and the powers of the complementary diagonals increase with k, starting from 0 for $k = 0$ and ending with $n(n^2/4)$ for $k = n/2$ (with the extra twist in this last value of k); they are all even integers (as they are divisible by n). Altogether the formula we obtain is

$$h_\Delta = \prod_{k=0}^{(n-2)/2} \prod_{i=-1}^{n-2} ([C_i, C_{i+k}][D_i, D_{i+k}])^{n[n^2/4 - k(n-k)]} ([C_i, D_{i+k}][D_i, C_{i+k}])^{nk(n-k)}$$

$$\times \prod_{i=-1}^{n-2} [D_i, C_{i+n/2}]^{n(n^2/4)}$$

(using the notation with the sets C_j and D_j with $j \geq n-1$), where the last product is what comes from the $(n/2)$th diagonal (recall that the main diagonal elements contribute nothing there). We give here the pictorial image of the matrix for $n = 4$, where the double vertical line has the usual meaning of the point to the right of which lie the columns labeled by "sets with two names".

16 0	4 12	0			
0 16	12 4	16 0			
	16 0	4 12	0		
	0 16	12 4	16 0		
		16 0	4 12	0	
		0 16	12 4	16 0	
			16 0	4 12	0
			0 16	12 4	16 0

In order to write h_Δ only with the usual sets C_j with $-1 \le j \le n-2$, and D_j with $-1 \le j \le n-2$, we do the usual trick, which allows us to write, for every $1 \le k \le (n-2)/2$, the product over i as

$$\prod_{j=-1}^{n-2-k} ([C_j,C_{j+k}][D_j,D_{j+k}])^{n[n^2/4-k(n-k)]} ([C_j,D_{j+k}][D_j,C_{j+k}])^{nk(n-k)}$$

$$\times \prod_{j=-1}^{k-2} ([C_j,C_{j+n-k}][D_j,D_{j+n-k}])^{n[n^2/4-k(n-k)]} ([C_j,D_{j+n-k}][D_j,C_{j+n-k}])^{nk(n-k)}.$$

The usual invariance under $k \mapsto n-k$ allows us to look at the second product as the product to $n-k$, which satisfies the inequalities $(n+2)/2 \le n-k \le n-1$ since $1 \le k \le (n-2)/2$. As for the remaining value $k = n/2$, the same argument allows us to write the corresponding product as

$$\prod_{i=-1}^{(n-4)/2} [D_i,C_{i+n/2}]^{n(n^2/4)} \prod_{i=(n-2)/2}^{n-2} [D_i,C_{i-n/2}]^{n(n^2/4)}.$$

Now, for every $(n-2)/2 \le i \le n-2$, we can write that $[D_i,C_{i-n/2}]$ equals (up to the usual sign, but $n(n^2/4)$ is even) $[C_{i-n/2},D_i]$, or equivalently $[C_j,D_{j+n/2}]$ with $j = i - n/2$ that satisfies $-1 \le j \le (n-2)/2$. We now recall that the main diagonal elements of the squares in the $(n/2)$th diagonal are supposed to appear to the power 0, so we add them trivially (as we did in the nonsingular case) and obtain the formula

$$h_\Delta = \prod_{k=0}^{n-1} \prod_{j=-1}^{n-2-k} ([C_j,C_{j+k}][D_j,D_{j+k}])^{n[n^2/4-k(n-k)]} ([C_j,D_{j+k}][D_j,C_{j+k}])^{nk(n-k)}.$$

$$(5.11)$$

When examining the expression for h_Δ in Eq. (4.8) for small values of even n, we obtain the following results. For $n = 2$, where k can be either 0 or 1, we recall that the power

$$n\left(\frac{n}{2}-k\right)^2 = 2(1-k)^2$$

vanishes for $k = 1$, leaving only $k = 0$ with the power 2. However, its complement with respect to $n(n^2/4) = 2$ is 2 for $k = 0$ and vanishes for $k = 1$. Looking at the pictorial description we see that the number of columns is $2n+n = 6$, 2 more than the number $2n = 4$ of rows; hence there are two long diagonals of squares, the zeroth (containing only main diagonals elements) and the first (containing only some complementary diagonals as the last). This simply gives the denominator

$$[C_{-1},C_{-1}]^2 [D_{-1},D_{-1}]^2 [C_0,C_0]^2 [D_0,D_0]^2 [C_{-1},D_0]^2 [D_{-1},C_0]^2,$$

which by Lemma 4.2 is simply $[E,E]^2 [F,F]^2$ for $E = C_{-1} \cup D_0$ and $F = D_{-1} \cup C_0$ from before and coincides with g_Δ. This is so since also here the operator N from Definition 2.18 is trivial for $n = 2$ by the same reasoning as in the nonsingular case.

We recall here the remark about these "singular Z_2 curves" being just nonsingular Z_2 curves with $r = m$ and with the sets E and F here in the roles of the nonsingular C_{-1} and C_0 respectively, and we deduce from this description that Theorem 5.6 below with $n = 2$ is consistent with Theorem 4.7 with $n = 2$.

In the case $n = 4$, where k can be 0, 1, 2 or 3, we have seen that the power

$$n\left(\frac{n}{2} - k\right)^2 = 4(2 - k)^2$$

vanishes for $k = 2$ and gives 16 for $k = 0$ and 4 for $k = 1$ and $k = 3$. Hence the complement of this power with respect to $n(n^2/4) = 16$ is 16 for $k = 2$, 12 for $k = 1$ and $k = 3$, and vanishes for $k = 0$. The reflection of this in pictorial description is that the number of columns is $2n + n = 12$, 4 more than the number $2n = 8$ of rows. Therefore there are three long diagonals of squares, the zeroth (with only elements from main diagonals) for $k = 0$, the first for $k = 1$ and $k = 3$, and the second (which as the last contains only elements from complementary diagonals) for $k = 2$. We did not do an example of a singular Z_4 curve, but the denominator h_Δ in this case is

$$([C_{-1}, C_{-1}][D_{-1}, D_{-1}][C_0, C_0][D_0, D_0][C_1, C_1][D_1, D_1][C_2, C_2][D_2, D_2]^{16})$$
$$\times ([C_{-1}, C_0][D_{-1}, D_0][C_0, C_1][D_0, D_1][C_1, C_2][D_1, D_2])^4 ([C_{-1}, C_2][D_{-1}, D_2])^4$$
$$\times ([C_{-1}, D_0][D_{-1}, C_0][C_0, D_1][D_0, C_1][C_1, D_2][D_1, C_2])^{12} ([C_{-1}, D_2][D_{-1}, C_2])^{12}$$
$$\times ([C_{-1}, D_1][D_{-1}, C_1][C_0, D_2][D_0, C_2])^{16}.$$

Also here we already have a special case with small m rather than small n, which is Theorem 3.10. The argument is identical to the one for odd n, but without the statements about stronger results and squaring. Thus, by the same considerations, we see that Theorem 3.10 from Section 3.3 is the special case with $m = 2$ of the general result composed of Theorems 5.6 and 5.7.

5.3.4 N-Invariance for Even n

We now prove

Theorem 5.6. *The expression $\theta^{2n^2}[\Delta](0, \Pi)/h_\Delta$ with h_Δ from Eq. (5.11) is invariant under the action of the operators N, T_R and T_S.*

Proof. The proof of the invariance under N goes word-for-word as the proof for odd n. The same extends to the remark about the equivalence of the N-invariance of the quotient to the fact that the expression for h_Δ in Eq. (5.11) itself is invariant under N, i.e., satisfies $h_{N(\Delta)} = h_\Delta$.

For the invariance under the operators T_R and T_S we again use the PMT from Proposition 5.3 as the key ingredient, hence we write, using the reduced g_Δ from Eq. (5.7) as n is even, that

$$\frac{\theta^{n^2}[\Delta](0,\Pi)}{h_\Delta} = \frac{\theta^{n^2}[\Delta](0,\Pi)}{g_\Delta} \bigg/ \frac{h_\Delta}{g_\Delta}.$$

The (reduced) PMT in Proposition 5.3 now reduces us to proving the equality $h_\Delta/g_\Delta = h_{T_R(\Delta)}/g_{T_R(\Delta)} = h_{T_S(\Delta)}/g_{T_S(\Delta)}$, or equivalently to proving the two equalities $h_\Delta/h_{T_R(\Delta)} = g_\Delta/g_{T_R(\Delta)}$ and $h_\Delta/h_{T_S(\Delta)} = g_\Delta/g_{T_S(\Delta)}$, for any divisor Δ (which determines the sets C_j and D_j, $-1 \le j \le n-2$) and branch points $R \in C_0$ and $S \in B = D_{-1}$. The splitting of h_Δ as $h_\Delta = s_\Delta t_\Delta$, as usual, is now with

$$s_\Delta = \prod_{k=0}^{n-3} \prod_{j=1}^{n-2-k} ([C_j, C_{j+k}][D_j, D_{j+k}])^{n[n^2/4-k(n-k)]} ([C_j, D_{j+k}][D_j, C_{j+k}])^{nk(n-k)}$$

and

$$t_\Delta = \prod_{k=0}^{n-2} ([C_0, C_k][D_0, D_k])^{n[n^2/4-k(n-k)]} ([C_0, D_k][D_0, C_k])^{nk(n-k)}$$

$$\times \prod_{k=-1}^{n-2} ([C_{-1}, C_k][D_{-1}, D_k])^{n[n^2/4-(k+1)(n-1-k)]} ([C_{-1}, D_k][D_{-1}, C_k])^{n(k+1)(n-1-k)}$$

(with the same index change taking k to $k+1$ for $j = -1$), and we wish to prove that $s_\Delta = s_{T_R(\Delta)} = s_{T_S(\Delta)}$. This can again be done as in the case of odd n, as we still have $\widetilde{Y}_j = \widehat{Y}_j = Y_{n-1-j}$ and $\widetilde{Z}_j = \widehat{Z}_j = Z_{n-1-j}$ for all $1 \le j \le n-2$ in this notation. Therefore we can again cancel s_Δ with $s_{T_R(\Delta)}$ and with $s_{T_S(\Delta)}$ in the quotients $h_\Delta/h_{T_R(\Delta)} = s_\Delta t_\Delta / s_{T_R(\Delta)} t_{T_R(\Delta)}$ and $h_\Delta/h_{T_S(\Delta)} = s_\Delta t_\Delta / s_{T_S(\Delta)} t_{T_S(\Delta)}$ respectively, and it remains to prove the equalities $t_\Delta/t_{T_R(\Delta)} = g_\Delta/g_{T_R(\Delta)}$ and $t_\Delta/t_{T_S(\Delta)} = g_\Delta/g_{T_S(\Delta)}$.

We again use the fact that the sets G and I are invariant under T_R and T_S, i.e., satisfy $G = \widetilde{G} = \widehat{G}$ and $I = \widetilde{I} = \widehat{I}$, so we collect as many expressions containing them in the expression for t_Δ as we can. Using the equality

$$(k+1)(n-1-k) = k(n-k) + (n-1-2k)$$

we split the products into 3 parts: the part with $k = -1$ and $k = 0$, the part with $1 \le k \le (n-2)/2$, and the part with $n/2 \le k \le n-2$ (no intermediate value here). In the part with $k = -1$ and $k = 0$ we again find that $[Y_{-1}, Z_{-1}]$ and $[Y_0, Z_0]$ with $Y \ne Z$ do not appear but all the expressions which are of one of the forms $[Y_{-1}, Z_{-1}]$, $[Y_0, Z_0]$ or $[Y_{-1}, Z_0]$ with $Y = Z$ appear there at least to the power $n(n^2 - 4n + 4)/4$. Hence we use the (up to sign) equalities $[C_{-1}, C_{-1}][C_0, C_0][C_{-1}, C_0] = [G, G]$ and $[D_{-1}, D_{-1}][D_0, D_0][D_{-1}, D_0] = [I, I]$ from Lemma 4.2 to deal with these expressions. As for the others, for every $1 \le k \le (n-2)/2$, the power to which $[Y_{-1}, Z_k]$ with $Y = Z$ appear is smaller than the power to which $[Y_0, Z_k]$ appear, while $[Y_0, Z_k]$ with $Y \ne Z$ appear to a power smaller than the one to which $[Y_{-1}, Z_k]$ appear. Finally, we have that for $n/2 \le k \le n-2$, the power to which $[Y_0, Z_k]$ with $Y = Z$ appear is smaller than the one to which $[Y_{-1}, Z_k]$ appear, while $[Y_{-1}, Z_k]$ with $Y \ne Z$ appear to a power smaller than the one to which $[Y_0, Z_k]$ appear. We again use the equalities

$$[C_{-1},C_k][C_0,C_k] = [G,C_k], \quad [D_{-1},D_k][D_0,D_k] = [I,D_k],$$

$$[C_{-1},D_k][C_0,D_k] = [G,D_k], \quad \text{and} \quad [D_{-1},C_k][D_0,C_k] = [I,C_k]$$

(up to the usual sign), to see that t_Δ is

$$([C_{-1},C_{-1}][D_{-1},D_{-1}][C_0,C_0][D_0,D_0][C_{-1},D_0][D_{-1},C_0])^{n(n-1)}$$

$$\times \prod_{k=1}^{(n-2)/2} ([G,C_k][I,D_k])^{n[n^2/4-(k+1)(n-1-k)]} ([C_0,C_k][D_0,D_k])^{n(n-1-2k)}$$

$$\times ([G,G][I,I])^{n(n^2-4n+4)/4} \prod_{k=1}^{(n-2)/2} ([G,D_k][I,C_k])^{nk(n-k)} ([C_{-1},D_k][D_{-1},C_k])^{n(n-1-2k)}$$

$$\times \prod_{k=n/2}^{n-2} ([G,C_k][I,D_k])^{n[n^2/4-k(n-k)]} ([C_{-1},C_k][D_{-1},D_k])^{n(2k-(n-1))}$$

$$\times \prod_{k=n/2}^{n-2} ([G,D_k][I,C_k])^{n(k+1)(n-1-k)} ([C_0,D_k][D_0,C_k])^{n(2k-(n-1))}$$

(with the G and I parts from the part with $k = -1$ and $k = 0$ written now in the third row). When combining to get the expressions

$$[G,C_j][G,C_{n-1-j}], \quad [I,D_j][I,D_{n-1-j}],$$

$$[G,D_j][G,D_{n-1-j}], \quad \text{and} \quad [I,C_j][I,C_{n-1-j}]$$

for every $1 \le j \le n-2$, we now see that for every $n/2 \le k \le n-2$, we have the equality $1 \le n-1-k \le (n-2)/2$, and the power to which $[G,C_{n-1-k}]$ and $[I,D_{n-1-k}]$ appear is n times the complement of

$$[(n-1-k)+1][n-1-(n-1-k)] = k(n-k)$$

with respect to $n^2/4$. Moreover, for $1 \le k \le (n-2)/2$ we again have the equality $n/2 \le n-1-k \le n-2$ and the power to which $[G,D_{n-1-k}]$ and $[I,C_{n-1-k}]$ appear is n times

$$[(n-1-k)+1][n-1-(n-1-k)] = k(n-k).$$

From all this we obtain that t_Δ can be written as

$$([G,G][I,I])^{n(n^2-4n+4)/4} \prod_{k=n/2}^{n-2} ([G,C_k][G,C_{n-1-k}][I,D_k][I,D_{n-1-k}])^{n[n^2/4-k(n-k)]}$$

$$\times \prod_{k=1}^{(n-2)/2} ([G,D_k][G,D_{n-1-k}][I,C_k][I,C_{n-1-k}])^{nk(n-k)}$$

$$\times ([C_{-1},C_{-1}][D_{-1},D_{-1}][C_0,C_0][D_0,D_0][C_{-1},D_0][D_{-1},C_0])^{n(n-1)}$$

$$\times \prod_{k=1}^{(n-2)/2} ([C_0,C_k][D_0,D_k])^{n(n-1-2k)} ([C_{-1},D_k][D_{-1},C_k])^{n(n-1-2k)}$$

$$\times \prod_{k=n/2}^{n-2} ([C_{-1},C_k][D_{-1},D_k])^{n(2k-(n-1))} ([C_0,D_k][D_0,C_k])^{n(2k-(n-1))}.$$

The last three lines here give again the reduced expression for g_Δ in Eq. (5.7), and putting a tilde or a hat on all the sets gives the expressions for $t_{T_R(\Delta)}$ and $t_{T_S(\Delta)}$ respectively.

Our conclusion is thus the same as it was for the case of odd n. We have in the numerators of these two quotients the expressions $[G,G]$ and $[I,I]$, the expressions $[G,C_k][G,C_{n-1-k}]$ and $[I,D_k][I,D_{n-1-k}]$ for $n/2 \le k \le n-2$, and $[G,D_k][G,D_{n-1-k}]$ and $[I,C_k][I,C_{n-1-k}]$ for $1 \le k \le (n-2)/2$. The same considerations allow us to cancel the expressions with the corresponding expressions from the denominators which are with a tilde or a hat on all the sets. Doing so leaves the numerators as just g_Δ and the denominators as $g_{T_R(\Delta)}$ and $g_{T_S(\Delta)}$ (since they are the same expressions but with a tilde or a hat on all the sets respectively). Therefore we have obtained that $t_\Delta/t_{T_R(\Delta)} = g_\Delta/g_{T_R(\Delta)}$ and $t_\Delta/t_{T_S(\Delta)} = g_\Delta/g_{T_S(\Delta)}$, which concludes the proof of the theorem. □

5.4 Thomae Formulae for Singular Z_n Curves

Now we can turn to prove the Thomae formulae for a singular Z_n curve with equation

$$w^n = \prod_{i=0}^{m-1} (z-\lambda_i) \prod_{i=1}^{m} (z-\mu_i)^{n-1}.$$

Before doing so though we note that the same remark about merging the formulae for odd and even n extends to this setting (and therefore also to the more general setting we present in Appendix A). From Theorem 5.7 below we see that the Thomae formulae is based on the denominators h_Δ from Eq. (5.10) when n is odd and from Eq. (5.11) when n is even. Thus it suffices to examine these expressions for this cause. The only difference here between the odd and even n cases is (apart from e being 2 and 1 respectively) the $n^2 - 1$ in comparison to the n^2 which appear in the powers of the expressions of the sort $[Y_j, Z_{j+k}]$ with $Y = Z$ as types. Therefore taking the 4th power of the denominator in Eq. (5.10) and the 8th power of the denominator in Eq. (5.11), and then multiplying the denominator in Eq. (5.10) by

$$\left[\bigcup_{j=-1}^{n-2} C_j, \bigcup_{j=-1}^{n-2} C_j \right]^{2n} = \prod_{0 \leq i < j \leq m-1} (\lambda_i - \lambda_j)^{2n}$$

and

$$\left[\bigcup_{j=-1}^{n-2} D_j, \bigcup_{j=-1}^{n-2} D_j \right]^{2n} = \prod_{1 \leq i < j \leq m} (\mu_i - \mu_j)^{2n}$$

we obtain a parity-independent Thomae formulae (after proving Theorem 5.7) stating that the quotient $\theta^{16n^2}[\Delta](0, \Pi)/h_\Delta$ with

$$h_\Delta = \prod_{k=0}^{n-1} \prod_{j=-1}^{n-2-k} ([C_j, C_{j+k}][D_j, D_{j+k}])^{2n(n-2k)^2} ([C_j, D_{j+k}][D_j, C_{j+k}])^{8nk(n-k)}$$

is independent of Δ. Nevertheless, as in the nonsingular case, we continue to work with the stronger and reduced Eqs. (5.10) and (5.11), even though they depend on the parity of n.

5.4.1 The Thomae Formulae

We have already seen in the example in Section 3.2 that when we find the quotient $\theta^{2en^2}[\Delta](0, \Pi)/h_\Delta$ that is invariant under the operators N, T_R, and T_S, we obtain the Thomae formulae, since one can get from any divisor to any other divisor using a finite sequence of applications of the operators N, T_R, and T_S. This again extends to our more general setting, a fact which we prove in

Theorem 5.7. *The expression* $\theta^{2en^2}[\Delta](0, \Pi)/h_\Delta$ *with* h_Δ *from Eq. (5.10) when n is odd and from Eq. (5.11) when n is even is independent of the divisor* Δ.

Proof. Since Theorems 5.5 and 5.6 show that this expression is invariant under the operators N, T_R and T_S, it suffices to show that one can get from any divisor to any other divisor using the operators N, T_R, and T_S. However, unlike the nonsingular case, Theorems 2.13 and 2.15 show that not all the non-special integral divisors of degree g supported on the branch points distinct from P_0 have the same form, since the cardinalities of the sets C_j and D_j, $-1 \leq j \leq n - 2$, may vary (they satisfy the conditions, but are not uniquely determined by them). This means that we can no longer use permutations (there are also two types of points here), and therefore we give a proof in the spirit of the second proof of Theorem 4.8. Note that we have chosen different notation for the divisors we construct on the way. This is since for the general family in Appendix A we combine the two proofs together, with notation for the divisors which is consistent with their roles in the proofs of Theorem 4.8 and this theorem.

We thus recall the ordering (4.9) of the sets C_j, $-1 \le j \le n - 2$, from the nonsingular case and the corresponding notation $k \prec l$ etc., and we shall use it here as well. One may assume that here in the singular case the set D_j goes together with C_j for every $-1 \le j \le n - 2$ in the ordering (4.9), but it is easier to work simply with the indices. Now the induction step for the stage corresponding to the sets C_k and D_k (or simply to k) is the following claim: Assume that Δ and Ξ are two divisors that are distinct only in the sets C_j and D_j with $j \preceq k$. Assume that one can get, using the operators N, T_R and T_S, from any divisor to any other divisor which is distinct from it only in sets C_j and D_j with $j \prec k$ (with strict "inequality"—this is the induction hypothesis). Then one can get from Δ to Ξ using only N, T_R and T_S. The proof is by this type of induction.

If $k = -1$, then Δ and Ξ can be distinct only on the sets C_{-1} and D_{-1}, but since they are the complements of the unions $\bigcup_{j=0}^{n-2} C_j$ and $\bigcup_{j=0}^{n-2} D_j$ to the sets of branch points P_i, $0 \le i \le m - 1$, and Q_i, $1 \le i \le m - 1$, respectively, we must have $\Delta = \Xi$ then and there is nothing to prove. If $k = 0$, then Ξ is obtained from Δ by mixing C_{-1} with C_0 and D_{-1} with D_0, but with the cardinality conditions preserved. For this we note again that if $R \in C_0$ and $P \in C_{-1}$, $P \ne P_0$, then T_P is permissible on the divisor $T_R(\Delta)$, and the resulting divisor is again the one obtained from Δ by interchanging the places of R and P. We have seen that this is the action of the combination $T_P T_R$ in the nonsingular case, and here the sets C_j with $-1 \le j \le n - 2$ behave just as in the nonsingular case. As for the sets D_j, $-1 \le j \le n - 2$, the fact that $\widetilde{D_j} = D_{n-1-j}$ for all $1 \le j \le n - 2$ and that $\widetilde{D_{-1}} = D_0$ and $\widetilde{D_0} = D_{-1}$ show that this is indeed the action of $T_P T_R$ also here (as double application leaves all the sets D_j, $-1 \le j \le n - 2$ fixed). Similarly, we can take $S \in D_{-1}$ and $Q \in D_0$, and find that T_Q is permissible on the divisor $T_S(\Delta)$, and the resulting divisor is the one obtained from Δ by interchanging the places of S and Q. Here the reason is that $\widehat{C_j} = C_{n-1-j}$ and $\widehat{D_j} = D_{n-1-j}$ for $1 \le j \le n - 2$ and this operation fixes all these sets when applied twice, and the same holds for $\widehat{A} = C_0$ and $\widehat{C_0} = A$. Since S is fixed by T_S and then taken to D_0 by T_Q and Q is taken by the action of T_S to D_{-1} where T_Q leaves it fixed, the action of $T_Q T_S$ is as stated. These are some of the mixing options involving only changes in C_{-1}, D_{-1}, C_0, and D_0.

These two pairs of operators fix the cardinalities of C_{-1}, D_{-1}, C_0, and D_0. We now introduce two pairs of operators that will change these cardinalities. For $R \in C_0$ and $Q \in D_0$, T_Q is permissible on $T_R(\Delta)$, and the resulting divisor is the one obtained from Δ by taking R and Q from C_0 and D_0 to C_{-1} and D_{-1} respectively; this is so since we have

$$\widetilde{\widehat{C_0}} = \widetilde{A} = C_0 \setminus R, \quad \widetilde{\widehat{D_0}} = \widetilde{B} \setminus Q = D_0 \setminus Q,$$

$$\widehat{\widetilde{A}} = \widetilde{C_0} = A \cup R, \quad \text{and} \quad \widehat{\widetilde{B}} = \widetilde{D_0} \cup Q = B \cup Q,$$

for the sets including P_0 put it where needed, and the other sets change as above. Similarly, for $S \in D_{-1}$ and $P \in C_{-1}$, $P \ne P_0$, we find that T_P is permissible on $T_S(\Delta)$, and the resulting divisor is the one obtained from Δ by taking P and S from C_{-1} and D_{-1} to C_0 and D_0 respectively; here we have

$$\widetilde{\widetilde{C_0}} = \widehat{A} \cup P = C_0 \cup P, \quad \widetilde{\widetilde{D_0}} = \widehat{B} = D_0 \cup S,$$

$$\widetilde{\widehat{A}} = \widehat{C_0} \setminus P = A \setminus P \ , \text{and} \ \ \widetilde{\widehat{B}} = \widehat{D_0} = B \setminus S,$$

with the same remark about P_0 and the other sets. Using these operations it is easy to see that one can get from Δ to any other divisor Ξ obtained only by mixing C_{-1}, D_{-1}, C_0, and D_0. The shortest way to explain this (but not to apply this) is by starting from Δ, and taking all of C_0 and D_0 to C_{-1} and D_{-1}, one from each set at a time using $T_P T_S$. Then one redistributes them as needed for Ξ, one from each set at a time using $T_Q T_R$ (note that the cardinality conditions on C_{-1}, D_{-1}, C_0, and D_0 imply that this process is feasible). This forms the basis for our induction.

We now turn to the induction step. We again have the operator N, which by Definition 2.18 interchanges C_j and D_j with C_{n-2-j} and D_{n-2-j} respectively, for every $0 \leq j \leq n-2$ (and fixes C_{-1} and D_{-1}). We also have the operators T_R and T_S, which interchange C_j and D_j with C_{n-1-j} and D_{n-1-j} respectively, for every $1 \leq j \leq n-2$ (and do some mixing in the remaining 4 sets, which is not important to us at this stage). This means that for our sets C_k and D_k with $1 \leq k \leq n-2$, the action of N interchanges them with the "neighbors" on one side in the ordering (4.9) and the action of T_R or T_S interchanges them with the "neighbors" on their other side there. We remark that the operator W we have defined in the nonsingular case can no longer be so easily defined, but any permissible operator T_R or T_S can fill its role in this proof. Therefore we assume that the "left neighbor" of k is h for some h (which means that the "left neighbors" of C_k and D_k in the ordering (4.9) are C_h and D_h for this h), and choose V to be the operator (either N or some T_R or T_S) which interchanges C_k and D_k with C_h and D_h respectively (in the case where V has to be some T_R or some T_S, the choice of exactly which one we take is not important and we simply assume that V is a permissible one at each application).

With these tools we can turn to the proof of the induction step. We do it like the induction step of the second proof of the nonsingular case, with 3 applications of the induction hypothesis, but here it is simpler. In the nonsingular case we needed to choose some set Y of cardinality r, and here we can take a different and simpler choice of the empty set (which can always be taken). Therefore the first induction step works like the other induction steps as well. Choose now a divisor Λ satisfying $(C_j)_\Lambda = (C_j)_\Delta$ and $(D_j)_\Lambda = (D_j)_\Delta$ for all $j \succeq k$, and with empty C_h and D_h (this can, of course, be done such that the cardinality conditions from Theorems 2.13 and 2.15 are satisfied). By the induction hypothesis we can get from Δ to Λ using the operators N, T_R and T_S, and therefore also to $V(\Lambda)$ by another single operation. If we define ε to be 2 if $V = N$, and to be 1 if V is some T_R or T_S, then we find by Definition 2.18 that the divisor $V(\Lambda)$ has empty C_k and D_k, and for $j \succ k$ we have that $(C_j)_{V(\Lambda)}$ is $(C_{n-\varepsilon-j})_\Delta = (C_{n-\varepsilon-j})_\Xi$ and $(D_j)_{V(\Lambda)}$ is $(D_{n-\varepsilon-j})_\Delta = (D_{n-\varepsilon-j})_\Xi$.

Now choose a divisor Ψ with empty C_k and D_k, such that $(C_h)_\Psi = (C_k)_\Xi$ and $(D_h)_\Psi = (D_k)_\Xi$, and with C_j and D_j like those corresponding to $V(\Lambda)$ for every

$j \succ k$ (again this can be done such that the cardinality conditions from Theorems 2.13 and 2.15 are satisfied, since the sets corresponding to Ξ satisfy them). Using the induction hypothesis again we find that we can get from $V(\Lambda)$ (and hence also from Δ) to Ψ using the operators N, T_R and T_S, and another application of V takes us to $V(\Psi)$. Using Definition 2.18 again we find that $V(\Psi)$ has the same C_k and D_k as Ξ, and also all the sets C_j and D_j for $j \succ k$ are like those corresponding to Ξ and Δ (and C_h and D_h are empty, but this does not matter). We can now apply the induction hypothesis yet another time and find that we can get from $V(\Psi)$, and hence from Δ, to Ξ as desired. This proves the theorem. $\qquad\square$

5.4.2 Changing the Basepoint

We would now like to present the basepoint change for this case, and combine the results for the different basepoints. As always, the Thomae formulae obtained from Theorems 5.5, 5.6 and 5.7 with the nice expressions appearing in Eqs. (5.10) and (5.11) are beautiful and symmetric, but they include some dependence on the choice of the basepoint P_0. We now have for every branch point S (either some point P_i, $0 \leq i \leq m-1$ or some Q_i, $1 \leq i \leq m$) the quotients $\theta[\varphi_S(\Delta) + K_S](0, \Pi)/h_\Delta$ with h_Δ as defined by Eqs. (5.10) and (5.11), and the value of this quotient is independent of the divisor Δ (whose support must now not contain S rather than P_0) but may still be different when letting the basepoint S vary. Moreover, we note again that here S can also be some point Q_i and is not restricted to be only some P_i, but choosing it to be such requires us to change the roles of the sets C_j and D_j. This is so since in the notations we have used, this will put us in the situation where the sets C_j, $-1 \leq j \leq n-2$, contain points Q_k and the sets D_j, $-1 \leq j \leq n-2$, contain points P_k, unlike what we always had.

We begin with changing the basepoint to some point P_i. This goes exactly as we changed basepoints in the nonsingular case with $r \geq 2$. Since $m \geq 2$ (we want positive genus) we do have some other point P_i, and we have divisors where the cardinality of C_{-1} (and hence also D_{-1}) is at least 2. Since Lemma 2.12 shows that the branch points satisfy the condition that $\varphi_{P_i}(P_j^n) = 0$, $\varphi_{P_i}(Q_j^n) = 0$, $\varphi_{Q_i}(P_j^n) = 0$ and $\varphi_{Q_i}(Q_j^n) = 0$ for every i and j, and since the divisors we have are integral, of degree g, supported on the branch points (all of them—P_0 is now allowed), contain no branch point to the power n or higher, and contain some points to the power $n-1$ (as we chose them to be), we can apply Corollary 1.13. Now, we start with our divisors and basepoint P_0, fix another basepoint P_i with some $1 \leq i \leq m-1$, and we want to show that the quotients obtained using the basepoint P_i give the same constant. This is done by the usual argument: choose a divisor Δ (with basepoint P_0) with $|C_{-1}| = |D_{-1}| \geq 2$ and such that P_i is in C_{-1} (or A), and then the divisor Γ_{P_i} from Corollary 1.13 is non-special by Theorem 2.13. This divisor Γ_{P_i} also has the same sets C_j and D_j with $-1 \leq j \leq n-2$, where as always the only difference is in A (which is replaced by $A \setminus P_i \cup P_0$) but when adding P_i instead of P_0 the set

C_{-1} is seen to be the same for Δ and for Γ_{P_i}. This gives that Eqs. (5.10) and (5.11) will give the same expression for $h_{\Gamma_{P_i}}$ as for h_Δ, and since Corollary 1.13 shows that $\varphi_{P_i}(\Gamma_{P_i}) + K_{P_i} = \varphi_{P_0}(\Delta) + K_{P_0}$ the comparison between the constant values obtained from P_0 and from P_i is complete (by these representatives).

We now turn to changing the basepoint to some point Q_i. We can choose a divisor where where the cardinality of D_{n-2} (and hence also C_{n-2}) is at least 1 together with the condition $|C_{-1}| = |D_{-1}| \geq 1$ (again since $m \geq 2$) and the other cardinality conditions from Theorems 2.13 and 2.15 are satisfied. Again Lemma 2.12 gives the conditions required for Corollary 1.13, and the divisors are of the sort on which it can be applied. We then choose such a divisor (with basepoint P_0) where we assume that the chosen new basepoint Q_i with some $1 \leq i \leq m$ is in D_{n-2}, and then we want to see that the divisor Γ_{Q_i} from Corollary 1.13 is again non-special and satisfies $h_{\Gamma_{Q_i}} = h_\Delta$.

The main argument for this is already in the proof of Proposition 1.15 for the case where Q_i there lies in D_{n-2}: as we said there, the set C_j with $0 \leq j \leq n-2$ plays the role of D_{n-3-j}, and the set D_j with $-1 \leq j \leq n-3$ plays the role of C_{n-3-j}, and the statements asserted there for A and D_{n-2} completes our statement for C_{-1} and D_{n-2}. We also saw there that the cardinality conditions hold for Γ_{Q_i} since they do for Δ, which means that Γ_{Q_i} is non-special by Theorem 2.13.

Now, we already have the equality $\varphi_{Q_i}(\Gamma_{Q_i}) + K_{Q_i} = \varphi_{P_0}(\Delta) + K_{P_0}$ from Corollary 1.13, and in order to see why the constant quotient obtained from the basepoint Q_i (represented by the divisor Γ_{Q_i}) is the same as the one from P_0 (represented by the divisor Δ) we have to verify that Eqs. (5.10) and (5.11) will give the same expression for $h_{\Gamma_{Q_i}}$ as for h_Δ. This is equivalent to saying that changing every C_j and D_j with $-1 \leq j \leq n-2$ to D_{n-3-j} and C_{n-3-j}, respectively, leaves these expressions invariant. This is indeed the case, as for every $0 \leq k \leq n-1$, the expressions $[Y_j, Z_{j+k}]$ (with the usual meaning) all appear to powers depending only on k and on whether $Y = Z$ as types of sets, and under this action every such expression transforms to $[\overline{Y}_{n-3-j}, \overline{Z}_{n-3-j-k}]$ where putting a bar on a type replaces it by the opposite one. As usual, we write this as $[\overline{Z}_i, \overline{Y}_{i+k}]$ for $i = n-3-j-k$ (up to sign), and since $Y = Z$ is equivalent to $\overline{Z} = \overline{Y}$ and when j runs through the numbers from -1 to $n-2-k$ so does i (we ignore the decreasing order again), the fact that all the powers are even shows that $h_{\Gamma_{Q_i}} = h_\Delta$. This concludes the proof of the claim that the constant quotients obtained from all the different $2m$ branch points are equal.

We now like to show that the sets of characteristics and thus the set of theta constants one obtains when going over all the divisors Theorems 2.13 and 2.15 give for any choice of basepoint are the same (and not only the constant quotients). Here the number of divisors is not as easy to calculate, but it is clear that it is the same for every choice of basepoint (and from what we now show, this is the number of characteristics appearing in the full Thomae formulae as well). We shall prove this using Proposition 1.15, which is applicable by the fact that $\prod_{i=0}^{m-1} P_i / \prod_{i=1}^{m} Q_i$ is principal and the fact that the divisors are as needed in the assumptions of Proposition 1.15. For any characteristic $\varphi_{P_0}(\Delta) + K_{P_0}$ obtained from the basepoint P_0 and any other branch point S Proposition 1.15 gives that we have a divisor Γ_S such that $\varphi_S(\Gamma_S) + K_S$ gives the same characteristic, and since Γ_S is of the same form as Δ,

Theorem 2.13 gives that Γ_S is non-special as well and with S not appearing in its support. The converse follows, as usual, either by an argument of symmetry or by the fact that the maps $\varphi_S(\cdot) + K_S$ are one-to-one on the relevant sets of divisors and the fact that there is the same number of divisors Γ with the basepoint S as divisors Δ with the basepoint P_0. Therefore the Thomae formulae from Theorems 5.5 and 5.6 (using Theorem 5.7) with the $2m$ distinct branch points P_i, $0 \le i \le m-1$, and Q_i, $1 \le i \le m$, involve the same theta constants and give the same constant.

As in the examples and in the nonsingular case, we would like to have new notation for the Thomae formulae from Theorems 5.5 and 5.6 which is independent of the choice of the basepoint. As always, for this we use the fact that if P_i appears to the $(n-1)$th power in Δ, then $P_0^{n-1}\Delta = P_i^{n-1}\Gamma_{P_i}$, and if Q_i appears to the $(n-1)$th power in Δ, then $P_0^{n-1}\Delta = Q_i^{n-1}\Gamma_{Q_i}$, and we denote this divisor by Ξ (hence Ξ equals $\prod_{j=-1}^{n-2} C_j^{n-2-j} D_j^{j+1}$, and we decide to continue taking the sets C_j, $-1 \le j \le n-2$, to contain the points P_i and the sets D_j, $-1 \le j \le n-2$, to contain the points Q_i, even though the results are correct with the opposite choice as well). Then we can replace the notation $\theta[\Delta]$ for a divisor Δ with basepoint P_0 by $\theta[\Xi]$, which is the theta function with characteristic $\varphi_S(\Xi) + K_S$ for some branch point S (either a P_i or a Q_i) and this does not depend on the choice of the basepoint S by the second statement in Corollary 1.13. We also denote the denominators from Eqs. (5.10) and (5.11) by h_Ξ (which is again a more suitable correspondence by the usual argument), and this almost proves

Theorem 5.8. *For every divisor $\Xi = \prod_{j=-1}^{n-2} C_j^{n-2-j} D_j^{j+1}$ which is based on a partition of the set of m branch points P_i, $0 \le i \le m-1$, into n sets C_j, $-1 \le j \le n-2$, and a partition of the set of m branch points Q_i, $1 \le i \le m$, into n sets D_j, $-1 \le j \le n-2$, such that the cardinality condition $|C_j| = |D_j|$ holds for every $-1 \le j \le n-2$, define $\theta[\Xi]$ to be the theta function with characteristic $\varphi_S(\Xi) + K_S$ branch point S (which is independent of the choice of S), and define h_Ξ to be the expression from Eq. (5.10) if n is odd and Eq. (5.11) if n is even. Then the quotient $\theta^{2en^2}[\Xi](0,\Pi)/h_\Xi$ is independent of the divisor Ξ.*

We have only nearly proved Theorem 5.8 as the proof from the basepoint P_0 applies only for divisors Ξ where where C_{-1} and D_{-1} are nonempty and C_{-1} contains P_0. We now want to see how to eliminate this extra assumption, which breaks the lovely symmetry of Theorem 5.8. Theorem 5.8 talks about the independence of the quotient $\theta^{2en^2}[\Xi](0,\Pi)/h_\Xi$ of the divisor Ξ, but for this paragraph it will be more convenient to interpret the phrase "Theorem 5.8 holds for the divisor Ξ" as "the value of quotient corresponding to Ξ is the same as the constant obtained from Theorems 5.5, 5.6 and 5.7 (with any basepoint)". In these terms we see that as usual, Theorem 5.8 applies to any divisor Ξ with nonempty C_{-1} and D_{-1} as it just comes from some point P_i from C_{-1}. Also divisors with nonempty C_{n-2} and D_{n-2} can be covered by Theorem 5.8 (as they come from some point Q_i from D_{n-2}), but this does not shorten the proof we give for the other cases, so we ignore it at this stage. We now see that the operator M from the remark after Proposition 1.15 allows us to extend the validity of Theorem 5.8 to all the divisors for which it is asserted.

We shall show that Theorem 5.8 holds for such a divisor Ξ by induction on the least (here in the usual order, not in (4.9)) index l such that C_l and D_l are nonempty. The basis of the induction, which is $l = -1$, is now known. Assume that $l \geq 0$ and that the statement is true for $l - 1$, and then notice that the action of M (which only multiplies Ξ by $\prod_{i=0}^{m-1} P_i / \prod_{i=1}^{m} Q_i$ as C_{-1} and D_{-1} are empty) takes Ξ to a divisor Υ with nonempty C_{l-1} and D_{l-1}. Since we know that the application of M leaves the theta constant $\theta[\Xi]$ invariant (as it does for the characteristic), we have to show that the application of M leaves also the denominator h_Ξ invariant in order to compare the quotients. Now, we have seen after Proposition 1.15 that the action of M is simply "rotating" the sets C_j and D_j, $-1 \leq j \leq n-2$ together, or more explicitly decreases each index by 1 (and takes the index -1 to $n-2$, which does not matter here since C_{-1} and D_{-1} are assumed to be empty). Since the powers to which expressions appear in Eqs. (5.10) and (5.11) are based only on the difference between indices, all the expressions are taken to expressions which appear to the same power (this also holds, though with another argument, for expressions containing C_{-1} and D_{-1}, but this does not matter here as these sets are empty). This shows that for $\Upsilon = M(\Xi)$ we also have $h_\Upsilon = h_\Xi$, and since Proposition 1.15 gives that $\varphi_S(\Xi) + K_S = \varphi_S(\Upsilon) + K_S$ for any branch point S, we find that $\theta^{2en^2}[\Xi](0, \Pi)/h_\Xi = \theta^{2en^2}[\Upsilon](0, \Pi)/h_\Upsilon$. Since by the induction hypothesis Theorem 5.8 holds for Υ, it now does also for Ξ, which proves it for every divisor Ξ as we wanted.

Theorem 5.8 is the most "symmetric" Thomae formulae for singular Z_n curves, and gives as a corollary the fact that the Thomae formulae from the $2m$ different basepoints give the same constant. As usual, unlike Theorems 5.5 and 5.6 in which every divisor gives different characteristics, Theorem 5.8 speaks about orbits of size n of M which gives these characteristics (and calculating their number is a nontrivial combinatorial mission with which we shall not deal here). It contains, of course, Theorem 3.7 from Section 3.2 as its special case with $n = 3$ and $m = 3$. It also gives the basepoint-independent formulae asserted in Section 3.3 for the case described there (after replacing every μ_1 by λ and omitting every expression involving μ_2, which is ∞).

Chapter 6

Some More Singular Z_n Curves

In this chapter we treat two other cases of families of singular curves and give a few examples. Contrary to the examples in Chapter 3, which were presented to demonstrate special cases of general families, here we have no general theory and we treat these cases separately.

The most general case of a Z_n curve is the Riemann surface corresponding to the algebraic equation

$$w^n = \prod_{i=1}^{t}(z - \mu_i)^{\alpha_i},$$

where the powers α_i are assumed (for simplicity of notation) to be written in increasing order, and with $1 \le \alpha_i \le n - 1$ for every $1 \le i \le t$. We shall usually be assuming for simplicity that $\sum_{i=1}^{t} \alpha_i$ is a multiple of n, a condition that is equivalent to the statement that ∞ is not a branch point. If the last assumption is not satisfied, then ∞ is considered as if appearing to the power $1 \le s \le n - 1$, which completes $\sum_{i=1}^{t} \alpha_i$ to a multiple of n. The assumption that ∞ is not a branch point does not cause any loss of generality, as an automorphism of the Riemann sphere can take us from one case to the other. We remark again that the results where ∞ is a branch point are obtained by simply removing every expression containing it from the denominators g_Δ and h_Δ we construct on the way, by the remark we gave in Section 2.6.

More importantly, we say that a Z_n curve is *fully ramified* if all the powers α_i are relatively prime to n, and in the case ∞ is a branch point we also demand that the complementary power s be relatively prime to n. Note that until now all the powers were 1 and $n - 1$ (including the complementary power s, which was $n - 1$ in the example in Section 3.3), making all the Z_n curves we dealt with until now fully ramified. We note that this information is enough for calculation of the genus. The total branching number is simply $(n - 1)t$ (or $(n - 1)(t + 1)$ if ∞ is also a branch point—let us assume that it is not), meaning that $2g - 2 = -2n + (n - 1)t$ and hence $g = (n - 1)(t - 2)/2$ (and hence t must be even if n is even). The reader can verify that the genera of all the Z_n curves we dealt with until now satisfied this rule.

Until this chapter, in all the Z_n curves we have dealt with only the powers 1 and $n - 1$ appeared. We note that by choosing other functions on the Z_n curve we can

H.M. Farkas and S. Zemel, *Generalizations of Thomae's Formula for Z_n Curves*,
Developments in Mathematics 21, DOI 10.1007/978-1-4419-7847-9_6,
© Springer Science+Business Media, LLC 2011

obtain other powers. For example, for any integer $1 \leq l \leq n-1$ which is prime to n, the Z_n curve described by

$$w^n = \prod_{i=0}^{rn-1} (z - \lambda_i)^l$$

is actually a nonsingular Z_n curve. This can be seen by replacing w by the function $u = w^k / \prod_{i=0}^{rn-1} (z - \lambda_i)^h$ with k and h satisfying $kl - hn = 1$ (integers), which gives the algebraic equation

$$u^n = \prod_{i=0}^{rn-1} (z - \lambda_i).$$

u (together with z) generates w as well since $w = u^l \prod_{i=0}^{rn-1} (z - \lambda_i)^{h(k-1)}$, as one can easily check. In this chapter we continue to consider only fully ramified Z_n curves, but also treat a few cases where powers other than 1 and $n-1$ appear. Namely, in Section 6.1 we allow the powers 2 and $n-2$ to appear together with 1 and $n-1$ (for general odd n) and in Section 6.2 we allow the power $n-3$ to appear together with 1 (for general n not divisible by 3). In light of the previous remark we state that the families of Z_n curves we deal with in this chapter are indeed new, i.e., not isomorphic (except maybe for some small values of n) in such a trivial way to our singular or nonsingular Z_n curves.

6.1 A Family of Z_n Curves with Four Branch Points and a Symmetric Equation

In this section we deal with Z_n curves with $t = 4$ and where the powers are $\alpha_1 = 1$, $\alpha_2 = 2$, $\alpha_3 = n-2$ and $\alpha_4 = n-1$. In order to keep the basepoint as P_0 (until the end) and appearing to the power 1 in the algebraic equation, the algebraic equation will be written as

$$w^n = (z - \lambda_0)(z - \lambda_1)^2 (z - \lambda_2)^{n-2}(z - \lambda_3)^{n-1},$$

and the point over λ_i will be denoted as usual by P_i. Note that since we are assuming full ramification the powers α_i need be relatively prime to n, so that we work only with odd n in this section. We can replace the function w by $\left(\prod_{i=0}(z - \lambda_i)\right)/w$ and obtain that the same Z_n curve can be described by the equation

$$\left(\frac{\prod_{i=0}(z - \lambda_i)}{w}\right)^n = (z - \lambda_0)^{n-1}(z - \lambda_1)^{n-2}(z - \lambda_2)^2 (z - \lambda_3),$$

showing that P_0 and P_3 are symmetric and also P_1 and P_2 are symmetric. However, the equation does not give any symmetry between these pairs of points, and we shall later see that (except for some small values of n) these pairs are indeed not symmetric to one another.

6.1.1 Functions, Differentials, Weierstrass Points and Abel–Jacobi Images

The Z_n curve thus described is a compact Riemann surface, and calculating its genus gives $g = (n-1)(4-2)/2 = n-1$. We note that $n = 3$ takes us back to the singular curve with $m = 2$, so we get new results only for $n \geq 5$.

In order to prove Thomae formulae for a Z_n curve, our technique requires us to know which characteristics (or equivalently divisors) appear. For this we need to find first the divisors of the defining functions and differentials and to construct bases of the holomorphic differentials on our Z_n curve X. We recall that instead of the function z it is usually more convenient to look at the functions $z - \lambda_i$, $0 \leq i \leq 3$. If we denote the n points over ∞ by ∞_h with $1 \leq h \leq n$ as usual we find that

$$\text{div}(z-\lambda_i) = \frac{P_i^n}{\prod_{h=1}^n \infty_h}, \quad \text{div}(w) = \frac{P_0 P_1^2 P_2^{n-2} P_3^{n-1}}{\prod_{h=1}^n \infty_h^2},$$

and

$$\text{div}(dz) = \frac{P_0^{n-1} P_1^{n-1} P_2^{n-1} P_3^{n-1}}{\prod_{h=1}^n \infty_h^2},$$

and it is easy to verify that the degree of $\text{div}(dz)$ is indeed $2g - 2$. Another useful fact is obtained by calculating

$$\text{div}\left(\frac{(z-\lambda_2)(z-\lambda_3)}{w}\right) = \frac{P_2^2 P_3}{P_0 P_1^2},$$

which also shows that our Riemann surface X can be represented as a 3-sheered cover of the sphere. We shall use this relation to construct more holomorphic differentials. We note that

$$\text{div}\left(\frac{dz}{w}\right) = P_0^{n-2} P_1^{n-3} P_2 = P_0^{n-2} P_1^{n-3} P_2^1 P_3^0,$$

which shows that dz/w is a holomorphic differential. It thus follows that the differential $(z-\lambda_2)^k (z-\lambda_3)^k dz/w^{k+1}$ (the product of dz/w by the kth power of $(z-\lambda_2)(z-\lambda_3)/w$) is a holomorphic differential for every $0 \leq k \leq (n-3)/2$ (recall that n is odd!), since it is easy to calculate that

$$\text{div}\left(\left(\frac{(z-\lambda_2)(z-\lambda_3)}{w}\right)^k \frac{dz}{w}\right) = P_0^{n-2-k} P_1^{n-3-2k} P_2^{2k+1} P_3^k.$$

This gives us $(n-1)/2$ holomorphic differentials, half of what we need.

It remains to produce another $(n-1)/2$ holomorphic differentials. The differential with $k = (n-3)/2$ has divisor $P_0^{(n-1)/2}P_2^{n-2}P_3^{(n-3)/2}$, and multiplying it again by $(z-\lambda_2)(z-\lambda_3)/w$ (which is the same as taking $k=(n-1)/2$ above) would give a differential with divisor $P_0^{(n-3)/2}P_2^n P_3^{(n-1)/2}/P_1^2$ which is no longer holomorphic. However, if we multiply this last differential by $(z-\lambda_1)/(z-\lambda_2)$ we multiply its divisor by P_1^n/P_2^n, which yields the differential

$$(z-\lambda_1)(z-\lambda_2)^{(n-3)/2}(z-\lambda_3)^{(n-1)/2}\frac{dz}{w^{(n+1)/2}}$$

whose divisor is $P_0^{(n-3)/2}P_1^{n-2}P_3^{(n-1)/2}$. Thus it is again holomorphic. We can multiply again by the same function $(z-\lambda_2)(z-\lambda_3)/w$ of degree 3 any number $0 \le l \le (n-3)/2$ of times and obtain that

$$\text{div}\left(\frac{z-\lambda_1}{z-\lambda_2}\left(\frac{(z-\lambda_2)(z-\lambda_3)}{w}\right)^k\frac{dz}{w}\right) = P_0^{n-2-k}P_1^{2n-3-2k}P_2^{2k+1-n}P_3^k$$

for any $(n-1)/2 \le k \le n-2$ (where we have substituted $k = l + (n-1)/2$) and this is also a holomorphic differential for every such k. Altogether we have constructed $n-1 = g$ holomorphic differentials, one for every $0 \le k \le n-2$, and these holomorphic differentials form a basis for the space of holomorphic differentials on X. In fact, more is true:

Lemma 6.1. *The differentials $(z-\lambda_2)^k(z-\lambda_3)^k dz/w^{k+1}$, $0 \le k \le (n-3)/2$, and the differentials $(z-\lambda_1)(z-\lambda_2)^{k-1}(z-\lambda_3)^k dz/w^{k+1}$, $(n-1)/2 \le k \le n-2$ (denoted by ω_k for any k), form a basis of the space of holomorphic differentials on X; this basis is adapted to any of the points P_i, $0 \le i \le 3$. The gap sequence at each of these points is simply the numbers between 1 and $n-1$, and hence none of them is a Weierstrass point.*

Proof. We will show that at each point P_i and power $0 \le l \le n-2$, there exists some $0 \le k \le n-2$ such that $\text{Ord}_{P_i}\omega_k = l$; then this k is unique because there are $n-1$ possible values of k and $n-1$ possible values of l. For P_3 and P_0 it is easy: simply take $k=l$ for the former and $k=n-2-l$ for the latter, as $\text{Ord}_{P_3}\omega_k = k$ and $\text{Ord}_{P_0}\omega_k = n-2-k$. For the other two branch points we have to split into cases according to the parity of l (recall that n is odd throughout). For P_1 we find that for even $0 \le l \le n-3$, we take $k=(n-3-l)/2$, and then $k \le (n-3)/2$ and therefore $\text{Ord}_{P_1}\omega_k = n-3-2k$; for odd $1 \le l \le n-2$, we take $k=n-(l+3)/2$, which satisfies $k \ge (n-1)/2$ and thus $\text{Ord}_{P_1}\omega_k = 2n-3-2k$. For P_2 we have that for odd $1 \le l \le n-2$, we take $k=(l-1)/2$, so that $k \le (n-3)/2$ and then $\text{Ord}_{P_2}\omega_k = 2k+1$; for even $0 \le l \le n-3$, we take $k=(n+l-1)/2$, which satisfies $k \ge (n-1)/2$ and therefore $\text{Ord}_{P_2}\omega_k = 2k+1-n$. This means that at every point P_i, $0 \le i \le 3$, the orders of the differentials ω_k are all distinct, which proves the first assertion. Also, these orders are the numbers between 0 and $n-2$, which proves the second assertion, since by Proposition 1.6 the gap sequence is obtained by adding 1 to each of these numbers. This concludes the proof of the lemma.   □

As an important corollary of Lemma 6.1 we see that since at every point P_i the orders of the differentials ω_k, $0 \leq k \leq n-2$, are all distinct, the proof of Lemma 2.7 holds equally here. Since the subspaces $\Omega_k(1)$, $0 \leq k \leq n-2$, are all 1-dimensional we therefore find that in order to find the space $\Omega(\Xi)$ for an integral divisor Ξ supported on the set of branch points, it suffices to see for each $0 \leq k \leq n-2$ whether $\Omega_k(\Xi)$ is 1-dimensional or 0-dimensional. This is equivalent to checking whether ω_k is a multiple of Ξ or not, and this gives us $\Omega(\Xi)$ and $i(\Xi)$ directly. In particular we shall apply this to integral divisors of degree $g = n-1$ supported on the set of branch points to see whether they are special or not.

In addition we have the following property, shared in different forms by the other Z_n curves we have dealt with.

Lemma 6.2. *For any $0 \leq i \leq 3$, the vector K_{P_i} of Riemann constants is a point of order $2n$ in the jacobian variety. Furthermore, the image of any other branch point P_j by the Abel–Jacobi map φ_{P_i} with basepoint P_i is a point of order n in the jacobian variety.*

Proof. The second statement follows as usual by the fact that the divisor P_j^n / P_i^n is the divisor of the meromorphic function $(z-\lambda_j)/(z-\lambda_i)$ and hence is principal. As for the first, we can choose any canonical divisor based on the branch points, which means that its image in $J(X)$ by the Abel–Jacobi map φ_{P_i} is of order n (as the sum of $2g-2$ elements of order n each). This means that $-2K_{P_i}$ is of order n in $J(X)$, and thus K_{P_i} itself is of order $2n$ in $J(X)$. This proves the lemma.   □

6.1.2 Non-Special Divisors in an Example of $n = 7$

As in all the previous cases, we would like to use this information to find all the integral divisors of degree $g = n-1$ which are supported on the set of branch points with the soon-to-be basepoint P_0 excluded and which are also non-special. In order to demonstrate how this is done we shall begin with an example.

The example we take is the case $n = 7$. This is the Z_7 curve with algebraic equation

$$w^7 = (z-\lambda_0)(z-\lambda_1)^2(z-\lambda_2)^5(z-\lambda_3)^6,$$

which is a compact Riemann surface of genus $g = 7-1 = 6$. The holomorphic differentials we have are, in increasing order of $0 \leq k \leq 5$, the differentials

$$\frac{dz}{w}, \quad \frac{(z-\lambda_2)(z-\lambda_3)dz}{w^2}, \quad \frac{(z-\lambda_2)^2(z-\lambda_3)^2dz}{w^3}, \quad \frac{(z-\lambda_1)(z-\lambda_2)^2(z-\lambda_3)^3dz}{w^4},$$

$$\frac{(z-\lambda_1)(z-\lambda_2)^3(z-\lambda_3)^4 dz}{w^4}, \quad \text{and} \quad \frac{(z-\lambda_1)(z-\lambda_2)^4(z-\lambda_3)^5 dz}{w^6},$$

with divisors

$$P_0^5 P_1^4 P_2^1 P_3^0, \quad P_0^4 P_1^2 P_2^3 P_3^1, \quad P_0^3 P_1^0 P_2^5 P_3^2, \quad P_0^2 P_1^5 P_2^0 P_3^3, \quad P_0^1 P_1^3 P_2^2 P_3^4, \quad \text{and} \quad P_0^0 P_1^1 P_2^4 P_3^5,$$

respectively, all of degree $2g-2=10$.

We now look at the divisors of degree $g=6$ supported on the set of branch points $\{P_1,P_2,P_3\}$ (with P_0 excluded). Throughout the following arguments of counting divisors we use repeatedly the two following well-known combinatorial facts. One is that the number of divisors of degree d supported on t given points, which is the same as the number of nonnegative integral solutions to the equation $\sum_{i=1}^t x_i = d$, and is thus the binomial coefficient $\binom{t+d-1}{t-1}$. The second fact is the corollary of the first stating that the number of such divisors that are multiples of a given divisor of degree c is just the binomial coefficient $\binom{t+d-c-1}{t-1}$ (as the quotient is based on the same t points but of degree $d-c$). Explicitly for our case, the degree d is $g=6$ and the number t of points allowed in the support is 3, which gives us $\binom{8}{2}=28$ divisors. However, in our search for the non-special divisors we exclude a subset of them. Since we know that $P_0 P_1^2/P_2^2 P_3$ is principal, we find that $r(1/P_2^2 P_3) \geq 2$ (actually this is equality, but this does not matter here). Thus for every divisor Δ of degree 6 which is a multiple of the divisor $P_2^2 P_3$, we have $r(1/\Delta) \geq r(1/P_2^2 P_3) \geq 2$ and hence $i(\Delta) \geq 1$ by the Riemann–Roch Theorem, meaning that Δ is special. Among our 28 divisors the multiples of $P_2^2 P_3$ are $\binom{5}{2}=10$ divisors, which are special and thus excluded. This leaves us with 18 divisors to examine.

The remaining divisors are not multiples of $P_2^2 P_3$, meaning either that P_3 does not appear at all, P_2 does not appear at all, or P_2 appears only to the first power. The first condition gives us the 7 divisors $P_1^l P_2^{6-l}$ with $0 \leq l \leq 6$. The second condition gives the 7 divisors $P_1^l P_3^{6-l}$ with $0 \leq l \leq 6$, and the third condition gives the 6 divisors $P_1^l P_2 P_3^{5-l}$ with $0 \leq l \leq 5$. Altogether this seems like 20 divisors, but the divisor P_1^6 appears both as $l=6$ in the first set and as $l=6$ in the second set, and the divisor $P_1^5 P_2$ appears both as $l=5$ in the first set and as $l=5$ in the third set. Since it is easy to verify that the second and third sets do not intersect and that the only intersections are the two we have already written, we do have 18 divisors. Thus we have a list of all the divisors that are possibly non-special. Our claim is that the non-special divisors among those are the 7 divisors $P_1^l P_2^{6-l}$ with $0 \leq l \leq 6$ from the first set, together with the divisors P_3^6 ($l=0$ in the second set) and $P_1^4 P_2 P_3$ ($l=4$ in the third set). In order to see that the first 7 divisors are non-special we note that the sum of powers to which P_1 and P_2 appear in any of them is 6, while for each holomorphic differential ω_k, $0 \leq k \leq 5$, we have $\mathrm{Ord}_{P_1}\omega_k + \mathrm{Ord}_{P_2}\omega_k = 5$ (check the list!). Therefore no ω_k can be a multiple of such a divisor, which makes all of them non-special. As for the divisor P_3^6, it is non-special since by Lemma 6.1 the point P_3 is not a Weierstrass point (this would also prove our assertion for P_1^6 and for P_2^6, with $l=6$ and $l=0$ in the first set—Lemma 6.1 actually shows that the orders of the holomorphic differentials on X at the branch points take value between 0 and 5, hence none of them can be a multiple of any divisor of the form P_i^6, $0 \leq i \leq 3$).

Finally for the divisor $P_1^4 P_2 P_3$ we see that the only holomorphic differentials ω_k with $\mathrm{Ord}_{P_1}\omega_k \geq 4$ are ω_0 and ω_3. Since ω_0 does not vanish at P_3 and ω_3 does not vanish at P_2 we see that this divisor is also non-special. Altogether this gives us 9 non-special divisors.

We now want to verify that the remaining 9 divisors are special. These divisors are the 5 divisors $P_1^l P_3^{6-l}$ with $1 \leq l \leq 5$ from the second set and the 4 divisors $P_1^l P_2 P_3^{5-l}$ with $0 \leq l \leq 3$ from the third set (the divisors in the intersections were seen to be non-special). Now, the divisor $P_0^0 P_1^1 P_2^4 P_3^5$ of ω_5 is a multiple of $P_1 P_3^5$ ($l = 1$ in the second set) and of $P_2 P_3^5$ and $P_1 P_2 P_3^4$ ($l = 0$ and $l = 1$ in the third set). The divisor $P_0^2 P_1^5 P_2^0 P_3^3$ of ω_3 is a multiple of $P_1^3 P_3^3$, $P_1^4 P_3^2$ and $P_1^5 P_3$ ($l = 3$, $l = 4$ and $l = 5$ in the second set), and the divisor $P_0^1 P_1^3 P_2^2 P_3^4$ of ω_4 is a multiple of the remaining divisors $P_1^2 P_3^4$ ($l = 2$ in the second set), $P_1^2 P_2 P_3^3$ and $P_1^3 P_2 P_3^2$ ($l = 2$ and $l = 3$ in the third set), together with $P_1^3 P_3^3$ and $P_1 P_2 P_3^4$ already mentioned. This shows that we have exactly 9 non-special divisors of degree $g = 6$ supported on the branch points P_1, P_2 and P_3, and that they are the ones listed in the previous paragraph.

We now know which are the divisors to take later as representatives of characteristics for the theta function when the basepoint is P_0. However, unlike the non-singular case and the singular case we dealt with in Chapters 4 and 5, here there is no symmetry between all the different basepoints. Therefore it will be useful, when we later want to compare what happens with different basepoints, to find all the non-special divisors of degree g supported on all the 4 branch points P_i, $0 \leq i \leq 3$. We continue with the example of $n = 7$. Dividing 6 powers between 4 points gives us altogether $\binom{9}{3} = 84$ divisors (of which 28 do not contain P_0 and these are the divisors we dealt with before). The fact that $P_0 P_1^2$ and $P_2^2 P_3$ are equivalent allows us to delete the multiples of both of them from the list of divisors we examine (as both $r(1/P_0 P_1^2)$ and $r(1/P_2^2 P_3)$ are at least 2). Each of them has $\binom{6}{3} = 20$ multiples that we have excluded from the list of 84 divisors when we look for non-special divisors, but doing so we see that we have taken out the divisor $P_0 P_1^2 P_2^2 P_3$ twice. Hence that the number of remaining divisors is $84 - 20 - 20 + 1 = 45$, of which 18 do not contain P_0 and were already analyzed. It remains to deal with 27 divisors, which all contain P_0 in their support. The fact that we insist P_0 appears in the support will be expressed by the fact that every power l below will start from 1, rather than 0 like before, in the sets we now define.

These 27 divisors can be described as the union of several sets, as we did with the previous 18. First, we use the fact that since P_0 appears in the support of all these divisors and they are not multiples of $P_0 P_1^2$, we see that either P_1 does not appear in these divisors, or it does, but only to the first power. Of course, the partition according to the appearance of P_1 splits these 27 divisors into 2 disjoint sets. Let us start with the divisors not containing P_1. The fact that these divisors are not multiples of $P_2^2 P_3$ as well again splits them into 3 sets according to P_3 not appearing, P_2 not appearing, or P_2 appearing to the first power. These conditions give us the 6 divisors $P_0^l P_2^{6-l}$ with $1 \leq l \leq 6$, the 6 divisors $P_0^l P_3^{6-l}$ with $1 \leq l \leq 6$, and the 5 divisors $P_0^l P_2 P_3^{5-l}$ with $1 \leq l \leq 5$ respectively. Although this looks as if these are 17 divisors, we see that the divisor P_0^6 appears both as $l = 6$ in the first set and as

$l = 6$ in the second set; the divisor $P_0^5 P_2$ appears both as $l = 5$ in the first set and as $l = 5$ in the third set, and these are the only intersections. Hence this gives 15 divisors. Examining the divisors containing P_1 in their support, since these divisors are not multiples of $P_2^2 P_3$, we split them again into 3 sets as before, which gives the 5 divisors $P_0^l P_1 P_2^{5-l}$ with $1 \leq l \leq 5$, the 5 divisors $P_0^l P_1 P_3^{5-l}$ with $1 \leq l \leq 5$, and the 4 divisors $P_0^l P_1 P_2 P_3^{4-l}$ with $1 \leq l \leq 4$ (with respect to the same 3 conditions). In this list of 14 divisors the only intersections are the divisor $P_0^5 P_1$ appearing both as $l = 5$ in the first set and as $l = 5$ in the second set, and the divisor $P_0^4 P_1 P_2$ appearing both as $l = 4$ in the first set and as $l = 4$ in the third set. Therefore these are 12 divisors, which together with the previous 15 are all the 27 new divisors we should examine.

We claim that among these divisors the ones that are non-special are the 6 divisors $P_0^l P_3^{6-l}$ with $1 \leq l \leq 6$ (all the second set with P_1 not appearing), together with the 4 divisors $P_0^2 P_2 P_3^3$ ($l = 2$ in the third set with P_1 not appearing), $P_0^3 P_1 P_3^2$ ($l = 3$ in the second set with P_1 appearing), $P_0 P_1 P_2^4$ ($l = 1$ in the first set with P_1 appearing), and $P_0^2 P_1 P_2 P_3^2$ ($l = 2$ in the third set with P_1 appearing). Let us first show that these divisors are non-special. For $P_0^l P_3^{6-l}$ with $1 \leq l \leq 6$ we use the fact that for every holomorphic differential ω_k, $0 \leq k \leq 5$, we have $\mathrm{Ord}_{P_0} \omega_k + \mathrm{Ord}_{P_3} \omega_k = 5$ (check the list again!) and the sum of the powers to which P_0 and P_3 appear in our divisor is 6. This gives the non-specialty of these divisors, as we obtained it for the divisors $P_1^l P_2^{6-l}$ with $0 \leq l \leq 6$. For the divisor $P_0^2 P_2 P_3^3$ we examine the list of holomorphic differentials and see that the only differentials ω_k such that $\mathrm{Ord}_{P_0} \omega_k \geq 2$ are those with $0 \leq k \leq 3$, while those satisfying $\mathrm{Ord}_{P_3} \omega_k \geq 2$ are those with $3 \leq k \leq 5$. The only differential satisfying both conditions is ω_3, and since it does not vanish at P_2, we conclude that $P_0^2 P_2 P_3^3$ is non-special. For the divisor $P_0^3 P_1 P_3^2$ a similar argument holds: The conditions imposed by P_0 and P_3 imply $0 \leq k \leq 2$ and $2 \leq k \leq 5$, leaving us only with the differential ω_2 which does not vanish at P_1. When considering the divisor $P_0 P_1 P_2^4$, we see from the list that the only holomorphic differentials ω_k with $\mathrm{Ord}_{P_2} \omega_k \geq 4$ are ω_2 and ω_5, and since ω_2 does not vanish at P_1 and ω_5 does not vanish at P_0, we conclude that this divisor is also non-special. Finally, for the divisor $P_0^2 P_1 P_2 P_3^2$ we use the fact (which we have already seen) that the conditions from P_0 and P_3 leave us only with the differentials ω_2 and ω_3 as possible multiples of this divisor, making it non-special since ω_2 does not vanish at P_1 and ω_3 does not vanish at P_2. Altogether this gives us another 10 non-special divisors. We note, before we continue, that only 9 of these divisors can serve us as representatives of characteristics for the theta function with any basepoint P_i, since we always demand that the basepoint not appear in the representing divisors and in the last divisor all the 4 branch points appear.

Before we turn to the proof of the general (odd) n case, we need to verify that we have indeed written all the non-special divisors. We have written 7 divisors not containing P_1 and 3 that do contain it, leaving 8 that do not contain P_1 and 9 that do. Explicitly, the first 8 are the 5 divisors $P_0^l P_2^{6-l}$ with $1 \leq l \leq 5$ from the first set and the 4 divisors $P_0^l P_3 P_2^{5-l}$ with l being 1, 3, 4 or 5 from the third set (with $P_0^5 P_2$ counted twice), and the other 9 are the 4 divisors $P_0^l P_1 P_2^{5-l}$ with $2 \leq l \leq 5$ from the first set, the 4 divisors $P_0^l P_1 P_3^{5-l}$ with l being 1, 2, 4 or 5 from the second set, and the

3 divisors $P_0^l P_1 P_2^{4-l}$ with l being 1, 3 or 4 from the third set (with $P_0^5 P_1$ and $P_0^4 P_1 P_2$ counted twice). Together these are the 17 remaining divisors, which we now show are all special. We begin with the first 8 divisors. The divisor $P_0^1 P_1^3 P_2^2 P_3^4$ of ω_4 is a multiple of $P_0 P_2 P_3^4$ ($l = 1$ in the third set). The divisor $P_0^5 P_1^4 P_2^1 P_3^0$ of ω_0 is a multiple of $P_0^5 P_2$ ($l = 5$ in the first and third sets). The divisor $P_0^4 P_1^2 P_2^3 P_3^1$ of ω_1 is a multiple of $P_0^4 P_2^2$ and $P_0^3 P_2^3$ ($l = 3$ and $l = 4$ in the first set) and of $P_0^4 P_2 P_3$ ($l = 4$ in the third set). Finally, the divisor $P_0^3 P_1^0 P_2^5 P_3^2$ of ω_2 is a multiple of $P_0 P_2^5$ and $P_0^2 P_2^4$ ($l = 1$ and $l = 2$ from the first set) and of $P_0^3 P_2 P_3^2$ ($l = 3$ in the third set), together with $P_0^3 P_2^3$ already mentioned. As for the 9 remaining divisors, we find that the divisor $P_0^2 P_1^5 P_2^0 P_3^3$ of ω_3 is a multiple of $P_0^2 P_1 P_3^3$ ($l = 2$ in the second set). The divisor $P_0^1 P_1^1 P_2^2 P_3^4$ of ω_4 is a multiple of $P_0 P_1 P_3^4$ ($l = 1$ in the second set) and of $P_0 P_1 P_2 P_3^3$ ($l = 1$ in the third set). The divisor $P_0^5 P_1^4 P_2^1 P_3^0$ of ω_0 is a multiple of $P_0^5 P_1$ ($l = 5$ in the first and second sets) and $P_0^4 P_1 P_2$ ($l = 4$ in the first and third sets). Completing the picture is the divisor $P_0^4 P_1^2 P_2^3 P_3^1$ of ω_1, which is a multiple of $P_0^4 P_1 P_3$ ($l = 4$ in the second set), of $P_0^2 P_1 P_2^3$ and $P_0^3 P_1 P_2^2$ ($l = 2$ and $l = 3$ in the first set), and of $P_0^3 P_1 P_2 P_3$ ($l = 3$ in the third set), together with $P_0^4 P_1 P_2$ with which we already dealt. This covers all the 17 divisors and shows that they are special.

6.1.3 Non-Special Divisors in the General Case

With the example of $n = 7$ in mind we can turn to the general odd n case. The result is as follows.

Theorem 6.3. *On the Z_n curve X there are, for every $n \geq 3$, exactly $n + 2$ non-special integral divisors of degree $g = n - 1$ supported on the branch points P_1, P_2 and P_3, namely $P_1^l P_2^{n-1-l}$ with $0 \leq l \leq n - 1$, P_3^{n-1}, and $P_1^{n-3} P_2 P_3$. If we allow also P_0 to be included in the support, then for $n \geq 5$ we obtain $n + 3$ more such non-special divisors, namely $P_0^l P_3^{n-1-l}$ with $1 \leq l \leq n - 1$, $P_0 P_1 P_2^{n-3}$, $P_0^{(n-1)/2} P_1 P_3^{(n-3)/2}$, $P_0^{(n-3)/2} P_2 P_3^{(n-1)/2}$, and $P_0^{(n-3)/2} P_1 P_2 P_3^{(n-3)/2}$, while for $n = 3$ the last two already appear and the two preceding them coincide. Altogether this gives together $2n + 5$ divisors if $n \geq 5$ and 8 if $n = 3$.*

Proof. We begin with the divisors where P_0 does not appear in the support. There are $\binom{n+1}{2} = n(n+1)/2$ total integral divisors of degree $n - 1$ supported on the set of 3 branch points. Since for every such divisor Δ which is a multiple of $P_2^2 P_3$ we have $r(1/\Delta) \geq r(1/P_2^2 P_3) \geq 2$ (since $P_0 P_1^2 / P_2^2 P_3$ is principal), the Riemann–Roch Theorem gives that every such Δ is special. In our search for non-special divisors we can thus delete all these divisors from the list. There are $\binom{n-2}{2} = (n-2)(n-3)/2$ such divisors, and removing them from the list of divisors leaves us with only $3n - 3$ divisors. As in the special case with $n = 7$, we see that the assumption that these divisors are not multiples of $P_2^2 P_3$ splits them into 3 sets, according to whether P_3 does not appear at all, P_2 does not appear at all, or P_2 appears but only to the first power. In the first set we have the n divisors $P_1^l P_2^{n-1-l}$ with $0 \leq l \leq n - 1$. In the second set there are the n divisors $P_1^l P_3^{n-1-l}$ with $0 \leq l \leq n-1$, and in the third set there are the $n-1$ divisors $P_1^l P_2 P_3^{n-2-l}$ with $0 \leq l \leq n-2$. Summing up the cardinalities of these divisors give $3n-1$, which is 2 more than what we need, but since the only intersections are easily seen to be just the two divisors P_1^{n-1} (which appears both as $l = n-1$ in the first set and as $l = n-1$ in the second set) and $P_1^{n-2} P_2$ (which appears both as $l = n-2$ in the first set and as $l = n-2$ in the third set) we have exactly the $3n-3$ divisors we are looking for.

The first assertion of the theorem is that the non-special among these divisors are the n divisors $P_1^l P_2^{n-1-l}$ with $0 \leq l \leq n-1$ from the first set, and the divisors P_3^{n-1} (which corresponds to $l=0$ in second set) and $P_1^{n-3} P_2 P_3$ (corresponding to $l = n-3$ in the third set). First, we show that these $n+2$ divisors are indeed non-special. For the n divisors of the first set we see that the sums of the powers to which P_1 and P_2 appear in them is is $n-1$. On the other hand, we have, for every holomorphic differential ω_k with $0 \leq k \leq n-2$, the equality $\mathrm{Ord}_{P_1} \omega_k + \mathrm{Ord}_{P_1} \omega_k = n-2$ (either as $(n-3-2k) + (2k+1)$ for $0 \leq k \leq (n-3)/2$ or as $(2n-3-2k) + (2k+1-n)$ for $(n-1)/2 \leq k \leq n-2$). This means that no ω_k can be a multiple of any of our divisors and thus they are all non-special. For the divisor P_3^{n-1} Lemma 6.1 gives that the point P_3 is not a Weierstrass point and that this divisor is non-special (as mentioned in the case $n=7$, the same argument also proves this for P_1^{n-1} and for P_2^{n-1} which are the elements with $l=n-1$ and $l=0$ in the first set). As for the remaining divisor $P_1^{n-3} P_2 P_3$, checking the orders of the holomorphic differential ω_k at P_1 gives that the only differentials ω_k with $\mathrm{Ord}_{P_1} \omega_k \geq n-3$ are ω_0 and $\omega_{(n-1)/2}$. The fact that ω_0 does not vanish at P_3 and $\omega_{(n-1)/2}$ does not vanish at P_2 shows that this divisor is non-special. This shows that our $n+2$ divisors are indeed non-special.

In order to show that there are no additional divisors which are non-special, one checks directly that the remaining $2n-5$ divisors are special. The remaining divisors are the divisors $P_1^l P_3^{n-1-l}$ with $1 \leq l \leq n-2$ from the second set and the divisors $P_1^l P_2 P_3^{n-2-l}$ with $0 \leq l \leq n-4$ from the third set (together $(n-2) + (n-3) = 2n-5$ divisors). Now, the reader can verify that for a divisor $P_1^l P_3^{n-1-l}$ with $0 \leq l \leq n-1$, some differential ω_k has a divisor that is a multiple of this divisor if and only if k satisfies both the inequalities $(n-1)/2 \leq k \leq n-2$ and $n-1-l \leq k \leq n-(l+3)/2$. Moreover, for every $1 \leq l \leq n-2$ such a choice of k is possible (and the cases $l=0$ and $l=n-1$ give the non-special divisors P_1^{n-1} and P_3^{n-1} respectively). Similarly, one can also verify that given a divisor $P_1^l P_2 P_3^{n-2-l}$ with $0 \leq l \leq n-2$, the divisor of the differential ω_k is a multiple of it if and only if k satisfies $(n-1)/2 \leq k \leq n-2$ and $n-2-l \leq k \leq n-(l+3)/2$. Here one can see that for every $0 \leq l \leq n-4$ one can find some k which satisfies these two inequalities (while $l=n-3$ and $l=n-2$ give the non-special divisors $P_1^{n-3} P_2 P_3$ and $P_1^{n-2} P_2$ respectively). Since this covers all the remaining $2n-5$ divisors, we know that our $n+2$ divisors are the only non-special ones.

Let us now allow P_0 to appear in the support of the divisors. Then the number of divisors we have in total is $\binom{n+2}{3} = n(n+1)(n+2)/6$, and of these $\binom{n+1}{2}$ do not include P_0 in their support. We want to delete from the full list of divisors both the multiples of $P_2^2 P_3$ and the multiples of $P_0 P_1^2$. The number of multiples of each of them is $\binom{n-1}{3} = (n-1)(n-2)(n-3)/6$, but there are divisors appearing as a multiple of both and then are counted twice this way. These divisors are the multiples of $P_0 P_1^2 P_2^2 P_3$ and the number of them is thus $\binom{n-4}{3} = (n-4)(n-5)(n-6)/6$. Therefore in the list remain $\binom{n+2}{3} - 2\binom{n-1}{3} + \binom{n-4}{3} = 9n - 18$ divisors, $3n - 3$ of which are those in which P_0 does not appear and have been already dealt with. Note that for $n = 3$ the number $\binom{n-4}{3}$ equals 0 and not -1 like $(n-4)(n-5)(n-6)/6$, so that there we have 10 divisors rather than $9n - 18 = 9$. In any case, to ensure that we now deal only with the $6n - 15$ (or 4 rather than $6n - 15 = 3$ if $n = 3$) divisors containing P_0 in their support we will again let the power l start from 1 rather than 0 in what follows.

Again these divisors, which contain P_0 in their support but are not multiples of $P_0 P_1^2$, split into 2 disjoint sets, one consisting of the divisors not containing P_1 in their support, and the other consisting of those which do but only to the first power. In each of these sets we have the usual partition, where the first subset of each set contains the divisors in which P_3 does not appear, the second subset contains those in which P_2 does not appear, and the third contains those in which P_2 appears to the first power. Let us see how many divisors we have in each set if we assume that P_1 is not contained in the support. This gives the $n - 1$ divisors $P_0^l P_2^{n-1-l}$ with $1 \le l \le n-1$, the $n - 1$ divisors $P_0^l P_3^{n-1-l}$ with $1 \le l \le n-1$, and the $n - 2$ divisors $P_0^l P_2 P_3^{n-2-l}$ with $1 \le l \le n-2$ respectively. Since we have the usual intersections (P_0^{n-1} appears both as $l = n - 1$ in the first set and as $l = n - 1$ in the second set, $P_0^{n-2} P_2$ appears both as $l = n - 2$ in the first set and as $l = n - 2$ in the third set, and nothing further), this sums up to $3n - 6$ divisors rather than $3n - 4$. As for the second part, we now have the $n - 2$ divisors $P_0^l P_1 P_2^{n-2-l}$ with $1 \le l \le n-2$, the $n - 2$ divisors $P_0^l P_1 P_3^{n-2-l}$ with $1 \le l \le n-2$, and the $n - 3$ divisors $P_0^l P_1 P_2 P_3^{n-3-l}$ with $1 \le l \le n-3$ respectively. The intersections behave as usual (the only two intersections are $P_0^{n-2} P_1$ appearing both as $l = n - 2$ in the first set and as $l = n - 2$ in the second set and $P_0^{n-3} P_1 P_2$ appearing both as $l = n - 3$ in the first set and as $l = n - 3$ in the third set), but here we have to be careful in the case $n = 3$, since $n - 3 = 0$ and we look only at divisors with $l \ge 1$. This means that for $n \ge 5$, these $3n - 7$ divisors are in fact $3n - 9$, while for $n = 3$, we only have 2 divisors which coincide. Together this sums up to $6n - 15$ divisors for $n \ge 5$, but to 4 divisors for $n = 3$, and these are the divisors we should examine.

Among these divisors, those which are asserted to be non-special are as follows. Of those not containing P_1 in their support, these are the $n - 1$ divisors $P_0^l P_3^{n-1-l}$ with $1 \le l \le n-1$ (which are all of the second set) and the divisor $P_0^{(n-3)/2} P_2 P_3^{(n-1)/2}$ (with $l = (n-3)/2$ in the third set). Of those which do contain P_1, there are only 3 such divisors, namely the divisor $P_0^{(n-1)/2} P_1 P_3^{(n-3)/2}$ (with $l = (n-1)/2$ from the second set), the divisor $P_0 P_1 P_2^{n-3}$ (with $l = 1$ from the first set), and the divisor $P_0^{(n-3)/2} P_1 P_2 P_3^{(n-3)/2}$ (with $l = (n-3)/2$ in the third set). First, we show that these $n + 3$ divisors (for $n = 3$ these are actually only 5 divisors as two of them coincide, and two other divisors do not include P_0 in their support, but for the moment this changes nothing) are non-special. In every divisor $P_0^l P_3^{n-1-l}$ with $1 \le l \le n-1$ the sum of the powers to which P_0 and P_3 appear is $n-1$, and since for each holomorphic differential ω_k with $0 \leq k \leq n-2$ we have $\operatorname{Ord}_{P_0}\omega_k + \operatorname{Ord}_{P_3}\omega_k = n-2$ (as the sum $(n-2-k)+k$) we see that ω_k cannot be a multiple of any such divisor. Hence all these $n-1$ divisors are non-special. The fact (already used a second ago) that $\operatorname{Ord}_{P_0}\omega_k = n-2-k$ and $\operatorname{Ord}_{P_3}\omega_k = k$ for every $0 \leq k \leq n-2$ will be very useful when we consider all the other divisors, as follows. For the divisor $P_0^{(n-3)/2}P_2P_3^{(n-1)/2}$ we obtain that $\operatorname{Ord}_{P_0}\omega_k \geq (n-3)/2$ implies $0 \leq k \leq (n-1)/2$ and $\operatorname{Ord}_{P_3}\omega_k \geq (n-1)/2$ implies $(n-1)/2 \leq k \leq 5$, and since the single divisor $\omega_{(n-1)/2}$ satisfying these assumptions does not vanish at P_2 we conclude that $P_0^{(n-3)/2}P_2P_3^{(n-1)/2}$ is non-special. Similar considerations can be done for the divisor $P_0^{(n-1)/2}P_1P_3^{(n-3)/2}$: The conditions on k giving high enough order of vanishing of ω_k at P_0 and P_3 are $0 \leq k \leq (n-3)/2$ and $(n-3)/2 \leq k \leq 5$ respectively and this leaves us only with the differential $\omega_{(n-3)/2}$ which does not vanish at P_1. When we examine the divisor $P_0P_1P_2^{n-3}$ and check the orders of the holomorphic differentials at P_2 we find that the only holomorphic differentials ω_k which vanish at P_2 at least to the order $n-3$ are $\omega_{(n-3)/2}$ and ω_{n-2}. Since $\omega_{(n-3)/2}$ does not vanish at P_1 and ω_{n-2} does not vanish at P_0 we obtain the non-specialty of this divisor as well. For the one remaining divisor $P_0^{(n-3)/2}P_1P_2P_3^{(n-3)/2}$ we collect the information we have already obtained that the only differentials satisfying the conditions at P_0 and P_3 are $\omega_{(n-3)/2}$ and $\omega_{(n-1)/2}$, and since we also saw that $\omega_{(n-3)/2}$ does not vanish at P_1 and $\omega_{(n-1)/2}$ does not vanish at P_2 this divisor is also non-special. This shows that all our divisors are indeed non-special.

For $n \geq 5$ we had $6n-15$ divisors, from which $n+3$ were non-special, and $5n-18$ remain to be seen to be special. For $n=3$ we had 4 divisors, and from the 5 divisors which we saw in the last paragraph that were non-special (there are 5 and not 6 since $P_0^{(n-1)/2}P_1P_3^{(n-3)/2} = P_0P_1P_2^{n-3}$ for $n=3$), only 3 count here since P_0 does not appear in $P_0^{(n-3)/2}P_2P_3^{(n-1)/2}$ and in $P_0^{(n-3)/2}P_1P_2P_3^{(n-3)/2}$ when $n=3$. Therefore for $n=3$ there is only one divisor (which is P_0P_2 and does not contain P_1) whose specialty needs verifying. In any case, let us write the list of the remaining divisors in order to show that they are special. From those not containing P_1 in their support these are the $n-2$ divisors $P_0^lP_2^{n-1-l}$ with $1 \leq l \leq n-2$ from the first set and the $n-3$ divisors $P_0^lP_2P_3^{n-2-l}$ with $l \neq (n-3)/2$ from the third set (with $P_0^{n-2}P_2$ appearing twice), together $2n-6$ divisors (or the one divisor P_0P_2 if $n=3$). From those containing P_1 we have the $n-3$ divisors $P_0^lP_1P_2^{n-2-l}$ with $2 \leq l \leq n-2$ from the first set, the $n-3$ divisors $P_0^lP_1P_3^{n-2-l}$ with $l \neq (n-1)/2$ from the third set, and the $n-4$ divisors $P_0^lP_1P_2P_3^{n-3-l}$ with $l \neq (n-3)/2$ from the third set (with $P_0^{n-2}P_1$ and $P_0^{n-3}P_1P_2$ counted twice), which are $3n-12$ divisors (and none for $n=3$).

The proof that all these divisors are special is the same as the proof of this assertion for the divisors not including P_0 in their support (and holds for the special case $n=3$ as well). We give it briefly and leave the verification of the details to the reader. For a divisor $P_0^lP_2^{n-1-l}$ with $0 \leq l \leq n-1$, the divisor of the holomorphic differential ω_k is a multiple of this divisor if and only if k satisfies both $0 \leq k \leq (n-3)/2$ and $(n-2-l)/2 \leq k \leq n-2-l$. Such k exists for $1 \leq l \leq n-2$

(and $l = 0$ and $l = n - 1$ give the non-special divisors P_2^{n-1} and P_0^{n-1} respectively). This completes the proof for $n = 3$, and a reader unhappy about empty statements can now assume $n \geq 5$ (even though everything remains true also for $n = 3$ as empty statements). Now, for a divisor $P_0^l P_2 P_3^{n-2-l}$ with $0 \leq l \leq n-2$, the divisor of the differential ω_{n-2-l} is a multiple of it if $l \neq (n-3)/2$ (and $l = (n-3)/2$ gives the non-special divisor $P_0^{(n-3)/2} P_2 P_3^{(n-1)/2}$). For a divisor $P_0^l P_1 P_2^{n-2-l}$ with $0 \leq l \leq n-2$, the equalities k must satisfy in order for the divisor of ω_k to be a multiple of this divisor are $0 \leq k \leq (n-5)/2$ and $(n-3-l)/2 \leq k \leq n-2-l$, and this can happen simultaneously for every $2 \leq l \leq n-2$ (the values $l = 0$ and $l = 1$ give the non-special divisors $P_1 P_2^{n-2}$ and $P_0 P_1 P_2^{n-3}$ respectively). For a divisor $P_0^l P_1 P_3^{n-2-l}$ with $0 \leq l \leq n-2$, the divisor of the differential ω_{n-2-l} is a multiple of it if $l \neq (n-1)/2$ (and $l = (n-1)/2$ gives the non-special divisor $P_0^{(n-1)/2} P_1 P_3^{(n-3)/2}$). Finally, for a divisor $P_0^l P_1 P_2 P_3^{n-3-l}$ with $0 \leq l \leq n-3$, the divisor of the differential ω_k is a multiple of this divisor for k being either $n-3-l$ or $n-2-l$, but this happens only if this k does not equal either $(n-3)/2$ or $(n-1)/2$. This covers all the values of l such that $l \neq (n-3)/2$ (and $l = (n-3)/2$ gives the non-special divisor $P_0^{(n-3)/2} P_1 P_2 P_3^{(n-3)/2}$). This covers all the remaining $5n - 18$ divisors, showing that our $n+3$ are the only additional non-special ones (these numbers are correct for $n \geq 5$—for $n = 3$ we had only one divisor $P_0 P_2$ which we showed was special, and 3 non-special divisors).

One can easily see that for $n \geq 5$ the divisors are all distinct since their supports divide them according to the way we have listed them, so that we indeed have $2n+5$ different non-special divisors. As for $n = 3$ we have seen at the beginning of the last paragraph that allowing P_0 to enter the support of the divisors adds only 3 new non-special divisors, which together with the first $n+2 = 5$ gives 8 different divisors. This finishes the proof of the theorem. $\square$

We recall that for $n = 3$ we obtain a singular Z_3 curve of the type we already dealt with and with $m = 2$. This is so since the expression $z - \lambda_2$ appears to the power $n - 2 = 1$ and the expression $z - \lambda_3$ appears to the power $n - 1 = 2$, indeed giving such a singular Z_3 curve. The points P_1 and P_3 here play the roles of Q_1 and Q_2 there and the point P_2 here plays the role of P_1 there. Let us thus write explicitly the divisors Theorem 6.3 gives us in this case in order to verify consistency. The first 5 divisors are indeed the non-special divisors P_1^2, $P_1 P_2$, P_2^2, P_3^2 and $P_2 P_3$, which we order as P_2^2, $P_2 P_1$, $P_2 P_3$, P_1^2 and P_3^2 in order to make the consistency verification easier. As for the additional 3 divisors, they are $P_0 P_3$, P_0^2 and $P_0 P_1$ (from the "list of 6" we should have gotten by the general assertion we have $P_2 P_3$ and $P_1 P_2$ appearing again and $P_0 P_1$ appearing twice there), which are indeed the additional 3 divisors P_0^2, $P_0 P_1$ and $P_0 P_3$ one obtains by symmetry when allowing P_0 in the supports.

We note that for $n \geq 5$, from the $2n+5$ non-special divisors only $2n+4$ can be used to represent characteristics for theta functions with a basepoint P_i. This is since the divisor $P_0^{(n-3)/2} P_1 P_2 P_3^{(n-3)/2}$ contains all the branch points in its support and for characteristics with basepoint P_i, $0 \leq i \leq 3$ we need that P_i will not appear in the support of the divisor. Note that this is a new phenomenon which did not happen before in either the nonsingular or singular cases already studied. Theorems 2.9 and 2.15, which showed us that the divisors we had obtained were all the non-special divisors, did not use the fact that P_0 had been excluded from the support. This shows that the Z_n curves we deal with here are indeed different from those treated before. In the case $n = 3$ this last divisor is simply $P_1 P_2$ appearing again, and indeed this is not a new Z_3 curve.

6.1.4 Operators

We remind ourselves that the Thomae formulae are proved for a given basepoint, and we continue choosing it to be P_0. Therefore we restrict our attention again to the first $n + 2$ non-special divisors from Theorem 6.3. We would like, as we did in all previous cases, to define operators N and T_R which will have the properties we require. Since we do not have a general form of presenting all the divisors in one notation, we shall define the operators ad hoc. We shall assume from now on that $n \geq 5$, and later we explain what changes for $n = 3$ and relate it to the usual singular Z_3 curve.

Definition 6.4. Let N be the operator defined by

$$N(P_1^l P_2^{n-1-l}) = \begin{cases} P_1^{n-3-l} P_2^{l+2} & 0 \leq l \leq n-3 \\ P_1^{2n-3-l} P_2^{l+2-n} & n-2 \leq l \leq n-1, \end{cases}$$

$$N(P_3^{n-1}) = P_1^{n-3} P_2 P_3, \quad \text{and} \quad N(P_1^{n-3} P_2 P_3) = P_3^{n-1}.$$

Let T_{P_3} be the operator defined only on the divisors $P_1^l P_2^{n-1-l}$ with $0 \leq l \leq n-1$ such that

$$T_{P_3}(P_1^l P_2^{n-1-l}) = \begin{cases} P_1^{n-5-l} P_2^{l+4} & 0 \leq l \leq n-5 \\ P_1^{2n-5-l} P_2^{l+4-n} & n-4 \leq l \leq n-1. \end{cases}$$

The operator T_{P_1} is defined only on the divisors $P_1^{n-3} P_2^2$ (with $l = n-3$) and $P_1^{n-3} P_2 P_3$ and is defined by

$$T_{P_1}(P_1^{n-3} P_2^2) = P_1^{n-3} P_2 P_3, \quad \text{and} \quad T_{P_1}(P_1^{n-3} P_2 P_3) = P_1^{n-3} P_2^2.$$

The operator T_{P_2} is defined only on the divisors $P_1^{n-2} P_2$ (with $l = n-2$) and $P_1^{n-3} P_2 P_3$ and is defined by

$$T_{P_2}(P_1^{n-2} P_2) = P_1^{n-3} P_2 P_3, \quad \text{and} \quad T_{P_2}(P_1^{n-3} P_2 P_3) = P_1^{n-2} P_2.$$

Before we prove that this definition of the operators allows them to play their usual role in the proof of the Thomae formulae, we first note the following important observations. The operator N is defined on all $n + 2$ divisors, and it is an involution.

This is clear since it interchanges $P_1^{n-2}P_2$ and P_1^{n-1}, and since when $0 \le l \le n-3$ also $0 \le n-3-l \le n-3$ (and on the last two divisors it is also clear). Since $n+2$ is odd, N must have a fixed point, and it is the divisor $P_1^{(n-3)/2}P_2^{(n+1)/2}$. P_3 appears only in the support of 2 divisors, P_3^{n-1} and $P_1^{n-3}P_2P_3$ (which N interchanges), and the corresponding operator T_{P_3} is defined on all the other divisors and is also an involution. This is also clear since when l satisfies an inequality (either $0 \le l \le n-5$ or $n-4 \le l \le n-1$) its "image" by T_{P_3} ($n-5-l$ or $2n-5-l$ respectively) satisfies the same inequality. Note that the divisor $P_1^{(n-5)/2}P_2^{(n+3)/2}$ is T_{P_3}-invariant. When examining the divisors $P_1^{n-3}P_2^2$, $P_1^{n-3}P_2P_3$ and $P_1^{n-2}P_2$ (with N-images P_2^{n-1}, P_3^{n-1} and P_1^{n-1} respectively) we see that the first two are the only divisors whose N-images do not contain P_1 and the last two are the only divisors whose N-images do not contain P_2 ($P_1^{n-3}P_2P_3$ satisfies both assertions). Since T_{P_1} interchanges the first two and T_{P_2} interchanges the last two, they are also involutions. Therefore the operators T_R are defined exactly on the divisors whose N-images do not contain R in their support. Note that on P_3^{n-1}, whose N-image is $P_1^{n-3}P_2P_3$ and contains all the three branch points distinct from P_0, no operator T_R is defined—again a situation we did not encounter in any of the previous Z_n curves.

We now show that the operators we have just defined have the properties making them worthy of their names. Δ is assumed, as usual, to be any of the $n+2$ non-special divisors listed above.

Proposition 6.5. *If $e = \varphi_{P_0}(\Delta) + K_{P_0}$, then $-e = \varphi_{P_0}(N(\Delta)) + K_{P_0}$. If R is some branch point P_i, $1 \le i \le 3$, such that T_R is defined on Δ, then*

$$e + \varphi_{P_0}(R) = -(\varphi_{P_0}(T_R(\Delta)) + K_{P_0}).$$

Proof. For the first statement we show that

$$\varphi_{P_0}(\Delta) + \varphi_{P_0}(N(\Delta)) = -2K_{P_0},$$

and it clearly suffices. We now observe that by Definition 6.4 we have

$$\varphi_{P_0}(\Delta) + \varphi_{P_0}(N(\Delta)) = \varphi_{P_0}(P_1^{n-3}P_2^{n+1}).$$

This is clear for $\Delta = P_1^l P_2^{n-1-l}$ with $0 \le l \le n-3$, while for the other two values of l we obtain $\varphi_{P_0}(P_1^{2n-3}P_2)$ and for the other two divisors we obtain $\varphi_{P_0}(P_1^{n-3}P_2P_3^n)$. We can now use Lemma 6.2 to move the nth power from P_1 or P_3 to P_2 and obtain this equality for these divisors as well. Applying Lemma 6.2 on the result and using the fact that $P_0P_1^2/P_2^2P_3$ is principal (and thus so is its square) and that $\varphi_{P_0}(P_0) = 0$ we find that this expression equals

$$\varphi_{P_0}(P_1^{n-3}P_2^{n+1}P_3^n) = \varphi_{P_0}(P_0^2P_1^{n+1}P_2^{n-3}P_3^{n-2}) = \varphi_{P_0}(P_1P_2^{n-3}P_3^{n-2}).$$

Since $P_1 P_2^{n-3} P_3^{n-2}$ is the divisor of the holomorphic differential ω_{n-2}, this expression indeed equals $-2K_{P_0}$.

As for the second statement, it suffices to show that for any such Δ and R we have

$$\varphi_{P_0}(\Delta) + \varphi_{P_0}(R) + \varphi_{P_0}(T_R(\Delta)) = -2K_{P_0}.$$

We begin with $R = P_3$, use Definition 6.4 to see that Δ has to be $P_1^l P_2^{n-1-l}$ for some $0 \le l \le n-1$, and then find that

$$\varphi_{P_0}(\Delta) + \varphi_{P_0}(P_3) + \varphi_{P_0}(T_{P_3}(\Delta)) = \varphi_{P_0}(P_1^{n-5} P_2^{n+3} P_3),$$

where for $0 \le l \le n-5$ this is clear, and for $n-4 \le l \le n-1$, we get $\varphi_{P_0}(P_1^{2n-5} P_2^3 P_3)$ and we use Lemma 6.2 to move the nth power from P_1 to P_2. We now use the fact that $P_0 P_1^2 / P_2^2 P_3$ is principal and that $\varphi_{P_0}(P_0) = 0$ to see that this expression equals

$$\varphi_{P_0}(P_0 P_1^{n-3} P_2^{n+1}) = \varphi_{P_0}(P_1^{n-3} P_2^{n+1});$$

we already saw that this expression equals $-2K_{P_0}$. As for the operators T_{P_1} and T_{P_2}, the divisors can be only $P_1^{n-3} P_2^2$ and $P_1^{n-3} P_2 P_3$ in the first case and $P_1^{n-2} P_2$ and $P_1^{n-3} P_2 P_3$ in the second case. Using Definition 6.4 we find that in all the four possibilities (two for $R = P_1$ and two for $R = P_2$) we have the equality

$$\varphi_{P_0}(\Delta) + \varphi_{P_0}(R) + \varphi_{P_0}(T_R(\Delta)) = \varphi_{P_0}(P_1^{2n-5} P_2^3 P_3);$$

we already saw that this expression equals $-2K_{P_0}$. This proves the proposition. $\quad\square$

6.1.5 First Identities Between Theta Constants

In the previous cases we have used the material from Section 2.6. Lemma 6.2 fills the role of Lemmas 2.4 and 2.12 in the proof of the statements there, which allows us to extend the validity of the use of Eqs. (1.6) and (1.7) to the case we are considering here. Similarly, Proposition 6.5 shows that the operators from Definition 6.4 play the same role as the operators from Definitions 2.16 and 2.18, and thus plays the role of Propositions 2.17 and 2.19 in the proof of the following analog of Propositions 2.20 and 2.21, where we write $e = 2$ since n is odd.

Proposition 6.6. *For any divisor Δ of the $n+2$ divisors of Theorem 6.3, the theta function $\theta[\Delta](\varphi_{P_0}(P), \Pi)$ for any lift of $\varphi_{P_0}(\Delta) + K_{P_0}$ does not vanish identically on the surface, and its divisor of zeros from Theorem 1.8 (the Riemann Vanishing Theorem) is the integral divisor of degree $g = n - 1$ denoted $N(\Delta)$ in Definition 6.4. For any two such divisors Δ and Ξ, the function*

$$f(P) = \frac{\theta[\Delta]^{2n^2}(\varphi_{P_0}(P), \Pi)}{\theta[\Xi]^{2n^2}(\varphi_{P_0}(P), \Pi)}$$

is a well-defined meromorphic function on the Z_n curve X, which is independent of the lifts of $\varphi_{P_0}(\Delta)$, $\varphi_{P_0}(\Xi)$, and K_{P_0} from $J(X)$ to $\mathbb{C}^g$. This function has the same divisor $[N(\Delta)/N(\Xi)]^{2n^2}$ as the function

$$\prod_{i=1}^{3}(z - z(P_i))^{2n[v_{P_i}(N(\Delta)) - v_{P_i}(N(\Xi))]},$$

and hence $f(P)$ is a nonzero constant multiple of this rational function of z.

As before, we will be evaluating these functions at branch points, and since $\varphi_{P_0}(R)$ is a point of order n for any branch point R, we can apply Eq. (1.7) with $\zeta = 0$ to see that the evaluation of this quotient at a branch point is the same as evaluating another quotient at 0. However, for our expressions to have meaning when evaluating these quotients at branch points, we have to choose only points where neither the numerator nor the denominator vanish, which by Proposition 6.6 are only points that do not appear either in the support of $N(\Delta)$ or in the support of $N(\Xi)$. In particular, P_0 will be a good point to substitute for any such Δ and Ξ, and this gives us the theta constants of the Riemann surface X. If now R is a point P_i with $i \neq 0$ which does not appear in both $N(\Delta)$ and $N(\Xi)$, then Eq. (1.7) with $\zeta = 0$ translates to

$$\frac{\theta^{2n^2}[\Delta](\varphi_{P_0}(R), \Pi)}{\theta^{2n^2}[\Xi](\varphi_{P_0}(R), \Pi)} = \frac{\theta^{2n^2}[\varphi_{P_0}(\Delta) + \varphi_{P_0}(R) + K_{P_0}](0, \Pi)}{\theta^{2n^2}[\varphi_{P_0}(\Xi) + \varphi_{P_0}(R) + K_{P_0}](0, \Pi)},$$

which by Proposition 6.5 equals

$$\frac{\theta^{2n^2}[-(\varphi_{P_0}(T_R(\Delta)) + K_{P_0})](0, \Pi)}{\theta^{2n^2}[-(\varphi_{P_0}(T_R(\Xi)) + K_{P_0})](0, \Pi)},$$

as in Section 2.6. Note that we have used the independence of the lifts; we have this independence, since we have taken the power of the theta functions to be $2n^2$. This means that we can extend the validity of Eq. (2.1) (with $e = 2$) to this situation (with the same proof). Since T_R is an involution and we already know that R does not appear in $N(T_R(\Delta))$ if it does not appear in $N(\Delta)$, the same holds for Eq. (2.3). We also remark that also in this situation, in the function

$$\frac{\theta^{2n^2}[\Delta](\varphi_{P_0}(P), \Pi)}{\theta^{2n^2}[T_R(\Delta)](\varphi_{P_0}(P), \Pi)}$$

for such Δ and R, the points P_0 and R are always the only branch points whose substitution gives a meaningful value.

We now use the quotient from Proposition 6.6 in order to obtain the PMT. We begin with the divisor $\Delta = P_1^l P_2^{n-1-l}$ with $0 \leq l \leq n-3$ and with $R = P_3$, giving that the quotient

$$\frac{\theta^{2n^2}[\Delta](\varphi_{P_0}(P),\Pi)}{\theta^{2n^2}[T_{P_3}(\Delta)](\varphi_{P_0}(P),\Pi)} = \frac{\theta^{2n^2}[P_1^l P_2^{n-1-l}](\varphi_{P_0}(P),\Pi)}{\theta^{2n^2}[P_1^{n-5-l}P_2^{l+4}\{P_1^{2n-5-l}P_2^{l+4-n}\}](\varphi_{P_0}(P),\Pi)}$$

is a meromorphic function on the Z_n curve X, where we have used Definition 6.4. In the quotient on the right-hand side (and in what follows) we mean that the divisor $P_1^{n-5-l}P_2^{l+4}$ appears if $0 \le l \le n-5$, and the divisor $P_1^{2n-5-l}P_2^{l+4-n}$ in the brackets appears if $n-4 \le l \le n-3$ (check Definition 6.4 again). However, we can do these two cases together and for this we use this notation. Since Definition 6.4 gives us also that $N(\Delta) = P_1^{n-3-l}P_2^{l+2}$ and $N(T_{P_3}(\Delta)) = P_1^{l+2}P_2^{n-3-l}$ both for $0 \le l \le n-5$ and for $n-4 \le l \le n-3$, Proposition 6.6 shows that in any of these cases the divisor of this function is $\left[P_1^{n-3-l}P_2^{l+2}/P_1^{l+2}P_2^{n-3-l}\right]^{2n^2}$. Hence this function is

$$c\frac{(z-\lambda_1)^{2n(n-3-l)}(z-\lambda_2)^{2n(l+2)}}{(z-\lambda_1)^{2n(l+2)}(z-\lambda_2)^{2n(n-3-l)}},$$

where c is some nonzero complex constant. This quotient is not reduced (except for the case $l = n-3$), but we prefer not to do the reduction as it will force us to split into cases according to the way l relates to $(n-5)/2$. We shall therefore obtain unreduced Thomae formulae at the end and will have to do some reduction then. We note that for this value $l = (n-5)/2$ this function reduces to be simply the constant function c, which we will soon seen to be 1; this is because the divisor $P_1^{(n-5)/2}P_2^{(n+3)/2}$ is T_{P_3}-invariant, as we have already seen. We shall later see where this fact matters, but for now we leave this function as is.

In order to calculate the constant c we do what we always do—substitute $P = P_0$, which gives

$$\frac{\theta^{2n^2}[P_1^l P_2^{n-1-l}](0,\Pi)}{\theta^{2n^2}[P_1^{n-5-l}P_2^{l+4}\{P_1^{2n-5-l}P_2^{l+4-n}\}](0,\Pi)} = c\frac{(\lambda_0-\lambda_1)^{2n(n-3-l)}(\lambda_0-\lambda_2)^{2n(l+2)}}{(\lambda_0-\lambda_1)^{2n(l+2)}(\lambda_0-\lambda_2)^{2n(n-3-l)}},$$

and therefore

$$\frac{\theta^{2n^2}[P_1^l P_2^{n-1-l}](\varphi_{P_0}(P),\Pi)}{\theta^{2n^2}[P_1^{n-5-l}P_2^{l+4}\{P_1^{2n-5-l}P_2^{l+4-n}\}](\varphi_{P_0}(P),\Pi)}$$

$$= \frac{\theta^{2n^2}[P_1^l P_2^{n-1-l}](0,\Pi)}{\theta^{2n^2}[P_1^{n-5-l}P_2^{l+4}\{P_1^{2n-5-l}P_2^{l+4-n}\}](0,\Pi)}$$

$$\times \frac{(\lambda_0-\lambda_1)^{2n(l+2)}(\lambda_0-\lambda_2)^{2n(n-3-l)}}{(\lambda_0-\lambda_1)^{2n(n-3-l)}(\lambda_0-\lambda_2)^{2n(l+2)}} \times \frac{(z-\lambda_1)^{2n(n-3-l)}(z-\lambda_2)^{2n(l+2)}}{(z-\lambda_1)^{2n(l+2)}(z-\lambda_2)^{2n(n-3-l)}}.$$

We now substitute the remaining point $P = P_3$, which gives us, on the one hand,

$$\frac{\theta^{2n^2}[P_1^l P_2^{n-1-l}](\varphi_{P_0}(P_3),\Pi)}{\theta^{2n^2}[P_1^{n-5-l}P_2^{l+4}\{P_1^{2n-5-l}P_2^{l+4-n}\}](\varphi_{P_0}(P_3),\Pi)}$$

$$=\frac{\theta^{2n^2}[P_1^l P_2^{n-1-l}](0,\Pi)}{\theta^{2n^2}[P_1^{n-5-l}P_2^{l+4}\{P_1^{2n-5-l}P_2^{l+4-n}\}](0,\Pi)}$$

$$\times\frac{(\lambda_0-\lambda_1)^{2n(l+2)}(\lambda_0-\lambda_2)^{2n(n-3-l)}}{(\lambda_0-\lambda_1)^{2n(n-3-l)}(\lambda_0-\lambda_2)^{2n(l+2)}}\times\frac{(\lambda_3-\lambda_1)^{2n(n-3-l)}(\lambda_3-\lambda_2)^{2n(l+2)}}{(\lambda_3-\lambda_1)^{2n(l+2)}(\lambda_3-\lambda_2)^{2n(n-3-l)}},$$

while on the other hand, Eq. (2.3) gives us

$$\frac{\theta^{2n^2}[P_1^l P_2^{n-1-l}](\varphi_{P_0}(P_3),\Pi)}{\theta^{2n^2}[P_1^{n-5-l}P_2^{l+4}\{P_1^{2n-5-l}P_2^{l+4-n}\}](\varphi_{P_0}(P_3),\Pi)}$$

$$=\frac{\theta^{2n^2}[P_1^{n-5-l}P_2^{l+4}\{P_1^{2n-5-l}P_2^{l+4-n}\}](0,\Pi)}{\theta^{2n^2}[P_1^l P_2^{n-1-l}](0,\Pi)}$$

(as $T_{P_3}(P_1^{n-5-l}P_2^{l+4})=P_1^l P_2^{n-1-l}$ for $0\le l\le n-5$, while for $n-4\le l\le n-3$ we have $T_{P_3}(P_1^{2n-5-l}P_2^{l+4-n})=P_1^l P_2^{n-1-l}$, since T_{P_3} is an involution). Therefore we find by combining the last two equations that

$$\frac{\theta^{2n^2}[P_1^{n-5-l}P_2^{l+4}\{P_1^{2n-5-l}P_2^{l+4-n}\}](0,\Pi)}{\theta^{2n^2}[P_1^l P_2^{n-1-l}](0,\Pi)}$$

$$=\frac{\theta^{2n^2}[P_1^l P_2^{n-1-l}](0,\Pi)}{\theta^{2n^2}[P_1^{n-5-l}P_2^{l+4}\{P_1^{2n-5-l}P_2^{l+4-n}\}](0,\Pi)}$$

$$\times\frac{(\lambda_0-\lambda_1)^{2n(l+2)}(\lambda_0-\lambda_2)^{2n(n-3-l)}}{(\lambda_0-\lambda_1)^{2n(n-3-l)}(\lambda_0-\lambda_2)^{2n(l+2)}}\times\frac{(\lambda_3-\lambda_1)^{2n(n-3-l)}(\lambda_3-\lambda_2)^{2n(l+2)}}{(\lambda_3-\lambda_1)^{2n(l+2)}(\lambda_3-\lambda_2)^{2n(n-3-l)}},$$

and hence

$$\frac{\theta^{4n^2}[P_1^l P_2^{n-1-l}](0,\Pi)}{(\lambda_0-\lambda_1)^{2n(n-3-l)}(\lambda_0-\lambda_2)^{2n(l+2)}(\lambda_3-\lambda_1)^{2n(l+2)}(\lambda_3-\lambda_2)^{2n(n-3-l)}}$$

$$=\frac{\theta^{4n^2}[P_1^{n-5-l}P_2^{l+4}\{P_1^{2n-5-l}P_2^{l+4-n}\}](0,\Pi)}{(\lambda_0-\lambda_1)^{2n(l+2)}(\lambda_0-\lambda_2)^{2n(n-3-l)}(\lambda_3-\lambda_1)^{2n(n-3-l)}(\lambda_3-\lambda_2)^{2n(l+2)}}.\qquad(6.1)$$

Note that for $l=(n-5)/2$ the last equation is trivial, as all the powers appearing there equal $2n(n-1)/2$ for this value of l and the divisor $P_1^{(n-5)/2}P_2^{(n+3)/2}$ is invariant under T_{P_3}, and that for any $0\le l\le n-5$, starting with the divisor $T_{P_3}(\Delta)=P_1^{n-5-l}P_2^{l+4}$ we obtain exactly the same equation. As for starting with $T_{P_3}(\Delta)=P_1^{2n-5-l}P_2^{l+4-n}$ for $n-4\le l\le n-3$, we have to see what changes if we let the index l be in the complementary set, i.e., to satisfy $n-2\le l\le n-1$. In this case the form of $T_{P_3}(\Delta)$ is the same as for $n-4\le l\le n-3$, but the powers in the polynomials change. $N(\Delta)$ becomes $P_1^{2n-3-l}P_2^{l+2-n}$ and $N(T_{P_3}(\Delta))$ becomes $P_1^{l+2-n}P_2^{2n-3-l}$, i.e., every power $n-3-l$ is replaced by $2n-3-l$ and every power

$l+2$ is replaced by $l+2-n$. Since the action of T_{P_3} takes l to $2n-5-l$, the reader can verify that starting with $n-2 \le l \le n-1$ gives the same two equations as $n-4 \le l \le n-3$ by the T_{P_3} correspondence. We remark again that Eq. (6.1) is not reduced (except for the case $l = n-3$, which also comes from $l = n-2$ in the way just described), and we shall give its reduced analog later.

6.1.6 The Poor Man's Thomae (Unreduced and Reduced)

We have completed obtaining the equations from the operator T_{P_3}. However, we do have two more operators (although with much smaller domains), T_{P_1} and T_{P_2}. We now return to the divisor $\Delta = P_1^{n-3}P_2^2$ (with $l = n-3$), and recall that on this divisor we can also apply the operator T_{P_1}. Then Proposition 6.6 gives, using Definition 6.4 to calculate that $T_{P_1}(\Delta) = P_1^{n-3}P_2P_3$, that the quotient

$$\frac{\theta^{2n^2}[\Delta](\varphi_{P_0}(P), \Pi)}{\theta^{2n^2}[T_{P_1}(\Delta)](\varphi_{P_0}(P), \Pi)} = \frac{\theta^{2n^2}[P_1^{n-3}P_2^2](\varphi_{P_0}(P), \Pi)}{\theta^{2n^2}[P_1^{n-3}P_2P_3](\varphi_{P_0}(P), \Pi)}$$

is a meromorphic function on the Z_n curve X. Here Definition 6.4 gives us that $N(\Delta) = P_2^{n-1}$ and $N(T_{P_1}(\Delta)) = P_3^{n-1}$, so that from Proposition 6.6 we obtain that the divisor of this function is $\left[P_2^{n-1}/P_3^{n-1}\right]^{2n^2}$ and therefore this function is the function $d(z-\lambda_2)^{2n(n-1)}/(z-\lambda_3)^{2n(n-1)}$ for some nonzero complex constant d. Note that this expression is already reduced. The value of d is found by the substitution $P = P_0$, which gives

$$\frac{\theta^{2n^2}[P_1^{n-3}P_2^2](0, \Pi)}{\theta^{2n^2}[P_1^{n-3}P_2P_3](0, \Pi)} = d\frac{(\lambda_0-\lambda_2)^{2n(n-1)}}{(\lambda_0-\lambda_3)^{2n(n-1)}},$$

and therefore

$$\frac{\theta^{2n^2}[P_1^{n-3}P_2^2](\varphi_{P_0}(P), \Pi)}{\theta^{2n^2}[P_1^{n-3}P_2P_3](\varphi_{P_0}(P), \Pi)} = \frac{\theta^{2n^2}[P_1^{n-3}P_2^2](0, \Pi)}{\theta^{2n^2}[P_1^{n-3}P_2P_3](0, \Pi)} \times \frac{(\lambda_0-\lambda_3)^{2n(n-1)}}{(\lambda_0-\lambda_2)^{2n(n-1)}}$$
$$\times \frac{(z-\lambda_2)^{2n(n-1)}}{(z-\lambda_3)^{2n(n-1)}}.$$

Substituting the remaining point $P = P_1$ yields, on the one hand,

$$\frac{\theta^{2n^2}[P_1^{n-3}P_2^2](\varphi_{P_0}(P_1), \Pi)}{\theta^{2n^2}[P_1^{n-3}P_2P_3](\varphi_{P_0}(P_1), \Pi)} = \frac{\theta^{2n^2}[P_1^{n-3}P_2^2](0, \Pi)}{\theta^{2n^2}[P_1^{n-3}P_2P_3](0, \Pi)} \times \frac{(\lambda_0-\lambda_3)^{2n(n-1)}}{(\lambda_0-\lambda_2)^{2n(n-1)}}$$
$$\times \frac{(\lambda_1-\lambda_2)^{2n(n-1)}}{(\lambda_1-\lambda_3)^{2n(n-1)}},$$

while Eq. (2.3) gives us, on the other hand, that

$$\frac{\theta^{2n^2}[P_1^{n-3}P_2^2](\varphi_{P_0}(P_1),\Pi)}{\theta^{2n^2}[P_1^{n-3}P_2P_3](\varphi_{P_0}(P_1),\Pi)} = \frac{\theta^{2n^2}[P_1^{n-3}P_2P_3](0,\Pi)}{\theta^{2n^2}[P_1^{n-3}P_2^2](0,\Pi)}$$

(as $T_{P_1}(P_1^{n-3}P_2P_3) = P_1^{n-3}P_2^2$). We can now combine these last two equations to obtain that

$$\frac{\theta^{2n^2}[P_1^{n-3}P_2P_3](0,\Pi)}{\theta^{2n^2}[P_1^{n-3}P_2^2](0,\Pi)} = \frac{\theta^{2n^2}[P_1^{n-3}P_2^2](0,\Pi)}{\theta^{2n^2}[P_1^{n-3}P_2P_3](0,\Pi)} \times \frac{(\lambda_0-\lambda_3)^{2n(n-1)}(\lambda_1-\lambda_2)^{2n(n-1)}}{(\lambda_0-\lambda_2)^{2n(n-1)}(\lambda_1-\lambda_3)^{2n(n-1)}},$$

or equivalently,

$$\frac{\theta^{4n^2}[P_1^{n-3}P_2^2](0,\Pi)}{(\lambda_0-\lambda_2)^{2n(n-1)}(\lambda_1-\lambda_3)^{2n(n-1)}} = \frac{\theta^{4n^2}[P_1^{n-3}P_2P_3](0,\Pi)}{(\lambda_0-\lambda_3)^{2n(n-1)}(\lambda_1-\lambda_2)^{2n(n-1)}}.$$

The reader can check that when starting with $T_{P_1}(\Delta) = P_1^{n-3}P_2P_3$ we obtain exactly the same equation.

In order to see what we obtain from the remaining operator T_{P_2} we go back to the divisor $\Delta = P_1^{n-2}P_2$ (with $l = n-2$), and recall that Definition 6.4 gives that $T_{P_2}(\Delta) = P_1^{n-3}P_2P_3$. Then from Proposition 6.6 we obtain that the quotient

$$\frac{\theta^{2n^2}[\Delta](\varphi_{P_0}(P),\Pi)}{\theta^{2n^2}[T_{P_2}(\Delta)](\varphi_{P_0}(P),\Pi)} = \frac{\theta^{2n^2}[P_1^{n-2}P_2](\varphi_{P_0}(P),\Pi)}{\theta^{2n^2}[P_1^{n-3}P_2P_3](\varphi_{P_0}(P),\Pi)}$$

is a meromorphic function on the Z_n curve X. Here from Definition 6.4 we see that $N(\Delta) = P_1^{n-1}$ and $N(T_{P_2}(\Delta)) = P_3^{n-1}$. Therefore we have from Proposition 6.6 that this function has divisor $\left[P_1^{n-1}/P_3^{n-1}\right]^{2n^2}$, which means that it is the function $a(z-\lambda_1)^{2n(n-1)}/(z-\lambda_3)^{2n(n-1)}$ with a being, as always, some nonzero complex constant. This expression is also already reduced. We find the value of a by the substitution $P = P_0$, which gives that

$$\frac{\theta^{2n^2}[P_1^{n-2}P_2](0,\Pi)}{\theta^{2n^2}[P_1^{n-3}P_2P_3](0,\Pi)} = a\frac{(\lambda_0-\lambda_1)^{2n(n-1)}}{(\lambda_0-\lambda_3)^{2n(n-1)}},$$

and thus this function is, explicitly,

$$\frac{\theta^{2n^2}[P_1^{n-2}P_2](\varphi_{P_0}(P),\Pi)}{\theta^{2n^2}[P_1^{n-3}P_2P_3](\varphi_{P_0}(P),\Pi)} = \frac{\theta^{2n^2}[P_1^{n-2}P_2](0,\Pi)}{\theta^{2n^2}[P_1^{n-3}P_2P_3](0,\Pi)} \times \frac{(\lambda_0-\lambda_3)^{2n(n-1)}}{(\lambda_0-\lambda_1)^{2n(n-1)}}$$

$$\times \frac{(z-\lambda_1)^{2n(n-1)}}{(z-\lambda_3)^{2n(n-1)}}.$$

We can now obtain, on the one hand, by substituting $P = P_2$, the equality

$$\frac{\theta^{2n^2}[P_1^{n-2}P_2](\varphi_{P_0}(P_2),\Pi)}{\theta^{2n^2}[P_1^{n-3}P_2P_3](\varphi_{P_0}(P_2),\Pi)} = \frac{\theta^{2n^2}[P_1^{n-2}P_2](0,\Pi)}{\theta^{2n^2}[P_1^{n-3}P_2P_3](0,\Pi)} \times \frac{(\lambda_0-\lambda_3)^{2n(n-1)}}{(\lambda_0-\lambda_1)^{2n(n-1)}}$$

$$\times \frac{(\lambda_2-\lambda_1)^{2n(n-1)}}{(\lambda_2-\lambda_3)^{2n(n-1)}},$$

and on the other hand, from Eq. (2.3), the equality

$$\frac{\theta^{2n^2}[P_1^{n-2}P_2](\varphi_{P_0}(P_2),\Pi)}{\theta^{2n^2}[P_1^{n-3}P_2P_3](\varphi_{P_0}(P_2),\Pi)} = \frac{\theta^{2n^2}[P_1^{n-3}P_2P_3](0,\Pi)}{\theta^{2n^2}[P_1^{n-2}P_2](0,\Pi)},$$

since we have $T_{P_2}(P_1^{n-3}P_2P_3) = P_1^{n-2}P_2$. Combining these equations yields

$$\frac{\theta^{4n^2}[P_1^{n-3}P_2P_3](0,\Pi)}{\theta^{2n^2}[P_1^{n-2}P_2](0,\Pi)} = \frac{\theta^{2n^2}[P_1^{n-2}P_2](0,\Pi)}{\theta^{2n^2}[P_1^{n-3}P_2P_3](0,\Pi)} \times \frac{(\lambda_0-\lambda_3)^{2n(n-1)}(\lambda_2-\lambda_1)^{2n(n-1)}}{(\lambda_0-\lambda_1)^{2n(n-1)}(\lambda_2-\lambda_3)^{2n(n-1)}},$$

and thus we have

$$\frac{\theta^{4n^2}[P_1^{n-2}P_2](0,\Pi)}{(\lambda_0-\lambda_1)^{2n(n-1)}(\lambda_2-\lambda_3)^{2n(n-1)}} = \frac{\theta^{4n^2}[P_1^{n-3}P_2P_3](0,\Pi)}{(\lambda_0-\lambda_3)^{2n(n-1)}(\lambda_2-\lambda_1)^{2n(n-1)}}.$$

As the reader can verify, when starting with $T_{P_2}(\Delta) = P_1^{n-3}P_2P_3$ we obtain exactly the same equation. Combining this equation with the one from the last paragraph (as their right-hand sides are the same) we obtain

$$\frac{\theta^{4n^2}[P_1^{n-3}P_2^2](0,\Pi)}{[(\lambda_0-\lambda_2)(\lambda_1-\lambda_3)]^{2n(n-1)}} = \frac{\theta^{4n^2}[P_1^{n-2}P_2](0,\Pi)}{[(\lambda_0-\lambda_1)(\lambda_2-\lambda_3)]^{2n(n-1)}} = \frac{\theta^{4n^2}[P_1^{n-3}P_2P_3](0,\Pi)}{[(\lambda_0-\lambda_3)(\lambda_2-\lambda_1)]^{2n(n-1)}}.$$

$$(6.2)$$

Note that Eq. (6.2) is consistent with Eq. (6.1) with $l = n - 3$, which compares exactly the first two terms here.

We have dealt with almost all the $n + 2$ divisors, but there is one missing: the divisor $\Delta = P_3^{n-1}$. Since on this divisor no operator T_R can act, the quotient $\theta^{4n^2}[\Delta](0,\Pi)/1$ can be considered as the part of the PMT corresponding to this divisor Δ. Therefore we have completed finding the PMT for our Z_n curve, as summarized in the following

Proposition 6.7. *The PMT for the Z_n curve X is composed of the following equations. Eq. (6.1) with $0 \leq l \leq n - 5$, which splits these $n - 4$ divisors into $(n-5)/2$ pairs and one set containing the single divisor $P_1^{(n-5)/2}P_2^{(n+3)/2}$; Eq. (6.1) with $l = n - 4$, which relates to the divisors $P_1^{n-4}P_3^3$ and P_1^{n-4}; Eq. (6.2), which relates 3 divisors; and the expression $\theta^{4n^2}[P_3^{n-1}](0,\Pi)/1$ involving the remaining divisor P_3^{n-1}.*

Proof. We have seen that starting from each divisor we can obtain the corresponding equation, and it is easy to see that the denominator g_Δ appearing under the theta constant $\theta^{4n^2}[\Delta](0,\Pi)$ in the corresponding equation depends only on Δ. This is so, since for $\Delta = P_1^l P_2^{n-1-l}$ it depends only on l, and the other two divisors are unique of their type.

It remains to show that all the sets appearing in these equations are closed under all the operators T_R. Now, except for the three divisors appearing in Eq. (6.2), the only operator T_R which can act is T_{P_3}, which takes $P_1^{(n-5)/2} P_2^{(n+3)/2}$ to itself and pairs all the others (being an involution). This gives the result for all the sets except for the one containing the three divisors appearing in Eq. (6.2), and the single divisor P_3^{n-1}. On the last divisor no operator T_R can act, meaning that for the set containing it alone there is nothing to prove. For the remaining set we see that on each of its divisors two operators can act (on $P_1^{n-3} P_2^2$ these are T_{P_1} and T_{P_3}, on $P_1^{n-2} P_2$ these are T_{P_2} and T_{P_3}, and on $P_1^{n-3} P_2 P_3$ these are T_{P_1} and T_{P_2}), and they take this divisor exactly to the other two (T_{P_1} pairs $P_1^{n-3} P_2^2$ and $P_1^{n-3} P_2 P_3$, T_{P_2} pairs $P_1^{n-2} P_2$ and $P_1^{n-3} P_2 P_3$, and T_{P_3} pairs $P_1^{n-3} P_2^2$ and $P_1^{n-2} P_2$). This proves the proposition. □

We note that Eq. (6.1) is not reduced, since for most values of l (except for the value $l = n-3$), powers of both the expression $(\lambda_0 - \lambda_1)(\lambda_3 - \lambda_2)$ and the expression $(\lambda_0 - \lambda_2)(\lambda_3 - \lambda_1)$ appear in the two denominators. This occurs, as we already remarked, since the expression

$$c \frac{(z - \lambda_1)^{2n(n-3-l)} (z - \lambda_2)^{2n(l+2)}}{(z - \lambda_1)^{2n(l+2)} (z - \lambda_2)^{2n(n-3-l)}}$$

we used to represent the quotient of theta functions was not reduced, its reduction being the expression $c(z - \lambda_1)^{2n(n-5-2l)} (z - \lambda_2)^{2n(2l+5-n)}$ (with one power nonnegative and one nonpositive). In order to obtain the reduced expression we take, for any pair $0 \le l \le n-5$ and $n-3-l$, the representative l satisfying $0 \le l \le (n-7)/2$, and then the reduced form of Eq. (6.1) becomes

$$\frac{\theta^{4n^2}[P_1^l P_2^{n-1-l}](0, \Pi)}{(\lambda_0 - \lambda_1)^{2n(n-5-2l)} (\lambda_3 - \lambda_2)^{2n(n-5-2l)}} = \frac{\theta^{4n^2}[P_1^{n-5-l} P_2^{l+4}](0, \Pi)}{(\lambda_0 - \lambda_2)^{2n(n-5-2l)} (\lambda_3 - \lambda_1)^{2n(n-5-2l)}}.$$
(6.3)

For the unpaired value $l = (n-5)/2$ we have just $\theta^{4n^2}[P_1^{(n-5)/2} P_2^{(n+3)/2}](0, \Pi)/1$ as the reduced PMT, and it is clear that this is the case $l = (n-5)/2$ of Eq. (6.3), as both sides are then really the same expression since the powers in the denominator vanish. For the complementary values $(n-3)/2 \le l \le n-5$, Eq. (6.3) still holds, but contains negative powers in the denominators; but after manipulating the formula to have positive powers we obtain the same Eq. (6.3) as for the value $n-5-l$, which satisfies the inequality $0 \le n-5-l \le (n-7)/2$. For the value $l = n-3$ we see that Eq. (6.1) is reduced, but for the PMT, Proposition 6.7 shows that it appears in Eq. (6.2), and the latter is reduced as it appears. Examining the remaining value $l = n-4$, we see that the expression

$$c \frac{(z - \lambda_1)^{2n} (z - \lambda_2)^{2n(n-2)}}{(z - \lambda_1)^{2n(n-2)} (z - \lambda_2)^{2n}}$$

we used to represent the theta function is not reduced, and its reduction is the expression $c(z - \lambda_2)^{2n(n-3)}/(z - \lambda_1)^{2n(n-3)}$. Therefore again some reduction can be done to the corresponding Eq. (6.1), giving the reduced equation

$$\frac{\theta^{4n^2}[P_1^{n-4}P_2^3](0,\Pi)}{(\lambda_0 - \lambda_2)^{2n(n-3)}(\lambda_3 - \lambda_1)^{2n(n-3)}} = \frac{\theta^{4n^2}[P_1^{n-1}](0,\Pi)}{(\lambda_0 - \lambda_1)^{2n(n-3)}(\lambda_3 - \lambda_2)^{2n(n-3)}}. \qquad (6.4)$$

Let us write Proposition 6.7 in the language we always used for PMT formulas. Define, for a divisor $\Delta = P_1^l P_2^{n-1-l}$, the expression g_Δ to be

$$(\lambda_0 - \lambda_1)^{2n(n-3-l)}(\lambda_0 - \lambda_2)^{2n(l+2)}(\lambda_3 - \lambda_1)^{2n(l+2)}(\lambda_3 - \lambda_2)^{2n(n-3-l)}$$

if $0 \le l \le n-3$ and

$$(\lambda_0 - \lambda_1)^{2n(2n-3-l)}(\lambda_0 - \lambda_2)^{2n(l+2-n)}(\lambda_3 - \lambda_1)^{2n(l+2-n)}(\lambda_3 - \lambda_2)^{2n(2n-3-l)}$$

if $n-2 \le l \le n-1$. For $\Delta = P_1^{n-3}P_2P_3$ define g_Δ to be

$$(\lambda_0 - \lambda_3)^{2n(n-1)}(\lambda_2 - \lambda_1)^{2n(n-1)},$$

and for $\Delta = P_3^{n-1}$ define $g_\Delta = 1$. Then the PMT in Proposition 6.7 states that the expression $\theta^{4n^2}[\Delta](0,\Pi)/g_\Delta$ is invariant under the three operators T_R. For the reduced PMT define, for a divisor $\Delta = P_1^l P_2^{n-1-l}$, the expression g_Δ to be

$$(\lambda_0 - \lambda_1)^{2n(n-5-2l)}(\lambda_3 - \lambda_2)^{2n(n-5-2l)}$$

if $0 \le l \le (n-5)/2$, and

$$(\lambda_0 - \lambda_2)^{2n(2l+5-n)}(\lambda_3 - \lambda_1)^{2n(2l+5-n)}$$

if $(n-5)/2 \le l \le n-3$. Note that both definitions coincide for $l = (n-5)/2$ to give $g_\Delta = 1$, and the reduced g_Δ equals the unreduced one for $l = n-3$. For $l = n-1$ define g_Δ to be

$$(\lambda_0 - \lambda_1)^{2n(n-3)}(\lambda_3 - \lambda_2)^{2n(n-3)},$$

and for the remaining divisors (the one with $l = n-2$ and the divisors $P_1^{n-3}P_2P_3$ and P_3^{n-1}) leave the old g_Δ. Then the reduced PMT states that the expression $\theta^{4n^2}[\Delta](0,\Pi)/g_\Delta$ with these expressions for g_Δ is invariant under the three operators T_R.

In order to obtain the full Thomae formulae for our Z_n curve X we will, as we did in all the previous cases, multiply the denominators g_Δ (either the unreduced or the reduced ones) by some expressions to obtain new denominators h_Δ. These denominators will preserve the property that $\theta^{4n^2}[\Delta](0,\Pi)/h_\Delta$ with these expressions for h_Δ is invariant under the three operators T_R, but will make this quotient also invariant under the operator N. As always, the fact that Eq. (1.5) gives that $\theta[N(\Delta)](0,\Pi) = \theta[\Delta](0,\Pi)$ up to the proper power shows that the last assertion is equivalent to the statement that $h_{N(\Delta)} = h_\Delta$. Before proving the general case we continue with an example to illustrate the process.

6.1.7 The Thomae Formulae in the Case $n = 7$

We shall continue with our example of $n = 7$, and we shall use Proposition 6.7 with the unreduced expressions for g_Δ (just for the sake of the example). Here Eq. (6.1) gives the equality

$$\frac{\theta^{196}[P_2^6](0, \Pi)}{(\lambda_0 - \lambda_1)^{56}(\lambda_0 - \lambda_2)^{28}(\lambda_3 - \lambda_1)^{28}(\lambda_3 - \lambda_2)^{56}}$$
$$= \frac{\theta^{196}[P_1^2 P_2^4](0, \Pi)}{(\lambda_0 - \lambda_1)^{28}(\lambda_0 - \lambda_2)^{56}(\lambda_3 - \lambda_1)^{56}(\lambda_3 - \lambda_2)^{28}}$$

for $l = 0$ and $l = 2$, the quotient

$$\frac{\theta^{196}[P_1 P_2^5](0, \Pi)}{(\lambda_0 - \lambda_1)^{42}(\lambda_0 - \lambda_2)^{42}(\lambda_3 - \lambda_1)^{42}(\lambda_3 - \lambda_2)^{42}}$$

for $l = 1$, and the equality

$$\frac{\theta^{196}[P_1^3 P_2^3](0, \Pi)}{(\lambda_0 - \lambda_1)^{14}(\lambda_0 - \lambda_2)^{70}(\lambda_3 - \lambda_1)^{70}(\lambda_3 - \lambda_2)^{14}}$$
$$= \frac{\theta^{196}[P_1^6](0, \Pi)}{(\lambda_0 - \lambda_1)^{70}(\lambda_0 - \lambda_2)^{14}(\lambda_3 - \lambda_1)^{14}(\lambda_3 - \lambda_2)^{70}}$$

for $l = 3$ (which includes $l = 6$). Equation (6.2) gives

$$\frac{\theta^{196}[P_1^4 P_2^2](0, \Pi)}{(\lambda_0 - \lambda_2)^{84}(\lambda_1 - \lambda_3)^{84}} = \frac{\theta^{196}[P_1^5 P_2](0, \Pi)}{(\lambda_0 - \lambda_1)^{84}(\lambda_2 - \lambda_3)^{84}} = \frac{\theta^{196}[P_1^4 P_2 P_3](0, \Pi)}{(\lambda_0 - \lambda_3)^{84}(\lambda_2 - \lambda_1)^{84}}$$

(including $l = 4$ and $l = 5$), and we also have the expression $\theta^{196}[P_3^6](0, \Pi)/1$. We would like to combine these equations into one big equality containing all these terms.

In order to do this combination we recall that N pairs P_2^6 with $P_1^4 P_2^2$, $P_1 P_2^5$ with $P_1^3 P_2^3$, $P_1^5 P_2$ with P_1^6, $P_1^4 P_2 P_3$ with P_3^6, and leaves $P_1^2 P_2^4$ invariant. This allows us to see that if we divide the first equation by

$$(\lambda_0 - \lambda_2)^{70}(\lambda_1 - \lambda_3)^{70},$$

the second by

$$(\lambda_0 - \lambda_1)^{42}(\lambda_2 - \lambda_3)^{42}(\lambda_0 - \lambda_2)^{28}(\lambda_1 - \lambda_3)^{28},$$

the third by
$$(\lambda_0 - \lambda_1)^{70}(\lambda_2 - \lambda_3)^{70},$$

the fourth by
$$(\lambda_0 - \lambda_1)^{56}(\lambda_2 - \lambda_3)^{56}(\lambda_0 - \lambda_2)^{14}(\lambda_1 - \lambda_3)^{14}$$

and the fifth by
$$(\lambda_0 - \lambda_1)^{56}(\lambda_2 - \lambda_3)^{56}(\lambda_0 - \lambda_2)^{14}(\lambda_1 - \lambda_3)^{14}(\lambda_0 - \lambda_3)^{84}(\lambda_2 - \lambda_1)^{84},$$

and use Eq. (1.5) in the form $\theta[N(\Delta)](0,\Pi) = \theta[\Delta](0,\Pi)$ to the proper power, we obtain the equality

$$\frac{\theta^{196}[P_1 P_2^5](0,\Pi)}{(\lambda_0 - \lambda_1)^{84}(\lambda_2 - \lambda_3)^{84}(\lambda_0 - \lambda_2)^{70}(\lambda_1 - \lambda_3)^{70}}$$

$$= \frac{\theta^{196}[P_1^3 P_2^3](0,\Pi)}{(\lambda_0 - \lambda_1)^{84}(\lambda_3 - \lambda_2)^{84}(\lambda_0 - \lambda_2)^{70}(\lambda_3 - \lambda_1)^{70}}$$

$$= \frac{\theta^{196}[P_1^6](0,\Pi)}{(\lambda_0 - \lambda_1)^{140}(\lambda_3 - \lambda_2)^{140}(\lambda_0 - \lambda_2)^{14}(\lambda_3 - \lambda_1)^{14}}$$

$$= \frac{\theta^{196}[P_1^5 P_2](0,\Pi)}{(\lambda_0 - \lambda_1)^{140}(\lambda_2 - \lambda_3)^{140}(\lambda_0 - \lambda_2)^{14}(\lambda_1 - \lambda_3)^{14}}$$

$$= \frac{\theta^{196}[P_1^4 P_2 P_3](0,\Pi)}{(\lambda_0 - \lambda_1)^{56}(\lambda_2 - \lambda_3)^{56}(\lambda_0 - \lambda_2)^{14}(\lambda_1 - \lambda_3)^{14}(\lambda_0 - \lambda_3)^{84}(\lambda_2 - \lambda_1)^{84}}$$

$$= \frac{\theta^{196}[P_1^4 P_2^2](0,\Pi)}{(\lambda_0 - \lambda_1)^{56}(\lambda_2 - \lambda_3)^{56}(\lambda_0 - \lambda_2)^{98}(\lambda_1 - \lambda_3)^{98}}$$

$$= \frac{\theta^{196}[P_2^6](0,\Pi)}{(\lambda_0 - \lambda_1)^{56}(\lambda_3 - \lambda_2)^{56}(\lambda_0 - \lambda_2)^{98}(\lambda_3 - \lambda_1)^{98}}$$

$$= \frac{\theta^{196}[P_1^2 P_2^4](0,\Pi)}{(\lambda_0 - \lambda_1)^{28}(\lambda_3 - \lambda_2)^{28}(\lambda_0 - \lambda_2)^{126}(\lambda_3 - \lambda_1)^{126}}$$

$$= \frac{\theta^{196}[P_3^6](0,\Pi)}{(\lambda_0 - \lambda_1)^{56}(\lambda_2 - \lambda_3)^{56}(\lambda_0 - \lambda_2)^{14}(\lambda_1 - \lambda_3)^{14}(\lambda_0 - \lambda_3)^{84}(\lambda_2 - \lambda_1)^{84}},$$

which includes all the 9 terms.

This last equation could have been the Thomae formulae for the Z_7 curve we are talking about. However, one can see that this formula is not reduced, as there are expressions appearing in all the denominators. Therefore we shall multiply the whole equation by

$$(\lambda_0 - \lambda_1)^{28}(\lambda_2 - \lambda_3)^{28}(\lambda_0 - \lambda_2)^{14}(\lambda_1 - \lambda_3)^{14},$$

and obtain the equality

$$\frac{\theta^{196}[P_1 P_2^5](0,\Pi)}{(\lambda_0 - \lambda_1)^{56}(\lambda_2 - \lambda_3)^{56}(\lambda_0 - \lambda_2)^{56}(\lambda_1 - \lambda_3)^{56}}$$

$$= \frac{\theta^{196}[P_1^3 P_2^3](0,\Pi)}{(\lambda_0 - \lambda_1)^{56}(\lambda_3 - \lambda_2)^{56}(\lambda_0 - \lambda_2)^{56}(\lambda_3 - \lambda_1)^{56}} = \frac{\theta^{196}[P_1^6](0,\Pi)}{(\lambda_0 - \lambda_1)^{112}(\lambda_3 - \lambda_2)^{112}}$$

$$= \frac{\theta^{196}[P_1^5 P_2](0,\Pi)}{(\lambda_0 - \lambda_1)^{112}(\lambda_2 - \lambda_3)^{112}} = \frac{\theta^{196}[P_1^4 P_2 P_3](0,\Pi)}{(\lambda_0 - \lambda_1)^{28}(\lambda_2 - \lambda_3)^{28}(\lambda_0 - \lambda_3)^{84}(\lambda_2 - \lambda_1)^{84}}$$

$$= \frac{\theta^{196}[P_1^4 P_2^2](0,\Pi)}{(\lambda_0 - \lambda_1)^{28}(\lambda_2 - \lambda_3)^{28}(\lambda_0 - \lambda_2)^{84}(\lambda_1 - \lambda_3)^{84}}$$

$$= \frac{\theta^{196}[P_1^6](0,\Pi)}{(\lambda_0 - \lambda_1)^{28}(\lambda_3 - \lambda_2)^{28}(\lambda_0 - \lambda_2)^{84}(\lambda_3 - \lambda_1)^{84}} = \frac{\theta^{196}[P_1^2 P_2^4](0,\Pi)}{(\lambda_0 - \lambda_2)^{112}(\lambda_3 - \lambda_1)^{112}}$$

$$= \frac{\theta^{196}[P_3^6](0,\Pi)}{(\lambda_0 - \lambda_1)^{28}(\lambda_2 - \lambda_3)^{28}(\lambda_0 - \lambda_3)^{84}(\lambda_2 - \lambda_1)^{84}}. \tag{6.5}$$

Since Eq. (6.5) contains all the non-special divisors of degree $g = 6$ supported on the branch points distinct from P_0 on our Z_7 curve (by Theorem 6.3), it is clear that

Theorem 6.8. *Equation* (6.5) *is the Thomae formulae for the Z_7 curve X with symmetric equation.*

The reduction we did in the end is needed, as usual, since we started in the PMT in Proposition 6.7 with the unreduced Eq. (6.1). The reader can check that if we start with the reduced PMT, i.e., with Eqs. (6.3) and (6.4), the appropriate manipulations are dividing the first equation by

$$(\lambda_0 - \lambda_2)(\lambda_1 - \lambda_3)^{84},$$

the second by

$$(\lambda_0 - \lambda_1)^{56}(\lambda_2 - \lambda_3)^{56}(\lambda_0 - \lambda_2)^{56}(\lambda_1 - \lambda_3)^{56},$$

the third by

$$(\lambda_0 - \lambda_1)^{56}(\lambda_2 - \lambda_3)^{56},$$

the fourth by

$$(\lambda_0 - \lambda_1)^{28}(\lambda_2 - \lambda_3)^{28},$$

and the fifth by

$$(\lambda_0 - \lambda_1)^{28}(\lambda_2 - \lambda_3)^{28}(\lambda_0 - \lambda_3)^{84}(\lambda_2 - \lambda_1)^{84},$$

and they bring us to Eq. (6.5) directly, without needing any reduction.

We remark at this point that once we remember that to the proper power we have $\theta[N(\Delta)](0,\Pi) = \theta[\Delta](0,\Pi)$ by Eq. (1.5), we could really write a shorter equation with only 5 terms, where the other 4 could be filled in using this relation. However, since the Thomae formulae is more complete when containing all the divisors, we preferred to give the longer (and partially redundant) but fuller equation in Eq. (6.5).

6.1.8 The Thomae Formulae in the General Case

We now turn to the general case. It turns out that even though we have restricted ourselves to odd values of n, we still have to split into two cases, according to whether n is congruent to 1 or to 3 modulo 4. The construction of the denominator h_Δ is similar in these two cases, but the details are different, and inside each case we have to separate again the values of odd l from the values of even l. The resulting denominators also depend on whether l is even or odd and the residue of n modulo 4, and this makes the construction a bit tedious. Therefore, as in the nonsingular and usual singular cases, we put this construction in Appendix B (we just state that it is quite similar to the proof of Theorem 3.10, but with more splitting into cases) and simply write explicitly the resulting denominators. Of course, for the sake of completeness we prove that these denominators do have all the desired properties. The result is as follows. For $n \equiv 1 \pmod 4$ and $\Delta = P_1^l P_2^{n-1-l}$ define h_Δ to be the expression

$$[(\lambda_0 - \lambda_1)(\lambda_2 - \lambda_3)]^{2n[(n^2+2n-3)/8-(l+2)(n-1-l)/2]}[(\lambda_0 - \lambda_2)(\lambda_1 - \lambda_3)]^{2n(l+2)(n-1-l)/2}$$

for even l and

$$[(\lambda_0 - \lambda_1)(\lambda_2 - \lambda_3)]^{2n[(n^2+2n-3)/8-(l+1)(n-2-l)/2]}[(\lambda_0 - \lambda_2)(\lambda_1 - \lambda_3)]^{2n(l+1)(n-2-l)/2}$$

for odd l. For the remaining divisors $\Delta = P_1^{n-3}P_2P_3$ or $\Delta = P_3^{n-1}$ define h_Δ to be

$$[(\lambda_0 - \lambda_1)(\lambda_2 - \lambda_3)]^{2n(n^2-6n+5)/8}[(\lambda_0 - \lambda_3)(\lambda_1 - \lambda_2)]^{2n(n-1)}.$$

When $n \equiv 3 \pmod 4$ define h_Δ for the divisor $\Delta = P_1^l P_2^{n-1-l}$ to be

$$[(\lambda_0 - \lambda_1)(\lambda_2 - \lambda_3)]^{2n[(n^2+2n+1)/8-(l+2)(n-1-l)/2]}[(\lambda_0 - \lambda_2)(\lambda_1 - \lambda_3)]^{2n(l+2)(n-1-l)/2}$$

for even l and

$$[(\lambda_0 - \lambda_1)(\lambda_2 - \lambda_3)]^{2n[(n^2+2n+1)/8-(l+1)(n-2-l)/2]}[(\lambda_0 - \lambda_2)(\lambda_1 - \lambda_3)]^{2n(l+1)(n-2-l)/2}$$

for odd l. If Δ is one of the remaining divisors $P_1^{n-3}P_2P_3$ and P_3^{n-1} we define h_Δ to be

$$[(\lambda_0 - \lambda_1)(\lambda_2 - \lambda_3)]^{2n(n^2-6n+9)/8}[(\lambda_0 - \lambda_3)(\lambda_1 - \lambda_2)]^{2n(n-1)}.$$

Note that for any fixed n the sum of the powers appearing in the expression for h_Δ is independent of Δ and equals $(n^2+2n-3)/8$ if $n \equiv 1 \pmod 4$ and $(n^2+2n+1)/8$ if $n \equiv 3 \pmod 4$. This is obvious for $\Delta = P_1^l P_2^{n-1-l}$ with any $0 \le l \le n-1$, and for the other two divisors it is also clear once one notices that

$$\frac{n^2+2n-3}{8} - \frac{n^2-6n+5}{8} = \frac{n^2+2n+1}{8} - \frac{n^2-6n+9}{8} = n-1.$$

This fact will be very useful in the proof of the following theorem.

With these expressions defined, we have

Theorem 6.9. *For any odd $n \geq 5$ the expression $\theta^{4n^2}[\Delta](0, \Pi)/h_\Delta$ with h_Δ defined to be the corresponding expression above is invariant under the three operators T_R and under N, and is consequently independent of the divisor Δ.*

Proof. We prove the theorem for the two cases $n \equiv 1 \pmod 4$ and $n \equiv 3 \pmod 4$ simultaneously, since the proofs are almost identical. We begin with the invariance under N, and recall that by the usual argument it suffices to show that $h_{N(\Delta)} = h_\Delta$. Since

$$(l+2)(n-1-l) = l(n-3-l) + 2(n-1)$$

and

$$(l+1)(n-2-l) = l(n-3-l) + (n-2),$$

we see that for the divisors $P_1^l P_2^{n-1-l}$ with $0 \leq l \leq n-3$, which N takes to $P_1^{n-3-l} P_2^{l+2}$, the powers of the expressions appearing in h_Δ for even l and odd l, respectively, are the same for l as for $n-3-l$ in both cases of n modulo 4. Therefore the expression for h_Δ for these divisors is invariant under taking l to $n-3-l$, which is exactly the action of N on these divisors. This gives the N-invariance of h_Δ on these divisors. We are left with two pairs to check: one is $P_1^{n-2} P_2$ and P_1^{n-1}, and the other is $P_1^{n-3} P_2 P_3$ and P_3^{n-1}. For the first pair it is clear that $(l+2)(n-1-l)$ vanishes when $l = n-1$ (which is even) and $(l+1)(n-2-l)$ vanishes when $l = n-2$ (which is odd), showing that the expressions for h_Δ coincide for these two divisors in both cases of n modulo 4. For the second pair we have simply defined the expressions for h_Δ to be the same in both cases of n modulo 4. This completes the proof of invariance under N.

The invariance under the operators T_R will be proved, as in the previous cases, by comparing it with the equations from the PMT in Proposition 6.7. As before we shall compare with the reduced expressions from Eqs. (6.3) and (6.4), together with Eq. (6.2) (which was already reduced). We write, as before,

$$\frac{\theta^{2n^2}[\Delta](0, \Pi)}{h_\Delta} = \frac{\theta^{2n^2}[\Delta](0, \Pi)}{g_\Delta} \Big/ \frac{h_\Delta}{g_\Delta},$$

and by the usual argument using the (reduced) PMT in Proposition 6.7 for the quotient in the numerator we obtain that it suffices to verify that the quotient in the denominator is also invariant under the three operators T_R (which is simpler). As always, for a divisor Δ and a point $R = P_i$, $1 \leq i \leq 3$, such that T_R can operate on Δ, the T_R-invariance of the quotient h_Δ/g_Δ, which is the same as the equality $h_\Delta/g_\Delta = h_{T_R(\Delta)}/g_{T_R(\Delta)}$, is equivalent to the equality $h_\Delta/h_{T_R(\Delta)} = g_\Delta/g_{T_R(\Delta)}$. This is the equality we shall check. This also has the advantage of giving us the same expression in the two cases $n \equiv 1 \pmod 4$ and $n \equiv 3 \pmod 4$, and we soon see that we can merge the odd l and even l cases as well.

We begin with the divisors $\Delta = P_1^l P_2^{n-1-l}$ with $0 < l < n-5$ and with the point $R = P_3$. Then we know that $T_R(\Delta) = P_1^{n-5-l} P_2^{l+4}$, and we begin by calculating

$g_\Delta/g_{T_R(\Delta)}$ for these divisors, in order to compare it to the expression obtained from $h_\Delta/h_{T_R(\Delta)}$. Now, for $0 \le l \le (n-7)/2$ Eq. (6.3) shows us that this expression is

$$\frac{\left[(\lambda_0 - \lambda_1)(\lambda_2 - \lambda_3)\right]^{2n(n-5-2l)}}{\left[(\lambda_0 - \lambda_2)(\lambda_3 - \lambda_1)\right]^{2n(n-5-2l)}}.$$

For $(n-3)/2 \le l \le n-5$ we get the reciprocals of both $g_\Delta/g_{T_R(\Delta)}$ and this expression obtained from $n-5-l$ (which satisfies $0 \le n-5-l \le (n-7)/2$), and for $l = (n-5)/2$ both expressions reduce to 1. Therefore we can extend the validity of this equality to every $0 \le l \le n-5$. Now, we already saw that the powers appearing in h_Δ are $l(n-3-l)/2$ (or its additive inverse) plus some expressions depending on the parity of l and on the residue of n modulo 4. Since l and $n-5-l$ have the same parity (n is odd!), we can reduce the expression $h_\Delta/h_{T_R(\Delta)}$ for such a divisor Δ to

$$\frac{\left[(\lambda_0 - \lambda_1)(\lambda_2 - \lambda_3)\right]^{-2nl(n-3-l)/2}\left[(\lambda_0 - \lambda_2)(\lambda_3 - \lambda_1)\right]^{2nl(n-3-l)/2}}{\left[(\lambda_0 - \lambda_1)(\lambda_2 - \lambda_3)\right]^{-2n(n-5-l)(l+2)/2}\left[(\lambda_0 - \lambda_2)(\lambda_3 - \lambda_1)\right]^{2n(n-5-l)(l+2)/2}}.$$

We now calculate that

$$(n-5-l)(l+2) - l(n-3-l) = 2(n-5-2l),$$

which gives that this quotient equals

$$\frac{\left[(\lambda_0 - \lambda_1)(\lambda_2 - \lambda_3)\right]^{2n(n-5-2l)}}{\left[(\lambda_0 - \lambda_2)(\lambda_3 - \lambda_1)\right]^{2n(n-5-2l)}},$$

just as we obtained from $g_\Delta/g_{T_R(\Delta)}$. This proves the T_{P_3}-invariance of the quotient $\theta^{4n^2}[\Delta](0,\Pi)/h_\Delta$ on these divisors.

Remaining with the point $R = P_3$, we now take the divisors $\Delta = P_1^l P_2^{n-1-l}$ with $n-4 \le l \le n-1$. Since here $T_R(\Delta) = P_1^{2n-5-l}P_2^{l+4-n}$, we compute $g_\Delta/g_{T_R(\Delta)}$ from Eq. (6.4) for $l = n-4$ and from Eq. (6.2) for $l = n-3$ to find that it is

$$\frac{\left[(\lambda_0 - \lambda_2)(\lambda_3 - \lambda_1)\right]^{2n(2l+5-n)}}{\left[(\lambda_0 - \lambda_1)(\lambda_2 - \lambda_3)\right]^{2n(2l+5-n)}}.$$

This power is $2n(n-3)$ for $l = n-4$ and $2n(n-1)$ for $l = n-3$. For both these values of l the expression for $h_{T_R(\Delta)}$ is only the power of $(\lambda_0 - \lambda_1)(\lambda_2 - \lambda_3)$; and since

$$(l+1)(n-2-l) \overset{l=n-4}{=} 2(n-3) \quad \text{and} \quad (l+2)(n-1-l) \overset{l=n-3}{=} 2(n-1)$$

we obtain the T_{P_3}-invariance also here. For the values $l = n-2$ and $l = n-1$ we use (both in the quotients $g_\Delta/g_{T_R(\Delta)}$ and $h_\Delta/h_{T_R(\Delta)}$) the reciprocals of the functions corresponding to $l = n-3$ and $l = n-4$ respectively; and we are done with the calculations concerning T_{P_3}. As for $R = P_1$ we take Δ with $l = n - 3$ again, where $T_R(\Delta) = P_1^{n-3}P_2P_3$ and Eq. (6.2) shows that $g_\Delta/g_{T_R(\Delta)}$ equals

$$\frac{\left[(\lambda_0 - \lambda_2)(\lambda_3 - \lambda_1)\right]^{2n(n-1)}}{\left[(\lambda_0 - \lambda_3)(\lambda_1 - \lambda_2)\right]^{2n(n-1)}}.$$

Then since we have $(l+2)(n-1-l) = 2(n-1)$ ($l = n - 3$ is even) we find that in either case of n modulo 4 the powers of $(\lambda_0 - \lambda_1)(\lambda_2 - \lambda_3)$ cancel out in $h_\Delta/h_{T_R(\Delta)}$ (this is easily seen using the fact that the sum of the powers appearing in the expressions for h_Δ is independent of Δ). Therefore it reduces to

$$\frac{\left[(\lambda_0 - \lambda_2)(\lambda_3 - \lambda_1)\right]^{2n(n-1)}}{\left[(\lambda_0 - \lambda_3)(\lambda_1 - \lambda_2)\right]^{2n(n-1)}}$$

and again equals $g_\Delta/g_{T_R(\Delta)}$, which proves the T_{P_1}-invariance of $\theta^{4n^2}[\Delta](0,\Pi)/h_\Delta$ on the divisors on which T_{P_1} is defined. For $R = P_2$ we take Δ with $l = n - 2$, where $T_R(\Delta) = P_1^{n-3}P_2P_3$ and by Eq. (6.2) again we see that the quotient $g_\Delta/g_{T_R(\Delta)}$ equals

$$\frac{\left[(\lambda_0 - \lambda_1)(\lambda_2 - \lambda_3)\right]^{2n(n-1)}}{\left[(\lambda_0 - \lambda_3)(\lambda_1 - \lambda_2)\right]^{2n(n-1)}}.$$

Here $(l+1)(n-2-l)$ vanishes ($l = n - 2$ is odd), and in either case of n modulo 4 only powers of $(\lambda_0 - \lambda_1)(\lambda_2 - \lambda_3)$ appear in h_Δ. Using again the fact that the sum of the powers appearing in the expressions for h_Δ is independent of Δ we see that canceling the powers of this expression from $h_{T_R(\Delta)}$ reduces this quotient to

$$\frac{\left[(\lambda_0 - \lambda_1)(\lambda_2 - \lambda_3)\right]^{2n(n-1)}}{\left[(\lambda_0 - \lambda_3)(\lambda_1 - \lambda_2)\right]^{2n(n-1)}}.$$

Since once again this equals $g_\Delta/g_{T_R(\Delta)}$, we obtain that $\theta^{4n^2}[\Delta](0,\Pi)/h_\Delta$ is invariant under T_{P_2} (when Δ is taken from the set of divisors on which T_{P_2} is defined). This completes the invariance of the quotient $\theta^{4n^2}[\Delta](0,\Pi)/h_\Delta$ under all the operators T_R.

The fact that this is the Thomae formulae for our Z_n curve now follows easily from the fact that one can get from any of our $n+2$ divisors to any other one by the operators N and T_R. Indeed, we have already seen that for $\Delta = P_1^l P_2^{n-1-l}$ with $0 \le l \le n - 3$ we have $NT_{P_3}(\Delta) = P_1^{l+2}P_2^{n-3-l}$, meaning that all these divisors with a given parity of l are in the same orbit of this action. The fact that all the divisors $P_1^l P_2^{n-1-l}$ (with both even and odd l) are in the same orbit can now be seen in three ways: Either observe that N pairs P_1^{n-1} and $P_1^{n-2}P_2$, or note that T_{P_3} pairs P_1^{n-1} and $P_1^{n-4}P_2^3$, or use the fact that T_{P_3} pairs $P_1^{n-3}P_2^2$ and $P_1^{n-2}P_2$. Applying either T_{P_1} on $P_1^{n-3}P_2^2$ or T_{P_2} on $P_1^{n-2}P_2$ adds $P_1^{n-3}P_2P_3$ into this orbit, and applying N on $P_1^{n-3}P_2P_3$

puts the remaining divisor P_3^{n-1} also in this orbit. This concludes the proof of the theorem.   □

Theorem 6.9 is the Thomae formulae for the Z_n curve X with $n \geq 5$ and the symmetric equation

$$w^n = (z - \lambda_0)(z - \lambda_1)^2 (z - \lambda_2)^{n-2}(z - \lambda_3)^{n-1},$$

and clearly contains Theorem 6.8 as its special case with $n = 7$.

We recall that in the nonsingular and singular cases we dealt with before we had two different formulae for odd and even n, but at the price of losing a bit of the strength of the formulae and also losing reduction in the odd n case we could show that the two cases were still the same in a sense. Here we deal only with odd n, but we obtain two different formulae for $n \equiv 1 (\mathrm{mod}\ 4)$ and $n \equiv 3 (\mathrm{mod}\ 4)$. Since the only difference between these cases is some number depending only on n which appears only in the powers of the expression $(\lambda_0 - \lambda_1)(\lambda_2 - \lambda_3)$, one might guess that also here this separation into cases is not "essential" in this sense. Indeed, one can verify that by taking the squares of the expressions for h_Δ in both cases and multiplying the expressions corresponding to $n \equiv 1 (\mathrm{mod}\ 4)$ by $(\lambda_0 - \lambda_1)^{2n}(\lambda_2 - \lambda_3)^{2n}$, we obtain unified Thomae formulae, stating that if we define h_Δ to be

$$[(\lambda_0 - \lambda_1)(\lambda_2 - \lambda_3)]^{2n[(n+1)^2/4 - (l+2)(n-1-l)]}[(\lambda_0 - \lambda_2)(\lambda_1 - \lambda_3)]^{2n(l+2)(n-1-l)}$$

for $\Delta = P_1^l P_2^{n-1-l}$ with even l,

$$[(\lambda_0 - \lambda_1)(\lambda_2 - \lambda_3)]^{2n[(n+1)^2/4 - (l+1)(n-2-l)]}[(\lambda_0 - \lambda_2)(\lambda_1 - \lambda_3)]^{2n(l+1)(n-2-l)}$$

for $\Delta = P_1^l P_2^{n-1-l}$ with odd l, and

$$[(\lambda_0 - \lambda_1)(\lambda_2 - \lambda_3)]^{2n(n-3)^2/4}[(\lambda_0 - \lambda_3)(\lambda_1 - \lambda_2)]^{4n(n-1)}$$

for $\Delta = P_1^{n-3} P_2 P_3$ and for $\Delta = P_3^{n-1}$, then the quotient $\theta^{16n^2}[\Delta](0, \Pi)/h_\Delta$ is independent of Δ. However, as usual we prefer to work with the strongest results possible even if it means we have to split into cases.

We now state what happens when $n = 3$, leaving the details for the reader to check. The only thing that changes in Definition 6.4 is the fact that now T_{P_1} can act also on the divisor $P_3^{n-1} = P_3^2$, and this divisor is T_{P_1}-invariant. This adds one single calculation to the proof of Proposition 6.5, which extends its validity to this divisor as well. However, since we added one action on one divisor and it is invariant under the new action, nothing changes later and Theorem 6.9 holds equally here. We also recall that this curve with $n = 3$ is the same as a singular Z_3 curve with $m = 2$, and we would like to verify consistency of all the results. Again we leave for the reader the details of verifying that if we take the square of the result of Theorem 3.13 with $n = 3$, or equivalently the results of the special case with $n = 3$ and $m = 2$ of Theorems 5.5 and 5.7 (together with the action of the operators), we obtain the same result as from Theorem 6.9 with $n = 3$. We only mention that the points denoted there by Q_1 and Q_2 (or by Q_λ and Q_∞ in Section 3.3) correspond here to P_1 and P_3, and the point denoted there by P_1 corresponds here to P_2. This means that μ_1 and μ_2 (or equivalently λ and the missing expressions with ∞) in the denominators h_Δ there should be replaced by λ_1 and λ_3 here, and λ_1 from there should be changed to λ_2 here.

We also remark that as the Z_n curves we discuss in this chapter can be considered as a different generalization of the corresponding Z_3 curve, one might expect that there is a proof for the Thomae formulae along the lines of the proof we gave for Theorem 3.13. This could be interesting, as Theorem 3.13 gave us a result which is a bit stronger than that of the general case, and also here a proof along the same lines is expected to give a stronger result. Such a proof indeed exists, but unfortunately for $n \geq 5$ it gives the same result as Theorem 6.9, without improving it. This gives another difference between $n = 3$ and $n \geq 5$. The details are left to the reader.

6.1.9 Changing the Basepoint

We now turn to changing the basepoint. Unlike the previous cases, where in the nonsingular case all the points were clearly symmetric and in the singular case we had the symmetry of replacing w by another function which showed that there as well all the points were symmetric, here this symmetry is only between P_0 and P_3 and between P_1 and P_2. P_1 and P_2 are indeed "different" from P_0 and P_3, as we shall soon see. We now describe briefly what happens with the other 3 basepoints P_i, $1 \leq i \leq 3$ and present the connections between them. For this we recall the $2n + 5$ (or 8 if $n = 3$) non-special integral divisors of degree $g = n - 1$ supported on all the branch points from Theorem 6.3, and use the part of them relevant for each basepoint. We shall not, however, write all the expressions for h_Δ here, and the interested reader can find them in Appendix B.

We begin with the basepoint P_3, which is symmetric to the basepoint P_0. Out of the $2n + 5$ (or 8 if $n = 3$) non-special integral divisors of degree $g = n - 1$ supported on the branch points from Theorem 6.3, there are again $n + 2$ which do not contain P_3 in their support. These divisors are the divisors $P_1^l P_2^{n-1-l}$ for $0 \leq l \leq n-1$, and the two divisors P_0^{n-1} and $P_0 P_1 P_2^{n-3}$ (note the symmetry to the basepoint P_0). The expressions for h_Δ in the Thomae formulae with basepoint P_3 are again separated into the cases $n \equiv 1 \pmod 4$ and $n \equiv 3 \pmod 4$, and in each case the expression for h_Δ with $\Delta = P_1^l P_2^{n-1-l}$ depends on the parity of l. We shall soon see that with the basepoints P_1 and P_2 this is not the case, which emphasizes the lack of symmetry between them.

Now we want to see what happens with the basepoints P_1 and P_2. We begin with the basepoint P_1, where Theorem 6.3 gives $n + 2$ non-special integral divisors of degree $g = n - 1$ supported on the branch points which do not contain P_1 in their support. Here these are the divisors $P_0^l P_3^{n-1-l}$ for $0 \leq l \leq n-1$, and the two divisors

P_2^{n-1} and $P_0^{(n-3)/2}P_2P_3^{(n-1)/2}$. A first indication for the lack of symmetry of this point to the points P_0 and P_3 is the fact that the second exceptional divisor no longer contains a point to the power $n-3$ (for $n \geq 7$ at least), but to smaller powers. As usual, the Thomae formulae with basepoint P_1 is based on expressions for h_Δ which are still defined slightly differently in the cases $n \equiv 1 \pmod 4$ and $n \equiv 3 \pmod 4$, but here the expression for h_Δ with $\Delta = P_0^l P_3^{n-1-l}$ does not depend on the parity of l but on the question whether $0 \leq l \leq (n-3)/2$ or $(n-1)/2 \leq l \leq n-1$. This different behavior emphasizes the lack of symmetry we are talking about.

As for the basepoint P_2, we see that from the divisors from Theorem 6.3 there are $n+2$ which do not contain P_2 in their support. The list of these divisors contains $P_0^l P_3^{n-1-l}$ for $0 \leq l \leq n-1$, and the two divisors P_1^{n-1} and $P_0^{(n-1)/2}P_1P_3^{(n-3)/2}$. Examining the powers of the second exceptional divisor relates this point to P_1, and not to P_0 and P_3, as expected. Then in the Thomae formulae with basepoint P_2 we have the usual splitting into the cases $n \equiv 1 \pmod 4$ and $n \equiv 3 \pmod 4$, and once again, like with the basepoint P_1 symmetric to P_2, the expression for h_Δ with $\Delta = P_0^l P_3^{n-1-l}$ depends on inequalities and not on the parity of l. The explicit inequalities in this case are whether $0 \leq l \leq (n-1)/2$ or $(n+1)/2 \leq l \leq n-1$. We just remark that this asymmetry appears only for $n \geq 7$ (look at the divisors to get a feeling of this), and indeed for $n = 3$ and $n = 5$ all four branch points are symmetric. For $n = 3$ we already saw this, and for $n = 5$ we give a few details at the end of Appendix B.

As in the previous cases, the Thomae formulae for the four possible basepoints P_i, $0 \leq i \leq 3$, give four constants, which in principal can be different, and we want to show that they are equal. Since we have not given the expressions for h_Δ with the basepoints P_i, $1 \leq i \leq 3$, we cannot do it in detail here, but the interested reader can find these expressions in Appendix B and do the necessary comparisons. We just describe the process that should be done for this. First note that Lemma 6.2 means that also for our Z_n curve X the conditions of Corollary 1.13 are satisfied, and it indeed can be applied to prove that some characteristics with the basepoint P_0 coincide with characteristics with some other base P_i, $1 \leq i \leq 3$. Since for applying Corollary 1.13 for this purpose we must have P_i appearing to the power $n-1$ in the representing divisor Δ, we must take $\Delta = P_i^{n-1}$ and this gives us that Γ_{P_i} from Corollary 1.13 is P_0^{n-1} for every $1 \leq i \leq 3$. The reader can now verify, using the expressions from Appendix B, that for every $0 \leq i \leq 3$, the expression $h_{\Gamma_{P_i}} = h_{P_0^{n-1}}$ with the basepoint P_i equals the expression $h_\Delta = h_{P_i^{n-1}}$ with the basepoint P_0. This shows that the Thomae formulae with the four different basepoints all give the same constants.

As in all the previous cases, we again would like to prove also that the four Thomae formulae involve the same characteristics of the theta constants. For this we give the corresponding analog of Propositions 1.14 and 1.15.

Proposition 6.10. *Let Δ be any of the $n+2$ divisors from Theorem 6.3 that do not contain P_0 in their support. Then for every point P_i, $1 \leq i \leq 3$, there is a divisor Γ_{P_i} appearing in Theorem 6.3, but with P_i not included in its support instead of P_0, and such that the equality $\varphi_{P_0}(\Delta) + K_{P_0} = \varphi_{P_i}(\Gamma_{P_i}) + K_{P_i}$ holds.*

We shall not write the details of the proof of Proposition 6.10 as it involves checking many cases, but just indicate how this should be done and give the relations thus obtained. We remember that the second statement of Corollary 1.13 allows us to change the basepoint freely for the divisors $P_0^{n-1}\Delta$, so we use it to write

$$\varphi_{P_0}(\Delta)+K_{P_0} = \varphi_{P_0}(P_0^{n-1}\Delta)+K_{P_0} = \varphi_{P_i}(P_0^{n-1}\Delta)+K_{P_i}$$

for any basepoint P_i to which we want to change, and then we search for the divisor Γ_{P_i} such that $P_0^{n-1}\Delta$ is equivalent (and hence gives the same value when φ_{P_i} is applied) to $P_i^{n-1}\Gamma_{P_i}$. The tools one applies for finding this divisor Γ_{P_i} are Lemma 6.2 (which allows us to put nth powers of branch points freely in and out of these divisors) and the fact that $P_0P_1^2/P_3P_2^2$ was seen to be principal. Then one multiplies by the relevant power of $P_0P_1^2/P_3P_2^2$ in order to obtain P_i to the power $n-1$ or -1, and then correct to integrality and powers not exceeding n by multiplying or dividing by expressions of the form P_j^n to get the desired results. The equalities one obtains when proving Proposition 6.10 are as follows. For even $0 \le l \le n-3$ we have

$$\varphi_{P_0}(P_1^l P_2^{n-1-l})+K_{P_0} = \varphi_{P_1}(P_0^{(n-3-l)/2}P_3^{(n+1+l)/2})+K_{P_1}$$
$$= \varphi_{P_2}(P_0^{n-1-l/2}P_3^{l/2})+K_{P_2} = \varphi_{P_3}(P_1^{l+2}P_2^{n-3-l})+K_{P_3},$$

for odd $1 \le l \le n-4$ we have

$$\varphi_{P_0}(P_1^l P_2^{n-1-l})+K_{P_0} = \varphi_{P_1}(P_0^{n-1-(l+1)/2}P_3^{(l+1)/2})+K_{P_1}$$
$$= \varphi_{P_2}(P_0^{(n-2-l)/2}P_3^{(n+l)/2})+K_{P_2} = \varphi_{P_3}(P_1^{l+2}P_2^{n-3-l})+K_{P_3},$$

and for the four remaining divisors we have

$$\varphi_{P_0}(P_1^{n-2}P_2)+K_{P_0} = \varphi_{P_1}(P_0^{(n-1)/2}P_3^{(n-1)/2})+K_{P_1} = \varphi_{P_2}(P_3^{n-1})+K_{P_2} = \varphi_{P_3}(P_2^{n-1})+K_{P_3},$$

$$\varphi_{P_0}(P_1^{n-1})+K_{P_0} = \varphi_{P_1}(P_0^{n-1})+K_{P_1} = \varphi_{P_2}(P_0^{(n-1)/2}P_3^{(n-1)/2})+K_{P_2} = \varphi_{P_3}(P_1P_2^{n-2})+K_{P_3},$$

$$\varphi_{P_0}(P_3^{n-1})+K_{P_0} = \varphi_{P_1}(P_0^{(n-3)/2}P_2P_3^{(n-1)/2})+K_{P_1}$$
$$= \varphi_{P_2}(P_0^{(n-1)/2}P_1P_3^{(n-3)/2})+K_{P_2} = \varphi_{P_3}(P_0^{n-1})+K_{P_3},$$

and

$$\varphi_{P_0}(P_1^{n-3}P_2P_3)+K_{P_0} = \varphi_{P_1}(P_2^{n-1})+K_{P_1} = \varphi_{P_2}(P_1^{n-1})+K_{P_2} = \varphi_{P_3}(P_0P_1P_2^{n-3})+K_{P_3}.$$

The reader is encouraged to check that for every divisor Δ with the basepoint P_0 and every basepoint P_i, the denominator $h_{\Gamma_{P_i}}$ with basepoint P_i from Appendix B equals the denominator h_Δ with basepoint P_0 from Theorem 6.9, and should note that this includes checking some inequalities for the divisors Γ_{P_1} and Γ_{P_2} and some parity conditions for the divisors Γ_{P_3}. More details of this verification can be found in Appendix B.

We conclude this discussion with the following remark. One could have proved only the Thomae formulae for the basepoint P_0 as proved in Theorem 6.9, and then used the relations in Proposition 6.10 to define the denominators $h_{\Gamma_{P_i}}$ for all the divisors with the other branch points. This proof, although shorter, misses the idea (which we wish to emphasize) that the Thomae formulae are equalities that can be obtained from any basepoint. The calculations in Proposition 6.10 and following it should be considered more as a consistency check than as a construction for the Thomae formulae for the other basepoints. We also state that one can write the Thomae formulae from Theorem 6.9 in a basepoint-independent notation, but since this does not give much insight or look any nicer, we shall not do it here.

6.2 A Family of Z_n Curves with Four Branch Points and an Asymmetric Equation

In this section we deal with Z_n curves again with $t = 4$ but with the powers $\alpha_1 = 1$, $\alpha_2 = 1$, $\alpha_3 = 1$ and $\alpha_4 = n - 3$. In this case we shall depart from our practice of taking the basepoint to appear to the power 1, so we leave the indexing as it is. However, in order to emphasize the different role of the point λ_4 we shall denote it simply by λ. Thus the algebraic equation is written as

$$w^n = (z - \lambda_1)(z - \lambda_2)(z - \lambda_3)(z - \lambda)^{n-3},$$

and we denote the point over λ_i with $1 \leq i \leq 3$, as usual, by P_i, while the point over $\lambda_4 = \lambda$ will be denoted by Q_λ (for more emphasis on its different behavior). Note that since we want full ramification the powers α_i must be relatively prime to n, so we shall assume throughout this section that n is not divisible by 3. The Z_n curve thus described is a compact Riemann surface, and by the general formula for totally ramified Z_n curves its genus is $g = (n-1)(4-2)/2 = n - 1$. We note that $n = 4$ takes us back to the nonsingular curve with $r = 1$ with which we dealt in Section 3.4, so that once again we get new results only for $n \geq 5$. We now proceed as usual to compute the divisors of the usual objects, where the points lying over ∞ will be denoted by ∞_h with $1 \leq h \leq n$ as always. The interesting aspect of this example will be that the number of divisors we obtain will be independent of n. Since the points P_1, P_2 and P_3 are clearly symmetric, we shall for convenience use the notation P_i, P_j and P_k where any time this notation appears we assume, as in previous cases, that i, j and k are 1, 2 and 3 and distinct.

6.2.1 An Example with $n = 10$

Let us start with an example. We take $n = 10$, so that the equation becomes

$$w^{10} = (z - \lambda_1)(z - \lambda_2)(z - \lambda_3)(z - \lambda)^7,$$

and the genus is $g = 10 - 1 = 9$. We calculate that

$$\operatorname{div}(z - \lambda) = \frac{Q_\lambda^{10}}{\prod_{h=1}^{10} \infty_h}, \quad \operatorname{div}(dz) = \frac{P_1^9 P_2^9 P_3^9 Q_\lambda^9}{\prod_{h=1}^{10} \infty_h^2}, \quad \text{and} \quad \operatorname{div}(w) = \frac{P_1 P_2 P_3 Q_\lambda^7}{\prod_{h=1}^{10} \infty_h},$$

from which we find that

$$\operatorname{div}\left(\frac{dz}{w}\right) = \frac{P_1^8 P_2^8 P_3^8 Q_\lambda^2}{\prod_{h=1}^{10} \infty_h} \quad \text{and} \quad \operatorname{div}\left(\frac{dz}{w^2}\right) = \frac{P_1^7 P_2^7 P_3^7}{Q_\lambda^5}.$$

We recall that in all the previous examples we found a meromorphic function on X of relatively small degree, and used it to construct more holomorphic differentials from a given one. Here we do not have a holomorphic differential yet, but this function will help us to obtain also the first holomorphic differential from the meromorphic differential dz/w^2. In this case we have

$$\operatorname{div}\left(\frac{z - \lambda}{w}\right) = \frac{Q_\lambda^3}{P_1 P_2 P_3},$$

and by multiplying the differential dz/w^2 by the lth power of this function we find that

$$\operatorname{div}\left((z - \lambda)^l \frac{dz}{w^{l+2}}\right) = P_1^{7-l} P_2^{7-l} P_3^{7-l} Q_\lambda^{3l-5}.$$

This gives us a holomorphic differential for every $2 \leq l \leq 7$, and explicitly

$$\operatorname{div}\left((z - \lambda)^2 \frac{dz}{w^4}\right) = P_2^5 P_2^5 P_3^5 Q_\lambda, \quad \operatorname{div}\left((z - \lambda)^3 \frac{dz}{w^5}\right) = P_1^4 P_2^4 P_3^4 Q_\lambda^4,$$

$$\operatorname{div}\left((z - \lambda)^4 \frac{dz}{w^6}\right) = P_1^3 P_2^3 P_3^3 Q_\lambda^7, \quad \operatorname{div}\left((z - \lambda)^5 \frac{dz}{w^7}\right) = P_1^2 P_2^2 P_3^2 Q_\lambda^{10},$$

$$\operatorname{div}\left((z - \lambda)^6 \frac{dz}{w^8}\right) = P_1 P_2 P_3 Q_\lambda^{13}, \quad \text{and} \quad \operatorname{div}\left((z - \lambda)^7 \frac{dz}{w^9}\right) = Q_\lambda^{16}.$$

This gives us 6 holomorphic differentials, but in order to obtain a basis for the space of holomorphic differentials (whose dimension is $g = 9$) we must first have 9 differentials. For this we note that in the last three divisors Q_λ appears to the power 10 or more, which means that by taking out one power of $z - \lambda$ we still have a holomorphic differential. This gives us also the 3 holomorphic differentials

$$\operatorname{div}\left((z - \lambda)^4 \frac{dz}{w^7}\right) = P_1^2 P_2^2 P_3^2 \prod_{h=1}^{10} \infty_h, \quad \operatorname{div}\left((z - \lambda)^5 \frac{dz}{w^8}\right) = P_1 P_2 P_3 Q_\lambda^3 \prod_{h=1}^{10} \infty_h,$$

and

$$\text{div}\left((z-\lambda)^6\frac{dz}{w^9}\right) = Q_\lambda^6 \prod_{h=1}^{10} \infty_h,$$

so together we indeed have $g = 9$ holomorphic differentials. These differentials are in fact linearly independent and compose a basis adapted to the point Q_λ. This basis, however, is not adapted to the points P_i, $1 \le i \le 3$. It is easy to verify (again by looking at the powers) that when replacing the last three differentials by

$$\text{div}\left((z-\lambda_i)(z-\lambda)^4\frac{dz}{w^7}\right) = P_i^{12}P_j^2P_k^2, \quad \text{div}\left((z-\lambda_i)(z-\lambda)^5\frac{dz}{w^8}\right) = P_i^{11}P_jP_kQ_\lambda^3,$$

and

$$\text{div}\left((z-\lambda_i)(z-\lambda)^6\frac{dz}{w^9}\right) = P_i^{10}Q_\lambda^6$$

we obtain, together with the first six differentials, a basis adapted to P_i, that is still also adapted to Q_λ.

A useful piece of information we draw from this is that Lemma 2.7 holds for this Z_{10} curve as well. We write $\Omega(1) = \bigoplus_{l=4}^9 \Omega_l(1)$ where in $\Omega_l(1)$ we put the divisors which are polynomials multiplied by dz/w^l. This means that $\Omega_l(1)$ is 1-dimensional for $4 \le l \le 6$ and 2-dimensional for $7 \le l \le 9$, and if one insists on decomposing $\Omega(1)$ into $n - 1 = 9$ subspaces, then $\Omega_l(1)$ is 0-dimensional for $1 \le l \le 3$. We note that for any divisor in $\Omega_l(1)$ the power to which P_1, P_2 and P_3 appear is either $9 - l$ or $19 - l$ and the power to which Q_λ appears is either $3l - 11$ or $3l - 21$. Since in any case the residues of these numbers modulo 10 determine l (as 3 is prime to 10), Lemma 2.7 holds for our Z_{10} curve with the same proof. We shall later see that this is true for general n.

Since we have bases adapted to all the branch points, Proposition 1.6 shows that we can obtain the Weierstrass gap sequences at the branch points by adding 1 to the orders of the differentials in the corresponding points. In this way we find that the Weierstrass gap sequence at the branch points P_1, P_2 and P_3 is 1,2,3,4,5,6,11,12,13 and the Weierstrass gap sequence at the branch point Q_λ is 1,2,4,5,7,8,11,14,17. In particular this shows that 3 is the smallest non-gap at Q_λ and that 7 is the smallest non-gap at the other three branch points. This means that there is a meromorphic function on X which has a pole only at Q_λ and of order 3, and that for $1 \le i \le 3$ there is a function which has a pole only at P_i and of order exactly 7. We have already seen the first function—it is $w/(z - \lambda)$, whose divisor is $P_1P_2P_3/Q_\lambda^3$. The function for the point P_i is $w^3/(z - \lambda_i)(z - \lambda)^2$ with divisor $P_j^3P_k^3Q_\lambda/P_i^7$. By multiplying these functions by $(z - \lambda)/w$ we also obtain the functions which have poles only at P_i and of orders 8, 9 and 10 (for the latter we obtain the function $(z - \lambda)/(z - \lambda_i)$, a linear combination of the constants and $1/(z - \lambda_i)$), but we shall not use them. These relations, $Q_\lambda^3 \equiv P_1P_2P_3$ and $P_i^7 \equiv P_j^3P_k^3Q_\lambda$ for $1 \le i \le 3$ (together with the relations obtained for P_i^8 and P_i^9 by the corresponding relations), can also be obtained by recalling that all the divisors of the holomorphic differentials are equivalent and finding these relations there. However, what will be important for us from this discussion is the fact that $r(1/Q_\lambda^3) \ge 2$ and for every $1 \le i \le 3$, $r(1/P_i^7) \ge 2$ as well.

We shall also make use of the fact that $r(1/P_1P_2P_3) \geq 2$, as seen from the divisor of the function $(z - \lambda)/w$.

6.2.2 Non-Special Divisors for $n = 10$

The Thomae formulae are based, as usual, on non-special integral divisors of degree $g = n - 1$ supported on the set of branch points (with one of them, the basepoint, omitted). Before turning to the general case we continue with our example of $n = 10$ and $g = 9$, and see what are these divisors in this case. The result we obtain is that there are exactly 18 non-special integral divisors of degree $g = 9$ supported on the branch points, and they are divided into three sets. The first contains the 6 divisors $P_i^6 P_j^3$, the second contains the 6 divisors $P_i^6 P_j^2 Q_\lambda$, and the third contains the 6 divisors $P_i^5 P_j^2 Q_\lambda^2$. In order to see why this is true we show explicitly that all these divisors are non-special and all the other integral divisors of degree 9 supported on the branch points are special.

First, we show that these divisors are all non-special. We shall use Lemma 2.7. If Δ is any of these divisors then $\Omega_l(\Delta) = 0$ for any $4 \leq l \leq 6$ since the divisor of the differential spanning $\Omega_l(1)$ for any such value of l is not a multiple of Δ. This is so, because either Δ contains the point P_i to the power 6, or only to the power 5 (which allows $\Omega_4(\Delta)$ to be nontrivial) but also Q_λ to the power 2 (which eliminates this possibility). Moreover, for the same reason we see that when checking $\Omega_l(\Delta)$ for $7 \leq l \leq 9$ we must stick to the 1-dimensional subspace of $\Omega_l(1)$ spanned by the differential which vanishes to high order (10, 11 or 12) at P_i. Looking at the divisors of these differentials we see that none of them vanishes at P_j to the 3rd order or more (which shows that the first 6 divisors are non-special), and if they vanish at Q_λ then they do not vanish at P_j to the second order or more (which shows that the remaining 12 divisors are non-special). This shows that all our divisors are non-special.

We now want to show that these are the only non-special divisors. First, By the remarks preceding the theorem we find that if Δ is an integral divisor of degree 9, which is divisible by a divisor Ξ that is either Q_λ^3, P_1^7, P_2^7, P_3^7, or $P_1P_2P_3$, then Δ is special. This is seen by considerations similar to those which allowed us to obtain that multiples of $P_0 P_1^2$ or $P_2^2 P_3$ were special in Section 6.1. This leaves us with only a few possibilities to find candidates for non-special divisors: Q_λ can appear to a power not exceeding 2, not all the points P_1, P_2, and P_3 can appear, and every one of them which appears can appear to a power not exceeding 6.

We thus split the possible divisors by the power (0, 1, or 2) to which Q_λ appears in them. If Q_λ does not appear in the divisor, then it contains only two other branch points, and they must appear to powers not exceeding 6 which sum to 9. This means either 6 and 3 or 5 and 4. This gives the 6 non-special divisors $P_i^6 P_j^3$ from the first set, together with the divisors $P_i^5 P_j^4$. In the case where Q_λ appears to the power 1 in the divisors we see that the two other points must appear to powers not exceeding 6 which sum to 8. This means either 6 and 2, 5 and 3 or 4 and 4. This gives the 6 non-special divisors $P_i^6 P_j^2 Q_\lambda$ from the second set, together with the 6 divisors $P_i^5 P_j^3 Q_\lambda$ and the 3 divisors $P_i^4 P_j^4 Q_\lambda$. After removing from this list all those divisors which we already know that are non-special, the remaining divisors are all special since the divisor $P_1^5 P_2^5 P_3^5 Q_\lambda$ of the holomorphic differential $(z-\lambda)^2 dz/w^4$ is a multiple of any one of them. It remains to see what happens in the case where Q_λ appears to the power 2 in the divisors.

If Q_λ appears to the power 2 in the divisors, then the two other points must appear to powers not exceeding 6 which sum to 7. This means either 6 and 1, 5 and 2 or 4 and 3. This gives us the 6 non-special divisors $P_i^5 P_j^2 Q_\lambda^2$ from the third set, together with the 6 divisors $P_i^6 P_j Q_\lambda^2$ and the 6 divisors $P_i^4 P_j^3 Q_\lambda^2$. Here the last 6 divisors are all special as the divisor $P_1^4 P_2^4 P_3^4 Q_\lambda^4$ of the holomorphic differential $(z-\lambda)^3 dz/w^5$ is a multiple of any one of them (and also of the 3 divisors $P_i^4 P_j^4 Q_\lambda$ from the previous paragraph). For the remaining 6 divisors, we see that the divisor $P_i^{11} P_j P_k Q_\lambda^3$ (with the corresponding i) of the holomorphic differential $(z-\lambda_i)(z-\lambda)^5 dz/w^8$ is a multiple of it. This means that the 18 divisors we have written at the start are indeed the only non-special divisors. In the general case we shall see (in Theorem 6.13 below) the interesting fact that this number 18 of the divisors will be the same for every n (except for some small values of n), unlike any of the previous cases.

6.2.3 The Basic Data for General n

Let us now turn to the case of general n. We shall see from the beginning that things look slightly different for $n \equiv 1 \pmod 3$ and $n \equiv 2 \pmod 3$. The best way to deal with both cases simultaneously is to divide n by 3 with remainder, i.e., write $n = 3s + t$ with $s \geq 1$ (as we want $n \geq 3$) integral, and t, the residue, being either 1 or 2. We shall use in the following both the notation n when convenient and the notation with s and t when it makes a difference.

As usual, we begin with finding the divisors of the useful functions and differentials on our Z_n curve X. We find that

$$\mathrm{div}(z-\lambda) = \frac{Q_\lambda^n}{\prod_{h=1}^n \infty_h}, \quad \mathrm{div}(dz) = \frac{P_1^{n-1} P_2^{n-1} P_3^{n-1} Q_\lambda^{n-1}}{\prod_{h=1}^n \infty_h^2},$$

and

$$\mathrm{div}(w) = \frac{P_1 P_2 P_3 Q_\lambda^{n-3}}{\prod_{h=1}^n \infty_h},$$

and we easily calculate that

$$\mathrm{div}\left(\frac{dz}{w}\right) = \frac{P_1^{n-2} P_2^{n-2} P_3^{n-2} Q_\lambda^2}{\prod_{h=1}^n \infty_h} \quad \text{and} \quad \mathrm{div}\left(\frac{dz}{w^2}\right) = \frac{P_1^{n-3} P_2^{n-3} P_3^{n-3}}{Q_\lambda^{n-5}}.$$

As we have seen in the example of $n = 10$, the fact that

$$\text{div}\left(\frac{z - \lambda}{w}\right) = \frac{Q_\lambda^3}{P_1 P_2 P_3}$$

(a formula that holds for general n here) was very useful and allowed us to construct holomorphic differentials from the meromorphic differentials that we have. Multiplying dz/w^2 by the lth power of this function yields

$$\text{div}\left((z - \lambda)^l \frac{dz}{w^{l+2}}\right) = P_1^{n-3-l} P_2^{n-3-l} P_3^{n-3-l} Q_\lambda^{3l-(n-5)}.$$

In order for this to be a holomorphic differential we must have that both $n - 3 - l$ and $3l - (n-5) = 3(l - (s-1)) + (2 - t)$ (recall that $n = 3s + t$) will be nonnegative. This happens exactly for $s - 1 \leq l \leq n - 3$ (recall that t is either 1 or 2, so that $2 - t$ is either 1 or 0—nonnegative but smaller than 3). Note that when $n = 10$ we have $s = 3$ and $t = 1$, and indeed the holomorphic differentials we had there were in the range $2 \leq l \leq 7$.

Substituting $l = m + s - 1$ we write the differential corresponding to l and its divisor as

$$\text{div}\left((z - \lambda)^{s-1+m} \frac{dz}{w^{s+1+m}}\right) = P_1^{2s+t-2-m} P_2^{2s+t-2-m} P_3^{2s+t-2-m} Q_\lambda^{3m+2-t}.$$

This is a holomorphic differential for every $0 \leq m \leq 2s + t - 2$. In particular we notice that for $m = 2s + t - 2$ (so that $m + s = n - 2$ and $3m = 2n + t - 6$) we obtain the holomorphic differential $(z - \lambda)^{n-3} dz/w^{n-1}$ whose divisor is Q_λ^{2n-4} (and indeed $2g - 2 = 2n - 4$). Later we shall also have use of the differential corresponding to $m = 0$, whose divisor is $P_1^{2s+t-2} P_2^{2s+t-2} P_3^{2s+t-2} Q_\lambda^{2-t}$. We also note that when $m \geq s + t - 1$ we find that

$$3m + 2 - t \geq 3s + 2t - 1 = n + t - 1 \geq n$$

(as $t \geq 1$), meaning that for these values of m the point Q_λ appears in the divisors with order n or more. Therefore by taking out one power of $z - \lambda$ we still have a holomorphic differential. Thus we get, for every $s + t - 1 \leq m \leq 2s + t - 2$,

$$\text{div}\left((z - \lambda)^{s-2+m} \frac{dz}{w^{s+1+m}}\right) = P_1^{2s+t-2-m} P_2^{2s+t-2-m} P_3^{2s+t-2-m} Q_\lambda^{3m+2-t-n} \prod_{h=1}^{n} \infty_h,$$

and this is a holomorphic differential. It will be easier to analyze these differentials if we write $m = s + t - 1 + r$, which means that for every $0 \leq r \leq s - 1$ we have

$$\text{div}\left((z - \lambda)^{2s+t-3+r} \frac{dz}{w^{2s+t+r}}\right) = P_1^{s-1-r} P_2^{s-1-r} P_3^{s-1-r} Q_\lambda^{3r+t-1} \prod_{h=1}^{n} \infty_h,$$

where we have used the equality $n = 3s + t$ in the power of Q_λ.

Let us now count how many holomorphic differentials we have constructed. We have first constructed $2s + t - 1$ differentials, and then added exactly s differentials (recall that for $n = 10$ we indeed had 6 divisors and added 3 more). Altogether we have $3s + t - 1 = n - 1 = g$ holomorphic differentials, and we claim that they are linearly independent and form a basis for the holomorphic differentials on X. In fact we claim even more:

Lemma 6.11. *The holomorphic differentials we have listed form a basis for the holomorphic differentials on X that is adapted to the branch point Q_λ. If we replace each of the last s differentials, i.e., with $s + t - 1 \leq m \leq 2s + t - 2$, by $(z - \lambda_i)(z - \lambda)^{s-2+m} dz / w^{s+1+m}$ for some $1 \leq i \leq 3$ and the corresponding m, we obtain a basis that is adapted to the branch point P_i (and that remains adapted also to Q_λ). The gap sequence at each of the points P_1, P_2, and P_3 is composed of the numbers between 1 and $2s + t - 1$ and the numbers between $n + 1$ and $n + s$, while the gap sequence at the branch point Q_λ is composed of the numbers $3k + (3 - t)$ with $0 \leq k \leq 2s + t - 2$ and the numbers $3k + t$ with $0 \leq k \leq s - 1$.*

Proof. Clearly the powers to which P_i appears in the differentials are the numbers between 0 and $2s + t - 2$ in the first set of differentials. For the last s differentials, writing them with the index $r = m - s - t + 1$ yields

$$\mathrm{div}\left((z - \lambda_i)(z - \lambda)^{2s+t-3+r} \frac{\mathrm{d}z}{w^{2s+t+r}} \right) = P_i^{s-1-r+n} P_j^{s-1-r} P_k^{s-1-r} Q_\lambda^{3r+t-1},$$

from which we obtain that the powers to which P_i appears in the second set of differentials are the numbers between n and $n + s - 1$. Clearly these numbers are all distinct, which means that these differentials form a basis adapted to P_i. As for the point Q_λ, we see that all the powers $3m + 2 - t$ to which it appears in the first set of differentials are congruent to $2 - t$ modulo 3 and distinct and all the powers $3r + t - 1$ in the second set of differentials (in either basis) are congruent to $t - 1$ modulo 3 and distinct. Since $2 - t \neq t - 1$ ($2 - 1 = 1$ while $1 - 1 = 0$ for $t = 1$ and $2 - 2 = 0$ while $2 - 1 = 1$ for $t = 2$), each of these bases is adapted to the point Q_λ. The gap sequence is obtained as usual from Proposition 1.6 by adding 1 to each of these orders, where for Q_λ we have put $k = m$ for $0 \leq m \leq 2s + t - 2$ in the first set and $k = r$ for $0 \leq r \leq s - 1$ in the second set. $\square$

Explicitly, the gap sequence at Q_λ is composed of all the numbers up to $3s = n - t$ which are not divisible by 3, followed by the numbers $3k + 2$ with $s \leq k \leq 2s - 1$ if $t = 1$ or by the numbers $3k + 1$ with $s \leq k \leq 2s$ if $t = 2$. Altogether this gives $2s + s = n - 1 = g$ numbers if $t = 1$ and $2s + s + 1 = n - 1 = g$ numbers if $t = 2$.

As an important corollary of Lemma 6.11 we now show that Lemma 2.7 holds here as well. We write $\Omega(1) = \bigoplus_{l=s+1}^{n-1} \Omega_l(1)$ where $\Omega_l(1)$ contains the divisors that are polynomials multiplied by $\mathrm{d}z / w^l$. Hence the dimension of $\Omega_l(1)$ is 1 for $s + 1 \leq l \leq 2s + t - 1$ and 2 for $2s + t \leq l \leq n - 1$ (and one might add 0 for $1 \leq l \leq s$). The power to which P_1, P_2 and P_3 appear in the divisor of any differential from $\Omega_l(1)$ is either $n - 1 - l$ or $2n - 1 - l$ and the power to which Q_λ appears in such a divisor is either $3l - n - 1$ or $3l - 2n - 1$ (this is clear from the calculation of the meromorphic differentials in the beginning). The residues of these numbers modulo n determine l for any branch point (for the points P_i, $1 \leq i \leq 3$, this is clear, and for the point Q_λ use the fact that 3 is prime to n), and therefore Lemma 2.7 holds for our Z_n curve with the same proof.

The information that we have here about elements of finite order in the jacobian $J(X)$ is as follows.

Lemma 6.12. *For any branch point S, the vector K_S of Riemann constants is a point of order $2n$ in the jacobian variety, while if $S = Q_\lambda$ it is of order 2 (recall that we are using the term "x is a point of order N" to denote that $Nx = 0$ without the minimality condition on N, so there is no contradiction between these statements for $S = Q_\lambda$). Furthermore, the image of any other branch point R by the Abel–Jacobi map φ_S with basepoint R is a point of order n in the jacobian variety.*

Proof. The second statement holds as in the previous cases since the divisors of the functions $(z - \lambda_i)/(z - \lambda_j)$, $(z - \lambda_i)/(z - \lambda)$, and $(z - \lambda)/(z - \lambda_i)$ are P_i^n/P_j^n, P_i^n/Q_λ^n, and Q_λ^n/P_i^n respectively, showing that all these divisors are principal. For the first we take any canonical divisor based on the branch points, and since its image in $J(X)$ by the Abel–Jacobi map φ_S, which is $-2K_S$, is of order n (being the sum of $2g - 2$ such elements), we conclude that K_S is of order $2n$ in $J(X)$. If $S = Q_\lambda$, then taking the differential $(z - \lambda)^{n-3} dz/w^{n-1}$ gives the divisor Q_λ^{2n-4}, whose image by φ_{Q_λ} vanishes. Therefore $-2K_{Q_\lambda} = 0$ and the lemma is proved. $\square$

We saw in the case $n = 10$ that knowing the smallest gaps at the branch points was useful to understanding what are the non-special divisors we are looking for. From Lemma 6.11 we find that 3 is the smallest non-gap at Q_λ and $2s + t$ is the smallest non-gap at the other three branch points, so that there is a function which has a pole only at Q_λ and to order 3 and for every $1 \leq i \leq 3$ there is a function which has a pole only at P_i and to order exactly $2s + t$. The first function is once again $w/(z - \lambda)$ with divisor $P_1 P_2 P_3/Q_\lambda^3$, and the function for the point P_i is $w^s/(z - \lambda_i)(z - \lambda)^{s-1}$ with divisor $P_j^s P_k^s Q_\lambda^t/P_i^{2s+t}$ (recall that $3s + t = n$ again). As in the case $n = 10$ we can obtain the function whose only pole at P_i is of any given order between $2s + t + 1$ and n (again the latter gives the familiar function $(z - \lambda)/(z - \lambda_i)$) from this function by multiplication by $(z - \lambda)/w$, and we can also find relations between these divisors using the holomorphic differentials. However, the only information that we shall use is the fact that $r(1/\Xi) \geq 2$ for Ξ being Q_λ^3, P_1^{2s+t}, P_2^{2s+t}, P_3^{2s+t} or $P_1 P_2 P_3$ (for the last Ξ the space $L(1/\Xi)$ includes the function $(z - \lambda)/w$ together with the constant ones, as we saw in the special case with $n = 10$). Recall that for $n = 10$, with $s = 3$ and $t = 1$, we have $2s + t = 7$ as we saw explicitly in that example.

6.2.4 Non-Special Divisors for General n

We now turn to finding the non-special integral divisors of degree $g = n - 1$ supported on the branch points. As in the example of $n = 10$, we shall be using the information from the previous paragraph. We can state and prove the results for both values of t, hence we do so.

Theorem 6.13. *The non-special integral divisors of degree $g = n - 1$ supported on the branch points are divided into three sets, where the first set contains the divisors $P_i^{2s+t-1}P_j^s$, the second set contains the divisors $P_i^{2s}P_j^{s+t-2}Q_\lambda$, and the third set contains the divisors $P_i^{2s+t-2}P_j^{s-1}Q_\lambda^2$. Altogether, for $n \geq 7$ (i.e., $s \geq 2$) we have 18 such divisors, while for $n = 5$ we have only 15 divisors and for $n = 4$ we have only 12.*

Proof. We first show, using Lemma 2.7, that all these divisors are non-special. We begin by showing that for every divisor Δ appearing in this list we have $\Omega_l(\Delta) = 0$ for any $s + 1 \leq l \leq 2s + t - 1$. Since in any case the power to which P_i appears is at least $2s + t - 2$ ($2s \geq 2s + t - 2$ as well since $t \leq 2$), the statement holds for every $s + 2 \leq l \leq 2s + t - 1$ (and thus is trivial for $n = 4$, but this should not bother us) and we only have to check $l = s + 1$. Since this power is $2s + t - 1$ in the divisors from the first set and in the case $t = 1$ also in those from the second set, we only have to verify that $\Omega_{s+1}(\Delta) = 0$ for the divisors from the third set and in the case $t = 2$ also for those from the second set. For this we note that in the divisor of any differential spanning $\Omega_{s+1}(1)$ the branch point Q_λ appears to the power $2 - t$. Since this power is at most 1 (since $t \geq 1$) the statement is true for the divisors from the third set in any case, and since for $t = 2$ this power is 0 we have it for the divisors from the second set in this case as well. This deals with the spaces $\Omega_l(\Delta) = 0$ for any $s + 1 \leq l \leq 2s + t - 1$.

We now show that $\Omega_l(\Delta) = 0$ also for $2s + t \leq l \leq n - 1$. The same considerations from the previous paragraph show that we only have to check the 1-dimensional subspace of $\Omega_l(1)$ in which P_i appears to a power at least n. Then the fact that P_j appears to the power at least $s - 1$ (the fact that $t \geq 1$ shows that $s + t - 2 \geq s - 1$ as well) proves that $\Omega_l(\Delta) = 0$ for every $2s + t + 1 \leq l \leq n - 1$ (this is trivial for $s = 1$, i.e., for $n = 4$ or $n = 5$, a fact which we can also ignore), leaving only the 1-dimensional subspace of $\Omega_{2s+t}(1)$ to verify. Since in the divisors from the first set, and also in the divisors from the second set if $t = 2$, the power to which P_j appears is actually s, these cases are covered as well. In order to see why $\Omega_{2s+t}(\Delta) = 0$ also in the remaining cases we note that every differential spanning the 1-dimensional subspace of $\Omega_{2s+t}(1)$ under consideration has a divisor in which Q_λ appears to the power $t - 1$. This power is again at most 1 (as $t \leq 2$), which covers the divisors from the third set in any case, and as for $t = 1$, this power is 0 and we are done with the divisors from the second set in this case as well. This completes the proof that all our divisors are non-special.

In order to show that these are the only non-special divisors we first use the usual trick to obtain that any integral divisor of degree $g = n - 1$ that is a multiple of Ξ where Ξ is either Q_λ^3, P_1^{2s+t}, P_2^{2s+t}, P_3^{2s+t}, or $P_1P_2P_3$, then Δ is special (since then

$r(1/\varDelta) \geq r(1/\varXi) \geq 2$, Riemann–Roch, etc.). The remaining possibilities we have to check are once again just a few, as they are subject to the restrictions that Q_λ can appear only to a power not exceeding 2, each of point the points P_1, P_2, and P_3 can appear to a power not exceeding $2s+t-1$, and one of them must not appear at all. As in the special case with $n = 10$ (where $2s+t-1$ is 6 as we saw in the proof there), we can write all the divisors satisfying these conditions and see why the only ones which are non-special are the ones written above.

Again the verification is easier if we split these divisors by the power to which Q_λ appears in them (which is either 0, 1, or 2). We start with the divisors in which this power is 0, meaning that the support of any such divisor consists of two branch points P_i and P_j which must appear to powers not exceeding $2s+t-1$ which sum to $3s+t-1 = n-1$. One possible such pair of numbers is $2s+t-1$ and s, which gives us the non-special divisors $P_i^{2s+t-1}P_j^s$ from the first set. In every other possibility the two branch points appear to powers not exceeding $2s+t-2$ (note that for $n=4$ this cannot happen as $2s+t-2 = 1$ and the sum cannot be $n-1 = 3$), and this gives a special divisor since the divisor $P_1^{2s+t-2}P_2^{2s+t-2}P_3^{2s+t-2}Q_\lambda^{2-t}$ of the holomorphic differential $(z-\lambda)^{s-1}dz/w^{s+1}$ is a multiple of every such divisor. This completes the proof for the divisors in which Q_λ does not appear at all.

Since when this power is 1 we shall have to split into the cases $t = 1$ and $t = 2$, we prefer to deal now with the case where this power is 2. The two other branch points must appear in such divisors to powers not exceeding $2s+t-1$ which sum to $3s+t-3 = n-3$. The pair $2s+t-2$ and $s-1$ gives us the non-special divisors $P_i^{2s+t-2}P_j^{s-1}Q_\lambda^2$ from the third set. There is also the pair $2s+t-1$ and $s-2$ (which cannot occur for $s=1$, i.e., for $n=4$ or $n=5$, since $s-2$ is negative then), and in all the remaining possibilities the two branch points both appear to powers not exceeding $2s+t-3$ (again impossible for $n=4$ since two 0s cannot sum to 1). All the latter divisors are special, as one easily sees that the divisor $P_1^{2s+t-3}P_2^{2s+t-3}P_3^{2s+t-3}Q_\lambda^{5-t}$ of the holomorphic differential $(z-\lambda)^s dz/w^{s+2}$ is a multiple of any one of them. The remaining divisors are the divisors $P_i^{2s+t-1}P_j^{s-2}Q_\lambda^2$, and for every such divisor we see that it is special since the divisor $P_i^{s-2+n}P_j^{s-2}P_k^{s-2}Q_\lambda^{t+2}$ of the holomorphic differential $(z-\lambda_i)(z-\lambda)^{2s+t-2}dz/w^{2s+t+1}$ (with the corresponding i) is a multiple of it. This proves the assertion if Q_λ appears to the power 2 in the divisors.

It remains to see what happens if Q_λ appears to the power 1 in the divisors. Then the possible pairs of powers to which the other branch points appear are pairs of numbers not exceeding $2s+t-1$ which sum to $3s+t-2 = n-2$. In any case one pair is the pair $2s$ and $s+t-2$, which gives us the non-special divisors $P_i^{2s}P_j^{s+t-2}Q_\lambda$ from the second set. Let us now split into the cases $t = 1$ and $t = 2$. If $t = 1$, then we have dealt with the pair including the maximal possible value $2s+t-1 = 2s$, and in any other pair the two numbers do not exceed $2s+t-2 = 2s-1$. This means that all these divisors are special, as one can see that the divisor $P_1^{2s-1}P_2^{2s-1}P_3^{2s-1}Q_\lambda$ of the holomorphic differential $(z-\lambda)^{s-1}dz/w^{s+1}$ is a multiple of each one of them. For $t = 2$ the pair which gave the non-special divisors is the one including the next-to-maximal value $2s = 2s+t-2$, leaving the pair $2s+t-1 = 2s+1$ and $s-1$ and pairs in which both numbers do not exceed $2s+t-3 = 2s-1$ (the latter possibility not existing for $n = 5$ as two numbers which are bounded by $2s - 1 = 1$ cannot give $n - 2 = 3$ as their sum). Since the divisor $P_1^{2s-1} P_2^{2s-1} P_3^{2s-1} Q_\lambda^3$ of the holomorphic differential $(z - \lambda)^s \mathrm{d}z / w^{s+2}$ is a multiple of any one of the latter divisors, they are all seen to be special. It thus remains to deal with the divisors $P_i^{2s+1} P_j^{s-1} Q_\lambda$ arising from the first pair, and for each such divisor we have the divisor $P_i^{s-1+n} P_j^{s-1} P_k^{s-1} Q_\lambda$ of the holomorphic differential $(z - \lambda_i)(z - \lambda)^{2s+t-3} \mathrm{d}z / w^{2s+t}$ (with the corresponding i) as a multiple of it, showing that it is special.

All this proves that the only non-special integral divisors of degree g supported on the branch points are those listed here. Now, clearly for $n \geq 7$ and $s \geq 2$ each set contains 6 divisors (the number of possibilities to choose i, j and k), since the powers to which P_i, P_j, and P_k appear are all distinct (the latter being 0). Since the three sets are disjoint, being distinguished by the power of Q_λ, this indeed gives 18 divisors. For $s = 1$ the sets are still disjoint, and the fact that the power to which P_i appears is higher than the one to which P_j appears is also valid, but some vanishing occurs. In the first set we have 6 divisors also for $s = 1$ by the same reasoning, but since the power $s - 1$ to which P_j appears in the last set vanishes for $s = 1$ we find that the divisors in this set depend only on the choice of i and are thus only 3. As for the second set with $s = 1$, we see that the power to which P_j appears is $s + t - 2 = t - 1$, which vanishes for $t = 1$ (and $n = 4$) and does not vanish for $t = 2$ (and $n = 5$). Therefore for $n = 5$ we have 15 divisors and for $n = 4$ we have only 12, which completes the proof of the theorem.   □

Since we shall soon have to work with the cases $t = 1$ and $t = 2$ differently, let us write the result of Theorem 6.13 explicitly in each case. For $t = 1$ we see that the first set contains the 6 divisors $P_i^{2s} P_j^s$, the second set contains the 6 divisors $P_i^{2s} P_j^{s-1} Q_\lambda$ (which are 3 divisors if $n = 4$), and the third set contains the 6 divisors $P_i^{2s-1} P_j^{s-1} Q_\lambda^2$ (which are also 3 divisors if $n = 4$). For $t = 2$ the first set contains the 6 divisors $P_i^{2s+1} P_j^s$, the second set contains the 6 divisors $P_i^{2s} P_j^s Q_\lambda$, and the third set contains the 6 divisors $P_i^{2s} P_j^{s-1} Q_\lambda^2$ (which are 3 divisors if $n = 5$).

Note that the fact that for $n = 4$ we have only 12 divisors is consistent with what we had in the case $r = 4$ in Section 3.4. Denoting Q_λ by P_0 in this case shows that Theorem 6.13 for $n = 4$ indeed gives the same 12 divisors, in which one branch points appears to the power 2 and another one appears to the power 1, as Theorems 2.6 and 2.9 give for $n = 4$ and $r = 1$. The reader can easily check this statement. In any case, we remember that we are interested, for any branch point, in those divisors which do not contain it in their support. For any value of $n = 3s + t$ (of course with $n \geq 4$ and $s \geq 1$) and any branch point we see that this gives us exactly 6 divisors. Indeed, for any branch point P_i with $1 \leq i \leq 3$ we have $P_j^{2s+t-1} P_k^s$ (which are 2 divisors), $P_j^{2s} P_k^{s+t-2} Q_\lambda$ (which are 2 divisors), and $P_j^{2s+t-2} P_k^{s-1} Q_\lambda^2$ (which are 2 divisors), and for the point Q_λ we have the 6 divisors $P_i^{2s+t-1} P_j^s$. The latter 6 divisors are $P_i^{2s} P_j^s$ if $t = 1$ and $P_i^{2s+1} P_j^s$ if $t = 2$.

6.2.5 Operators and Theta Quotients

We shall now depart from the practice of working with a basepoint appearing to the power 1 in the algebraic equation of the Z_n curve, and take our basepoint to be the point Q_λ. The reason for this choice is that it is the best one to make in order to maximize our results using the soon to be defined operators N and T_R. With Q_λ as our basepoint, we now define the operators N and T_R on the set of 6 divisors $P_i^{2s+t-1}P_j^s$. In order for the usual process to work we have to make different definitions, depending on whether $t=1$ or $t=2$.

Definition 6.14. For $t=1$ every divisor is of the form $\Delta = P_i^{2s}P_j^s$, and on such a divisor we define the operators N and T_{P_i} (with the corresponding index i) by

$$N(\Delta) = P_k^{2s}P_j^s \quad \text{and} \quad T_{P_i}(\Delta) = P_i^{2s}P_k^s.$$

For $t=2$ the divisors are of the form $\Delta = P_i^{2s+1}P_j^s$, and on every such divisor the operators which are defined are N and T_{P_j} (with the corresponding j) by

$$N(\Delta) = P_i^{2s+1}P_k^s \quad \text{and} \quad T_{P_j}(\Delta) = P_k^{2s+1}P_j^s.$$

As usual, we would like to show that the operators we have just defined in Definition 6.14 have the same role as they do in all the previous examples.

Proposition 6.15. *If* $e = \varphi_{Q_\lambda}(\Delta) + K_{Q_\lambda}$, *then* $-e = \varphi_{Q_\lambda}(N(\Delta)) + K_{Q_\lambda}$. *If* R *is some branch point* P_i, $1 \le i \le 3$, *such that* T_R *is defined on* Δ, *then*

$$e + \varphi_{Q_\lambda}(R) = -(\varphi_{Q_\lambda}(T_R(\Delta)) + K_{Q_\lambda}).$$

Proof. Since Definition 6.14 is different for $t=1$ and $t=2$ we shall have to split into cases when doing calculations, but the main ideas remain the same. We recall that by Lemma 6.12 we have $-2K_{Q_\lambda} = 0$ (so that φ_{Q_λ} vanishes on divisors of meromorphic differentials).

Now, for the first statement it suffices to show that for every divisor Δ we have

$$\varphi_{Q_\lambda}(\Delta) + \varphi_{Q_\lambda}(N(\Delta)) = -2K_{Q_\lambda} = 0.$$

Definition 6.14 gives that for $t=1$ we have for any divisor Δ the equality

$$\varphi_{Q_\lambda}(\Delta) + \varphi_{Q_\lambda}(N(\Delta)) = \varphi_{Q_\lambda}(P_1^{2s}P_2^{2s}P_3^{2s}).$$

This vanishes because φ_{Q_λ} vanishes either on the divisor $P_1^{2s}P_2^{2s}P_3^{2s}/Q_\lambda^2$ of the meromorphic differential $(z-\lambda)^{s-2}dz/w^s$, or on the divisor $P_1^{2s}P_2^{2s}P_3^{2s}/Q_\lambda^{6s}$ of the meromorphic function $w^{?s}/(z-\lambda)^{2s}$. As for $t=2$, then Definition 6.14 shows that for $\Delta = P_i^{2s+1}P_j$ we have

$$\varphi_{Q_\lambda}(\Delta) + \varphi_{Q_\lambda}(N(\Delta)) = \varphi_{Q_\lambda}(P_i^{4s+2}P_j^s P_k^s),$$

and since $n = 3s + 2$ Lemma 6.12 shows that for every Δ this expression equals $\varphi_{Q_\lambda}(P_1^s P_2^s P_3^s)$. This expression vanishes because φ_{Q_λ} vanishes either on the divisor $P_1^s P_2^s P_3^s Q_\lambda^{n-2}$ of the holomorphic differential $(z-\lambda)^{2s-1}dz/w^{2s+1}$, or on the divisor $P_1^s P_2^s P_3^s / Q_\lambda^{3s}$ of the meromorphic function $w^s/(z-\lambda)^s$.

For the second statement, as usual, we prove the equivalent equality stating that for any such Δ and R one has

$$\varphi_{Q_\lambda}(\Delta) + \varphi_{Q_\lambda}(R) + \varphi_{Q_\lambda}(T_R(\Delta)) = -2K_{Q_\lambda} = 0.$$

Now Definition 6.14 gives that for $t = 1$ we have for $\Delta = P_i^{2s}P_j$ and $R = P_i$ the equality

$$\varphi_{Q_\lambda}(\Delta) + \varphi_{Q_\lambda}(R) + \varphi_{Q_\lambda}(T_R(\Delta)) = \varphi_{Q_\lambda}(P_i^{4s+1}P_j^s P_k^s),$$

and since $n = 3s + 1$ we can use Lemma 6.12 to see that for any Δ and R this expression equals $\varphi_{Q_\lambda}(P_1^s P_2^s P_3^s)$. This expression vanishes because φ_{Q_λ} vanishes either on the divisor $P_1^s P_2^s P_3^s Q_\lambda^{n-3}$ of the holomorphic differential $(z-\lambda)^{2s-2}dz/w^{2s}$, or on the divisor $P_1^s P_2^s P_3^s / Q_\lambda^{3s}$ of the meromorphic function $w^s/(z-\lambda)^s$. In the case $t = 2$, Definition 6.14 shows that for any divisor Δ and branch point R we have the equality

$$\varphi_{Q_\lambda}(\Delta) + \varphi_{Q_\lambda}(R) + \varphi_{Q_\lambda}(T_R(\Delta)) = \varphi_{Q_\lambda}(P_1^{2s+1}P_2^{2s+1}P_3^{2s+1}).$$

This vanishes because φ_{Q_λ} vanishes either on the divisor $P_1^{2s+1}P_2^{2s+1}P_3^{2s+1}/Q_\lambda^3$ of the meromorphic differential $(z-\lambda)^{s-2}dz/w^s$, or on $P_1^{2s+1}P_2^{2s+1}P_3^{2s+1}/Q_\lambda^{6s+3}$, the divisor of the meromorphic function $w^{2s+1}/(z-\lambda)^{2s+1}$. This concludes the proof. $\qquad\square$

All the arguments used in each of the previous cases were based on the material from Section 2.6. We would like to extend its validity to the current case as well. For the role of Lemmas 2.4 and 2.12 in the proof there we now have Lemma 6.12, and therefore we can use Eqs. (1.6) and (1.7) also here. We recall that n can be either odd or even, so we again adopt the notation $e = 2/\gcd(2,n)$, which equals 1 for even n and 2 for odd n. As for the operators, where in the previous cases we used Propositions 2.17, 2.19, and 6.5, our current Proposition 6.15 shows that the operators from Definition 6.14 operate in a way corresponding to those from Definitions 2.16, 2.18, and 6.4. Therefore the appropriate analog of Propositions 2.20, 2.21, and 6.6 is

Proposition 6.16. *For any divisor Δ of the 6 divisors of Theorem 6.3 that do not include Q_λ in their support, the theta function $\theta[\Delta](\varphi_{Q_\lambda}(P), \Pi)$ for any lift of $\varphi_{Q_\lambda}(\Delta) + K_{Q_\lambda}$ does not vanish identically on the surface, and its divisor of zeros from Theorem 1.8 (the Riemann Vanishing Theorem) is the integral divisor of degree $g = n - 1$ denoted $N(\Delta)$ in Definition 6.14. For any two such divisors Δ and Ξ, the function*

$$f(P) = \frac{\theta[\Delta]^{en^2}(\varphi_{Q_\lambda}(P), \Pi)}{\theta[\Xi]^{en^2}(\varphi_{Q_\lambda}(P), \Pi)}$$

is a well-defined meromorphic function on the Z_n curve X, which is independent of the lifts of $\varphi_{Q_\lambda}(\Delta)$, $\varphi_{Q_\lambda}(\Xi)$, and K_{Q_λ} from $J(X)$ to $\mathbb{C}^g$. This function has the same divisor $[N(\Delta)/N(\Xi)]^{en^2}$ as the function

$$\prod_{i=1}^{3} (z - \lambda_i)^{en[v_{P_i}(N(\Delta)) - v_{P_i}(N(\Xi))]}.$$

Hence $f(P)$ is a nonzero constant multiple of this rational function of z.

Moreover, by the same considerations we showed in Section 6.1 we find that Eqs. (2.1) and (2.3) extend to this case as well.

We now use the quotient from Proposition 6.16 in order to obtain the PMT. This is done by the usual considerations, by which the divisor Ξ we take in the quotient in Proposition 6.16 is $T_R(\Delta)$ for some permissible branch point $R = P_i$ with some $1 \le i \le 3$. We recall from Definition 6.14 that the 6 relevant divisors split into 3 pairs, where each pair is connected via the operator T_{P_i} for different $1 \le i \le 3$. At that stage we remember that Definition 6.14 is different for $t = 1$ and $t = 2$, which means that the proof in the two cases will be different. Thus we now split into these two cases. We remark that the final Thomae formulae will indeed be different for these two cases, and this difference will be essential, i.e., there will be no way to merge these cases together.

6.2.6 Thomae Formulae for $t = 1$

We begin with the case $t = 1$. We take the divisor $\Delta = P_i^{2s} P_j^s$ and the branch point $R = P_i$, and using Definition 6.14 we obtain that the quotient

$$\frac{\theta^{en^2}[\Delta](\varphi_{Q_\lambda}(P), \Pi)}{\theta^{en^2}[T_{P_i}(\Delta)](\varphi_{Q_\lambda}(P), \Pi)} = \frac{\theta^{en^2}[P_i^{2s} P_j^s](\varphi_{Q_\lambda}(P), \Pi)}{\theta^{en^2}[P_i^{2s} P_k^s](\varphi_{Q_\lambda}(P), \Pi)}$$

is a meromorphic function on the Z_n curve X. Since Definition 6.14 also shows that $N(\Delta) = P_k^{2s} P_j^s$ and $N(T_{P_i}(\Delta)) = P_j^{2s} P_k^s$, we obtain from Proposition 6.16 that the divisor of this function is $[P_k^s / P_j^s]^{en^2}$, and therefore this function is just the function $c(z - \lambda_k)^{ens} / (z - \lambda_j)^{ens}$, where c is some nonzero complex constant. In order to calculate c we now set $P = Q_\lambda$, which gives us

$$\frac{\theta^{en^2}[P_i^{2s} P_j^s](0, \Pi)}{\theta^{en^2}[P_i^{2s} P_k^s](0, \Pi)} = c \frac{(\lambda - \lambda_k)^{ens}}{(\lambda - \lambda_j)^{ens}},$$

and hence

$$\frac{\theta^{en^2}[P_i^{2s}P_j^s](\varphi_{Q_\lambda}(P),\Pi)}{\theta^{en^2}[P_i^{2s}P_k^s](\varphi_{Q_\lambda}(P),\Pi)} = \frac{\theta^{en^2}[P_i^{2s}P_j^s](0,\Pi)}{\theta^{en^2}[P_i^{2s}P_k^s](0,\Pi)} \times \frac{(\lambda-\lambda_j)^{ens}}{(\lambda-\lambda_k)^{ens}} \times \frac{(z-\lambda_k)^{ens}}{(z-\lambda_j)^{ens}}.$$

Substituting the only other reasonable point $P = P_i$ gives us, on the one hand,

$$\frac{\theta^{en^2}[P_i^{2s}P_j^s](\varphi_{Q_\lambda}(P_i),\Pi)}{\theta^{en^2}[P_i^{2s}P_k^s](\varphi_{Q_\lambda}(P_i),\Pi)} = \frac{\theta^{en^2}[P_i^{2s}P_j^s](0,\Pi)}{\theta^{en^2}[P_i^{2s}P_k^s](0,\Pi)} \times \frac{(\lambda-\lambda_j)^{ens}}{(\lambda-\lambda_k)^{ens}} \times \frac{(\lambda_i-\lambda_k)^{ens}}{(\lambda_i-\lambda_j)^{ens}},$$

while on the other hand, we get from Eq. (2.3) the equality

$$\frac{\theta^{en^2}[P_i^{2s}P_j^s](\varphi_{Q_\lambda}(P_i),\Pi)}{\theta^{en^2}[P_i^{2s}P_k^s](\varphi_{Q_\lambda}(P_i),\Pi)} = \frac{\theta^{en^2}[P_i^{2s}P_k^s](0,\Pi)}{\theta^{en^2}[P_i^{2s}P_j^s](0,\Pi)}$$

(as $T_{P_i}(P_i^{2s}P_k^s) = P_i^{2s}P_j^s$—also here T_{P_i} is an involution). By combining the last two equations we obtain that

$$\frac{\theta^{en^2}[P_i^{2s}P_k^s](0,\Pi)}{\theta^{en^2}[P_i^{2s}P_j^s](0,\Pi)} = \frac{\theta^{en^2}[P_i^{2s}P_j^s](0,\Pi)}{\theta^{en^2}[P_i^{2s}P_k^s](0,\Pi)} \times \frac{(\lambda-\lambda_j)^{ens}}{(\lambda-\lambda_k)^{ens}} \times \frac{(\lambda_i-\lambda_k)^{ens}}{(\lambda_i-\lambda_j)^{ens}},$$

and finally that

$$\frac{\theta^{2en^2}[P_i^{2s}P_j^s](0,\Pi)}{(\lambda_i-\lambda_j)^{ens}(\lambda-\lambda_k)^{ens}} = \frac{\theta^{2en^2}[P_i^{2s}P_k^s](0,\Pi)}{(\lambda_i-\lambda_k)^{ens}(\lambda-\lambda_j)^{ens}}. \tag{6.6}$$

The reader can check that starting with the divisor $\Delta = P_i^{2s}P_k^s$ and $R = P_i$, we obtain exactly the same Eq. (6.6). We thus obtain three Eqs. (6.6), one for every value of $1 \le i \le 3$.

Since the only action of each operator T_{P_i} is to pair the two divisors in Eq. (6.6), it is now clear that

Proposition 6.17. *The PMT for the Z_n curve X with $t = 1$ is composed of Eqs. (6.6) for the three values of i.*

In order to put Proposition 6.17 in the usual language we use for PMT we describe it as follows: For the divisor $\Delta = P_i^{2s}P_j^s$, define the denominator g_Δ to be $(\lambda_i - \lambda_j)^{ens}(\lambda - \lambda_k)^{ens}$, i.e., the (ens)th power of the difference between the z-values of the points appearing in Δ, multiplied by the (ens)th power of the difference between λ and the z-value of the point not appearing in Δ. Then the quotient $\theta^{2en^2}[\Delta](0,\Pi)/g_\Delta$ is invariant under the operators T_R. These are, of course, the reduced denominators, as no common factor appear in them.

We now multiply the denominators g_Δ by appropriate expressions in order to obtain new denominators h_Δ, such that the quotient $\theta^{2en^2}[\Delta](0,\Pi)/h_\Delta$ will be invariant under the three operators T_R and also under the operator N. Since N

pairs $P_i^{2s}P_j^s$ with $P_k^{2s}P_j^s$, we do this by dividing Eq. (6.6) for P_i by the expression $(\lambda_k - \lambda_j)^{ens}(\lambda - \lambda_i)^{ens}$ (note that this is the same expression as one obtains when starting with $\Delta = P_i^{2s}P_k^s$), and we denote the expression thus obtained by h_Δ (which is equivalent to taking $h_\Delta = g_\Delta g_{N(\Delta)}$). Since it is clear that this h_Δ satisfies $h_{N(\Delta)} = h_\Delta$ (as the product of two expressions, one corresponding to Δ and the other corresponding to $N(\Delta)$), using what we already know and the identity $\theta[N(\Delta)](0, \Pi) = \theta[\Delta](0, \Pi)$ from Eq. (1.5) to the proper power we obtain that

$$\frac{\theta^{2en^2}[\Delta](0, \Pi)}{h_\Delta} = \frac{\theta^{2en^2}[P_i^{2s}P_j^s](0, \Pi)}{(\lambda_i - \lambda_j)^{ens}(\lambda - \lambda_k)^{ens}(\lambda_k - \lambda_j)^{ens}(\lambda - \lambda_i)^{ens}} \qquad (6.7)$$

is invariant under the operators T_R and N. Since it is clear that one can get from any non-special divisor of degree $g = n - 1$ on our Z_n curve from Theorem 6.13 to any other such divisor using the operators N and T_R (seen explicitly as

$$P_1^{2s}P_2^s \xoverset{T_{P_1}}{\longleftrightarrow} P_1^{2s}P_3^s \xoverset{N}{\longleftrightarrow} P_2^{2s}P_3^s \xoverset{T_{P_2}}{\longleftrightarrow} P_2^{2s}P_1^s \xoverset{N}{\longleftrightarrow} P_3^{2s}P_1^s \xoverset{T_{P_1}}{\longleftrightarrow} P_3^{2s}P_2^s,$$

and N also connects the two sides), we conclude that

Theorem 6.18. *The expression in Eq. (6.7) is independent of the divisor Δ, and is the Thomae formulae for our Z_n curve X, in the case $t = 1$.*

6.2.7 Thomae Formulae for $t = 2$

We now turn to the case $t = 2$. In this case Definition 6.14 shows that by taking the divisor $\Delta = P_i^{2s+1}P_j^s$ and the branch point $R = P_j$ we again obtain that the quotient

$$\frac{\theta^{en^2}[\Delta](\varphi_{Q_\lambda}(P), \Pi)}{\theta^{en^2}[T_{P_j}(\Delta)](\varphi_{Q_\lambda}(P), \Pi)} = \frac{\theta^{en^2}[P_i^{2s+1}P_j^s](\varphi_{Q_\lambda}(P), \Pi)}{\theta^{en^2}[P_k^{2s+1}P_j^s](\varphi_{Q_\lambda}(P), \Pi)}$$

is a meromorphic function on the Z_n curve X. However, now Definition 6.14 gives that $N(\Delta) = P_i^{2s+1}P_k^s$ and $N(T_{P_j}(\Delta)) = P_k^{2s+1}P_i^s$. Therefore Proposition 6.16 gives that the divisor of this function is $[P_i^{s+1}/P_k^{s+1}]^{en^2}$, meaning that this function is simply the function $c(z - \lambda_i)^{en(s+1)}/(z - \lambda_k)^{en(s+1)}$, where c is some nonzero complex constant. Setting $P = Q_\lambda$, for calculating c, yields

$$\frac{\theta^{en^2}[P_i^{2s+1}P_j^s](0, \Pi)}{\theta^{en^2}[P_k^{2s+1}P_j^s](0, \Pi)} = c\frac{(\lambda - \lambda_i)^{en(s+1)}}{(\lambda - \lambda_k)^{en(s+1)}},$$

and thus

$$\frac{\theta^{en^2}[P_i^{2s+1}P_j^s](\varphi_{Q_\lambda}(P),\Pi)}{\theta^{en^2}[P_k^{2s+1}P_j^s](\varphi_{Q_\lambda}(P),\Pi)}=\frac{\theta^{en^2}[P_i^{2s+1}P_j^s](0,\Pi)}{\theta^{en^2}[P_k^{2s+1}P_j^s](0,\Pi)}\times\frac{(\lambda-\lambda_k)^{en(s+1)}}{(\lambda-\lambda_i)^{en(s+1)}}\times\frac{(z-\lambda_i)^{en(s+1)}}{(z-\lambda_k)^{en(s+1)}}.$$

The only reasonable point to substitute now is $P=P_j$, which gives us the equality

$$\frac{\theta^{en^2}[P_i^{2s+1}P_j^s](\varphi_{Q_\lambda}(P_j),\Pi)}{\theta^{en^2}[P_k^{2s+1}P_j^s](\varphi_{Q_\lambda}(P_j),\Pi)}=\frac{\theta^{en^2}[P_i^{2s+1}P_j^s](0,\Pi)}{\theta^{en^2}[P_k^{2s+1}P_j^s](0,\Pi)}\times\frac{(\lambda-\lambda_k)^{en(s+1)}}{(\lambda-\lambda_i)^{en(s+1)}}\times\frac{(\lambda_j-\lambda_i)^{en(s+1)}}{(\lambda_j-\lambda_k)^{en(s+1)}},$$

while on the other hand, from Eq. (2.3) we obtain

$$\frac{\theta^{en^2}[P_i^{2s+1}P_j^s](\varphi_{Q_\lambda}(P_j),\Pi)}{\theta^{en^2}[P_k^{2s+1}P_j^s](\varphi_{Q_\lambda}(P_j),\Pi)}=\frac{\theta^{en^2}[P_k^{2s+1}P_j^s](0,\Pi)}{\theta^{en^2}[P_i^{2s+1}P_j^s](0,\Pi)}$$

(as $T_{P_j}(P_k^{2s+1}P_j^s)=P_i^{2s+1}P_j^s$—$T_{P_j}$ is an involution). The combination of these two equations gives that

$$\frac{\theta^{en^2}[P_k^{2s+1}P_j^s](0,\Pi)}{\theta^{en^2}[P_i^{2s+1}P_j^s](0,\Pi)}=\frac{\theta^{en^2}[P_i^{2s+1}P_j^s](0,\Pi)}{\theta^{en^2}[P_k^{2s+1}P_j^s](0,\Pi)}\times\frac{(\lambda-\lambda_k)^{en(s+1)}}{(\lambda-\lambda_i)^{en(s+1)}}\times\frac{(\lambda_j-\lambda_i)^{en(s+1)}}{(\lambda_j-\lambda_k)^{en(s+1)}},$$

and hence

$$\frac{\theta^{2en^2}[P_i^{2s+1}P_j^s](0,\Pi)}{(\lambda_j-\lambda_k)^{en(s+1)}(\lambda-\lambda_i)^{en(s+1)}}=\frac{\theta^{2en^2}[P_k^{2s+1}P_j^s](0,\Pi)}{(\lambda_j-\lambda_i)^{en(s+1)}(\lambda-\lambda_k)^{en(s+1)}}. \tag{6.8}$$

As in the case $t=1$, it is easily verified that if one starts with the divisor $\Delta=P_k^{2s+1}P_j^s$ and $R=P_j$, then one reaches exactly the same Eq. (6.8). Therefore for every value of $1\le j\le 3$ we get one Eq. (6.8), together three equations.

Every operator T_{P_j} pairs the two divisors in Eq. (6.8) and does nothing more, which clearly gives

Proposition 6.19. *The PMT for the Z_n curve X with $t=2$ is composed of Eqs. (6.8) for the three values of j.*

In the usual language of the PMT Proposition 6.19 can be described as follows: For $\Delta=P_i^{2s+1}P_j^s$, define g_Δ to be $(\lambda_j-\lambda_k)^{en(s+1)}(\lambda-\lambda_i)^{en(s+1)}$, or, in a language more based on Δ, the $en(s+1)$th power of the difference between the z-values of the branch point appearing in Δ to the low power and the branch point not appearing in Δ at all, multiplied by the $en(s+1)$th power of the difference between λ and the z-value of the branch point appearing in Δ to the high power. With this definition, the quotient $\theta^{2en^2}[\Delta](0,\Pi)/g_\Delta$ is invariant under the operators T_R. Once again these are the reduced expressions for g_Δ.

As usual, we would like to replace the denominators g_Δ by new denominators h_Δ which will also satisfy $h_{N(\Delta)}=h_\Delta$, and therefore by Eq. (1.5) in the form of the identity $\theta[N(\Delta)](0,\Pi)=\theta[\Delta](0,\Pi)$ to the proper power the quotient

$\theta^{2en^2}[\Delta](0,\Pi)/h_\Delta$ will be invariant under all the operators T_R and also under N. In this case however no further manipulations are necessary, as the denominators g_Δ already have this property: N pairs $P_i^{2s+1}P_j^s$ with $P_i^{2s+1}P_k^s$, and the denominators g_Δ for these two divisors are the same. This means that we can define $h_\Delta = g_\Delta$ for every divisor Δ, and then obtain that

$$\frac{\theta^{2en^2}[\Delta](0,\Pi)}{h_\Delta} = \frac{\theta^{2en^2}[P_i^{2s+1}P_j^s](0,\Pi)}{(\lambda_j-\lambda_k)^{en(s+1)}(\lambda-\lambda_i)^{en(s+1)}} \tag{6.9}$$

is invariant under the operators T_R and N. Also here one easily sees that it is possible to get from any non-special divisor of degree $g = n - 1$ on our Z_n curve from Theorem 6.13 to any other such divisor using the operators N and T_R (we can draw the diagram

$$P_1^{2s+1}P_2^s \overset{T_{P_2}}{\longleftrightarrow} P_3^{2s+1}P_2^s \overset{N}{\longleftrightarrow} P_3^{2s+1}P_1^s \overset{T_{P_1}}{\longleftrightarrow} P_2^{2s+1}P_1^s \overset{N}{\longleftrightarrow} P_2^{2s+1}P_3^s \overset{T_{P_3}}{\longleftrightarrow} P_1^{2s+1}P_3^s,$$

with N also connecting the two sides, similarly to the case $t = 1$), so that we obtain

Theorem 6.20. *The expression in Eq. (6.9) is independent of the divisor Δ, and is the Thomae formulae for our Z_n curve X, in the case $t = 2$.*

6.2.8 Changing the Basepoint

Let us now say a few words about changing the basepoint. For this we briefly present what happens when the basepoint is any other point P_i, $1 \le i \le 3$. For such a basepoint Theorem 6.13 gives 6 non-special integral divisors of degree $g = n - 1$ supported on the branch points distinct from P_i, which were seen to be $P_j^{2s+t-1}P_k^s$ (2 divisors), $P_j^{2s}P_k^{s+t-2}Q_\lambda$ (2 divisors), and $P_j^{2s+t-2}P_k^{s-1}Q_\lambda^2$ (2 divisors). On these divisors we define the operator N, which pairs $P_j^{2s+t-1}P_k^s$ with $P_k^{2s+t-2}P_j^{s-1}Q_\lambda^2$ (2 pairs) and $P_j^{2s}P_k^{s+t-2}Q_\lambda$ with $P_k^{2s}P_j^{s+t-2}Q_\lambda$. The reader can check that this operator has the desired properties as in the proof of Proposition 6.15 (the product of each pair is $P_j^{n-2}P_k^{n-2}Q_\lambda^2$ and the divisor $P_j^{n-2}P_k^{n-2}Q_\lambda^2/P_i^2$ is the divisor of the meromorphic differential $dz/(z-\lambda_i)w$—but recall that $-2K_{P_i}$ no longer vanishes so no proof using meromorphic functions is available). From the operators T_R we can define (for $n \ge 7$ and $s \ge 2$, at least) just the operator T_{Q_λ}, which acts only on the two divisors $P_j^{2s+t-2}P_k^{s-1}Q_\lambda^2$ and pairs them. The reader can check again that this is the proper definition as in the proof of Proposition 6.15 (their product with one another and with Q_λ is $P_j^{n-3}P_k^{n-3}Q_\lambda^5$, which is equivalent to $P_iP_j^{n-2}P_k^{n-2}Q_\lambda^2$ and continue as before). The other branch points appear in all the divisors, meaning that no other operator T_R can be defined. Therefore we will not be able to reach from any divisor to any other divisor using the operators N and T_R (we obtain two orbits, one containing the two divisors $P_j^{2s+t-1}P_k^s$ and the two divisors $P_j^{2s+t-2}P_k^{s-1}Q_\lambda^2$, and the other containing the two divisors $P_j^{2s} P_k^{s+t-2} Q_\lambda$). This is why we have taken the basepoint for our Z_n curve to be Q_λ.

What we said above holds only for $n \geq 7$, i.e., for $s \geq 2$. For $s = 1$ the powers $s - 1$ vanish and some more operators T_R can be defined. What happens in the case $n = 5$ (with $s = 1$ and $t = 2$) is that the two operators T_{P_j} with $j \neq i$ can be defined, but each only on the one divisor $P_j^3 P_k^1$ (with corresponding i, j and k) and to take it to itself. This means that the orbits for $n = 5$ are also as said in the previous paragraph. As for $n = 4$, we find that for the basepoint P_i the two operators T_{P_j} with $j \neq i$ pair respectively the divisors $P_j^2 P_k$ and $P_j^2 Q_\lambda$. We remember, however, that for $n = 4$ we are back to the situation in Section 3.4, where we obtain Thomae formulae for any basepoint and they were all symmetric. Therefore the restrictions on the formulae with the basepoint P_i will hold only for $n \geq 5$ but the process of basepoint change works equally for all $n \geq 4$.

When doing the process with the basepoint P_i, $1 \leq i \leq 3$, we do not get the full Thomae formulae (for $n \geq 5$). What we do get, is that the set of 6 divisors splits into two sets, one containing 4 divisors and the other containing 2 divisors, and on these sets the quotient $\theta^{2en^2}[\Delta](0, \Pi)/h_\Delta$ is constant for properly defined denominators h_Δ. Explicitly the result is

$$\frac{\theta^{2en^2}[P_k^{2s+t-1} P_j^s](0, \Pi)}{(\lambda_i - \lambda_k)^{en(s+t-1)}(\lambda - \lambda_j)^{en(s+t-1)}} = \frac{\theta^{2en^2}[P_j^{2s+t-2} P_k^{s-1} Q_\lambda^2](0, \Pi)}{(\lambda_i - \lambda_k)^{en(s+t-1)}(\lambda - \lambda_j)^{en(s+t-1)}}$$

$$= \frac{\theta^{2en^2}[P_k^{2s+t-2} P_j^{s-1} Q_\lambda^2](0, \Pi)}{(\lambda_i - \lambda_j)^{en(s+t-1)}(\lambda - \lambda_k)^{en(s+t-1)}} = \frac{\theta^{2en^2}[P_j^{2s+t-1} P_k^s](0, \Pi)}{(\lambda_i - \lambda_j)^{en(s+t-1)}(\lambda - \lambda_k)^{en(s+t-1)}}$$

as the large orbit, and

$$\frac{\theta^{2en^2}[P_j^{2s} P_k^{s+t-2} Q_\lambda](0, \Pi)}{1} = \frac{\theta^{2en^2}[P_k^{2s} P_j^{s+t-2} Q_\lambda](0, \Pi)}{1}$$

as the small orbit. What we now obtain is that by dividing these equations by appropriate expressions we can merge these equations to be the Thomae formulae for these basepoints, and that the 4 constants we obtain from the Thomae formulae for the 4 basepoints are the same. We shall, however, only give the correspondence on the divisors, and the reader can check the details of verifying the consistency of these equations with Theorems 6.18 and 6.20. All this, of course, holds only for $n \geq 5$. As for $n = 4$, we already know from Section 3.4 that all the points are symmetric and present the same behavior. This is also a good place to emphasize that for $n \geq 5$ the point Q_λ is indeed not symmetric to the other points P_i, $1 \leq i \leq 3$ (they are symmetric to one another, of course), so that this example is again different from the nonsingular and singular families we dealt with in Chapters 4 and 5. The reader can also verify that what we obtain for $n = 4$ in this notation is consistent with what we obtained in Section 3.4.

We now indicate how one shows that the expressions we obtained for the basepoints P_i can be merged (after division by appropriate expressions) into Thomae formulae, and that the 4 constants we obtain, which in principal can be different, are in fact equal. We shall not give the expressions for h_Δ with the basepoints P_i, $1 \le i \le 3$, but just state by what one has to divide the equations. For $t = 1$ one divides the long equation by

$$(\lambda_j - \lambda_k)^{ens}(\lambda - \lambda_i)^{ens}$$

and the short equation by

$$(\lambda_i - \lambda_k)^{ens}(\lambda - \lambda_j)^{ens}(\lambda_i - \lambda_j)^{ens}(\lambda - \lambda_k)^{ens}.$$

For $t = 2$ one leaves the long equation as it is (i.e., divide it by 1) and divides the short equation by

$$(\lambda_j - \lambda_k)^{en(s+1)}(\lambda - \lambda_i)^{en(s+1)}.$$

The reader can obtain the expressions and do the necessary comparisons (but be careful with the indices!). We just describe the process that should be done for the proof: First, note that Lemma 6.12 means that also for our Z_n curve X, the conditions of Corollary 1.13 are satisfied, but here it cannot be applied to relate characteristics with the basepoint Q_λ with characteristics with the basepoint P_i, as no branch point appears to the power $n - 1$ in any of the divisors in Theorem 6.13. However, the second statement of Corollary 1.13 can also here be used to prove the corresponding analog of Propositions 1.14, 1.15 and 6.10.

Proposition 6.21. *Let Δ be any of the 6 divisors from Theorem 6.13 that do not contain Q_λ in their support. Then for every point P_i, $1 \le i \le 3$, there is a divisor Γ_{P_i} appearing in Theorem 6.13, but with P_i not included in its support instead of Q_λ, such that the equality $\varphi_{Q_\lambda}(\Delta) + K_{Q_\lambda} = \varphi_{P_i}(\Gamma_{P_i}) + K_{P_i}$ holds.*

As in all the previous cases, Proposition 6.10 shows that not only do we have Thomae formulae for every basepoint and they give the same constants, but also that the four Thomae formulae involve the same characteristics of the theta constants.

We leave the details of the proof of Proposition 6.21 to the reader, who at this point should be able to fill them in. One uses the second statement of Corollary 1.13 to change the basepoint freely for the divisors $Q_\lambda^{n-1}\Delta$ and write

$$\varphi_{Q_\lambda}(\Delta) + K_{Q_\lambda} = \varphi_{Q_\lambda}(Q_\lambda^{n-1}\Delta) + K_{Q_\lambda} = \varphi_{P_i}(Q_\lambda^{n-1}\Delta) + K_{P_i}$$

for the basepoint P_i to which we want to change, and then multiplies the divisor $Q_\lambda^{n-1}\Delta$ by some power of the principal divisor $P_1 P_2 P_3 / Q_\lambda^3$ in order to obtain a (equivalent) divisor in which the desired point P_i appears either to the power $n - 1$ or to the power -1. Then one uses Lemma 6.12 in order to move nth powers of branch points in these divisors to the right direction and obtain an integral divisor in which P_i appears to the power $n - 1$. Dividing by P_i^{n-1} one finds the desired divisor Γ_{P_i}. The equalities one obtains when proving Proposition 6.21 can be summarized in one equation (as all the 6 divisors Δ have the same form $P_i^{2s+t-1}P_j^s$), and even though one can do the calculations for $t = 1$ and $t = 2$ simultaneously, we shall separate these cases for things to be clearer. For $t = 1$ we have

$$\varphi_{Q_\lambda}(P_i^{2s}P_j^s) + K_{Q_\lambda} = \varphi_{P_i}(P_j^{2s}P_k^s) + K_{P_i}$$
$$= \varphi_{P_j}(P_k^{2s}P_i^{s-1}Q_\lambda) + K_{P_j} = \varphi_{P_k}(P_i^{2s-1}P_j^{s-1}Q_\lambda^2) + K_{P_k},$$

while for $t = 2$ we have

$$\varphi_{Q_\lambda}(P_i^{2s+1}P_j^s) + K_{Q_\lambda} = \varphi_{P_i}(P_j^{2s}P_k^s Q_\lambda) + K_{P_i}$$
$$= \varphi_{P_j}(P_k^{2s+1}P_i^s) + K_{P_j} = \varphi_{P_k}(P_i^{2s}P_j^{s-1}Q_\lambda^2) + K_{P_k}.$$

It is easy to verify that for every basepoint P_t, $1 \leq t \leq 3$, each divisor from Theorem 6.13 which does not contain P_t in its support appears here with the basepoint P_t exactly for one choice of divisor Δ with the basepoint Q_λ. The reader is encouraged to check that for every divisor Δ with the basepoint Q_λ and every basepoint P_i, the denominator $h_{\Gamma_{P_i}}$ with basepoint P_i (after the corrections) equals the denominator h_Δ with basepoint P_0 from Theorems 6.18 and 6.20 .

Once again we would like to remark that it is better to use Proposition 6.21 as a merging tool and a consistency check, rather than a way to define the denominators for the basepoints P_i, $1 \leq i \leq 3$ from the denominators for the basepoint Q_λ. We also note that Theorem 6.18 with $n = 4$ and $s = 1$ is consistent with Theorem 3.14 from Section 3.4 and Theorem 5.6 with $n = 4$, as seen by replacing every Q_λ by P_0 and every λ by λ_0. As for basepoint-independent notation for Theorems 6.18 and 6.20, it does exist but we shall not give it.

We close this chapter and volume with some remarks for the reader. The purpose of this chapter was to give some examples of families of singular Z_n curves for which we do not have a general theory but yet succeeded in deriving Thomae formulae. Our reason for doing this was to encourage a new generation of scientists to continue this project by indicating a way. We should however mention that new ideas will be required for the general singular case since our methods are strongly based on finding non-special integral divisors of degree g supported on the set of branch points and Gonzalez-Diez and Torres have exhibited in [GDT] families of singular curves for which one cannot find any such divisors. We do however feel that the ideas of this chapter can be generalized to many other cases as well without the introduction of new ideas.

Appendix A

Constructions and Generalizations for the Nonsingular and Singular Cases

In this appendix we describe how one constructs the expressions for h_Δ appearing in Eqs. (4.7), (4.8), (5.10), and (5.11), which are invariant under the operator N. The construction begins with the expressions for g_Δ in Proposition 4.4 and Eq. (4.2) or in Proposition 5.3 and Eq. (5.5), and is executed while keeping all the properties g_Δ already had. We wish to present the actual process of obtaining the expressions for h_Δ, rather than just stating them and proving that they have the needed properties. While the proofs given do not require one to know the process of how the results were obtained, we feel that the ideas may be useful in other contexts, and therefore we present these ideas here in an appendix.

We shall first present the construction in the nonsingular case, and then in the singular case, where the former is in a way a "sub-construction" of the latter. Since in each of these cases the final results are slightly different for odd and even n, the process will also be different in these two cases, but it will resemble the construction for the case $n = 5$ described in Sections 4.2 and 5.2 in the nonsingular and singular cases respectively. We recall that in the process described there we divided the PMT (or equivalently multiplied g_Δ) by certain expressions at each stage, but the main idea was to preserve the properties of the PMT by making sure that each expression by which we divide remains the same after putting a tilde, and in the singular case also a hat, on all the sets appearing in it.

As we have mentioned throughout the book, both the nonsingular and general singular families of Z_n curves dealt with are special cases of a more general family. In the last two sections of this appendix we introduce this family in detail, and show that the final results of the singular case (i.e., Theorems 5.7 and 5.8, but also most of the results along the way) hold equally for this family.

H.M. Farkas and S. Zemel, *Generalizations of Thomae's Formula for Z_n Curves*,
Developments in Mathematics 21, DOI 10.1007/978-1-4419-7847-9,
© Springer Science+Business Media, LLC 2011

A.1 The Proper Order to do the Corrections in the Nonsingular Case

We start with the nonsingular case, and we first present the idea behind the specific order in which we do the operations. First, recall that by Proposition 4.4 and Eq. (4.2) we have

$$g_\Delta = [C_{-1}, C_{-1}]^{en(n-1)} [C_0, C_0]^{en(n-1)} \prod_{j=1}^{n-2} [C_{-1}, C_j]^{enj} [C_0, C_j]^{en(n-1-j)}.$$

Since by Definition 2.16 the operator N acts on the sets C_j, $-1 \le j \le n-2$, by fixing the set C_{-1} and interchanging each C_j, $0 \le j \le n-2$ with C_{n-2-j}, we find that

$$g_{N(\Delta)} = [C_{-1}, C_{-1}]^{en(n-1)} [C_{n-2}, C_{n-2}]^{en(n-1)} \prod_{j=0}^{n-3} [C_{-1}, C_j]^{en(n-2-j)} [C_{n-2}, C_j]^{en(j+1)},$$

where we replaced j by $n-2-j$ as the index of the product. We now recall that the expressions by which we divide must remain the same when putting a tilde on all the sets. As we have already seen, the form of these expressions depends on whether the expressions needing corrections involve some expression $[C_{-1}, C_j]^{ent}$ or $[C_0, C_j]^{ent}$ with $1 \le j \le n-2$ (and some number t), or not. We begin with those which involve one of these expressions (but with $j \ne (n-1)/2$ in the odd n case), and recall that the set $G = C_{-1} \cup C_0$ satisfies $\widetilde{G} = G$ and that $\widetilde{C}_j = C_{n-1-j}$ and $\widetilde{C_{n-1-j}} = C_j$; the value $j = (n-1)/2$ in the odd n case is different simply since $n-1-j = j$ there. Therefore, to correct such an expression we see that the expression by which we multiply g_Δ has to be

$$[C_{-1}, C_j]^{ent} [C_0, C_j]^{ent} [C_{-1}, C_{n-1-j}]^{ent} [C_0, C_{n-1-j}]^{ent}.$$

This means that $g_{N(\Delta)}$ will be multiplied by

$$[C_{-1}, C_{n-2-j}]^{ent} [C_{n-2}, C_{n-2-j}]^{ent} [C_{-1}, C_{j-1}]^{ent} [C_{n-2}, C_{j-1}]^{ent};$$

replace any index $0 \le k \le n-2$ by its complement with respect to $n-2$, and leave $k = -1$ invariant. On the other hand, an expression that does not involve the sets C_{-1} and C_0 is $[C_i, C_j]^{ent}$ with i and j being between 1 and $n-2$ and some number t. Under the assumption that i and j do not sum to $n-1$, we see that in this case the expression by which we have to multiply g_Δ has to be

$$[C_i, C_j]^{ent} [C_{n-1-i}, C_{n-1-j}]^{ent}$$

(the case where $i + j = n-1$ is different because the two expressions here coincide). Accordingly, $g_{N(\Delta)}$ will be multiplied by

$$[C_{n-2-i}, C_{n-2-j}]^{ent} [C_{i-1}, C_{j-1}]^{ent}.$$

We shall call corrections involving expressions including C_{-1} *corrections of the first type*, and corrections involving expressions not including C_{-1} *corrections of the second type*.

This begins to explain why we do the corrections in the order we do them, i.e., first correcting the expressions $[C_{-1}, C_j]$ with $1 \le j \le (n-4)/2$ for even n or with $1 \le j \le (n-3)/2$ for odd n by decreasing order of j, and then with all the expressions not involving C_{-1} at all. From the form of g_Δ, $g_{N(\Delta)}$, and the corrections of the first type for them, we find that all the expressions appearing in $g_{N(\Delta)}$ and not in g_Δ, and not including C_{-1}, involve the set C_{n-2}. Therefore they lead to corrections of the second type, whose corrections do not affect the powers to which any expressions including C_{-1} appear, either in g_Δ or in $g_{N(\Delta)}$. However, the corrections of the first type do affect the powers to which expressions $[C_{n-2}, C_j]$ appear in $g_{N(\Delta)}$ and expressions $[C_0, C_j]$ appear in g_Δ. This suggests that we should begin with the corrections of the first type, and then go on to the proper corrections of the second type. Finally, when dealing with the corrections of the first type, we see that the correction of $[C_{-1}, C_j]$ in g_Δ changes the power to which $[C_{-1}, C_{j-1}]$ appears in $g_{N(\Delta)}$. This explains why we have to start with large values of j and deal with these expressions in decreasing order, having the proper power to correct at each stage.

We now turn to the process itself. As we already mentioned, the cases of even n and odd n are slightly different, so we treat them separately.

A.2 Nonsingular Case, Odd n

Assume that n is odd, hence $e = 2$, and by what we have said before, we should start by correcting the expressions involving the set C_{-1}. We start by multiplying g_Δ by expressions of the form

$$[C_{-1}, C_j]^{2nt} [C_0, C_j]^{2nt} [C_{-1}, C_{n-1-j}]^{2nt} [C_0, C_{n-1-j}]^{2nt},$$

with decreasing j and the proper t at every stage. We thus begin by looking for the largest value of j such that $[C_{-1}, C_j]$ appears in $g_{N(\Delta)}$ to a power larger than the one to which it appears in g_Δ. This j turns out to be $(n-3)/2$, with the difference $2n(n-1)/2 - 2n(n-3)/2 = 2n$. From there we work by decreasing induction on j to find out that when we deal with the expression $[C_{-1}, C_j]$, the proper multiplying correction for g_Δ has to be

$$([C_{-1}, C_j][C_0, C_j][C_{-1}, C_{n-1-j}][C_0, C_{n-1-j}])^{2n[(n-1)^2/4 - j(n-1-j)]}.$$

This means that $g_{N(\Delta)}$ will be multiplied by

$$([C_{-1},C_{n-2-j}][C_{n-2},C_{n-2-j}][C_{-1},C_{j-1}][C_{n-2},C_{j-1}])^{2n[(n-1)^2/4-j(n-1-j)]}.$$

Indeed, for $j = (n-3)/2$, we have $(n-1)^2/4 - j(n-1-j) = 1$, and we start with the power $2n$ as stated. After correcting the expression $[C_{-1},C_j]$, we find that the next expression $[C_{-1},C_{j-1}]$ now appears in g_Δ to the power $2n(j-1)$, but in $g_{N(\Delta)}$ summing the powers from the original $g_{N(\Delta)}$ and from correcting $[C_{-1},C_j]$ gives

$$2n(n-1-j)+2n\left(\frac{(n-1)^2}{4}-j(n-1-j)\right) = 2n\left(\frac{(n-1)^2}{4}-(j-1)(n-1-j)\right).$$

The difference between these two powers is $2n[(n-1)^2/4 - (j-1)(n-j)]$, as needed. After doing all this, we see, by the same argument we used for the induction, that also for $j = 0$, the power to which $[C_{-1},C_0]$ appears in $g_{N(\Delta)}$ is $2n(n-1)^2/4$ and in g_Δ it does not appear at all. Now, since the expression involving $[C_{-1},C_0]$ that remains the same after putting a tilde on all the sets is $[G,G]$ (as $\tilde{G} = G$), we find (by expanding this expression) that the further correction is to multiply g_Δ by

$$[G,G]^{2n(n-1)^2/4} = [C_{-1},C_{-1}]^{2n(n-1)^2/4}[C_{-1},C_0]^{2n(n-1)^2/4}[C_0,C_0]^{2n(n-1)^2/4}.$$

The corresponding action on $g_{N(\Delta)}$ is multiplication by

$$[C_{-1},C_{-1}]^{2n(n-1)^2/4}[C_{-1},C_{n-2}]^{2n(n-1)^2/4}[C_{n-2},C_{n-2}]^{2n(n-1)^2/4}.$$

In order to verify that we have completed the corrections of the first type and that we can go on to the corrections of the second type, we have to examine to which powers the expressions involving C_{-1} appear in g_Δ and in $g_{N(\Delta)}$. For every $0 \le j \le (n-3)/2$ we find that the expression $[C_{-1},C_j]$ appears in g_Δ to the power

$$2nj+2n\left(\frac{(n-1)^2}{4}-j(n-1-j)\right) = 2n\left(\frac{(n-1)^2}{4}-j(n-2-j)\right)$$

(summing powers from the original g_Δ and from the corresponding correction). Note that this includes $j = 0$, by the fact that we could continue the induction to this case as well. However, this statement extends to every $-1 \le j \le n-2$, as we now show. For $j = -1$ the corresponding power is

$$2n(n-1) + 2n\frac{(n-1)^2}{4} = 2n\frac{n^2+2n-3}{4},$$

and substituting $j = -1$ before gives us exactly this value. For $j = (n-1)/2$, the power that comes from the correction vanishes (in correspondence with the fact that $[C_{-1},C_j]$ with this value of j does not appear in any correction for g_Δ), and substituting $j = (n-1)/2$ above indeed gives us the original power $2n(n-1)/2$. Finally, for $(n+1)/2 \le j \le n-2$ this holds, since the corresponding correction is the one involving $[C_{-1},C_{n-1-j}]$, and the power that comes from the correction is invariant under replacing j by $n-1-j$. Hence the same calculation applies. As for the powers to which the expression $[C_{-1}, C_j]$ with $-1 \leq j \leq n-2$ appears in $g_{N(\Delta)}$, there are two ways to see that they are the same as the powers to which they appear in g_Δ. One way is by the fact that we kept changing $g_{N(\Delta)}$ according to the action of N from Definition 2.16, which is fixing C_{-1} and interchanging C_j with $0 \leq j \leq n-2$ with C_{n-2-j}, and noticing that the obtained power is invariant under replacing j by $n-2-j$. The other way is by following the previous induction, and noticing that the expressions $[C_{-1}, C_j]$ with $(n-1)/2 \leq j \leq n-2$ (which appear in the original g_Δ to powers larger than those to which they appear in the original $g_{N(\Delta)}$) are also corrected in this process. First, the expression $[C_{-1}, C_{(n-1)/2}]$ appeared in g_Δ to a power larger by $2n$ than the power to which it appeared in $g_{N(\Delta)}$. Then, after the correction of $g_{N(\Delta)}$ corresponding to the one involving the expression $[C_{-1}, C_j]$ with $1 \leq j \leq (n-3)/2$ in g_Δ corrected the difference of powers also for $[C_{-1}, C_{n-2-j}]$, we have that the difference between the powers to which $[C_{-1}, C_{n-1-j}]$ appeared in g_Δ and $g_{N(\Delta)}$ was

$$2n\left(\frac{(n-1)^2}{4} - (j-1)(n-1-j)\right) - 2n(j-1) = 2n\left(\frac{(n-1)^2}{4} - (j-1)(n-j)\right).$$

This difference is dealt with in the next correction. Finally the appearance of $[C_{-1}, C_{n-2}]$ only in $g_{N(\Delta)}$ and to the power $2n(n-1)^2/4$ is dealt with in the last correction, and $[C_{-1}, C_{-1}]$ now appears also in $g_{N(\Delta)}$ to the power

$$2n(n-1) + 2n\frac{(n-1)^2}{4} = 2n\frac{n^2 + 2n - 3}{4}.$$

We have now completed all the corrections of the first type; in order to see what the corrections of the second type are that need to be done, we have to see what the relevant expressions are (i.e., those which do not involve the set C_{-1}) which now appear in g_Δ and $g_{N(\Delta)}$. We find that the only expressions of that sort which appear in g_Δ are the expressions $[C_0, C_j]$ with $0 \leq j \leq n-2$; summing the powers to which they appeared in the original g_Δ and in the relevant correction gives that now they appear to the power

$$2n(n-1-j) + 2n\left(\frac{(n-1)^2}{4} - j(n-1-j)\right) = 2n\left(\frac{n^2 + 2n - 3}{4} - j(n-j)\right);$$

to see this, write $(j-1)(n-1-j)$ as $j(n-j) - (n-1)$. Note that this holds not only for $1 \leq j \leq (n-3)/2$ but also for $j = 0$ and for $(n-1)/2 \leq j \leq n-2$ by what we said in the previous paragraph. At this point, in order to be more similar to the desired expression h_Δ from Eq. (4.7), we change the index j to k (so that $[C_0, C_k]$ appears in g_Δ to the power $2n[(n^2 + 2n - 3)/4 - k(n-k)]$); it is good to note at this point that by changing j to $k-1$ in the previous paragraph we find that the expression $[C_{-1}, C_{k-1}]$ appears now in both g_Δ and $g_{N(\Delta)}$ to exactly the same power $2n[(n^2 + 2n - 3)/4 - k(n-k)]$ (just substitute $j = k-1$ there and do the same trick as above). This will be useful when examining the final expressions. As for the new $g_{N(\Delta)}$, the expressions not involving C_{-1} that appear in it are the expressions

$[C_{n-2}, C_j]$ with $0 \leq j \leq n-2$. The power to which such expression appears is (by the same summation of the original power and the one from the correction)

$$2n(j+1) + 2n\left(\frac{(n-1)^2}{4} - (j+1)(n-2-j)\right) = 2n\left(\frac{(n-1)^2}{4} - (j+1)(n-3-j)\right);$$

again note that this statement holds not only for the usual corrections, i.e., for those corresponding to $(n-1)/2 \leq j \leq n-3$, but also for the value $j = n-2$ and for $0 \leq j \leq (n-1)/2$ by what we said above. Here the change of index done is to change j to $n-2-k$, which gives that the expression $[C_{n-2}, C_{n-2-k}]$ appears in $g_{N(\Delta)}$ also to the power $2n[(n^2 + 2n - 3)/4 - k(n-k)]$ (substitute $j = n-2-k$ in the relevant expression and see that you can use the same trick).

We can now describe the corrections of the second type. Fix $0 \leq k \leq n-3$ (as for $k = n-2$ there is nothing to correct—the expression $[C_0, C_{n-2}]$ appear in both g_Δ and $g_{N(\Delta)}$ to the same power $2n(n^2 - 6n + 9)/4)$, and we see that $[C_{n-2}, C_{n-2-k}]$ appears in $g_{N(\Delta)}$ and not in g_Δ. From here we see by induction on i that we shall have to multiply g_Δ by

$$[C_{n-2-i}, C_{n-2-i-k}]^{2n[(n^2+2n-3)/4 - k(n-k)]}[C_{i+1}, C_{i+1+k}]^{2n[(n^2+2n-3)/4 - k(n-k)]},$$

and therefore $g_{N(\Delta)}$ will be multiplied by

$$[C_i, C_{i+k}]^{2n[(n^2+2n-3)/4 - k(n-k)]}[C_{n-3-i}, C_{n-3-i-k}]^{2n[(n^2+2n-3)/4 - k(n-k)]}.$$

This is true except for the case where $n-3-k$ is even and $i = (n-3-k)/2$, since then $n-2-i$ and $n-2-i-k$ sum to $n-1$ and thus we multiply g_Δ only by

$$[C_{n-2-i}, C_{n-2-i-k}]^{2n[(n^2+2n-3)/4 - k(n-k)]} = [C_{(n-1+k)/2}, C_{(n-1-k)/2}]^{2n[(n^2+2n-3)/4 - k(n-k)]}$$

and $g_{N(\Delta)}$ is multiplied only by

$$[C_i, C_{i+k}]^{2n[(n^2+2n-3)/4 - k(n-k)]} = [C_{(n-3-k)/2}, C_{(n-3+k)/2}]^{2n[(n^2+2n-3)/4 - k(n-k)]}.$$

Indeed, in order to correct the fact that $[C_{n-2}, C_{n-2-k}]$ appears in $g_{N(\Delta)}$ and not in g_Δ, we have to do the correction with $i = 0$; since after doing the ith correction we see that $[C_{n-3-i}, C_{n-3-i-k}]$ now appears in $g_{N(\Delta)}$ and not in g_Δ, we must proceed with the $(i+1)$th correction. The way this process ends depends on the parity of $n-3-k$. If it is odd, then after the correction corresponding to $i = (n-4-k)/2$ we see that the expression $[C_{n-3-i}, C_{n-3-i-k}]$ now appears also in g_Δ as $[C_{i+1}, C_{i+1+k}]$ (as these expressions are $[C_{(n-2+k)/2}, C_{(n-2-k)/2}]$ and $[C_{(n-2-k)/2}, C_{(n-2+k)/2}]$ respectively, which are equal up to the usual sign). If it is even, then after doing the shorter correction corresponding to $i = (n-3-k)/2$, the new expression $[C_{(n-3-k)/2}, C_{(n-3+k)/2}]$ from $g_{N(\Delta)}$ already appears in g_Δ as $[C_{i+1}, C_{i+1+k}]$ for $i = (n-5-k)/2$. In any case we find that after doing all these corrections we have actually multiplied g_Δ by

$$\prod_{j=1}^{n-2-k} [C_j, C_{j+k}]^{2n[(n^2+2n-3)/4-k(n-k)]}$$

and $g_{N(\Delta)}$ by

$$\prod_{j=0}^{n-3-k} [C_j, C_{j+k}]^{2n[(n^2+2n-3)/4-k(n-k)]}.$$

To see this, merge all these expressions together and write $j = n - 2 - i - k$ in the first expressions correcting g_Δ, $j = i + 1$ in the second expressions correcting g_Δ, $j = n - 3 - i - k$ in the first expressions correcting $g_{N(\Delta)}$, and $j = i$ in the second expressions correcting $g_{N(\Delta)}$. Then follow the indices (for both even and odd values of $n - 3 - k$), use the fact that $[Y, Z] = [Z, Y]$ (taken to even powers), and verify that these are the results.

Now it is easy to see that the obtained expressions for g_Δ and $g_{N(\Delta)}$ are the same: For every $0 \leq k \leq n - 1$, we have now that the expression $[C_j, C_{j+k}]$ appears in both g_Δ and $g_{N(\Delta)}$ to the power $2n[(n^2 + 2n - 3)/4 - k(n - k)]$ for any $-1 \leq j \leq n - 2 - k$. Indeed, for $j = -1$ we already know that for both of them. For the other values of j, we see that for g_Δ the value $j = 0$ was present before, and we now added the remaining values. As for $g_{N(\Delta)}$, the value $j = n - 2 - k$ already appeared (recall that $[C_{n-2}, C_{n-2-k}]$ is $[C_{n-2-k}, C_{n-2}]$ up to sign), and the last action we just did put in the remaining values. Therefore we obtain both for g_Δ and for $g_{N(\Delta)}$ the same expression

$$\prod_{k=0}^{n-1} \prod_{j=-1}^{n-2-k} [C_j, C_{j+k}]^{2n[(n^2+2n-3)/4-k(n-k)]}$$

as the nominee for h_Δ, which is very similar to the formula for h_Δ in Eq. (4.7). The only difference is the fact that the number independent of k appearing in the power is $(n^2 + 2n - 3)/4$ here and $(n^2 - 1)/4$ there. This comes from the fact that our expression is not reduced. Since the number $k(n - k)$ is maximal for $k = (n - 1)/2$ or $k = (n + 1)/2$ and then equals $(n^2 - 1)/4$, we see that every expression $[C_i, C_j]$ appears at least to the power $2n(n - 1)/2$. This means that the expression

$$\left[\bigcup_{j=-1}^{n-2} C_j, \bigcup_{j=-1}^{n-2} C_j \right]^{2n(n-1)/2} = \prod_{0 \leq i < j \leq nr-1} (\lambda_i - \lambda_j)^{2n(n-1)/2}$$

(which is clearly invariant under everything) is a "common divisor" for all of these nominees for h_Δ. By dividing all the nominees for h_Δ by this invariant "common divisor", we obtain the expression for h_Δ given in Eq. (4.7).

The fact that we encounter this common divisor and have to do this reduction comes from the fact that we have started with the unreduced expression for g_Δ in Eq. (4.2). Had we started with the reduced formula for g_Δ with n odd given in Eq. (4.3), we could have gotten directly to the formula for h_Δ given in Eq. (4.7), without needing any reduction. The reader can verify that doing so would be in the same order of stages, as follows. In the jth correction of the first type the power will be $2n[(n-1)(n-3)/4 - j(n-2-j)]$; in the correction with $[G,G]$ it will thus be $2n(n-1)(n-3)/4$; and in the corrections of the second type with the index k, it will be $2n[(n^2-1)/4 - k(n-k)]$. Note that the latter power is the one which appears in h_Δ in Eq. (4.7) and thus we do obtain this h_Δ without any reduction. We conclude this section by remarking that the special case $n = 5$ here recovers exactly the process described in Section 4.2.

A.3 Nonsingular Case, Even n

Now assume that n is even, hence $e = 1$, and once again we start by correcting the expressions involving the set C_{-1}. We start by multiplying g_Δ by the expressions

$$[C_{-1},C_j]^{nt}[C_0,C_j]^{nt}[C_{-1},C_{n-1-j}]^{nt}[C_0,C_{n-1-j}]^{nt}$$

with decreasing order of j and the corresponding t. Here the largest value of j, such that $[C_{-1},C_j]$ appears in $g_{N(\Delta)}$ to a power larger than the one to which it appears in g_Δ, is $(n-4)/2$; for $j = (n-2)/2$ these powers are the same, which is not surprising since N fixes both the sets C_{-1} and $C_{(n-2)/2}$—note that this does not happen for any set when n is odd. The corresponding difference is $n(n/2) - n(n-4)/2 = 2n$. In the process of decreasing induction on j, we now find that when we deal with the expression $[C_{-1},C_j]$ the proper multiplying correction for g_Δ has to be

$$([C_{-1},C_j][C_0,C_j][C_{-1},C_{n-1-j}][C_0,C_{n-1-j}])^{n[n(n-2)/4-j(n-1-j)]},$$

so that $g_{N(\Delta)}$ will be multiplied by

$$([C_{-1},C_{n-2-j}][C_{n-2},C_{n-2-j}][C_{-1},C_{j-1}][C_{n-2},C_{j-1}])^{n[n(n-2)/4-j(n-1-j)]}.$$

To see this note that for $j = (n-4)/2$ we have $n(n-2)/4 - j(n-1-j) = 2$ and we indeed start with the power $2n$. Then after correcting the expression $[C_{-1},C_j]$ the next expression $[C_{-1},C_{j-1}]$ still appears in g_Δ to the power $n(j-1)$ but appears in $g_{N(\Delta)}$ to the power

$$n(n-1-j) + n\left(\frac{n(n-2)}{4} - j(n-1-j)\right) = n\left(\frac{n(n-2)}{4} - (j-1)(n-1-j)\right),$$

by the usual summation argument. It is easy to see that the difference between these two powers is the required difference $n[n(n-2)/4 - (j-1)(n-j)]$. The power analyzing argument again extends to $j = 0$ and shows that the power to which $[C_{-1},C_0]$ appears in $g_{N(\Delta)}$ is $n[n(n-2)/4]$ and in g_Δ it does not appear at all. Hence (recall that the corresponding correction is based on the expression $[G,G]$) we multiply g_Δ by

$$[G,G]^{n[n(n-2)/4]} = [C_{-1},C_{-1}]^{n[n(n-2)/4]}[C_{-1},C_0]^{n[n(n-2)/4]}[C_0,C_0]^{n[n(n-2)/4]},$$

and thus multiply $g_{N(\Delta)}$ by

$$[C_{-1},C_{-1}]^{n[n(n-2)/4]}[C_{-1},C_{n-2}]^{n[n(n-2)/4]}[C_{n-2},C_{n-2}]^{n[n(n-2)/4]}.$$

The verification that we have completed the corrections of the first type now goes as follows. For $0 \le j \le (n-4)/2$, we now find that the expression $[C_{-1},C_j]$ appears in g_Δ to the power which is the sum

$$nj + n\left(\frac{n(n-2)}{4} - j(n-1-j)\right) = n\left(\frac{n(n-2)}{4} - j(n-2-j)\right).$$

Here as well, this holds also for $j = 0$, by continuing the induction as before, and for $j = -1$, as the calculation

$$n(n-1) + n\frac{n(n-2)}{4} = n\frac{n^2+2n-4}{4}$$

and substitution show. For j being $(n-2)/2$ and $n/2$, it holds as there is no correction for g_Δ and substitution gives back the original values $n(n-2)/2$ and $n^2/2$. For $(n+2)/2 \le j \le n-2$, it holds by the same calculations, as the corresponding correction is the one involving $[C_{-1},C_{n-1-j}]$ and the power which comes from the correction is invariant under replacing j by $n-1-j$. Hence, as before, this statement holds for every $-1 \le j \le n-2$. As for the powers to which $[C_{-1},C_j]$ with $-1 \le j \le n-2$ appears in $g_{N(\Delta)}$, we find that they are the same as the powers to which they appear in g_Δ. This is again seen either by the fact that the changes in $g_{N(\Delta)}$ corresponded via the action of N from Definition 2.16 to those in g_Δ, or by following the induction we did and verifying that the expressions $[C_{-1},C_j]$ with $n/2 \le j \le n-2$ (which appear in the original g_Δ to powers larger than those to which they appear in the original $g_{N(\Delta)}$) are also corrected in this process. The details of the latter verification are as follows. First, the expression $[C_{-1},C_{n/2}]$ appeared in g_Δ to a power larger by $2n$ than the power to which it appeared in $g_{N(\Delta)}$. Then, in the inductive process, we see that correcting the expression $[C_{-1},C_j]$ with $1 \le j \le (n-4)/2$ in g_Δ also dealt with the difference of powers for $[C_{-1},C_{n-2-j}]$, and after that, the difference between the powers to which $[C_{-1},C_{n-1-j}]$ appeared in g_Δ and $g_{N(\Delta)}$ was

$$n\left(\frac{n(n-2)}{4} - (j-1)(n-1-j)\right) - n(j-1) = n\left(\frac{n(n-2)}{4} - (j-1)(n-j)\right).$$

The next correction deals with this difference. At the last stage, $[C_{-1},C_{n-2}]$ did not appear in g_Δ and appeared in $g_{N(\Delta)}$ to the power $n[n(n-2)/4]$; after the last correction, we see that $[C_{-1},C_{-1}]$ now appears also in $g_{N(\Delta)}$ to the power

$$n(n-1) + n\frac{n(n-2)}{4} = n\frac{n^2+2n-4}{4},$$

We again want to write the expressions not involving C_{-1} that appear in g_Δ and $g_{N(\Delta)}$, in order to see what are the corrections of the second type to be done now. Also here we find that the only expressions of that sort that appear in g_Δ are the expressions $[C_0, C_j]$ with $0 \le j \le n-2$, which now appear to the power

$$n(n-1-j) + n\left(\frac{n(n-2)}{4} - j(n-1-j)\right) = n\left(\frac{n^2+2n-4}{4} - j(n-j)\right)$$

(expand $(j-1)(n-1-j)$ as $j(n-j) - (n-1)$ again). Also here, note that what we said in the previous paragraph extends the validity of this from $1 \le j \le (n-4)/2$ also to $j=0$ and to $(n-2)/2 \le j \le n-2$. We now again change the index j to k (and see that $[C_0, C_k]$ appears in g_Δ to the power $n[(n^2+2n-4)/4 - k(n-k)]$), and once again substituting $j = k-1$ in the expressions from the previous paragraph shows that the expression $[C_{-1}, C_{k-1}]$ appears now in both g_Δ and $g_{N(\Delta)}$ to exactly the same power. This resembles more the expressions appearing in the desired expression h_Δ from Eq. (4.8), and will again be useful when we examine the expressions we get in the end. In the new $g_{N(\Delta)}$, the expressions not involving C_{-1} that appear in it are once again the expressions $[C_{n-2}, C_j]$ with $0 \le j \le n-2$, and the power to which such an expression appears is

$$n(j+1) + n\left(\frac{n(n-2)}{4} - (j+1)(n-2-j)\right) = n\left(\frac{n(n-2)}{4} - (j+1)(n-3-j)\right)$$

(the relevant correction is the one we did for the expression $[C_{-1}, C_{n-2-j}]$ in g_Δ). Again note that this holds not only for $n/2 \le j \le n-3$, but for all $0 \le j \le n-2$ by what we said above. We again change the index j to $n-2-k$, and do the usual trick, in order to obtain that the expression $[C_{n-2}, C_{n-2-k}]$ appears in $g_{N(\Delta)}$ to the same power $n[(n^2+2n-4)/4 - k(n-k)]$.

The corrections of the second type that we do now are very similar to what we did in Section A.2. We fix some $0 \le k \le n-3$ (again for $k = n-2$ there is nothing to correct), and we find that $[C_{n-2}, C_{n-2-k}]$ appears in $g_{N(\Delta)}$ and not in g_Δ. From here apply the same induction on i as in Section A.2, to see that what we do here is to multiply g_Δ by

$$[C_{n-2-i}, C_{n-2-i-k}]^{n[(n^2+2n-4)/4 - k(n-k)]} [C_{i+1}, C_{i+1+k}]^{n[(n^2+2n-4)/4 - k(n-k)]}$$

and $g_{N(\Delta)}$ by

$$[C_i, C_{i+k}]^{n[(n^2+2n-4)/4 - k(n-k)]} [C_{n-3-i}, C_{n-3-i-k}]^{n[(n^2+2n-4)/4 - k(n-k)]}$$

repeatedly (while for even $n-3-k$ and $i = (n-3-k)/2$ we reduce this action to multiplication of g_Δ only by

$$[C_{n-2-i}, C_{n-2-i-k}]^{n[(n^2+2n-4)/4 - k(n-k)]} = [C_{(n-1+k)/2}, C_{(n-1-k)/2}]^{n[(n^2+2n-4)/4 - k(n-k)]}$$

and $g_{N(\Delta)}$ only by

$$[C_i,C_{i+k}]^{n[(n^2+2n-4)/4-k(n-k)]} = [C_{(n-3-k)/2},C_{(n-3+k)/2}]^{n[(n^2+2n-4)/4-k(n-k)]}$$

as before), and exactly the same argument as in Section A.2 shows that this sums up to multiplying g_Δ by

$$\prod_{j=1}^{n-2-k} [C_j,C_{j+k}]^{n[(n^2+2n-4)/4-k(n-k)]}$$

and $g_{N(\Delta)}$ by

$$\prod_{j=0}^{n-3-k} [C_j,C_{j+k}]^{n[(n^2+2n-4)/4-k(n-k)]}.$$

The same considerations as in Section A.2 show that the obtained expressions for g_Δ and $g_{N(\Delta)}$ are the same expression; one can see that for each $0 \le k \le n-1$, we now have that all the expressions $[C_j,C_{j+k}]$ appear in both g_Δ and $g_{N(\Delta)}$ to the power $n[(n^2+2n-4)/4-k(n-k)]$ for every $-1 \le j \le n-2-k$, and thus both the expressions for g_Δ and $g_{N(\Delta)}$ are equal to

$$\prod_{k=0}^{n-1}\prod_{j=-1}^{n-2-k} [C_j,C_{j+k}]^{n[(n^2+2n-4)/4-k(n-k)]}.$$

This nominee for h_Δ is again very similar to the formula for h_Δ in Eq. (4.8), but the number, independent of k, appearing in the power is $(n^2+2n-4)/4$ here and $n^2/4$ there. This difference comes again from the fact that our expression is not reduced. The expression $k(n-k)$ for even n is maximal for $k=n/2$ with value $n^2/4$, so that every expression $[C_i,C_j]$ appears at least to the power $n(n-2)/2$. Hence we can divide all the nominees for h_Δ by the "common divisor"

$$\left[\bigcup_{j=-1}^{n-2}C_j,\bigcup_{j=-1}^{n-2}C_j\right]^{n(n-2)/2} = \prod_{0\le i<j\le nr-1}(\lambda_i-\lambda_j)^{n(n-2)/2}$$

(an expression which is clearly invariant under everything), and after doing so we obtain exactly the expression for h_Δ from Eq. (4.8).

This common divisor and this reduction occur since we have started with the unreduced expression for g_Δ in Eq. (4.2). We can, as before, start with the reduced formula for g_Δ with even n given in Eq. (4.4) and get directly to the formula for h_Δ given in Eq. (4.8), without reductions. The reader can check that this is done by the same stages (in the same order), but with the following differences. In the jth correction of the first type, replace the power by $n[(n-2)^2/4-j(n-2-j)]$; in the correction with $[G,G]$ replace it by $n(n-2)^2/4$; and in the corrections of the second type with the index k, it will be replaced by $n[n^2/4-k(n-k)]$. This power is the same as the one appearing in h_Δ in Eq. (4.8), so that we obtain this h_Δ without any reduction.

A.4 The Proper Order to do the Corrections in the Singular Case

We now turn to describing the construction in the singular case. Once again we start by explaining why we do the operations in a specific order. For this we invoke Proposition 5.3 and Eq. (5.5) to see that

$$g_{\Delta} = ([C_{-1},C_{-1}][C_{-1},D_0][D_0,D_0][C_0,C_0][C_0,D_{-1}][D_{-1},D_{-1}])^{en(n-1)}$$

$$\times \prod_{j=1}^{n-2}([C_{-1},C_j][D_0,C_j][C_0,D_j][D_{-1},D_j])^{enj}$$

$$\times \prod_{j=1}^{n-2}([C_{-1},D_j][D_0,D_j][C_0,C_j][D_{-1},C_j])^{en(n-1-j)},$$

Since by Definition 2.18 the action of the operator N is to fix the sets C_{-1} and D_{-1} and interchange each C_j and each D_j with $0 \le j \le n-2$ by C_{n-2-j} and D_{n-2-j} respectively, we see that

$$g_{N(\Delta)} = ([C_{-1},C_{-1}][C_{-1},D_{n-2}][D_{n-2},D_{n-2}][C_{n-2},C_{n-2}][C_{n-2},D_{-1}][D_{-1},D_{-1}])^{en(n-1)}$$

$$\times \prod_{j=1}^{n-2}([C_{-1},C_j][D_{n-2},C_j][C_{n-2},D_j][D_{-1},D_j])^{en(n-2-j)}$$

$$\times \prod_{j=1}^{n-2}([C_{-1},D_j][D_{n-2},D_j][C_{n-2},C_j][D_{-1},C_j])^{en(j+1)}$$

(again replacing the index j of the product by $n-2-j$). The expressions by which we divide must now remain the same when putting a tilde or a hat on all of the sets. Once again we have to separate these expressions into those involving the sets C_{-1}, D_{-1}, C_0 and D_0 and to those which do not; we start with the former ones. We recall the sets $G = C_{-1} \cup C_0$ and $I = D_{-1} \cup D_0$, and we know that we have the equalities

$$\widetilde{G} = \widehat{G} = G, \quad \widetilde{C_j} = \widehat{C_j} = C_{n-1-j}, \quad \widetilde{C_{n-1-j}} = \widehat{C_{n-1-j}} = C_j,$$

$$\widetilde{I} = \widehat{I} = I, \quad \widetilde{D_j} = \widehat{D_j} = D_{n-1-j}, \quad \text{and} \quad \widetilde{D_{n-1-j}} = \widehat{D_{n-1-j}} = D_j,$$

where j is assumed to satisfy $1 \le j \le n-2$ here. We adopt again the notation Y_j and Z_j for either C_j or D_j, and continue using the notation $Y = Z$ or $Y \ne Z$ there. All this again means that if some expression $[Y_{-1},Z_j]^{ent}$ or $[Y_0,Z_j]^{ent}$ with $1 \le j \le n-2$ (and $j \ne (n-1)/2$ in the odd n case, since this value is exceptional as in the nonsingular case) and some number t is involved, then the corresponding expression by which we have to multiply g_{Δ} has to be

$$[Y_{-1},Z_j]^{ent}[Y_0,Z_j]^{ent}[Y_{-1},Z_{n-1-j}]^{ent}[Y_0,Z_{n-1-j}]^{ent}.$$

Therefore $g_{N(\Delta)}$ will be multiplied by

$$[Y_{-1},Z_{n-2-j}]^{ent}[Y_{n-2},Z_{n-2-j}]^{ent}[Y_{-1},Z_{j-1}]^{ent}[Y_{n-2},Z_{j-1}]^{ent}$$

(as seen by replacing any index $0 \leq k \leq n-2$ by its complement with respect to $n-2$ and leaving $k = -1$ fixed).

When examining the expressions with no set C_{-1}, D_{-1}, C_0, and D_0 involved, we see that when correcting some expression $[Y_i, Z_j]^{ent}$ with i and j being between 1 and $n-2$ and some number t (the exceptional case is again $i + j = n - 1$ if $Y = Z$, but only $i = j = (n-1)/2$ for odd n if $Y \neq Z$) then, as in the nonsingular case, the expression by which we have to multiply g_Δ has to be

$$[Y_i, Z_j]^{ent}[Y_{n-1-i}, Z_{n-1-j}]^{ent}.$$

This means that $g_{N(\Delta)}$ will be multiplied by

$$[Y_{n-2-i}, Z_{n-2-j}]^{ent}[Y_{i-1}, Z_{j-1}]^{ent}.$$

We shall continue calling corrections involving expressions including C_{-1} and D_{-1} *corrections of the first type* and corrections involving expressions not including C_{-1} and D_{-1} *corrections of the second type*. In addition, corrections that are based on expressions $[Y_i, Z_j]$ will be called *corrections of pure type* if $Y = Z$, and *corrections of mixed type* if $Y \neq Z$.

The first thing we note is that the expressions of pure type and the expressions of mixed type can be dealt with separately, since correcting one does not affect the other. The other thing we note is why, in each of these types, we have to do the corrections of the first type with decreasing order of j, and only after this, deal with the corrections of second type. This is because the assertion we gave in the nonsingular case, about corrections of the second type having no effect on the expressions dealt with in corrections of the first type (since again any expression not involving C_{-1} and D_{-1} in $g_{N(\Delta)}$ involves C_{n-2} and D_{n-2}), holds equally here, as the previous paragraph shows. Furthermore, the same holds for the assertion about the corrections dealing with $[Y_{-1}, Z_j]$ changing the power to which $[Y_{-1}, Z_{j-1}]$ appears in $g_{N(\Delta)}$ (which implies working in decreasing order of j in both pure and mixed types). This is why we work in this order here as well.

We can now describe the details of the process, again splitting into the cases of even n and odd n.

A.5 Singular Case, Odd n

We start with the odd n case, hence $e = 2$, and we know that the corrections of the pure type should be started with correcting the expressions involving the sets C_{-1} and D_{-1}. Since in this part the behavior of the expressions of pure type is different from the behavior of those of mixed type, we do them separately. We start with those of pure type. These corrections resemble very much the corrections of the first type we did in Section A.2. The first expressions by which we multiply g_Δ are of the form

$$[C_{-1}, C_j]^{2nt} [C_0, C_j]^{2nt} [C_{-1}, C_{n-1-j}]^{2nt} [C_0, C_{n-1-j}]^{2nt}$$

and

$$[D_{-1}, D_j]^{2nt} [D_0, D_j]^{2nt} [D_{-1}, D_{n-1-j}]^{2nt} [D_0, D_{n-1-j}]^{2nt},$$

and we do this in decreasing order of j and taking care of the power t every time. This is again done by decreasing induction on j, starting with the largest value $j = (n-3)/2$, and we see that when we deal with the expressions $[C_{-1}, C_j]$ and $[D_{-1}, D_j]$ the proper multiplying corrections for g_Δ have to be

$$([C_{-1}, C_j][C_0, C_j][C_{-1}, C_{n-1-j}][C_0, C_{n-1-j}])^{2n[(n-1)^2/4 - j(n-1-j)]}$$

and

$$([D_{-1}, D_j][D_0, D_j][D_{-1}, D_{n-1-j}][D_0, D_{n-1-j}])^{2n[(n-1)^2/4 - j(n-1-j)]}.$$

Hence $g_{N(\Delta)}$ will be multiplied by

$$([C_{-1}, C_{n-2-j}][C_{n-2}, C_{n-2-j}][C_{-1}, C_{j-1}][C_{n-2}, C_{j-1}])^{2n[(n-1)^2/4 - j(n-1-j)]}$$

and

$$([D_{-1}, D_{n-2-j}][D_{n-2}, D_{n-2-j}][D_{-1}, D_{j-1}][D_{n-2}, D_{j-1}])^{2n[(n-1)^2/4 - j(n-1-j)]}.$$

The reason is exactly as in Section A.2, which applies just as well to the sets D_j, $-1 \le j \le n-2$ since in any expression (g_Δ, h_Δ, etc.) in Chapter 5 an expression $[D_i, D_j]$ appears to the same power to which the corresponding expression $[C_i, C_j]$ appears. Just as when dealing with expressions involving $[C_{-1}, C_0]$ we had to work with powers of $[G, G]$, we see that when dealing with expressions involving $[D_{-1}, D_0]$, we have to work with powers of $[I, I]$ (as we know that $\widetilde{G} = \widehat{G} = G$ and $\widetilde{I} = \widehat{I} = I$). Hence, we see that after doing all this, the further corrections to be done are to multiply g_Δ by

$$[G, G]^{2n(n-1)^2/4} = [C_{-1}, C_{-1}]^{2n(n-1)^2/4} [C_{-1}, C_0]^{2n(n-1)^2/4} [C_0, C_0]^{2n(n-1)^2/4}$$

and

$$[I, I]^{2n(n-1)^2/4} = [D_{-1}, D_{-1}]^{2n(n-1)^2/4} [D_{-1}, D_0]^{2n(n-1)^2/4} [D_0, D_0]^{2n(n-1)^2/4},$$

and thus multiply $g_{N(\Delta)}$ by

$$[C_{-1}, C_{-1}]^{2n(n-1)^2/4} [C_{-1}, C_{n-2}]^{2n(n-1)^2/4} [C_{n-2}, C_{n-2}]^{2n(n-1)^2/4}$$

and

$$[D_{-1}, D_{-1}]^{2n(n-1)^2/4} [D_{-1}, D_{n-2}]^{2n(n-1)^2/4} [D_{n-2}, D_{n-2}]^{2n(n-1)^2/4}.$$

The fact that this completes the corrections of the first pure type is now verified as in Section A.2. The same argument holds for the expressions $[D_{-1}, D_j]$ as it holds for $[C_{-1}, C_j]$, by the remark in the previous paragraph. This also holds for the extension of the statement that the power to which the expression $[C_{-1}, C_j]$, and here also the expression $[D_{-1}, D_j]$, appears in both g_Δ and $g_{N(\Delta)}$ is $2n \left[(n-1)^2/4 - j(n-2-j) \right]$ for every $-1 \le j \le n-2$, and not only for $0 \le j \le (n-3)/2$. Therefore in the expressions of pure type we can go on to the corrections of second type.

Before doing so, we want to begin doing also the corrections of the mixed type, which, by what we said above, should also be started with correcting the expressions involving the sets C_{-1} and D_{-1} (i.e., the first type). These corrections start by multiplying g_Δ by

$$[C_{-1}, D_j]^{2nt} [C_0, D_j]^{2nt} [C_{-1}, D_{n-1-j}]^{2nt} [C_0, D_{n-1-j}]^{2nt}$$

and

$$[D_{-1}, C_j]^{2nt} [D_0, C_j]^{2nt} [D_{-1}, C_{n-1-j}]^{2nt} [D_0, C_{n-1-j}]^{2nt},$$

again with decreasing j and the proper t at every stage, and they have no analog in the nonsingular case. The largest value of j such that $[C_{-1}, D_j]$ and $[D_{-1}, C_j]$ appear in $g_{N(\Delta)}$ is now $n-2$, with difference $2n(n-1) - 2n \times 1 = 2n(n-2)$. From there we work by decreasing induction on j, and we obtain that at the stage corresponding to the expressions $[C_{-1}, D_j]$ and $[D_{-1}, C_j]$, the corrections are multiplying g_Δ by

$$([C_{-1}, D_j][C_0, D_j][C_{-1}, D_{n-1-j}][C_0, D_{n-1-j}])^{2nj(n-1-j)}$$

and

$$([D_{-1}, C_j][D_0, C_j][D_{-1}, C_{n-1-j}][D_0, C_{n-1-j}])^{2nj(n-1-j)}.$$

By the usual argument, $g_{N(\Delta)}$ will be multiplied by

$$([C_{-1}, D_{n-2-j}][C_{n-2}, D_{n-2-j}][C_{-1}, D_{j-1}][C_{n-2}, D_{j-1}])^{2nj(n-1-j)}$$

and

$$([D_{-1}, C_{n-2-j}][D_{n-2}, C_{n-2-j}][D_{-1}, C_{j-1}][D_{n-2}, C_{j-1}])^{2nj(n-1-j)}.$$

This is easily verified for $j = n-2$, as $j(n-1-j) = n-2$, and we start with the power $2n(n-2)$ as needed. For the following values of j, after correcting the expressions $[C_{-1}, D_j]$ and $[D_{-1}, C_j]$, we have that the next expressions $[C_{-1}, D_{j-1}]$ and $[D_{-1}, C_{j-1}]$ now appear in g_Δ to the power $2nj(n-j)$; in $g_{N(\Delta)}$, summing the powers from the original $g_{N(\Delta)}$ and from the last correction, gives that these expressions now appear there to the power

$$2nj + 2nj(n-1-j) = 2nj(n-j).$$

The difference between these two powers is clearly $2n(j-1)(n-j)$, which allows going on to the next step of the induction. We continue with this process until we reach the value $j = (n-1)/2$, where the corrections for g_Δ are only

$$[C_{-1}, D_{(n-1)/2}]^{2nj(n-1-j)} [C_0, D_{(n-1)/2}]^{2nj(n-1-j)}$$

and

$$[D_{-1}, C_{(n-1)/2}]^{2nj(n-1-j)} [D_0, C_{(n-1)/2}]^{2nj(n-1-j)},$$

and the corrections for $g_{N(\Delta)}$ are thus

$$[C_{-1}, D_{(n-3)/2}]^{2nj(n-1-j)} [C_{n-2}, D_{(n-3)/2}]^{2nj(n-1-j)}$$

and

$$[D_{-1}, C_{(n-3)/2}]^{2nj(n-1-j)} [D_{n-2}, C_{(n-3)/2}]^{2nj(n-1-j)}.$$

We now stop, since for the preceding value $(n+1)/2$ of j we already had multiplication of g_Δ by powers of $[C_{-1}, D_{(n-3)/2}]$ and $[D_{-1}, C_{(n-3)/2}]$.

Let us now show that we have completed the corrections of the first mixed type, which allows us to move on to corrections of the second type (for both pure and mixed expressions). This is done as follows. For $(n-1)/2 \le j \le n-2$ we find that the expressions $[C_{-1}, D_j]$ and $[D_{-1}, C_j]$ appear in g_Δ to the power

$$2n(n-1-j) + 2nj(n-1-j) = 2n(j+1)(n-1-j)$$

(again presented as the sum of the power from the original g_Δ and the power from the corresponding correction). Note that this includes $j = (n-1)/2$ by the fact that we could continue the induction to this case. However, also here the usual arguments show that this holds for every $-1 \le j \le n-2$. Indeed, for $1 \le j \le (n-3)/2$, the same calculation holds since the corresponding corrections are those involving $[C_{-1}, D_{n-1-j}]$ and $[D_{-1}, C_{n-1-j}]$, and the power that comes from the correction is invariant under replacing j by $n-1-j$. For $j = -1$ and $j = 0$ there is no correction, and we do see that $(j+1)(n-1-j)$ vanishes for the former and gives $n-1$ for the latter, in correspondence with the fact that in the original g_Δ, the expressions $[C_{-1}, D_0]$ and $[D_{-1}, C_0]$ appeared to the power $2n(n-1)$, while the expression $[C_{-1}, D_{-1}]$ (which is the same as $[D_{-1}, C_{-1}]$ up to the usual sign) indeed did not appear at all. When examining the powers to which the expressions $[C_{-1}, D_j]$ and $[D_{-1}, C_j]$ with $-1 \le j \le n-2$ appear in $g_{N(\Delta)}$, we again have two ways to show that they are the same as the powers to which they appear in g_Δ. As before, the fact that the manipulations on $g_{N(\Delta)}$ are obtained by applying the operator N from Definition 2.18 (fixing every index -1 and interchanging every index $0 \le j \le n-2$ with $n-2-j$) to those done on g_Δ, and the fact that each of the final powers we have obtained is invariant under replacing j by $n-2-j$ is one way to see this. The other way is by verifying that in the inductive process we also have corrected the expressions $[C_{-1}, D_j]$ and $[D_{-1}, C_j]$ with $0 \le j \le (n-3)/2$ (which appear in the original g_Δ to powers larger than those to which they appear in the original $g_{N(\Delta)}$) on the way. Indeed, at first the expressions $[C_{-1}, D_0]$ and $[D_{-1}, C_0]$ appeared in g_Δ to a power larger by $2n(n-2)$ than the power to which they appeared in $g_{N(\Delta)}$. In the following steps we see that after the corrections corresponding to some $(n+1)/2 \leq j \leq n-2$ in g_Δ fixed the difference of powers also for $[C_{-1}, D_{n-2-j}]$ and $[D_{-1}, C_{n-2-j}]$, the next expressions $[C_{-1}, D_{n-1-j}]$ and $[D_{-1}, C_{n-1-j}]$ now appear in g_Δ and $g_{N(\Delta)}$ to powers whose difference is

$$2nj(n-j) - 2n(n-j) = 2n(j-1)(n-j).$$

This difference is easily seen to be fixed by the next correction. This of course concludes the case $j = (n-3)/2$, which corresponds to the simpler corrections that were the last we did of the first and mixed type.

We are now done with all the corrections of the first type, so we write the expressions not involving the sets C_{-1} and D_{-1} that appear now in g_Δ and $g_{N(\Delta)}$ to see how to move on. We find that the only expressions that appear in g_Δ are $[Y_0, Z_j]$ with $0 \leq j \leq n-2$. The power to which the expressions appear is $2n[(n^2 + 2n - 3)/4 - j(n-j)]$ if $Y = Z$ (as in Section A.2), while if $Y \neq Z$ it is

$$2nj + 2nj(n-1-j) = 2nj(n-j),$$

by the usual summation argument. We have already explained why this holds for every $0 \leq j \leq n-2$. Now change the index j to k in all these expressions, and change j to $k-1$ in all the preceding expressions (those involve the sets C_{-1} and D_{-1}, both pure and mixed). When doing so, we see that the power to which any expression $[Y_{-1}, Z_{k-1}]$ appears is the same power to which $[Y_0, Z_k]$ does, and that this power is $2n[(n^2 + 2n - 3)/4 - k(n-k)]$ (as in Section A.2) if $Y = Z$ and $2nk(n-k)$ if $Y \neq Z$ (recall that $[C_{-1}, D_j]$ and $[D_{-1}, C_j]$ now appear to the power $2n(j+1)(n-1-j)$, and substitute $j = k-1$ there). This puts our expressions in a form more similar to the desired expression h_Δ from Eq. (5.10) and will make it easier to compare with it later. When we examine the new $g_{N(\Delta)}$ we see that the expressions not involving C_{-1} and D_{-1} that appear in it are $[Y_{n-2}, Z_j]$ with $0 \leq j \leq n-2$, which appear to the power $2n[(n-1)^2/4 - (j+1)(n-3-j)]$ if $Y = Z$ (again as in Section A.2), and if $Y \neq Z$ to the power

$$2n(n-2-j) + 2n(j+1)(n-2-j) = 2n(j+2)(n-2-j).$$

Note that the usual extension of these statements to every $0 \leq j \leq n-2$ holds also here. Changing the index j to $n-2-k$ gives (by the usual substitutions and manipulations) that the expression $[Y_{n-2}, Z_{n-2-k}]$ appears also in $g_{N(\Delta)}$ to the same power $2n[(n^2 + 2n - 3)/4 - k(n-k)]$ if $Y = Z$, or to the power $2nk(n-k)$ if $Y \neq Z$.

We can now say what the corrections of the second type are that have to be done. We formulate them in a way where we do the pure and mixed types together. For any fixed $0 \leq k \leq n-3$ (as for $k = n-2$ there is nothing to correct—in the pure type it is as in Section A.2 and in the mixed type the expressions $[C_0, D_{n-2}]$ and $[D_0, C_{n-2}]$ appear in both g_Δ and $g_{N(\Delta)}$ to the same power $2n(2n-4)$) define α_k to be $2n[(n^2 + 2n - 3)/4 - k(n-k)]$ if $Y = Z$ and $2nk(n-k)$ if $Y \neq Z$. Then the process goes just as in Section A.2, with the same inductive process and by the same reasoning. Explicitly, at the ith stage of the induction, we have to multiply g_Δ by the expression

$$[Y_{n-2-i}, Z_{n-2-i-k}]^{\alpha_k} [Y_{i+1}, Z_{i+1+k}]^{\alpha_k},$$

and thus $g_{N(\Delta)}$ will be multiplied by

$$[Y_i, Z_{i+k}]^{\alpha_k} [Y_{n-3-i}, Z_{n-3-i-k}]^{\alpha_k}.$$

In the intermediate case of $n-3-k$ even and $i = (n-3-k)/2$ this action reduces to multiplying g_Δ only by

$$[Y_{n-2-i}, Z_{n-2-i-k}]^{\alpha_k} = [Y_{(n-1+k)/2}, Z_{(n-1-k)/2}]^{\alpha_k},$$

and multiplying $g_{N(\Delta)}$ only by

$$[Y_i, Z_{i+k}]^{\alpha_k} = [Y_{(n-3-k)/2}, Z_{(n-3+k)/2}]^{\alpha_k}.$$

Indeed, by the same considerations as in Section A.2 (with more types of expressions and two different powers, but this does not make any significant changes in the process) we conclude that the process ends with the correction corresponding to $i = (n-4-k)/2$ for $n-3-k$ odd, and with the shorter correction corresponding to $i = (n-3-k)/2$ for $n-3-k$ even. Altogether, we get that we have actually multiplied g_Δ by

$$\prod_{j=1}^{n-2-k} [C_j, C_{j+k}]^{2n[(n^2+2n-3)/4-k(n-k)]} \prod_{j=1}^{n-2-k} [D_j, D_{j+k}]^{2n[(n^2+2n-3)/4-k(n-k)]}$$

$$\times \prod_{j=1}^{n-2-k} [C_j, D_{j+k}]^{2nk(n-k)} \prod_{j=1}^{n-2-k} [D_j, C_{j+k}]^{2nk(n-k)},$$

and $g_{N(\Delta)}$ by

$$\prod_{j=0}^{n-3-k} [C_j, C_{j+k}]^{2n[(n^2+2n-3)/4-k(n-k)]} \prod_{j=0}^{n-3-k} [D_j, D_{j+k}]^{2n[(n^2+2n-3)/4-k(n-k)]}$$

$$\times \prod_{j=0}^{n-3-k} [C_j, D_{j+k}]^{2nk(n-k)} \prod_{j=0}^{n-3-k} [D_j, C_{j+k}]^{2nk(n-k)}$$

(with the same remarks about the substitutions of indices as in Section A.2).

We now want to compare the obtained expressions for g_Δ and $g_{N(\Delta)}$ and see that they are the same. Indeed, for every $0 \le k \le n-1$, we now have that all the expressions $[Y_j, Z_{j+k}]$ with $-1 \le j \le n-2-k$ appear in both g_Δ and $g_{N(\Delta)}$ to the power $2n[(n^2+2n-3)/4 - k(n-k)]$ if $Y = Z$ and $2nk(n-k)$ if $Y \ne Z$ (this is just the power α_k from the previous paragraph). For $j = -1$, this was already verified for both of them. In g_Δ we already had the expressions corresponding to $j = 0$, and the last operation put in all the other values. In $g_{N(\Delta)}$ the expression with $j = n-2-k$

already appeared (use the fact that up to sign $[Y_{n-2}, Z_{n-2-k}]$ equals $[Z_{n-2-k}, Y_{n-2}]$, and the separation between $Y = Z$ and $Y \neq Z$ is of course preserved in this presentation), and we now added the remaining values. We then get for both g_Δ and $g_{N(\Delta)}$ the same expression

$$\prod_{k=0}^{n-1} \prod_{j=-1}^{n-2-k} ([C_j, C_{j+k}][D_j, D_{j+k}])^{2n[(n^2+2n-3)/4-k(n-k)]} ([C_j, D_{j+k}][D_j, C_{j+k}])^{2nk(n-k)}$$

to be the nominee for h_Δ, which is already very close to the formula for h_Δ in Eq. (5.10). The difference is only in the expressions of pure type, and it summarizes to the fact that the number, independent of k, appearing in the powers of the pure expressions, is $(n^2 + 2n - 3)/4$ here and $(n^2 - 1)/4$ there. This happens because of the fact that our expression is not reduced, and as in Section A.2, we see that every expression $[Y_i, Z_j]$ with $Y = Z$ appears at least to the power $2n(n-1)/2$. This means that the (clearly invariant) expressions

$$\left[\bigcup_{j=-1}^{n-2} C_j, \bigcup_{j=-1}^{n-2} C_j \right]^{2n(n-1)/2} = \prod_{0 \le i < j \le m-1} (\lambda_i - \lambda_j)^{2n(n-1)/2}$$

and

$$\left[\bigcup_{j=-1}^{n-2} D_j, \bigcup_{j=-1}^{n-2} D_j \right]^{2n(n-1)/2} = \prod_{1 \le i < j \le m} (\mu_i - \mu_j)^{2n(n-1)/2}$$

are "common divisors" for all of these nominees for h_Δ. Dividing all the nominees for h_Δ by these invariant "common divisors" gives us exactly the expression for h_Δ appearing in Eq. (5.10).

As always, this happens because we have started the calculations using the unreduced expression for g_Δ in Eq. (5.5). Here the reduced formula for g_Δ with n odd is Eq. (5.6), and starting from it brings us directly to the formula for h_Δ given in Eq. (5.10), without any reduction. Leaving the details for the reader to verify, we state that the stages of this construction are the same (including their ordering), but with different powers. In the jth correction of the first pure type the power will be $2n[(n-1)(n-3)/4 - j(n-2-j)]$, and in the corrections with $[G, G]$ and $[I, I]$, it will thus be $2n(n-1)(n-3)/4$. In the jth correction of the first mixed type the power will be $2n(j+1)(n-1-j)$, and in the corrections of the second type with the index k, it will be $2n[(n^2-1)/4 - k(n-k)]$ for the pure type, and $2nk(n-k)$ for the mixed type. Since these powers are exactly those which appear in h_Δ in Eq. (5.10), we see that this process indeed gives this h_Δ, without any reduction. Also in the singular case, the special case $n = 5$ of this construction is exactly the process described in Section 5.2. We also remark that as we saw, the mixed part of the process already gives us (for general odd n) the reduced part of h_Δ even when we start with the unreduced expression for g_Δ in Eq. (5.5).

A.6 Singular Case, Even n

We now turn to the even n case, hence $e = 1$, and as before we start with correcting the expressions of pure type that involve the sets C_{-1} and D_{-1}. This part is similar to what we did in Section A.3. Once again we start by multiplying g_Δ by

$$[C_{-1}, C_j]^{nt} [C_0, C_j]^{nt} [C_{-1}, C_{n-1-j}]^{nt} [C_0, C_{n-1-j}]^{nt}$$

and

$$[D_{-1}, D_j]^{nt} [D_0, D_j]^{nt} [D_{-1}, D_{n-1-j}]^{nt} [D_0, D_{n-1-j}]^{nt}$$

with decreasing j and some t. We do the decreasing induction on j as in Section A.3, starting with the largest value $j = (n-4)/2$ (again for $j = (n-2)/2$ there is nothing to do), and then in the correction corresponding to the expressions $[C_{-1}, C_j]$ and $[D_{-1}, D_j]$ the proper expressions by which we multiply g_Δ must be

$$([C_{-1}, C_j][C_0, C_j][C_{-1}, C_{n-1-j}][C_0, C_{n-1-j}])^{n[n(n-2)/4 - j(n-1-j)]}$$

and

$$([D_{-1}, D_j][D_0, D_j][D_{-1}, D_{n-1-j}][D_0, D_{n-1-j}])^{n[n(n-2)/4 - j(n-1-j)]}.$$

Therefore we multiply $g_{N(\Delta)}$ by

$$([C_{-1}, C_{n-2-j}][C_{n-2}, C_{n-2-j}][C_{-1}, C_{j-1}][C_{n-2}, C_{j-1}])^{n[n(n-2)/4 - j(n-1-j)]}$$

and

$$([D_{-1}, D_{n-2-j}][D_{n-2}, D_{n-2-j}][D_{-1}, D_{j-1}][D_{n-2}, D_{j-1}])^{n[n(n-2)/4 - j(n-1-j)]}.$$

The reason is exactly as in Section A.3, which again can be applied as well to the sets D_j, $-1 \le j \le n-2$, as in Section A.5. Once again we work with the set I as we do with the set G (as $\widetilde{G} = \widehat{G} = G$ and $\widetilde{I} = \widehat{I} = I$), and we see that after doing all this, the further corrections to be done are to multiply g_Δ by

$$[G, G]^{n[n(n-2)/4]} = [C_{-1}, C_{-1}]^{n[n(n-2)/4]} [C_{-1}, C_0]^{n[n(n-2)/4]} [C_0, C_0]^{n[n(n-2)/4]}$$

and

$$[I, I]^{n[n(n-2)/4]} = [D_{-1}, D_{-1}]^{n[n(n-2)/4]} [D_{-1}, D_0]^{n[n(n-2)/4]} [D_0, D_0]^{n[n(n-2)/4]}.$$

The corresponding action on $g_{N(\Delta)}$ is multiplying it by

$$[C_{-1}, C_{-1}]^{n[n(n-2)/4]} [C_{-1}, C_{n-2}]^{n[n(n-2)/4]} [C_{n-2}, C_{n-2}]^{n[n(n-2)/4]}$$

and

$$[D_{-1}, D_{-1}]^{n[n(n-2)/4]} [D_{-1}, D_{n-2}]^{n[n(n-2)/4]} [D_{n-2}, D_{n-2}]^{n[n(n-2)/4]}.$$

We now verify as in Section A.3 that this completes the corrections of the first pure type. By the argument from Section A.5 the statements about the expressions $[C_{-1}, C_j]$ holds for the expressions $[D_{-1}, D_j]$ just as well, and the same holds for the extendibility of the validity of the statement about the power to which the expressions $[C_{-1}, C_j]$ and $[D_{-1}, D_j]$ appear in g_Δ and $g_{N(\Delta)}$ to every $-1 \leq j \leq n - 2$. Therefore the corrections of the expressions of the first pure type are complete.

We now turn to correcting the expressions of mixed type, which again should be started by multiplying g_Δ by expressions of the form

$$[C_{-1}, D_j]^{nt} [C_0, D_j]^{nt} [C_{-1}, D_{n-1-j}]^{nt} [C_0, D_{n-1-j}]^{nt}$$

and

$$[D_{-1}, C_j]^{nt} [D_0, C_j]^{nt} [D_{-1}, C_{n-1-j}]^{nt} [D_0, C_{n-1-j}]^{nt},$$

again with decreasing j and certain t at every stage. The largest value of j such that $[C_{-1}, D_j]$ and $[D_{-1}, C_j]$ appear in $g_{N(\Delta)}$ is once again $n - 2$, and with difference $n(n - 1) - n \times 1 = n(n - 2)$. In the process of decreasing induction on j, we now find that at the stage corresponding to the expressions $[C_{-1}, D_j]$ and $[D_{-1}, C_j]$, the proper multiplying corrections for g_Δ have to be

$$([C_{-1}, D_j][C_0, D_j][C_{-1}, D_{n-1-j}][C_0, D_{n-1-j}])^{nj(n-1-j)}$$

and

$$([D_{-1}, C_j][D_0, C_j][D_{-1}, C_{n-1-j}][D_0, C_{n-1-j}])^{nj(n-1-j)},$$

and then for $g_{N(\Delta)}$ they are

$$([C_{-1}, D_{n-2-j}][C_{n-2}, D_{n-2-j}][C_{-1}, D_{j-1}][C_{n-2}, D_{j-1}])^{nj(n-1-j)}$$

and

$$([D_{-1}, C_{n-2-j}][D_{n-2}, C_{n-2-j}][D_{-1}, C_{j-1}][D_{n-2}, C_{j-1}])^{nj(n-1-j)}.$$

The arguments are as in Section A.5, but with the powers having coefficient n rather than $2n$ (since e is now 1 and not 2). The process here ends when we reach the value $j = n/2$, where now the last correction is not shorter and also involves multiplication of g_Δ by powers of $[C_{-1}, D_{(n-2)/2}]$ and $[D_{-1}, C_{(n-2)/2}]$.

The verification that we are done with the corrections of the first mixed type is very similar to Section A.5, and goes as follows. For $n/2 \leq j \leq n - 2$, the expressions $[C_{-1}, D_j]$ and $[D_{-1}, C_j]$ now appear in g_Δ to the power

$$n(n - 1 - j) + nj(n - 1 - j) = n(j + 1)(n - 1 - j),$$

and the same calculations we did in Section A.5 show that the validity of this extends to $-1 \leq j \leq n - 2$. The usual considerations show that in $g_{N(\Delta)}$ the same expressions appear and to the same powers (again both ways work). As the reader can check, the calculations of following the inductive process in this case conclude with the value $j = (n - 2)/2$ and give the desired result.

We now examine the expressions not containing C_{-1} and D_{-1} which appear in g_Δ and $g_{N(\Delta)}$. Also here we find that in g_Δ only $[Y_0, Z_j]$ with $0 \leq j \leq n-2$ appear, and to the power $n[(n^2 + 2n - 4)/4 - j(n - j)]$ if $Y = Z$ (as in Section A.3), or

$$nj + nj(n - 1 - j) = nj(n - j)$$

if $Y \neq Z$ (as in Section A.5, with coefficient n instead of $2n$). Note again that this holds for every $0 \leq j \leq n-2$ as before. We now do the usual change of index from j to k here and from j to $k-1$ in the expressions containing C_{-1} and D_{-1} above, and obtain that both $[Y_0, Z_k]$ and $[Y_0, Z_{k-1}]$ now appear in g_Δ to the same power $n[(n^2 + 2n - 4)/4 - k(n - k)]$ if $Y = Z$ (follow the considerations of Section A.3) or $nk(n - k)$ if $Y \neq Z$ (as in Section A.5, with the same substitution for the latter). This is again more easily compared with the desired expression h_Δ from Eq. (5.11) since the form is similar. Now, the expressions not involving C_{-1} and D_{-1} which appear in $g_{N(\Delta)}$ are $[Y_{n-2}, Z_j]$ with $0 \leq j \leq n-2$, and such an expression appears there to the power $n[n(n - 2)/4 - (j + 1)(n - 3 - j)]$ if $Y = Z$ (again as in Section A.3) or to the power

$$n(n - 2 - j) + n(j + 1)(n - 2 - j) = n(j + 2)(n - 2 - j)$$

if $Y \neq Z$ (similar to Section A.5). Again note that this continues to hold for every $0 \leq j \leq n-2$. The considerations, as we always did after changing the index j to $n - 2 - k$ at this point, now yield that any expression $[Y_{n-2}, Z_{n-2-k}]$ now appears in $g_{N(\Delta)}$ to the same power, i.e., $n[(n^2 + 2n - 4)/4 - k(n - k)]$ if $Y = Z$ and $nk(n - k)$ if $Y \neq Z$.

The corrections of the second type are done by the usual argument. Given some $0 \leq k \leq n-3$ (as always, for $k = n-2$ there is nothing to correct) we can do the usual inductive process, with the same argument as in Section A.5 allowing us to do the corrections of pure and mixed types together. Here we define β_k to be the expression $n[(n^2 + 2n - 4)/4 - k(n - k)]$ if $Y = Z$ and $nk(n - k)$ if $Y \neq Z$, and then this argument shows us that the ith induction step is multiplying g_Δ by the expressions

$$[Y_{n-2-i}, Z_{n-2-i-k}]^{\beta_k} [Y_{i+1}, Z_{i+1+k}]^{\beta_k}$$

and $g_{N(\Delta)}$ by the expressions

$$[Y_i, Z_{i+k}]^{\beta_k} [Y_{n-3-i}, Z_{n-3-i-k}]^{\beta_k}.$$

As usual, in the intermediate case where $n - 3 - k$ is even $i = (n - 3 - k)/2$, this is just multiplying g_Δ by

$$[Y_{n-2-i}, Z_{n-2-i-k}]^{\beta_k} = [Y_{(n-1+k)/2}, Z_{(n-1-k)/2}]^{\beta_k}$$

and multiplying $g_{N(\Delta)}$ by

$$[Y_i, Z_{i+k}]^{\beta_k} = [Y_{(n-3-k)/2}, Z_{(n-3+k)/2}]^{\beta_k}.$$

All this sums up to multiplying g_Δ by

$$\prod_{j=1}^{n-2-k} [C_j, C_{j+k}]^{n[(n^2+2n-4)/4-k(n-k)]} \prod_{j=1}^{n-2-k} [D_j, D_{j+k}]^{n[(n^2+2n-4)/4-k(n-k)]}$$

$$\times \prod_{j=1}^{n-2-k} [C_j, D_{j+k}]^{nk(n-k)} \prod_{j=1}^{n-2-k} [D_j, C_{j+k}]^{nk(n-k)},$$

and $g_{N(\Delta)}$ by

$$\prod_{j=0}^{n-3-k} [C_j, C_{j+k}]^{n[(n^2+2n-4)/4-k(n-k)]} \prod_{j=0}^{n-3-k} [D_j, D_{j+k}]^{n[(n^2+2n-4)/4-k(n-k)]}$$

$$\times \prod_{j=0}^{n-3-k} [C_j, D_{j+k}]^{nk(n-k)} \prod_{j=0}^{n-3-k} [D_j, C_{j+k}]^{nk(n-k)},$$

by the usual changes of indices.

As in all the previous constructions, the usual considerations show that the obtained expressions for g_Δ and $g_{N(\Delta)}$ are the same (since for every $0 \le k \le n-1$ and $-1 \le j \le n-2-k$, the expression $[Y_j, Z_{j+k}]$ appears in both g_Δ and $g_{N(\Delta)}$ to the power β_k, i.e., $n[(n^2+2n-4)/4 - k(n-k)]$ if $Y = Z$ and $nk(n-k)$ if $Y \ne Z$, for every $-1 \le j \le n-2-k$), and the common expression is

$$\prod_{k=0}^{n-1} \prod_{j=-1}^{n-2-k} ([C_j, C_{j+k}][D_j, D_{j+k}])^{n[(n^2+2n-4)/4-k(n-k)]} ([C_j, D_{j+k}][D_j, C_{j+k}])^{nk(n-k)}.$$

This expression is very similar to the formula for h_Δ in Eq. (5.11), and is our nominee for h_Δ. Once again the difference (in the expressions of pure type here appears the number $(n^2+2n-4)/4$ and there appears $n^2/4$) comes from the fact that this expression is not reduced, as every expression $[Y_i, Z_j]$ with $Y = Z$ appears here at least to the power $n(n-2)/2$ (see the calculations in Section A.3). Dividing our nominees for h_Δ by the (invariant) "common divisors"

$$\left[\bigcup_{j=-1}^{n-2} C_j, \bigcup_{j=-1}^{n-2} C_j \right]^{n(n-2)/2} = \prod_{0 \le i < j \le m-1} (\lambda_i - \lambda_j)^{n(n-2)/2}$$

and

$$\left[\bigcup_{j=-1}^{n-2} D_j, \bigcup_{j=-1}^{n-2} D_j \right]^{n(n-2)/2} = \prod_{1 \le i < j \le m} (\mu_i - \mu_j)^{n(n-2)/2}$$

yields the expression for h_Δ from Eq. (5.11).

We conclude by describing briefly the reduced process, i.e., the one that starts with the reduced formula for g_Δ with n even from Eq. (5.7) rather than the unreduced one from Eq. (5.5). This process ends with the formula for h_Δ given in Eq. (5.11) without any reduction, and it goes by the same stages with the following powers at every stage. In the jth correction of the first and pure type, the power becomes $n[(n-2)^2/4 - j(n-2-j)]$, and in the corrections involving $[G,G]$ and $[I,I]$, it is $n(n-2)^2/4$. In the jth correction of the first and mixed type, the power has to be $n(j+1)(n-1-j)$, and in the corrections of the second type with the index k, it will be $n[n^2/4 - k(n-k)]$ in the pure type, and $nk(n-k)$ in the mixed type. As always, the fact that these powers are those that appear in h_Δ from Eq. (5.11) indeed shows that no reduction is needed here. Again we note that in the mixed part of the process there is no difference between the reduced and unreduced processes.

We end our discussion about these constructions with a remark about merging the odd and even n cases. We recall from Sections 4.4 and 5.4 that in both the singular and nonsingular cases one could (at some expense) merge the formulae for odd and even n into combined Thomae formulae. We recall that there we saw that in this merging we touched the pure part (the expressions in the nonsingular case were touched since they correspond to expressions of pure type), but did not touch the mixed part (except for taking the power). Examining the constructions in Sections A.5 and A.6 gives more insight into this, as the differences between the odd and even n cases were much more evident in the pure part than in the mixed one.

A.7 The General Family

We now present the larger family generalizing the nonsingular and singular cases. In this family the two parameters, r from the nonsingular case and m from the singular one, both appear (for this reason we denoted them differently in the first place). Assume that $m \geq 0$ and take r such that $m+nr$ is positive (this allows r to be negative if m is large enough!), and let X be the Riemann surface defined by the equation

$$w^n = \prod_{i=0}^{m+nr-1} (z-\lambda_i) \prod_{i=1}^{m} (z-\mu_i)^{n-1}$$

with λ_i, $0 \leq i \leq m+nr-1$, and μ_i, $1 \leq i \leq m$, distinct complex numbers. The Riemann surface X is clearly a fully ramified Z_n curve which is not branched over ∞. One may take the first product only up to $m+rn-2$, or the second product only up to $m-1$, and obtain a Z_n curve that has a branch point (either P_∞ or Q_∞ respectively) over ∞, but for convenience we shall assume that this is not the case. For the reader insisting on doing so, we recall that in Section 2.6 we showed that if there is branching over ∞, then the Thomae formulae remain the same, but with every expression including ∞ omitted. Note that by taking $m=0$ (and thus $r \geq 1$) we obtain a nonsingular Z_n curve, and by taking $r=0$ we obtain a singular Z_n curve of the sort we already dealt with; hence this family does generalize both cases. We also recall that for $n=2$ the singular curves we have dealt with are simply nonsingular Z_2 curves with $r=m$. This extends to this general case as well, as one can easily see that for $n = 2$ here we obtain yet again a nonsingular Z_2 curve (with the new parameter r being $r + m$ from here). In fact, every Z_2 curve can be seen to be nonsingular in this way. We also mention that this general family covers all the possible Z_3 curves, and also all the fully ramified Z_4 and Z_6 curves.

We note that the equation of our Z_n curve can be replaced by

$$\left(\frac{\prod_{i=0}^{m+nr-1}(z - \lambda_i) \prod_{i=1}^{m}(z - \mu_i)}{w} \right)^n = \prod_{i=0}^{m+nr-1} (z - \lambda_i)^{n-1} \prod_{i=1}^{m}(z - \mu_i),$$

which means that the pair (m, r) describing the Z_n curve can be equally replaced by the pair $(m + nr, -r)$. This means that we can restrict ourselves to nonnegative r, but we prefer not to impose this restriction. This is so since throughout the proof we choose our basepoint to be P_0, and there is no reason to assume that we chose the basepoint from the larger set. Nevertheless, the demand that $m + nr$ is strictly positive is made in order to make sure that we have at least one point P_0 in this set.

Our claim is that in this more general situation the Thomae formulae are the same formulae from Theorems 5.5, 5.6 and 5.7, and are proved in the same way except for a few minor changes. What does change in the result is the cardinality conditions on the sets C_j and D_j, $-1 \le j \le n - 2$, composing a divisor Δ, which are equivalent to Δ being non-special, as we later see. However, the final Thomae formulae will have the same form. The reason why the general formulae look like the singular case and not the nonsingular one is simply because in the singular case we have two kinds of sets, i.e., the sets C_j with $-1 \le j \le n - 2$ and the sets D_j with $-1 \le j \le n - 2$ like here, while in the nonsingular case we only have one type of sets. We now present the proof in this general setting, with emphasis on the differences from the special cases already shown. We shall omit from this presentation proofs that are identical or very close to proofs already given. The reader is encouraged to verify that substituting $m = 0$ and $r = 0$ gives the familiar nonsingular and singular formulae respectively at every stage.

As usual, we begin by calculating the genus of the Z_n curve X and by calculating some useful divisors of differentials and functions. First, the fact that X is a fully ramified Z_n curve and the number of branch points on it is $t = 2m + nr$ shows that the genus of X is $g = (n - 1)(2m + nr - 2)/2$. Next, we find that the divisors of the functions $\text{div}(z - \lambda_j[\mu_j])$ look as usual, and also the divisor $\text{div}(dz)$ of the differential dz looks as in the singular case (but counting all the points P_i means taking the corresponding index until $m + nr - 1$). The degree of the latter is indeed seen to be $2g - 2$, as it should be. As for the function w, its divisor here is

$$\text{div}(w) = \frac{\prod_{i=0}^{m+nr-1} P_i \prod_{i=1}^{m} Q_i^{n-1}}{\prod_{h=1}^{n} \infty_h^{m+r}},$$

and it is again useful to calculate that

$$\text{div}\left(\frac{\prod_{t=1}^{m}(z - \mu_t)}{w} \right) = \frac{\prod_{i=1}^{m} Q_i \prod_{h=1}^{n} \infty_h^{r}}{\prod_{i=0}^{m} | {}^{n_i - 1} P_i}.$$

As before one sees that dz/w is a holomorphic differential, and more generally we have the analog of Lemmas 2.2 and 2.10 which gives bases for the holomorphic differentials that are adapted to the branch points. Explicitly we have, for every $0 \leq k \leq n-2$ and $0 \leq l \leq m+r(k+1)-2$, that

$$\mathrm{div}\left[(z-\lambda_i)^l\left(\frac{\prod_{t=1}^{m}(z-\mu_t)}{w}\right)^k \frac{dz}{w}\right] = P_i^{n(l+1)-2-k}\prod_{j\neq i}P_j^{n-2-k}\prod_{j=1}^{m}Q_j^k\prod_{h=1}^{n}\infty_h^{m+r(k+1)-2-l}$$

for every $0 \leq i \leq m+rn-1$ and

$$\mathrm{div}\left[(z-\mu_i)^l\left(\frac{\prod_{t=1}^{m}(z-\mu_t)}{w}\right)^k \frac{dz}{w}\right] = \left(\prod_{j=0}^{m+nr-1}P_j^{n-2-k}\right)Q_i^{nl+k}\prod_{j\neq i}Q_j^k\prod_{h=1}^{n}\infty_h^{m+r(k+1)-2-l}$$

for every $1 \leq i \leq m$; these differentials form a basis for the holomorphic differentials that is adapted to the branch point P_i or Q_i respectively. Clearly the assumptions on k and l assure that all these differentials are indeed holomorphic, and the calculation

$$\sum_{k=0}^{n-2}\sum_{l=0}^{m+rk-2}1 = \sum_{k=0}^{n-2}(m+r(k+1)-1) = (n-1)(m-1)+r\frac{n(n-1)}{2} = g$$

shows the we indeed have g holomorphic differentials in every such basis. It is clear that their orders at the chosen point P_i or Q_i are all distinct, which shows that this gives us bases for the holomorphic differentials that are adapted to the points P_i and Q_i respectively.

The analog of Lemmas 2.3 and 2.11 concerning the gap sequences at the branch points in this case is as follows. Similar to the singular case, Proposition 1.6 shows that $n(l+1)-1-k$ is a gap at every branch point P_i and $nl+k+1$ is a gap at every branch point Q_i, and this holds for every $0 \leq k \leq n-2$ and $0 \leq l \leq m+r(k+1)-2$. In order to see what the gap sequence is explicitly, one has to separate into the cases of $r=0$, $r>0$ and $r<0$. For $r=0$ we already saw (see Lemma 2.11) that for every $0 \leq l \leq m-2$ the numbers between $nl+1$ and $nl+n-1$ are gaps at both a branch point P_i and a branch point Q_i, and this exhausts the gap sequences there. For $r>0$ and $r<0$ the statements split according to the value of l. We begin with $r>0$, where for every $0 \leq l \leq m+r-2$ the numbers between $nl+1$ and $nl+n-1$ are gaps at every branch point (P_i or Q_i). For $m+r-1 \leq l \leq m+r(n-1)-2$, however, we have that the gaps at P_i that lie between nl and $nl+n-1$ are exactly the numbers between $nl+1$ and $nl+n-1-\lfloor(l+1-m)/r\rfloor$, and the gaps at Q_i that lie in that range are exactly the numbers between $nl+1+\lfloor(l+1-m)/r\rfloor$ and $nl+n-1$ (this integral value comes from the inequality $k \geq (l+2-m-r)/r$). For $r<0$ we find that if $0 \leq l \leq m+r(n-1)-2$, then the numbers between $nl+1$ and $nl+n-1$ are gaps at every branch point. When $m+r(n-1)-1 \leq l \leq m+r-2$ we see that the gaps at a point P_i are the numbers between $nl+1+\lfloor(m+r-2-l)/|r|\rfloor$ and $nl+n-1$, while the gaps at a point Q_i are the numbers between $nl+1$ and $nl+n-1-\lfloor(m+r-2-l)/|r|\rfloor$. We note that in this general case the gap sequence at any point P_i, $0 \leq i \leq m+rn-1$ (which is the same for every such i) and the gap sequence at any point Q_i, $1 \leq i \leq m$ (which is again the same for every such i) are not the same.

In the nonsingular case it is quite clear why we did not encounter this phenomenon: there is only one type of branch point. To see why in the singular case we dealt with we did not observe it, recall we saw that we could replace w by another function and this changed the roles of the branch points P_i and Q_i, so there was no intrinsic way to distinguish between the two types. In this more general setting we "can" distinguish between them, since there are $m + rn$ of one type and m of the other. One can check that the gap sequences are consistent with the action of replacing w by $\prod_{i=0}^{m+nr-1}(z - \lambda_i)\left(\prod_{i=1}^{m}(z - \mu_i)\right)/w$, and hence with replacing the role of the branch points P_i with the role of the branch points Q_i, and the pair (m, r) with the pair $(m + nr, -r)$. Note, however, that this change also involves replacing every k by $n - 2 - k$. One needs to be careful with signs when checking these statements.

The next object we examine is the Abel–Jacobi map. Lemma 2.12 holds here as stated (and not Lemma 2.4, as K_R can be shown to be of order $2n$ but not necessarily 2 as in the special case of a nonsingular curve), and the equivalent of Lemma 2.5 is the statement that $\varphi_R\left(\prod_{i=0}^{m+nr-1} P_i\right) = \varphi_R\left(\prod_{i=1}^{m} Q_i\right)$ for every branch point R. Indeed this gives Lemma 2.5 if $m = 0$, and one now sees that the special case of this lemma for the singular case with $r = 0$ reduces to the statement that $\prod_{i=1}^{m} Q_i / \prod_{i=0}^{m-1} P_i$ is principal.

We can now discuss our first main object of interest. We choose P_0 as the basepoint, and we look for integral divisors of degree g supported on the branch points distinct from P_0 that are non-special. If we assume that no branch point appears to a power exceeding $n - 1$, we can write such a divisor Δ in the usual way $A^{n-1} \prod_{j=0}^{n-2} D_j^{j+1} C_j^{n-2-j}$, and thus define the sets A, B, and C_j and D_j with $0 \leq j \leq n - 2$, as we did in the singular case. Note that here $|B| + \sum_{j=0}^{n-2}|D_j| = m$, but $|A| + \sum_{j=0}^{n-2}|C_j| = m + rn - 1$ (which in the special case of a nonsingular curve reduces to B and D_j being empty and the cardinalities of the other sets summing to $rn - 1$). The analog of Theorems 2.6, 2.9, 2.13, and 2.15 (we put the two directions in one theorem) is

Theorem A.1. *An integral divisor Δ of degree g supported on the branch points distinct from P_0 is non-special if and only if $\Delta = A^{n-1} \prod_{j=0}^{n-2} D_j^{j+1} C_j^{n-2-j}$, where A and C_j, $0 \leq j \leq n - 2$, is a partition of the set of branch points P_i, $1 \leq i \leq m + nr - 1$, and B and D_j, $0 \leq j \leq n - 2$, is a partition of the branch points Q_i, $1 \leq i \leq m$, with $|A| = |B| + r - 1$ and $|C_j| = |D_j| + r$ for all $0 \leq j \leq n - 2$.*

The proof is very similar to the proofs of Theorems 2.6, 2.9, 2.13 and 2.15. First, in order to see that this gives a divisor of degree g we calculate that

$$(n-1)|A| + \sum_{j=0}^{n-2}\left((j+1)|D_j| + (n-2-j)|C_j|\right) = (n-1)\left(|A| + \sum_{j=0}^{n-2}|C_j|\right) - \sum_{j=0}^{n-2} r(j+1)$$

$$= (n-1)(m + rn - 1) - r\frac{n(n-1)}{2} = (n-1)\left(m - 1 + \frac{rn}{2}\right) = g.$$

Now, for the first direction (generalizing Theorems 2.6 and 2.13) one verifies that the decomposition according to k in each adapted basis shows that Lemma 2.7 holds also here. Then one sees, as before, that a holomorphic differential in $\Omega_k(\Delta)$ must be based on a polynomial with too many roots, forcing the polynomial, and hence the divisor, to vanish. As for the other direction, one shows, as in Theorems 2.9 and 2.15, that in such a divisor no branch point can appear to power n or more. Then one uses again the knowledge of the holomorphic differentials on X to obtain inequalities on the cardinalities of the sets A, C_j, D_j, and B, and finishes with a lemma analogous to Lemmas 2.8 and 2.14. This lemma states that if A and C_j, $0 \le j \le n-2$, form a partition of the $m+nr-1$ branch points P_i, $1 \le i \le m+nr-1$, into n sets, and B and D_j, $0 \le j \le n-2$, form a partition of the m branch points Q_i, $1 \le i \le m$, into n sets, such that the equality

$$(n-1)|A| + \sum_{j=0}^{n-3}(n-2-j)|C_j| + \sum_{j=0}^{n-2}(j+1)|D_j| = \frac{(n-1)(2m+rn-2)}{2}$$

holds, then either $|A| = |B|+r-1$ and $|C_j| = |D_j|+r$ for every $0 \le j \le n-2$, or the inequality

$$|A| + \sum_{j=0}^{k-1}|C_j| \le |B| + \sum_{j=0}^{k-1}|D_j| + r(k+1) - 2$$

must be satisfied for some $0 \le k \le n-2$, where the inequality corresponding to $k=0$ is just $|A| \le |B|+r-2$. This lemma is proved by considerations similar to those we applied for its special cases Lemmas 2.8 and 2.14. Clearly taking $r=0$ in Theorem A.1 returns us to Theorems 2.13 and 2.15, and since taking $m=0$ in Theorem A.1 makes the sets B and D_j, $0 \le j \le n-2$ empty, this does give us back Theorems 2.6 and 2.9. In this way we have found for $r \ge 0$ all the non-special divisors on X, since as in the special cases we examined the fact that the function $\left(\prod_{t=1}^{m}(z-\mu_t)\right)/w$ lies in $L\left(1/\prod_{i=0}^{m+nr-1}P_i\right)$ shows that no non-special divisor can contain all the points P_i, $0 \le i \le m+nr-1$, in its support (with the case $n=2$ being trivial as usual, and except for the nonsingular cases with $n=3$, or with $n=4$ and $r=1$, the only new trivial case is $n=3$ and $m=1$ with any r). In the case $r \le 0$ we find that the inverse of this function lies in $L\left(1/\prod_{i=1}^{m}Q_i\right)$, and then for any non-special divisor, there is some point Q_i not appearing in its support, which also means that we obtained all the divisors (with the only trivial cases being $n=2$ or $n=3$ and $m+rn=1$).

From this point on we notice that the whole construction of the Thomae formulae is based on the form of the divisor Δ as $A^{n-1}\prod_{j=0}^{n-2}D_j^{j+1}C_j^{n-2-j}$, without any use of the cardinality conditions themselves. This means that from now on we can copy all the propositions and all the proofs, and they will hold equally for this much more general setting. Indeed, one can define the operators N, T_R, and T_S here just as in Definition 2.18, and since here the divisors $P_0^{n(m+r-1)-2}\prod_{i=1}^{m+nr-1}P_i^{n-2}$ (which is the divisor of the differential $(z-\lambda_0)^{m+r-2}\mathrm{d}z/w$) and $P_0^{n(m+2r-1)-3}\prod_{i=1}^{m+nr-1}P_i^{n-3}\prod_{i=1}^{m}Q_i$ (which is the divisor of the differential $(z-\lambda_0)^{m+2r-2}\left(\prod_{t=1}^{m}(z-\mu_t)\right)\mathrm{d}z/w^2$) are canonical, we find that the proof of Proposition 2.19 holds almost word-for-word (up to the right choice of differentials and powers of P_0 inside φ_{P_0}). The theta function theory from Section 2.6 (including the index $e = 2/\gcd(2,n)$ of course) extends to this case; it is much more general, as we saw that it held also in the examples in Chapter 6. So do the constructions of the PMT, and later of the denominators h_Δ (either the formulae we have given for them in Chapter 5 or the construction in Sections A.5 and A.6), and the proof that the quotients we obtain are invariant under all the operators N, T_R, and T_S (including the new notation with C_{-1} and D_{-1} and with the auxiliary sets E, F and H). Therefore one obtains that Theorems 5.5 and 5.6 with Eqs. (5.10) and (5.11) hold for this general family, and so does the remark from the beginning of Section 5.4 about combining the formulae for the odd and even n cases. Since putting $m = 0$ here makes all the sets D_j, $-1 \leq j \leq n-2$, empty, we now fully understand why the process in the nonsingular case looks like this "partial process" of the one in the singular case.

It only remains to see why in this general setting we obtain the full Thomae formulae in this way. Explicitly, we need some assertion that one can get from any non-special divisor of degree g supported on the branch points distinct from P_0 (i.e., a divisor described in Theorem A.1) to any other such divisor (i.e., we look for some extension of Theorems 4.8 and 5.7 to this case). Clearly one cannot expect such a theorem in total generality, since one can take $m = 0$, $r = 1$ and $n \geq 5$ and obtain the Z_n curves from Section 3.4 for which we know that such a statement is not true. However, this is apparently the only problematic case, since what we have in the general setting is

Theorem A.2. *For any pair (m,r) of integers with $m \geq 0$ and $m + rn > 0$, except for the pair $r = 1$ and $m = 0$, one can get from any non-special divisor of degree g supported on the branch points distinct from P_0 to any other such divisor. This means that the expression $\theta^{2en^2}[\Delta](0,\Pi)/h_\Delta$ with h_Δ from Eq. (5.10) when n is odd and from Eq. (5.11) when n is even is independent of the divisor Δ, and is thus the Thomae formulae for the Riemann surface X.*

The proof of Theorem A.2 is the only thing that takes more work in this general case than in the singular case with which we already dealt. We therefore postpone the proof by a few paragraphs, to finish stating all the results that do not require much extra work.

We conclude our assertions about this family by showing that the idea of basepoint change and the basepoint-independent Thomae formulae hold here as well (using Theorem A.2 freely since we prove it in a moment). For this we first observe that Corollary 1.13 holds just as well in this case, and we need a result generalizing Propositions 1.14 and 1.15. What we have here (in the notation of Propositions 1.14 and 1.15) is $t + 1 = 2m + nr$ with the last m points denoted Q_i, $1 \leq i \leq m$, and with $\varphi_S\left(\prod_{i=0}^{m+nr-1} P_i / \prod_{i=1}^{m} Q_i\right) = 0$ for any relevant point S, and the divisors have the form as in Theorem A.1 (but with B already denoted by D_{-1}). For this setting we want to prove the same assertion as in Propositions 1.14 and 1.15, i.e., that for every point P_i there is a divisor Γ_{P_i} of the same form as Δ but with P_i removed instead of P_0 such that the equality $\varphi_{P_0}(\Delta) + K_{P_0} = \varphi_{P_i}(\Gamma_{P_i}) + K_{P_i}$ holds, and for every point Q_i there is a divisor Γ_{Q_i} of the same form as Δ but with Q_i removed instead of P_0 which satisfies the equality $\varphi_{P_0}(\Delta) + K_{P_0} = \varphi_{Q_i}(\Gamma_{Q_i}) + K_{Q_i}$. The proof is like the one of Proposition 1.15 (again we need the one which deals with two types of sets of branch points), where we take into consideration the fact that when changing to a point Q_i we also replace the pair (m, r) by the pair $(m + nr, -r)$ as described above. This is what shows that the proper cardinality conditions hold in this operation. Similarly, we again add $k + 1$ (for the same k) times the expression $\varphi_S\left(\prod_{i=0}^{m+nr-1} P_i / \prod_{i=1}^{m} Q_i\right)$ (with S being the corresponding branch point), and since there are nr more points P_i than points Q_i (also when r is negative!), the cardinality conditions from Theorem A.1 allow us to continue by the same considerations as in the proof of Proposition 1.15 and obtain the desired result. The operator M we have here multiplies our divisor $P_0^{n-1}\Delta$ by $\left(\prod_{i=0}^{m+nr-1} P_i / \prod_{i=1}^{m} Q_i\right)(D_{-1}^n / C_{-1}^n)$ (which is again of degree 0 by the cardinality conditions) and is still an operation of order n, and the action we did is applying this M successively $k + 1$ times. The same uniqueness assertion we had there of course extends to here by the same reasoning.

Knowing all this, we conclude that the basepoint-independent Thomae formulae Theorem 5.8 (including the definitions $\Xi = \prod_{j=-1}^{n-2} C_j^{n-2-j} D_j^{j+1}$ with the corresponding $\theta[\Xi]$ and h_Ξ) given for the singular case also holds here (under the cardinality conditions $|C_j| = |D_j| + r$ for $-1 \leq j \leq n - 2$, of course). Actually for $r \neq 0$ this is easy, since for $r > 0$ the set C_{-1} is never empty and for $r < 0$ the set D_{n-2} is never empty. This means that we can prove this theorem directly from Theorem A.2, and this already reaches all these divisors Ξ without needing to apply the operator M. In the remaining case $r = 0$, this is Theorem 5.8 itself, which we already proved. Of course, taking $m = 0$ and empty sets D_j, $-1 \leq j \leq n - 2$, gives back Theorem 4.9 if $r \geq 2$ and Theorem 4.11 if $r = 1$ (note that the former is indeed a special case here by all means, but the latter is a special case only by assertion and not by proof, as we recall that the assertion of Theorem A.2 excludes the pair $r = 1$ and $m = 0$). Therefore the nonsingular case (both with $r \geq 2$ and $r = 1$, interpreted correctly) is a special case of this general family until the end.

A.8 Proof of Theorem A.2

In order to prove Theorem A.2 we shall have to combine the tools from the second proof of Theorem 4.8 (permutations cannot help us here) and the proof of Theorem 5.7. We do this by the same sort of induction based on the ordering (4.9) of the sets C_j, $-1 \leq j \leq n - 2$ (where D_j is assumed to lie in the same place as the corresponding C_j—or simply using the indices themselves) and we adopt again the notation $k \prec l$, etc. More precisely, what we prove is the following inductive assertion: Assume that Δ and Ξ are two divisors that are distinct only in the sets C_j and D_j with $j \preceq k$. Assume that one can get, using the operators N, T_R, and T_S, from any divisor to any other divisor, that is distinct from it only in sets C_j and D_j with

$j \prec k$ (strict inequality, for the induction hypothesis). Then one can get from Δ to Ξ using only N, T_R and T_S. Since here the details do differ from both the special cases already proven (Theorems 4.8 and 5.7) and are a bit more subtle, we give the complete proof.

The very beginning of the proof, i.e., the basis for this induction for the values $k = -1$ (trivial) and $k = 0$, goes exactly as in the proof of Theorem 5.7 (using the pairs of operators $T_P T_R$, $T_P T_S$, $T_Q T_R$, and $T_Q T_S$ for $R \in C_0$, $S \in D_{-1}$, $P \in C_{-1}$, and $Q \in D_0$). We also keep the notation V for an operator (either N or some T_R or T_S) which interchanges C_k and D_k with C_h and D_h respectively, where C_h and D_h are the "left neighbors" of C_k and D_k in the ordering (4.9) (again in the case where V has to be some T_R or some T_S any permissible choice suffices for every application).

We now recall that in the proof of Theorem 5.7 in the singular case the passing from Δ to Ξ went through a divisor Γ, which had empty sets C_h and D_h (where h is the "left neighbor" of k), and to which one got from Δ using the induction hypothesis. Then we went through another divisor Υ, which also had empty sets C_h and D_h, such that one can get to it from $V(\Gamma)$, and such that one can get from $V(\Upsilon)$ to Ξ. Here we cannot take such divisors (as the cardinality conditions force a difference of r between the cardinalities of C_h and D_h, and in general $r \neq 0$), but also here we find that divisors with small C_k and D_k are easier to work with.

What was said in the last paragraph suggests that one should first move from Δ to a divisor Λ satisfying $(C_j)_\Lambda = (C_j)_\Delta$ and $(D_j)_\Lambda = (D_j)_\Delta$ every $j \succeq k$, and where where C_h and D_h have minimal cardinality (i.e., one of them is empty and the other has cardinality $|r|$, chosen properly). Such a divisor Λ always exists (we are free to mix the elements of the sets C_j and D_j with $j \prec k$ as we want as long as the cardinality conditions are satisfied), and clearly the induction hypothesis allows us to go from Δ to Λ using the operators N, T_R, and T_S. Then another application of V takes us to $V(\Lambda)$, which has sets C_k and D_k of minimal cardinality. Similarly, if we assume (and demonstrating this claim will require most of the work of proving the theorem) that one can get from $V(\Lambda)$ to a divisor Ψ, which also has sets C_k and D_k of minimal cardinality, which satisfies $(C_j)_\Psi = (C_j)_{V(\Lambda)}$ and $(D_j)_\Psi = (D_j)_{V(\Lambda)}$ for every $j \succ k$, and also satisfies $(C_h)_\Psi = (C_k)_\Xi$ and $(D_h)_\Psi = (D_k)_\Xi$, then one can also get to $V(\Psi)$. Now, $V(\Psi)$ satisfies $(C_j)_{V(\Psi)} = (C_j)_\Xi$ and $(D_j)_{V(\Psi)} = (D_j)_\Xi$ for $j \succeq k$. This is clear for $j = k$, and for $j \succ k$, we recall that V acts like an involution on these sets, and we applied it twice on the way: one from Λ to $V(\Lambda)$ and one from Ψ to $V(\Psi)$. The conclusion is that one can use the induction hypothesis to get from $V(\Psi)$ to Ξ, meaning that one can get from Δ to Ξ. A divisor Ψ "close" to Ξ in this manner always exists (construct $V(\Psi)$ as we constructed Λ, and then $\Psi = V(V(\Psi))$ as V is an involution), so we have to show that we can get from our $V(\Lambda)$ to our Ψ.

The last paragraph reduces us to proving the following claim: Assume that $V(\Lambda)$ and Ψ are two divisors, which are distinct only in the sets C_j and D_j with $j \preceq k$, and that both have sets C_k and D_k of minimal cardinality. Assume that one can get, using the operators N, T_R, and T_S, from any divisor to any other divisor, which is distinct from it only on sets C_j and D_j with $j \prec k$ (the inequality again being strict, in order to serve as the induction hypothesis). Then one can get from Δ to Ξ using only N,

T_R, and T_S. What we shall actually prove is that under these assumptions we can get from $V(\Lambda)$ to Ψ, but the previous paragraph shows that this gives what we want for Δ and Ξ as well. This allows us to put the extra assumption on the cardinalities of C_k and D_k only in the hypothesis; this is nice to know, even though it will not be used. Note that this statement is the same assertion we wanted to prove in the first place, but now we have the extra assumption that C_k and D_k are small in our divisors. This extra assumption will be very helpful in the proof. We note that in the nonsingular case we have done nothing (as in this case the cardinality to every set $(C_j)_\Delta$ is independent of Δ), but in the general case this is of course not true.

We now prove this claim. In the special case with $r = 0$ (the singular case which we already did), this was very easy: Both sets C_k and D_k were empty in both $V(\Lambda)$ and Ψ, and these divisors are also assumed to have the same sets C_j and D_j with $j \succ k$. Therefore the induction hypothesis immediately allows us to get from $V(\Lambda)$ to Ψ, and we are done (in fact, this is how we proved Theorem 5.7). However, for $r \neq 0$ only one of the sets C_k and D_k is empty in $V(\Lambda)$ and Ψ and we cannot assume that the nonempty sets coincide. Hence we have to work a bit harder. We thus apply the considerations used in the second proof of Theorem 4.8 (and thus proving it again in the same way as a special case). We shall make a choice of a set Y of branch points which allows us to define a divisor Γ and later a divisor Υ through which we can pass on the way from $V(\Lambda)$ to Ψ (and hence from Δ to Ξ), and this will finish the proof. We divide the proof into three cases, in increasing degree of difficulty. First we deal with $r < 0$ (and arbitrary $m > -nr$), then with $r \geq 2$ (and arbitrary $m \geq 0$), and finally with the case $r = 1$ and arbitrary $m \geq 1$ (we exclude the case $r = 1$ and $m = 0$, for which we know the assertion of the theorem is not true in general).

We begin with the easiest case $r < 0$. Then we adopt the notation U for the union of the "problematic" sets either for $V(\Lambda)$ or for Ψ, which in the case $r < 0$ are the sets D_j for $j \preceq k$ (or explicitly $U = \bigcup_{j \preceq k}(D_j)_{V(\Lambda)} = \bigcup_{j \preceq k}(D_j)_\Psi$). Since we assume that k is neither -1 nor 0 and the cardinality conditions clearly imply that $|D_j| \geq |r|$ for every j, we find that the cardinality of U is at least $3|r|$. Now, since the cardinalities of $(D_k)_{V(\Lambda)}$ and $(D_k)_\Psi$ are exactly $-r = |r|$ (since $(C_k)_{V(\Lambda)}$ and $(C_k)_\Psi$ are assumed to be empty), we find that one can take a subset Y of U of cardinality $|r|$ that is disjoint from both $(D_k)_{V(\Lambda)}$ and $(D_k)_\Psi$. We can now choose a divisor Γ such that $(C_j)_\Gamma = (C_j)_{V(\Lambda)}$ and $(D_l)_\Gamma = (D_l)_{V(\Lambda)}$ for every $j \succeq k$, and with $(D_h)_\Gamma$ chosen to be Y and $(C_h)_\Gamma$ empty. This is possible since $Y \cap (D_k)_{V(\Lambda)} = \emptyset$. Then, by the induction hypothesis, we can get from $V(\Lambda)$ to Γ, and then also to $V(\Gamma)$. We now choose a divisor Υ such that $(C_j)_\Upsilon = (C_j)_{V(\Gamma)}$ and $(D_j)_\Upsilon = (D_j)_{V(\Gamma)}$ for every $j \succeq k$, and with $(D_h)_\Upsilon$ chosen to be $(D_k)_\Psi$ and $(C_h)_\Upsilon$ empty. This is also possible since $Y \cap (D_k)_{V(\Lambda)} = \emptyset$. Then, by the induction hypothesis, we can get from $V(\Gamma)$ to Υ, and then also to $V(\Upsilon)$; and we now see that since V acts on the sets with indices different from -1 and 0 as an involution (V is not exactly an involution since when $V \neq N$ we may take, in any application, a different operator T_R or T_S, but the exact action on the sets C_{-1}, D_{-1}, C_0, and D_0 is insignificant at this point), the divisor $V(\Upsilon)$ satisfies $(C_j)_{V(\Upsilon)} = (C_j)_{V(\Lambda)} = (C_j)_\Psi$ and $(D_j)_{V(\Upsilon)} = (D_j)_{V(\Lambda)} = (D_j)_\Psi$ for every $j \succ k$ (we interchanged these sets once when we applied V to Γ, and once again when we applied V to Υ). However, by our choice of Υ we also have that $(C_k)_{V(\Upsilon)} = (C_h)_\Upsilon$ is empty like $(C_k)_\Psi$, and $(D_k)_{V(\Upsilon)} = (D_h)_\Upsilon$ and thus equals $(D_k)_\Psi$, and thus we can use the induction hypothesis once again to get from $V(\Upsilon)$ to Ψ. Altogether, this shows that we can get from $V(\Lambda)$ to Ψ (and hence from the original Δ to the original Ξ) using the operators N, T_R, and T_S. This finishes the proof for the case $r < 0$.

We now turn to the cases with $r > 0$, where $V(\Lambda)$ and Ψ have empty D_k, the corresponding sets C_k are of cardinality r, and U is $\bigcup_{j \prec k}(C_j)_{V(\Lambda)} = \bigcup_{j \prec k}(C_j)_\Psi$. One sees that once one can choose the subset Y of U of cardinality r that is disjoint from both $(C_k)_{V(\Lambda)}$ and $(C_k)_\Psi$, we can continue word-for-word as in the previous paragraph (up to replacing every C_j by D_j and every D_j by C_j) and complete the proof. This is what we actually did in the second proof of Theorem 4.8. However, the difference between the cases $r < 0$ and $r > 0$ here is the presence of the point P_0 in U when $r > 0$. The fact that this point P_0 is unusable for our purposes is what makes the case $r \geq 2$ more difficult in comparison to $r < 0$, and the case $r = 1$ restricted to $m \geq 1$ even more tricky. It also explains why we it does not suffice to use the symmetry taking (m, r) to $(m + nr, -r)$ and dispense with the case $r > 0$—our choice to fix P_0 forces us to deal with these cases separately and differently.

We now show how this is done in the case $r \geq 2$. This case is proved in the same way as Theorem 4.8. We know that we have to take the subset Y of cardinality r out of the set U with $(C_k)_{V(\Lambda)}$ and $(C_k)_\Psi$ (both of cardinality r) omitted and also with P_0 omitted, meaning that at most $2r + 1$ points were removed from U. Therefore we see that once the first induction step is proved, in the following steps the cardinality of U is at least $4r$, and then taking out $2r + 1$ points still leaves us with the possibility of taking a subset Y of cardinality r (since $2r - 1 \geq r$ for $r \geq 1$). Thus we only have to show that the first induction step, with $k = n - 2$ and $h = 0$, can also be done. At this stage U does not have to be of cardinality $3r$, but can be bigger; but we cannot assume that it is bigger than $3r$, and we may still find ourselves in the situation where taking out $2r + 1$ points off U leaves us with less than r branch points to put in Y. However, we solve this just as we did in the proof of Theorem 4.8. If $(C_{n-2})_{V(\Lambda)}$ and $(C_{n-2})_\Psi$ have nonempty intersection, then they contain together at most $2r - 1$ points, hence taking them and P_0 from U means taking at most $2r$ points out of a set of cardinality at least $3r$. Then we can choose the subset Y and continue as usual. In the remaining case, where the intersection of $(C_{n-2})_{V(\Lambda)}$ and $(C_{n-2})_\Psi$ is empty and $|U| = 3r$ (for $|U| > 3r$ we have no problem even with empty intersection of $(C_{n-2})_{V(\Lambda)}$ and $(C_{n-2})_\Psi$), we can choose (since $r \geq 2$) a subset Z of U of cardinality r not containing P_0 which intersects both $(C_{n-2})_{V(\Lambda)}$ and $(C_{n-2})_\Psi$ nontrivially. We can then choose a divisor Σ such that $(C_j)_\Sigma = (C_j)_{V(\Gamma)} = (C_j)_\Psi$ and $(D_j)_\Sigma = (D_j)_{V(\Gamma)} = (D_j)_\Psi$ for every $j \succ n - 2$, with empty D_{-1}, D_0 and D_{n-2}, with $C_{n-2} = Z$, and with C_{-1} and C_0 being some partition of $U \setminus Z$ (which is of cardinality $2r$ and contains P_0) into two sets of cardinality r where C_{-1} contains P_0. Then by what we have just shown one can get from $V(\Lambda)$ to Σ and from Σ to Ψ (recall that $(C_{n-2})_\Sigma = Z$ intersects both $(C_{n-2})_{V(\Lambda)}$ and $(C_{n-2})_\Psi$), and hence from $V(\Lambda)$ to Ψ. This completes the first induction step, and hence finishes the proof for the case $r \geq 2$ (repeating the proof of Theorem 4.8 if $m = 0$).

It remains to deal with the case $r = 1$ and $m \geq 1$. In this case we cannot prove the statement as we did for $r \geq 2$ (as no such set Z can be defined in order to overcome the problem in the first induction step), so that the claim we prove is slightly different and goes as follows. Assume that $V(\Lambda)$ and Ψ are two divisors that are distinct only in the sets C_j and D_j with $j \preceq k$, that both have empty sets D_k and sets C_k of cardinality 1, and that the set U defined above is of cardinality larger than the minimal one (i.e., more than just 1 for every $j \preceq k$—this is equivalent to the assertion that one of the sets D_j with $j \prec k$ is nonempty, the last inequality being strict because of the minimal cardinality assumption on C_k and D_k). Assume that one can get, using the operators N, T_R and T_S, from any divisor to any other divisor which is distinct from it only on sets C_j and D_j with $j \prec k$ (with the usual strict inequality for the induction hypothesis) under the extra assumption that the cardinality of the corresponding set U is not the minimal one. Then one can get from Δ to Ξ using only N, T_R, and T_S. The same remark about proving the assertion for $V(\Lambda)$ and Ψ and using the material from above to get it for Δ and Ξ (and the same remark about not using this) applies to this case as well. We will actually prove this statement without any limitation on m, but for $m = 0$ this is an empty statement since the assumptions can never be satisfied.

Let us now prove this last claim, and then see why it finishes the proof of the theorem for the case $r = 1$ and $m \geq 1$. The proof is simply by verifying that the argument we used for $r \geq 2$ works also here, but now we have to be more careful since the induction hypothesis can only be applied if the extra assumption it now contains is also satisfied. Now, the extra assumption on U assures us that at every induction step (including the first one) we can choose the subset Y (which is actually choosing one branch point P_i in U with $1 \leq i \leq m+n-1$, and note that $i \neq 0$ as we cannot take P_0) of U (with the usual disjointness properties) and hence define the divisors Γ and later Υ. This means that the first induction step holds since the basis of the induction was independent of the value of r or any other assumptions. As for any following induction steps, we see that since we assume that the cardinality of the set $(C_k)_{V(\Lambda)} = (C_k)_\Gamma$ is just 1 and the set $(D_k)_{V(\Lambda)} = (D_k)_\Gamma$ is empty, the assumption on U for $V(\Lambda)$ and Ψ implies the assumption on U for $V(\Lambda)$ and Γ (the second U is obtained from the first U by taking out one set of cardinality 1). This shows that the induction hypothesis can be applied for the stage of passing from $V(\Lambda)$ to Γ. Later, since the set $(C_k)_{V(\Gamma)} = (C_k)_\Upsilon = Y$ has cardinality 1 and the set $(D_k)_{V(\Gamma)} = (D_k)_\Upsilon$ is empty, similar considerations show that we can apply the induction hypothesis at the stage of passing from $V(\Gamma)$ to Υ as well. Finally, since the set $(C_k)_{V(\Upsilon)} = (C_k)_\Psi$ has cardinality 1 and the set $(D_k)_{V(\Upsilon)} = (D_k)_\Psi$ is empty we find by the same reasoning that we can apply the induction hypothesis also at the stage of passing from $V(\Upsilon)$ to Ψ. This means that the proof we used above also holds for this case once we add the extra assumption on U. Now it is clear that for $m \geq 1$ and for the rightmost sets C_k and D_k, i.e., with k maximal with respect to the order (4.9) (this means explicitly $k = (n-2)/2$ for even n and $k = (n-1)/2$ for odd n), the set U contains all the branch points P_i, $0 \leq i \leq m+n-1$. Therefore it is of cardinality $m+n > n$, and hence our claim finishes the proof for the case $r = 1$ and $m \geq 1$. As mentioned above, for $r = 1$ and $m = 0$ the proof still holds but proves an empty statement, and therefore the theorem itself does not extend to this case (and we already know that it is false then). In any case this concludes the proof of Theorem A.2.

Appendix B

The Construction and Basepoint Change Formulae for the Symmetric Equation Case

In this appendix we give some extra material for the symmetric equation case presented in Section 6.1. In the first three sections we describe the construction of the (N-invariant) expression for h_Δ appearing in Theorem 6.9. As in Appendix A, we start with the expression for g_Δ from Proposition 6.7 and Eqs. (6.1) and (6.2), and manipulate then, without losing any important properties of g_Δ, until we obtain the desired expressions. By a reasoning similar to the one explained in Appendix A, we want to present the construction of these expressions, rather than just unloading them on the reader. The slight difference between the cases $n \equiv 1 \pmod 4$ and $n \equiv 3 \pmod 4$ suggests that the process will not be exactly the same in these two cases. In any case, they are done by similar means: One merges Eq. (6.2) and the different Eqs. (6.1) (after some manipulations) into one large equation, with the operator N as the connecting operation. In the last two sections we present the operators and denominators h_Δ with the other basepoints, and say a few words about how to relate them to one another.

B.1 Description of the Process

We start with presenting the construction process. It resembles the process we did for proving Theorem 6.8 for the case $n = 7$, but this value of n is too small to get a true picture of how it goes in the general case. Now, each Eq. (6.1) relates only two divisors (except for the intermediate case $l = (n-5)/2$, whose role in the process we shall see), and Eq. (6.2) relates three divisors such that N takes two of them to divisors appearing in some Eqs. (6.1) and the third to the isolated divisor P_3^{n-1}. Therefore the best way to understand this construction is to write only the first two divisors appearing in Eq. (6.2), and then add, to each side, the Eq. (6.1) with which N connects. Continuing in this manner, we can follow the multiplications that have to be done in order to combine all these equations into one large equation. We complete the task by adding the remaining divisor of Eq. (6.2) accordingly, and the isolated divisor P_3^{n-1} is trivially done after that.

Since we are going to write equalities and relations between many terms now, we should work with shortened notation to make things clearer. Therefore, until further notice, anytime we write $\theta[\Delta]$ for some divisor Δ we mean the theta constant $\theta^{4n^2}[\Delta](0,\Pi)$, and we denote

$$\alpha = (\lambda_0 - \lambda_1)^{2n}(\lambda_2 - \lambda_3)^{2n}, \quad \beta = (\lambda_0 - \lambda_2)^{2n}(\lambda_1 - \lambda_3)^{2n}$$

and

$$\gamma = (\lambda_0 - \lambda_3)^{2n}(\lambda_1 - \lambda_2)^{2n}$$

(the latter appearing only in the part of Eq. (6.2) we leave to the end). Therefore Eq. (6.1) for $0 \leq l \leq n-5$ now reads

$$\frac{\theta[P_1^l P_2^{n-1-l}]}{\alpha^{n-3-l}\beta^{l+2}} = \frac{\theta[P_1^{n-5-l}P_2^{l+4}]}{\alpha^{l+2}\beta^{n-3-l}};$$

Eq. (6.1) for $l = n-4$ and $l = n-1$ now reads

$$\frac{\theta[P_1^{n-4}P_2^3]}{\alpha\beta^{n-2}} = \frac{\theta[P_1^{n-1}]}{\alpha^{n-2}\beta},$$

and Eq. (6.2) is written as

$$\frac{\theta[P_1^{n-2}P_2]}{\alpha^{n-1}} = \frac{\theta[P_1^{n-3}P_2P_3]}{\gamma^{n-1}} = \frac{\theta[P_1^{n-3}P_2^2]}{\beta^{n-1}}$$

(note the change in order from the original form!). We omit the middle term of this Eq. (6.2) (which makes it look like Eq. (6.1) for $l = n-3$ and $l = n-2$), and bring it back at the end of the process. If we now write the equations in the "proper order", where a symbol of $\leftrightarrow$ connects two theta constants between which N interchanges (and are hence equal by the form $\theta[N(\Delta)](0,\Pi) = \theta[\Delta](0,\Pi)$ of Eq. (1.5)), then what we get in the middle part (including the relevant part of Eq. (6.2)) is

$$\ldots \leftrightarrow \frac{\theta[P_1^{n-6}P_2^5]}{\alpha^3\beta^{n-4}} = \frac{\theta[P_1P_2^{n-2}]}{\alpha^{n-4}\beta^3} \leftrightarrow \frac{\theta[P_1^{n-4}P_2^3]}{\alpha\beta^{n-2}} = \frac{\theta[P_1^{n-1}]}{\alpha^{n-2}\beta}$$

$$\leftrightarrow \frac{\theta[P_1^{n-2}P_2]}{\alpha^{n-1}} = \frac{\theta[P_1^{n-3}P_2^2]}{\beta^{n-1}} \leftrightarrow \frac{\theta[P_2^{n-1}]}{\alpha^{n-3}\beta^2} = \frac{\theta[P_1^{n-5}P_2^4]}{\alpha^2\beta^{n-3}} \leftrightarrow \ldots \quad \text{(B.1)}$$

(the equations appearing here are Eq. (6.1) with values $l = n-6$ together with $l = 1$, then $l = n-4$ together with $l = n-1$, then the part of Eq. (6.2), and then Eq. (6.1) with values $l = 0$ together with $l = n-5$, in this order). We continue to the left with the expressions

$$\cdots \leftrightarrow \frac{\theta[P_1^{n-7-l}P_2^{l+6}]}{\alpha^{l+4}\beta^{n-5-l}} = \frac{\theta[P_1^{l+2}P_2^{n-3-l}]}{\alpha^{n-5-l}\beta^{l+4}} \leftrightarrow \frac{\theta[P_1^{n-5-l}P_2^{l+4}]}{\alpha^{l+2}\beta^{n-3-l}} = \frac{\theta[P_1^l P_2^{n-1-l}]}{\alpha^{n-3-l}\beta^{l+2}} \leftrightarrow \cdots$$

(B.2)

for small odd l and to the right with the expressions

$$\cdots \leftrightarrow \frac{\theta[P_1^l P_2^{n-1-l}]}{\alpha^{n-3-l}\beta^{l+2}} = \frac{\theta[P_1^{n-5-l}P_2^{l+4}]}{\alpha^{l+2}\beta^{n-3-l}} \leftrightarrow \frac{\theta[P_1^{l+2}P_2^{n-3-l}]}{\alpha^{n-5-l}\beta^{l+4}} = \frac{\theta[P_1^{n-7-l}P_2^{l+6}]}{\alpha^{l+4}\beta^{n-5-l}} \leftrightarrow \cdots$$

(B.3)

for small even l, until it ends on each side. The way in which it ends in either side and the resulting multiplications we have to do depend on whether $n \equiv 1 \pmod 4$ or $n \equiv 3 \pmod 4$.

Before presenting explicitly the way this process ends and the resulting expressions, we want to describe generally how it is done. The first observations about this process is that we work on each side separately, then combine them, and only finally add the two exceptional divisors. The second remark is that on either side, we shall start with the equation appearing in the middle (i.e., with Eq. (6.1) with the value $l = n - 4$ together with $l = n - 1$ on the left side and with Eq. (6.2) on the right side), and go toward the one lying at the extreme. Therefore, on each side, an expression closer to the middle will be referred to as *old*, while one more to the side will be called *new*. Now, at every stage we see (in either side) that to the sides of every $\leftrightarrow$, the power of α is larger in the denominator appearing under the newer expression, and the power of β is larger in the one appearing under the older expression. This means that in every correction, the newer part of the $\leftrightarrow$ has to be multiplied by some power of β, and the older one (together with all the ones preceding it) must be multiplied by a power of α, in order to do the merging. Note that in the passage from one stage to the next, the new equation of a stage becomes the old equation of the next one. Since this new equation was multiplied at that stage only by powers of β, we find that it is easier to follow the powers of α, and at every stage we can use the initial powers of α, since they remain untouched until after this stage is completed. Altogether, we find that the expressions in the very middle are multiplied in this part of the process only by powers of α.

We can yet give more general statements about these operations before separating into the cases $n \equiv 1 \pmod 4$ and $n \equiv 3 \pmod 4$, and into the left and right sides of each. Examining Eqs. (B.2) and (B.3) shows that at the lth stage described in any of them, the expression α appears in the denominator under the newer expression to a power larger by

$$n - 5 - l - (l+2) = n - 7 - 2l$$

(note that since Eq. (B.2) describes the left side and Eq. (B.3) describes the right side, in Eq. (B.2) the newer expression lies to the left of the $\leftrightarrow$ and the older one to its right side while in Eq. (B.3) it is the other way around). Clearly, the smallest (odd) value of l appearing in Eq. (B.2) is $l = 1$, while the smallest (even) value of l appearing in Eq. (B.3) is $l = 0$. Now, another look at Eq. (B.1) gives that the left side of it (seen as the upper row here) behaves exactly by the same law corresponding to the (odd) value $l = -1$. The other side of Eq. (B.1) (the lower row) exhibits the same behavior corresponding to the (even) value $l = -2$. All that remains to figure out in the explicit cases, is by which power of α every expression (on either side) has to be multiplied. We emphasize again that in order to calculate this, we can simply take the powers of α appearing in the original Eqs. (6.1) and (6.2). Since clearly the sum of powers to which α and β appears in every denominator is the same (it is $n - 1$ for every expression at the beginning), the powers of β by which we multiply every equation can be obtained simply by complementing those of β to a certain fixed number.

The exact details depend on whether $n \equiv 1 \pmod 4$ or $n \equiv 3 \pmod 4$. We now present them in every case and obtain the concrete expressions.

B.2 The Case $n \equiv 1 \pmod 4$

We start with the case $n \equiv 1 \pmod 4$. Calculating the powers of α by which a given expression is multiplied in the process depends on how the corresponding side ends. Let us examine the left side. We see that we can go with Eq. (B.2) with increasing small odd l until we reach the value $l = (n - 7)/2$, where in the corresponding Eq. (6.1) we have the quotient corresponding to the divisor $P_1^{n-5-l}P_2^{l+4} = P_1^{(n-3)/2}P_2^{(n+1)/2}$ (with denominator $\alpha^{(n-3)/2}\beta^{(n+1)/2}$). Since this divisor is invariant under N, we find that after the $\leftrightarrow$ there must appear the same divisor itself. Hence we do not put this $\leftrightarrow$ and the left side ends there. As for the right side, here we find that continuing with Eq. (B.3) with small even values of l is possible until the value $l = (n - 5)/2$ is reached. This is so since the corresponding Eq. (6.1) consists of only one quotient which is based on the T_{P_3}-invariant divisor $P_1^{(n-5)/2}P_2^{(n+3)/2}$ (with denominator $\alpha^{(n-1)/2}\beta^{(n-1)/2}$). Hence we cannot continue after this equation, and this is the end of the right side.

With this information at hand, we can now calculate the power of α by which we divide the equations in the middle, i.e., the two sides of Eq. (B.1), altogether in all these operations. In the left side of it we find, by what we said in Section B.1, that Eq. (6.1) with values $l = n - 4$ and $l = n - 1$ is divided in total by α taken to the power which is the sum of the numbers $n - 7 - 2l$ over the odd values of l between -1 and the one-before-last value $(n - 11)/2$ (recall that the first stage behaves like the value $l = -1$). In order to write this sum more conveniently, we use the fact that for $n \equiv 1 \pmod 4$ and odd l the number $k = (n - 7 - 2l)/4$ is an integer, and the number $n - 7 - 2l$ is just $4k$ for the corresponding k. Since for odd l between -1 and $(n - 11)/2$ this k goes (in decreasing order) over the integers from 1 to $(n - 5)/4$, we find that this power of α (by which we divide Eq. (6.1) with $l = n - 4$ and $l = n - 1$) is simply

$$\sum_{k=1}^{(n-5)/4} 4k = \frac{n^2 - 6n + 5}{8}.$$

The fact that this equation is not divided by any power of β implies that every equation on the left side must be divided by α and β to powers summing to this number $(n^2 - 6n + 5)/8$. Hence knowing the power of α (which is again easier to calculate) gives us immediately the power of β.

We now take Eq. (6.1) with some small odd l (and the complementary $n - 5 - l$) and see by which expression it was divided during these manipulations. When we reach this equation we find that the first $(l+1)/2$ stages were already done; hence they give to this equation powers of β rather than α. Since k decreases when l increases, we find that these stages correspond to the maximal values of k. Subtracting this $(l+1)/2$ from the $(n-5)/4$ values of k shows that the power of α by which this equation is divided is

$$\sum_{k=1}^{(n-7-2l)/4} 4k = \frac{(n-7-2l)(n-3-2l)}{8},$$

which we prefer to write as

$$\frac{[n-5-(2l+2)][n-1-(2l+2)]}{8} = \frac{n^2-6n+5}{8} - \frac{(l+1)(2n-6)}{4} + \frac{(l+1)^2}{2}.$$

The total power of β by which we divided this equation is the complement of this number with respect to $(n^2 - 6n + 5)/8$, which is easily seen from the last expansion to equal $(l+1)(n-4-l)/2$. Note that these assertions hold also for the final value $l = (n-7)/2$.

Bearing all this in mind, we want to see how the left side looks after that. Under each theta constant we have to put the denominator, that is the product of the old denominator under this theta constant, and the correction corresponding to the relevant equation. For a given odd $1 \le l \le (n-7)/2$, we find that the power of β which is now in the denominator appearing under $\theta[P_1^l P_2^{n-1-l}]$ is

$$(l+1) + 1 + \frac{(l+1)(n-4-l)}{2} = \frac{(l+1)(n-2-l)}{2} + 1.$$

For the powers of α we write $n - 3 - l$ as $n - 2 - (l+1)$, and then the same calculation and the fact that

$$n - 2 + \frac{n^2 - 6n + 5}{8} = \frac{n^2 + 2n - 11}{8}$$

show that this denominator is

$$\alpha^{(n^2+2n-11)/8-(l+1)(n-2-l)/2} \beta^{(l+1)(n-2-l)/2+1}$$

in total. In the denominator appearing under $\theta[P_1^{n-5-l} P_2^{l+4}]$ with the same value of l, we find that the power of β is now

$$(n-4-l)+1+\frac{(l+1)(n-4-l)}{2}=\frac{(l+3)(n-4-l)}{2}+1.$$

Finding the power of α is done using the fact that $l+2=n-2-(n-4-l)$ together with the previous two calculations, and we obtain that the corresponding denominator is

$$\alpha^{(n^2+2n-11)/8-(l+3)(n-4-l)/2}\beta^{(l+3)(n-4-l)/2+1}.$$

Replacing in the last assertion l by $n-5-l$ (giving another odd number since n is odd) we obtain that the new denominator under $\theta[P_1^l P_2^{n-1-l}]$ is

$$\alpha^{(n^2+2n-11)/8-(l+1)(n-2-l)/2}\beta^{(l+1)(n-2-l)/2+1}$$

also for odd $(n-3)/2 \leq l \leq n-6$, and the remaining odd value is $l=n-4$. In this case we see that since the denominator under $\theta[P_1^{n-4}P_2^3]$ was multiplied only by powers of α, it is now $\alpha^{(n^2-6n+13)/8}\beta^{n-2}$. Since

$$\frac{n^2+2n-11}{8}-(n-3)=\frac{n^2-6n+13}{8},$$

this gives the same value as substituting $l=n-4$ in the previous formula. The same considerations show that the new denominator under the remaining theta constant $\theta[P_1^{n-1}]$ is $\alpha^{(n^2+2n-11)/8}\beta$, and can be seen as if obtained by substituting $l=-1$ in the general expression (since then the l-dependent part vanishes). This finishes with the left side in the case $n \equiv 1 \pmod 4$.

We now turn to the right side of the case $n \equiv 1 \pmod 4$. The middle equation here is (the relevant part of) Eq. (6.2), and finding the power of α by which it was divided altogether is also done by summing the powers appearing at the end of Section B.1. This gives us the sum of the expressions $n-7-2l$ for even l between -2 and the one-before-last value $(n-9)/2$ (recall that here it is as if we start with $l=-2$), and for $n \equiv 1 \pmod 4$ and even l the integer we take is $k=(n-5-2l)/4$. Then $n-7-2l$ is $4k-2$, and on the range of even l between -2 and $(n-9)/2$, the index k covers (again in decreasing order) the integers from 1 to $(n-1)/4$. Combining all this yields that Eq. (6.2) is divided in total by α to the power

$$\sum_{k=1}^{(n-1)/4}(4k-2)=\frac{n^2-2n+1}{8}.$$

This equation is also not multiplied by any power of β; hence the powers of α and β by which any equation on the right side was divided must in total be $(n^2-2n+1)/8$.

Turning to Eq. (6.1) with some small even l (and also $n-5-l$), we find that we reach it after already doing the first $(l+2)/2$ stages. Therefore in these stages this equation was multiplied by powers of β; hence when searching for the total power of α by which we divided this equation, we must take (k decreases with l again) the $(l+2)/2$ largest values of k out of the total $(n-1)/4$ values. This gives the sum

$$\sum_{k=1}^{(n-5-2l)/4} (4k-2) = \frac{(n-5-2l)^2}{8},$$

alternatively written as

$$\frac{[n-1-(2l+4)]^2}{8} = \frac{n^2-2n+1}{8} - \frac{(l+2)(n-1)}{2} + \frac{(l+2)^2}{2}.$$

In order to find the total power of β by which we divided this equation, we just take the complement of this number with respect to $(n^2-2n+1)/8$, which equals just $(l+2)(n-3-l)/2$. Also here this holds for the final value $l = (n-5)/2$ as well.

In the new right side we now also have, under each theta constant, the denominator that is the product of the old denominator under this theta constant, and the expression coming from the correction involving the corresponding equation. We start by finding the power of β now appearing under the theta constant $\theta[P_1^l P_2^{n-1-l}]$ with even $0 \le l \le (n-5)/2$, which is the sum

$$(l+2) + \frac{(l+2)(n-3-l)}{2} = \frac{(l+2)(n-1-l)}{2}.$$

The power of α appearing under the same theta constant is calculated using the fact that $n-3-l$ is $n-1-(l+2)$, and hence the same calculation and the equality

$$n-1+\frac{n^2-2n+1}{8} = \frac{n^2+6n-7}{8}$$

yield that the new denominator there is

$$\alpha^{(n^2+6n-7)/8-(l+2)(n-1-l)/2} \beta^{(l+2)(n-1-l)/2}.$$

With the same value of l, the power of β appearing now under $\theta[P_1^{n-5-l} P_2^{l+4}]$ (note that for the value $l = (n-5)/2$ this is the same as the last constant, but consistency will be verified soon) is obtained by summing

$$(n-3-l) + \frac{(l+2)(n-3-l)}{2} = \frac{(l+4)(n-3-l)}{2}.$$

For the power of α here we write $l+2 = n-1-(n-3-l)$ and use the previous two calculations, which concludes that the denominator in question is

$$\alpha^{(n^2+6n-7)/8-(l+4)(n-3-l)/2} \beta^{(l+4)(n-3-l)/2}.$$

Once again replacing l by (the even number) $n-5-l$ in the last assertion shows the new denominator under $\theta[P_1^l P_2^{n-1-l}]$ is

$$\alpha^{(n^2+6n-7)/8-(l+2)(n-1-l)/2} \beta^{(l+2)(n-1-l)/2}.$$

for even $(n-5)/2 \leq l \leq n-5$ as well (which also verifies the consistency of the two assertions for $l = (n-5)/2$), and it remain to check the remaining even value $l = n-3$. The denominator under the corresponding theta constant $\theta[P_1^{n-3}P_2^2]$ was multiplied only by powers of α, yielding the expression is $\alpha^{(n^2-2n+1)/8}\beta^{n-1}$. The equality

$$\frac{n^2+6n-7}{8} - (n-1) = \frac{n^2-2n+1}{8}$$

again shows that this denominator is consistent with the expression obtained by substituting $l = n-3$ in the formula above. Similarly, one obtains that under $\theta[P_1^{n-2}P_2]$ now lies the new denominator $\alpha^{(n^2+6n-7)/8}$, and this gives the same value as the substitution $l = -2$ there. With this we conclude the right side in the case $n \equiv 1 \pmod 4$.

We now want to combine the two sides, where the linking point is the $\leftrightarrow$ between the two quotients that are now $\theta[P_1^{n-1}]/\alpha^{(n^2+2n-11)/8}\beta$ from Eq. (6.1) with values $l = n-4$ and $l = n-1$ from the left side, and $\theta[P_1^{n-2}P_2]/\alpha^{(n^2+6n-7)/8}$ from Eq. (6.2) from the right side. They are connected by a $\leftrightarrow$ since their numerators are equal, and therefore, in order to make the quotients equal, we have to make the denominators equal as well. For this we have to divide the quotient on the left (and hence the whole left side) by $\alpha^{(n+1)/2}$, and the quotient on the right (and hence the whole right side) by β. After doing so we obtain a large equality, which in particular state that the quotients $\theta[P_1^{n-1}]/\alpha^{(n^2+6n-7)/8}\beta$ and $\theta[P_1^{n-2}P_2]/\alpha^{(n^2+6n-7)/8}\beta$ are equal. To this common value also equal all the quotients with numerators $\theta[P_1^l P_2^{n-1-l}]$ with odd $1 \leq l \leq n-3$ and denominators

$$\alpha^{(n^2+6n-7)/8-(l+1)(n-2-l)/2}\beta^{(l+1)(n-2-l)/2+1},$$

and all the quotients with numerators $\theta[P_1^l P_2^{n-1-l}]$ with even $0 \leq l \leq n-4$ and denominators

$$\alpha^{(n^2+6n-7)/8-(l+2)(n-1-l)/2}\beta^{(l+2)(n-1-l)/2+1}.$$

Note that the first quotient is simply the case $l = n-1$ of the formula for even l, and the second quotient is just the case $l = n-2$ of the formula for odd l. Therefore we can now unite them this way in our statements. One can now see that since Eq. (6.2) was divided in total by $\alpha^{(n^2-2n+1)/8}\beta$, we also have that the quotient $\theta[P_1^{n-3}P_2P_3]/\alpha^{(n^2-2n+1)/8}\beta\gamma^{n-1}$ equals this common value. Finally, since by Definition 6.4 the operator N takes this extra divisor $P_1^{n-3}P_2P_3$ to the isolated divisor P_3^{n-1}, we can use again the fact that Eq. (1.5) gives us $\theta[N(\Delta)] = \theta[\Delta]$ to get that the quotient $\theta[P_3^{n-1}]/\alpha^{(n^2-2n+1)/8}\beta\gamma^{n-1}$, which deals with the remaining divisor, also equals the common constant value.

This now gives us a good candidate for the Thomae formulae, but, as usual, we see that the expressions we have constructed are not reduced. First, we see that β appears at least to the power 1 in all these expressions. As for the powers of α, we find that $(l+1)(n-2-l)$ attains its maximal value for odd l at $l = (n-3)/2$, giving $(n^2-2n+1)/4$, and $(l+2)(n-1-l)$ attains its maximal value for even l either at

$l = (n-5)/2$ or at $l = (n-1)/2$, giving $(n^2 + 2n - 3)/4$. Hence α appears in all these expressions at least to the power

$$\frac{n^2 + 6n - 7}{8} - \frac{n^2 + 2n - 3}{8} = \frac{n-1}{2}.$$

Therefore we can multiply the whole large equality by $\alpha^{(n-1)/2}\beta$, and obtain that $\theta[\Delta]/h_\Delta$ is independent of Δ where h_Δ is

$$\alpha^{(n^2 + 2n - 3)/8 - (l+1)(n-2-l)/2} \beta^{(l+1)(n-2-l)/2}$$

for $\Delta = P_1^l P_2^{n-1-l}$ with odd $1 \leq l \leq n-2$,

$$\alpha^{(n^2 + 2n - 3)/8 - (l+2)(n-1-l)/2} \beta^{(l+2)(n-1-l)/2}$$

for the same Δ with even $0 \leq l \leq n-1$, and $\alpha^{(n^2 - 6n + 5)/8}\gamma^{n-1}$ for the divisors $\Delta = P_1^{n-3}P_2P_3$ and $\Delta = P_3^{n-1}$. We recall that $\theta[\Delta]$ denotes the theta constant $\theta^{4n^2}[\Delta](0,\Pi)$, α equals $(\lambda_0 - \lambda_1)^{2n}(\lambda_2 - \lambda_3)^{2n}$, β equals $(\lambda_0 - \lambda_2)^{2n}(\lambda_1 - \lambda_3)^{2n}$, and γ equals $(\lambda_0 - \lambda_3)^{2n}(\lambda_1 - \lambda_2)^{2n}$, and we see that we obtain exactly the formulae stated for the case $n \equiv 1 \pmod 4$ in Theorem 6.9.

As always, the reason we obtain unreduced denominators is the fact that we have started with the unreduced Eqs. (6.1). As in Appendix A, had we started with the reduced Eqs. (6.3) and (6.4), we could have obtained the formula for h_Δ given in Theorem 6.9 directly, without any reduction needed. The reader can verify that the construction would have been the same, but with the details changed as follows. First, for every l (in both sides, i.e., in both Eqs. (B.2) and (B.3)), the difference between the powers of α is now $n - 9 - 2l$, while the difference between the powers of β is $n - 5 - 2l$; hence the sum of powers is not the same for every l, and the powers of both α and β have to be calculated. In fact, one could have calculated the powers of β also in the unreduced process, since we implicitly stated what the necessary summation was. There we could bypass doing it (even though it is not difficult), but here we cannot. However, the statements about extending the validity of the calculations to $l = -1$ in the left side of Eq. (B.1) and to $l = -2$ in the right side of Eq. (B.1) hold equally here, and we continue to use the index k corresponding to every side. Thus on the left side, the total power of α by which we divide Eq. (6.1) with $l = n-4$ and $l = n-1$ is

$$\sum_{k=1}^{(n-5)/4} (4k-2) = \frac{n^2 - 10n + 25}{8},$$

the total power of α by which we divide Eq. (6.1) with small odd l is

$$\sum_{k=1}^{(n-7-2l)/4} (4k-2) = \frac{[n-5-(2l+2)]^2}{8},$$

and the power of β by which we divide this equation is

$$\sum_{k=(n-3-2l)/4}^{(n-5)/4} (4k+2) = \frac{(l+1)(n-2-l)}{2}$$

(they no longer sum to a value independent of l!). This yields a denominator of $\alpha^{(n^2-2n+1)/8}$ under $\theta[P_1^{n-1}]$, and a denominator of

$$\alpha^{(n^2-2n+1)/8-(l+1)(n-2-l)/2}\beta^{(l+1)(n-2-l)/2}$$

under $\theta[P_1^l P_2^{n-1-l}]$ with odd l (either $1 \le l \le (n-7)/2$ or $(n-3)/2 \le l \le n-4$). On the right side, we divide Eq. (6.2) by the total power of α which is

$$\sum_{k=1}^{(n-1)/4} (4k-4) = \frac{n^2-6n+5}{8},$$

and divide Eq. (6.1) with small even l by α to the total power

$$\sum_{k=1}^{(n-5-2l)/4} (4k-4) = \frac{[n-1-(2l+4)][n-5-(2l+4)]}{8},$$

and β to the total power

$$\sum_{k=(n-1-2l)/4}^{(n-1)/4} 4k = \frac{(l+2)(n-1-l)}{2}$$

(again they no longer sum to a value independent of l). This gives a denominator of $\alpha^{(n^2+2n-3)/8}$ under $\theta[P_1^{n-2}P_2]$, and a denominator of

$$\alpha^{(n^2+2n-3)/8-(l+2)(n-1-l)/2}\beta^{(l+2)(n-1-l)/2}$$

under $\theta[P_1^l P_2^{n-1-l}]$ with even l (either $0 \le l \le (n-5)/2$ or $(n-5)/2 \le l \le n-3$). When merging the two sides, we divide only the left side by $\alpha^{(n-1)/2}$, and then we merge in the quotients corresponding to the divisors $P_1^{n-3}P_2P_3$ and P_3^{n-1}. After all this we obtain the $n \equiv 1(\mathrm{mod}\ 4)$ part of Theorem 6.9 directly.

B.3 The Case $n \equiv 3(\mathrm{mod}\ 4)$

We now turn to the case $n \equiv 3(\mathrm{mod}\ 4)$. We again have the two sides, but it turns out that they switch their behaviors when compared to the $n \equiv 1(\mathrm{mod}\ 4)$ case. However, since some calculations are different here, we prefer to give the proof of this case as well with all the details. We again start by examining how the left side ends, and this shows us that the value of (small odd) l which terminates the process of the left side is now $l = (n-5)/2$, with the corresponding Eq. (6.1) consisting of only one quotient. On the right side, we see that the terminal (small even) value of l is $l = (n-7)/2$, for which Eq. (6.1) involves the quotient corresponding to the N-invariant divisor. These values mark the ends of the corresponding sides by the same reasoning as in Section B.2.

Here the considerations from Section B.1 show that the power of α by which we have to divide the left part of Eq. (B.1), or equivalently Eq. (6.1) with $l = n-4$ and $l = n-1$, is the sum of the expressions $n - 7 - 2l$ for odd l between -1 and $(n-9)/2$. For $n \equiv 3 \pmod 4$ and odd l, we take $k = (n-5-2l)/4$ (which is integral), hence $n - 7 - 2l$ is again $4k - 2$, and the sum is

$$\sum_{k=1}^{(n-3)/4} (4k-2) = \frac{n^2 - 6n + 9}{8}.$$

For Eq. (6.1) with some small odd l (and $n - 5 - l$), the power of α by which we divide this equation is obtained by removing the maximal $(l+1)/2$ values of k from the sum (k again runs in decreasing order with respect to l, and as in Section B.2, this is the number of stages already done before getting to this equation), giving us that this power is

$$\sum_{k=1}^{(n-5-2l)/4} (4k-2) = \frac{(n-5-2l)^2}{8}.$$

We write this number as

$$\frac{[n-3-(2l+2)]^2}{8} = \frac{n^2 - 6n + 9}{8} - \frac{(l+1)(n-3)}{2} + \frac{(l+1)^2}{2},$$

which makes it easier to see that its complement with respect to $(n^2 - 6n + 9)/8$ (and is thus the total power of β by which we divided this equation) equals the number $(l+1)(n-4-l)/2$. As before, this holds also for the final value $l = (n-5)/2$.

For each theta constant, taking now the old denominator appearing under it, and multiplying it by the corresponding correction, completes the picture of the new left side. The same calculations and considerations we did in Section B.2 (including the merging of $1 \leq l \leq (n-5)/2$ and $(n-5)/2 \leq l \leq n-6$, with the multiple definition for $l = (n-5)/2$ appearing now on the left side) can be applied also here to get these denominators, except that for the powers of α we now calculate that

$$n - 2 + \frac{n^2 - 6n + 9}{8} = \frac{n^2 + 2n - 7}{8}.$$

The results are as follows. The new denominator under $\theta[P_1^l P_2^{n-1-l}]$ with any odd $1 \leq l \leq n - 6$ is

$$\alpha^{(n^2+2n-7)/8-(l+1)(n-2-l)/2} \beta^{(l+1)(n-2-l)/2+1},$$

and the new denominators under $\theta[P_1^{n-4}P_2^3]$ and $\theta[P_1^{n-1}]$ are $\alpha^{(n^2-6n+17)/8}\beta^{n-2}$ and $\alpha^{(n^2+2n-7)/8}\beta$, respectively. The calculation

$$\frac{n^2+2n-7}{8}-(n-3)=\frac{n^2-6n+17}{8},$$

and the vanishing of the l-dependent part for $l=-1$, show that these denominators can be obtained by simply substituting $l=n-4$ and $l=-1$, respectively, in the general formula. This completes the left side in the case $n\equiv 3(\mathrm{mod}\ 4)$.

On the right side of the case $n\equiv 3(\mathrm{mod}\ 4)$, we have divided the right part of Eq. (B.1) (which is part of Eq. (6.2)) by the following power of α, which Section B.1 gives us. The expressions are $n-7-2l$ as always, and they are summed over even l between -2 and $(n-11)/2$. Since $n\equiv 3(\mathrm{mod}\ 4)$ and l is even, the number $k=(n-7-2l)/4$ is integral, and this substitution turns $n-7-2l$ simply into $4k$. This presents the sum in question as

$$\sum_{k=1}^{(n-3)/4}4k=\frac{n^2-2n-3}{8}.$$

When we look for the power of α by which we divided Eq. (6.1) with small even l (and $n-5-l$), we find, by taking out the $(l+2)/2$ largest values of k (by the usual decreasing order and preceding stages counting considerations, as in Section B.2) from the sum, that this power is

$$\sum_{k=1}^{(n-7-2l)/4}4k=\frac{(n-7-2l)(n-3-2l)}{8}.$$

When we present it in the form

$$\frac{[n-3-(2l+4)][n+1-(2l+4)]}{8}=\frac{n^2-2n-3}{8}-\frac{(l+2)(2n-2)}{4}+\frac{(l+2)^2}{2},$$

it is clear, by taking the complement with respect to $(n^2-2n-3)/8$, that the total power of β by which we divided this equation is $(l+2)(n-3-l)/2$. As usual, this holds also for the final value $l=(n-7)/2$.

We now put, under each theta constant, the product of the old denominator that appeared under it, and the expression coming from the corresponding correction. This gives us the new right side. Again, most of what we did in Section B.2 (with the merging of $0\le l\le(n-7)/2$ and $(n-3)/2\le l\le n-5$ included) does the work here as well, but the calculation for the powers of α is now

$$n-1+\frac{n^2-2n-3}{8}=\frac{n^2+6n-11}{8}.$$

We thus obtain that the denominator under $\theta[P_1^l P_2^{n-1-l}]$ with $0\le l\le(n-5)/2$ and even is

$$\alpha^{(n^2+6n-11)/8-(l+2)(n-1-l)/2}\beta^{(l+2)(n-1-l)/2},$$

and under the theta constants $\theta[P_1^{n-3}P_2^2]$ and $\theta[P_1^{n-2}P_2]$ appear the denominators $\alpha^{(n^2-2n-3)/8}\beta^{n-1}$ and $\alpha^{(n^2+6n-11)/8}$, respectively. We now calculate that

$$\frac{n^2+6n-11}{8}-(n-1)=\frac{n^2-2n-3}{8},$$

and notice that the l-dependent part in the formula vanishes for $l=-2$. This allows us to look at the last denominators as the special cases of that formula when substituting $l=n-3$ or $l=-2$, respectively. We are thus done with the right side in the case $n \equiv 3 \pmod 4$.

The point linking now between the two sides is the $\leftrightarrow$ between the quotients which are now $\theta[P_1^{n-1}]/\alpha^{(n^2+2n-7)/8}\beta$ from Eq. (6.1) with values $l=n-4$ and $l=n-1$ from the left side, and $\theta[P_1^{n-2}P_2]/\alpha^{(n^2+6n-11)/8}$ from Eq. (6.2) from the right side. In order to make the denominators in them (and thus the quotients) equal, we now have to divide the left side by $\alpha^{(n-1)/2}$, and the right side by β. The large equality we obtain relates the quotients $\theta[P_1^{n-1}]/\alpha^{(n^2+6n-11)/8}\beta$ and $\theta[P_1^{n-2}P_2]/\alpha^{(n^2+6n-11)/8}\beta$, and to them also all the quotients with numerators $\theta[P_1^lP_2^{n-1-l}]$ with odd $1 \leq l \leq n-3$ and denominators

$$\alpha^{(n^2+6n-11)/8-(l+1)(n-2-l)/2}\beta^{(l+1)(n-2-l)/2+1},$$

and all the quotients with numerators $\theta[P_1^lP_2^{n-1-l}]$ with even $0 \leq l \leq n-4$ and denominators

$$\alpha^{(n^2+6n-11)/8-(l+2)(n-1-l)/2}\beta^{(l+2)(n-1-l)/2+1}$$

(again the connecting quotients are the even $l=n-1$ and odd $l=n-2$ cases, respectively). The extra term from Eq. (6.2), which was divided by $\alpha^{(n^2-2n-3)/8}\beta$ in total, is the quotient $\theta[P_1^{n-3}P_2P_3]/\alpha^{(n^2-2n-3)/8}\beta\gamma^{n-1}$, and once again, since Definition 6.4 shows that N takes the divisor $P_1^{n-3}P_2P_3$ to P_3^{n-1}, we obtain that this quotient, and the quotient $\theta[P_3^{n-1}]/\alpha^{(n^2-2n-3)/8}\beta\gamma^{n-1}$, also equal this common constant value.

This candidate for the Thomae formulae is again not reduced, since β appears at least to the power 1 in all these expressions. For the powers of α, one checks that $(l+1)(n-2-l)$ attains its maximal value for odd l either at $l=(n-5)/2$ or at $l=(n-1)/2$, giving $(n^2-2n-3)/4$, and $(l+2)(n-1-l)$ attains its maximal value for even l at $l=(n-3)/2$, giving $(n^2+2n+1)/4$, so that α appears to at least the power

$$\frac{n^2+6n-11}{8}-\frac{n^2+2n+1}{8}=\frac{n-3}{2}$$

in all these expressions. Thus the reduction is made by multiplying the whole large equality by $\alpha^{(n-3)/2}\beta$, and we obtain that the quotient $\theta[\Delta]/h_\Delta$ is independent of Δ where h_Δ is

$$\alpha^{(n^2+2n+1)/8-(l+1)(n-2-l)/2}\beta^{(l+1)(n-2-l)/2}$$

for $\Delta = P_1^l P_2^{n-1-l}$ with odd $1 \leq l \leq n-2$,

$$\alpha^{(n^2+2n+1)/8-(l+2)(n-1-l)/2}\beta^{(l+2)(n-1-l)/2}$$

for $\Delta = P_1^l P_2^{n-1-l}$ with even $0 \leq l \leq n-1$, and $\alpha^{(n^2-6n+9)/8}\gamma^{n-1}$ for the remaining divisors $\Delta = P_1^{n-3}P_2P_3$ and $\Delta = P_3^{n-1}$. Expanding again the notation $\theta[\Delta]$ to $\theta^{4n^2}[\Delta](0,\Pi)$, α to $(\lambda_0-\lambda_1)^{2n}(\lambda_2-\lambda_3)^{2n}$, β to $(\lambda_0-\lambda_2)^{2n}(\lambda_1-\lambda_3)^{2n}$, and γ to $(\lambda_0-\lambda_3)^{2n}(\lambda_1-\lambda_2)^{2n}$, yields exactly the formulae stated for the case $n \equiv 3 \pmod 4$ in Theorem 6.9.

Once again we have obtained unreduced denominators since we have started with the unreduced Eqs. (6.1), and again starting with the reduced Eqs. (6.3) and (6.4) sends us directly to the formula for h_Δ given in Theorem 6.9, with no reduction necessary. The details for this construction would have been the same, but with the following change in details, as the reader can verify. The assertions from Section B.2 about the difference between the powers of α and β being $n-9-2l$ and $n-5-2l$ respectively, together with the extension to $l = -1$ and $l = -2$ and the usability of the index k, extend to this case of $n \equiv 3 \pmod 4$ just as well (as they are independent of the ending of the sides). Hence we obtain that we divide Eq. (6.1) with $l = n-4$ and $l = n-1$ by α to the total power of

$$\sum_{k=1}^{(n-3)/4} (4k-4) = (n^2-10n+21)/8,$$

and that we divide Eq. (6.1) with small odd l by α to the power

$$\sum_{k=1}^{(n-5-2l)/4} (4k-4) = \frac{[n-3-(2l+2)][n-7-(2l+2)]}{8},$$

and β to the power

$$\sum_{k=(n-1-2l)/4}^{(n-3)/4} 4k = \frac{(l+1)(n-2-l)}{2}$$

(the sum no longer independent of l!). From this we obtain a denominator of $\alpha^{(n^2-2n-3)/8}$ under $\theta[P_1^{n-1}]$, and a denominator of

$$\alpha^{(n^2-2n-3)/8-(l+1)(n-2-l)/2}\beta^{(l+1)(n-2-l)/2}$$

under $\theta[P_1^l P_2^{n-1-l}]$ with odd l (either $1 \leq l \leq (n-5)/2$ or $(n-5)/2 \leq l \leq n-4$). On the right side, Eq. (6.2) was divided by a the total power of α which is

$$\sum_{k=1}^{(n-3)/4} (4k-2) = \frac{n^2 - 6n + 9}{8},$$

and Eq. (6.1) with small even l is divided by a total power of α which is

$$\sum_{k=1}^{(n-7-2l)/4} (4k-2) = \frac{[n-3-(2l+4)]^2}{8},$$

and a total power of β which is

$$\sum_{k=(n-3-2l)/4}^{(n-3)/4} (4k+2) = \frac{(l+2)(n-1-l)}{2}$$

(again with sum not independent of l). These calculations yield a denominator of $\alpha^{(n^2+2n+1)/8}$ under $\theta[P_1^{n-2}P_2]$, and a denominator of

$$\alpha^{(n^2+2n+1)/8-(l+2)(n-1-l)/2}\beta^{(l+2)(n-1-l)/2}$$

under $\theta[P_1^l P_2^{n-1-l}]$ with even l (either in the range $0 \le l \le (n-7)/2$ or in the range $(n-3)/2 \le l \le n-3$). When we merge the two sides, only the left side is manipulated, and it is divided by $\alpha^{(n+1)/2}$. After that, we can put in the quotients corresponding to the exceptional divisors $P_1^{n-3}P_2P_3$ and P_3^{n-1}, as before. This leads us directly to the $n \equiv 3 \pmod 4$ part of Theorem 6.9.

B.4 The Operators for the Other Basepoints

We have promised to give, in this appendix, the explicit expressions for the denominators h_Δ when one chooses the basepoint to be any other branch point P_i, $1 \le i \le 3$. However, in order to understand them better, we have to see what are the operators N and T_R for these basepoints. Throughout all the following discussion, we keep in mind that by replacing the function w by the function $\left(\prod_{i=0}^{3}(z-\lambda_i)\right)/w$, we interchange the roles of P_0 and P_3, and the roles of P_1 and P_2. This implies that there is symmetry between the two points in each pair, but there is no reason to expect any symmetry between the two pairs. While presenting the operators, we put our emphasis on a specific combination. We recall that in the heart of the construction for the denominators h_Δ with the basepoint P_0, was the fact that applying the operator T_{P_3}, and then the operator N, took the divisor $P_1^l P_2^{n-1-l}$ with $0 \le l \le n-3$ to the divisor $P_1^{l+2}P_2^{n-3-l}$, i.e., increased the index l by 2. The construction of the denominators h_Δ with any other branch point P_i, $1 \le i \le 3$, and obtaining the formulas we now present for them, are based again on the composition of T_{P_j}, where P_j is the branch point symmetric to P_i (i.e., $j = 3 - i$), with N.

We start with describing the action of the operators with the basepoint P_3, the one symmetric to P_0. As we have seen, the $n+2$ non-special divisors of degree $g = n-1$ supported on the branch points with P_3 omitted are $P_1^l P_2^{n-1-l}$ with $0 \le l \le n-1$, and the two divisors P_0^{n-1} and $P_0 P_1 P_2^{n-3}$. The operator N with the basepoint P_3 is defined by the equalities

$$N(P_1^l P_2^{n-1-l}) = \begin{cases} P_1^{n+1-l} P_2^{l-2} & 2 \le l \le n-1 \\ P_1^{1-l} P_2^{l-2+n} & 0 \le l \le 1, \end{cases}$$

and by pairing P_0^{n-1} with $P_0 P_1 P_2^{n-3}$. The operator T_{P_0} with the basepoint P_3 is defined by

$$T_{P_0}(P_1^l P_2^{n-1-l}) = \begin{cases} P_1^{n+3-l} P_2^{l-4} & 4 \le l \le n-1 \\ P_1^{3-l} P_2^{l-4+n} & 0 \le l \le 3. \end{cases}$$

As for the remaining operators, T_{P_1} pairs $P_1 P_2^{n-2}$ with $P_0 P_1 P_2^{n-3}$, and T_{P_2} pairs $P_1^2 P_2^{n-3}$ with $P_0 P_1 P_2^{n-3}$ and also takes $P_0^{n-1} = P_0^2$ to itself if $n = 3$. Then the combination NT_{P_0} is easily seen to take the divisor $P_1^l P_2^{n-1-l}$ with $2 \le l \le n-1$ to $P_1^{l-2} P_2^{n+1-l}$ (i.e., decreases l by 2). This indeed means that working with the basepoint P_3 gives results similar to those we have with the basepoint P_0.

We now turn to the other pair, and examine the effect of taking the basepoint to be P_1 or P_2. We look at the basepoint P_1 first. The $n+2$ divisors not containing P_1 in their support are $P_0^l P_3^{n-1-l}$ with $0 \le l \le n-1$, together with the two divisors P_2^{n-1} and $P_0^{(n-3)/2} P_2 P_3^{(n-1)/2}$. We define the operator N with the basepoint P_1 by

$$N(P_0^l P_3^{n-1-l}) = \begin{cases} P_0^{(n-3)/2-l} P_3^{(n+1)/2+l} & 0 \le l \le (n-3)/2 \\ P_0^{(3n-3)/2-l} P_3^{l-(n-1)/2} & (n-1)/2 \le l \le n-1, \end{cases}$$

and to pair P_2^{n-1} with $P_0^{(n-3)/2} P_2 P_3^{(n-1)/2}$. We define the operator T_{P_2} with the basepoint P_1 to take any divisor $P_0^l P_3^{n-1-l}$ with $0 \le l \le n-2$ to $P_0^{n-2-l} P_3^{l+1}$, and P_0^{n-1} to itself. When examining the remaining two operators, we see that the operator T_{P_3} pairs $P_0^{(n-1)/2} P_3^{(n-1)/2}$ with $P_0^{(n-3)/2} P_2 P_3^{(n-1)/2}$, while T_{P_0} pairs $P_0^{(n-3)/2} P_3^{(n+1)/2}$ with $P_0^{(n-3)/2} P_2 P_3^{(n-1)/2}$ and also takes $P_2^{n-1} = P_2^2$ to itself if $n = 3$. Here the behavior of the combined operation NT_{P_2} is different, as it takes the divisor $P_0^l P_3^{n-1-l}$ to $P_0^{l+(n+1)/2} P_3^{(n-3)/2-l}$ if $0 \le l \le (n-3)/2$, and to the divisor $P_0^{l-(n-1)/2} P_3^{(3n-3)/2-l}$ if $(n-1)/2 \le l \le n-1$. This is not very comfortable for understanding the results, but doing this operation twice leads to an action with which it is easier to work—it is easy to verify that the multiple operation $NT_{P_2} NT_{P_2}$ takes the divisor $P_0^l P_3^{n-1-l}$ with $0 \le l \le n-2$ to $P_0^{l+1} P_3^{n-2-l}$ (i.e., increases l by 1).

As for the basepoint P_2, the picture is of course similar to the one we have with the basepoint P_1 (and not P_0 or P_3). The $n+2$ corresponding divisors are $P_0^l P_3^{n-1-l}$ for $0 \le l \le n-1$, together with P_1^{n-1} and $P_0^{(n-1)/2} P_1 P_3^{(n-3)/2}$. The action of the operator N with the basepoint P_2 is described by

$$N(P_0^l P_3^{n-1-l}) = \begin{cases} P_0^{(n-1)/2-l} P_3^{(n-1)/2+l} & 0 \le l \le (n-1)/2 \\ P_0^{(3n-1)/2-l} P_3^{l-(n+1)/2} & (n+1)/2 \le l \le n-1, \end{cases}$$

and the pairing of P_1^{n-1} with $P_0^{(n-1)/2} P_1 P_3^{(n-3)/2}$. The action of the operator T_{P_1} with the basepoint P_2 is to take the divisor $P_0^l P_3^{n-1-l}$ with $1 \le l \le n-1$ to $P_0^{n-l} P_3^{l-1}$, and P_3^{n-1} to itself. The action of the remaining operators is described by T_{P_0} pairing $P_0^{(n-1)/2} P_3^{(n-1)/2}$ with $P_0^{(n-1)/2} P_1 P_3^{(n-3)/2}$, and T_{P_3} pairing $P_0^{(n+1)/2} P_3^{(n-3)/2}$ with $P_0^{(n-1)/2} P_1 P_3^{(n-3)/2}$ and also taking $P_1^{n-1} = P_1^2$ to itself if $n = 3$. The double operation NT_{P_1} again resembles the corresponding one from the basepoint P_1, as it takes the divisor $P_0^l P_3^{n-1-l}$ to $P_0^{l+(n-1)/2} P_3^{(n-1)/2-l}$ if $0 \le l \le (n-1)/2$ and to $P_0^{l-(n+1)/2} P_3^{(3n-1)/2-l}$ if $(n+1)/2 \le l \le n-1$. Once again doing this operation twice leads to an action with which it is easier to work, as the composite operation $NT_{P_1} NT_{P_1}$ takes $P_0^l P_3^{n-1-l}$ with $1 \le l \le n-1$ to $P_0^{l-1} P_3^{n-l}$ (i.e., decreases l by 1).

When imagining how the process works with the basepoints P_1 and P_2, one might be intimidated by the fact that the combination which is easy to work with is composed of 4 operators. However, this will be compensated by the fact that there will be only one side there. We recall that with the basepoint P_0 (and similar things happen with the symmetric basepoint P_3), Eq. (6.2), the PMT relating three divisors which are related also via the two other operators T_{P_1} and T_{P_2}, was "in the middle" of our diagram with the connecting symbols ↔. When looking at the equivalent to Eq. (6.2) with the basepoint P_1, we find that except for the exceptional divisor $P_0^{(n-3)/2} P_2 P_3^{(n-1)/2}$, whose N-image is the isolated divisor P_2^{n-1}, this equation concerns also the divisor $P_0^{(n-1)/2} P_3^{(n-1)/2}$. Since the N-image of this divisor is the T_{P_2}-invariant divisor P_0^{n-1}, we conclude that here Eq. (6.2) is located "on one side" of this diagram. The difference in the expressions for h_Δ does not depend on the side in the diagram in which the quotient corresponding to Δ lies, but rather on the parity of the number of applications of T_{P_2} on the way from the analog of Eq. (6.2) to the divisor Δ. We have a similar image with the basepoint P_2, since then the corresponding equivalent relates the exceptional divisor $P_0^{(n-1)/2} P_2 P_3^{(n-3)/2}$ (with the isolated divisor P_1^{n-1} as its N-image) to the divisor $P_0^{(n-1)/2} P_3^{(n-1)/2}$. The N-image of the latter is the T_{P_1}-invariant divisor P_3^{n-1}, giving us the same picture as with the basepoint P_1. The interested reader can check the details, and also see the difference between the $n \equiv 1 \pmod 4$ and $n \equiv 3 \pmod 4$ cases (which again differ according to the exact way this process ends on the longer side).

B.5 The Expressions for h_Δ for the Other Basepoints

We now state the results. The Thomae formulae with basepoint P_3 is as follows. For $n \equiv 1 \pmod 4$ and $\Delta = P_1^l P_2^{n-1-l}$, define h_Δ to be

$$[(\lambda_0 - \lambda_1)(\lambda_2 - \lambda_3)]^{2n[(n^2+2n-3)/8-l(n+1-l)/2]}[(\lambda_0 - \lambda_2)(\lambda_1 - \lambda_3)]^{2nl(n+1-l)/2}$$

for even l, and

$$[(\lambda_0 - \lambda_1)(\lambda_2 - \lambda_3)]^{2n[(n^2+2n-3)/8-(l-1)(n-l)/2]}[(\lambda_0 - \lambda_2)(\lambda_1 - \lambda_3)]^{2n(l-1)(n-l)/2}$$

for odd l. For the remaining divisors $\Delta = P_0 P_1 P_2^{n-3}$ or $\Delta = P_0^{n-1}$, define h_Δ to be the expression

$$[(\lambda_0 - \lambda_1)(\lambda_2 - \lambda_3)]^{2n(n^2-6n+5)/8}[(\lambda_0 - \lambda_3)(\lambda_1 - \lambda_2)]^{2n(n-1)}.$$

When $n \equiv 3 \pmod 4$ and $\Delta = P_1^l P_2^{n-1-l}$, define h_Δ to be the expression

$$[(\lambda_0 - \lambda_1)(\lambda_2 - \lambda_3)]^{2n[(n^2+2n+1)/8-l(n+1-l)/2]}[(\lambda_0 - \lambda_2)(\lambda_1 - \lambda_3)]^{2nl(n+1-l)/2}$$

for even l, and

$$[(\lambda_0 - \lambda_1)(\lambda_2 - \lambda_3)]^{2n[(n^2+2n+1)/8-(l-1)(n-l)/2]}[(\lambda_0 - \lambda_2)(\lambda_1 - \lambda_3)]^{2n(l-1)(n-l)/2}$$

for odd l. If Δ is one of the remaining divisors $P_0 P_1 P_2^{n-3}$ or P_0^{n-1}, then define h_Δ as

$$[(\lambda_0 - \lambda_1)(\lambda_2 - \lambda_3)]^{2n(n^2-6n+9)/8}[(\lambda_0 - \lambda_3)(\lambda_1 - \lambda_2)]^{2n(n-1)}.$$

The Thomae formulae with the basepoint P_3 states that then (for any odd n) the expression $\theta^{4n^2}[\Delta](0, \Pi)/h_\Delta$ is independent of the divisor Δ. One way to prove this is by using the symmetry argument from the basepoint P_0 case (but remember to interchange also P_1 with P_2!). Alternatively, one can follow the process for the basepoint P_3 (which is quite similar to the one with the basepoint P_0), either proving the analog of Theorem 6.9, or executing the constructions analogous to Sections B.2 and B.3, and obtain these results. In order to check the consistency of these results with Theorem 6.9 for the basepoint P_0, we use the fact that Proposition 6.10 relates the characteristic $\varphi_{P_0}(P_1^l P_2^{n-1-l}) + K_{P_0}$ with $\varphi_{P_3}(P_1^{l+2} P_2^{n-3-l}) + K_{P_3}$ for every $0 \le l \le n-3$. Then substituting $l+2$ in the place of l here gives us the correct expressions from Theorem 6.9. As for the remaining divisors, Proposition 6.10 relates $P_1^{n-2} P_2$, P_1^{n-1}, $P_1^{n-3} P_2 P_3$, and P_3^{n-1} with the basepoint P_0 with P_2^{n-1}, $P_1 P_2^{n-2}$, $P_0 P_1 P_2^{n-3}$, and P_0^{n-1} with the basepoint P_3, respectively. The reader can easily complete the consistency check from here.

We now present the results concerning the other two basepoints, and recall that in any case the Thomae formulae again splits into two cases according to the residue of n modulo 4. We begin with the basepoint P_1. When $n \equiv 1 \pmod 4$ and $\Delta = P_0^l P_3^{n-1-l}$, the expression h_Δ is defined to be

$$[(\lambda_0 - \lambda_1)(\lambda_2 - \lambda_3)]^{2n[(n^2+2n-3)/8-(l+1)(n-1-2l)]}[(\lambda_0 - \lambda_2)(\lambda_1 - \lambda_3)]^{2n(l+1)(n-1-2l)}$$

for $0 \le l \le (n-3)/2$, and

$$[(\lambda_0-\lambda_1)(\lambda_2-\lambda_3)]^{2n[(n^2+2n-3)/8-(n-1-l)(2l+1-n)]}[(\lambda_0-\lambda_2)(\lambda_1-\lambda_3)]^{2n(n-1-l)(2l+1-n)}$$

for $(n-1)/2 \le l \le n-1$. For the extra two divisors, $\Delta = P_0^{(n-3)/2}P_2P_3^{(n-1)/2}$ or $\Delta = P_2^{n-1}$, define h_Δ to be

$$[(\lambda_0 - \lambda_1)(\lambda_2 - \lambda_3)]^{2n(n^2-6n+5)/8}[(\lambda_0 - \lambda_3)(\lambda_1 - \lambda_2)]^{2n(n-1)}.$$

For $n \equiv 3 \pmod 4$ and $\Delta = P_0^l P_1^{n-1-l}$, let h_Δ be

$$[(\lambda_0 - \lambda_1)(\lambda_2 - \lambda_3)]^{2n[(n^2+2n+1)/8-(l+1)(n-1-2l)]}[(\lambda_0 - \lambda_2)(\lambda_1 - \lambda_3)]^{2n(l+1)(n-1-2l)}$$

for $0 \le l \le (n-3)/2$, and

$$[(\lambda_0-\lambda_1)(\lambda_2-\lambda_3)]^{2n[(n^2+2n+1)/8-(n-1-l)(2l+1-n)]}[(\lambda_0-\lambda_2)(\lambda_1-\lambda_3)]^{2n(n-1-l)(2l+1-n)}$$

for $(n-1)/2 \le l \le n-1$. For Δ being either $\Delta = P_0^{(n-3)/2}P_2P_3^{(n-1)/2}$ or $\Delta = P_2^{n-1}$, let h_Δ be

$$[(\lambda_0 - \lambda_1)(\lambda_2 - \lambda_3)]^{2n(n^2-6n+9)/8}[(\lambda_0 - \lambda_3)(\lambda_1 - \lambda_2)]^{2n(n-1)}.$$

Then the Thomae formulae with the basepoint P_1 asserts that (for any odd n) the expression $\theta^{4n^2}[\Delta](0,\Pi)/h_\Delta$ is independent of Δ. This cannot be obtained using any symmetry argument from the previous cases, and has to be proved by following the process for the basepoint P_1 (unless one insists on using the consistency checks as proofs). Once again one can either prove that these expressions work as we did in Theorem 6.9, or use a construction in the general form of Sections B.2 and B.3 (once again the difference between the cases $n \equiv 1 \pmod 4$ and $n \equiv 3 \pmod 4$ will occur since the interesting side of the diagram ends differently). Now use the fact that Proposition 6.10 relates the characteristic $\varphi_{P_0}(P_1^l P_2^{n-1-l}) + K_{P_0}$ with $\varphi_{P_1}(P_0^{(n-3-l)/2}P_3^{(n+1+l)/2}) + K_{P_1}$ for even $0 \le l \le n-3$, and with $\varphi_{P_1}(P_0^{n-1-(l+1)/2}P_3^{(l+1)/2}) + K_{P_1}$ for odd $1 \le l \le n-2$, and substitute $(n-3-l)/2$ in the place of l in $(l+1)(n-1-2l)$, and $n-1-(l+1)/2$ in the place of l in $(n-1-l)(2l+1-n)$. This finishes the consistency check for these divisors, and as Proposition 6.10 relates P_1^{n-1}, $P_1^{n-3}P_2P_3$, and P_3^{n-1} with the basepoint P_0 with P_0^{n-1}, P_2^{n-1}, and $P_0^{(n-3)/2}P_2P_3^{(n-1)/2}$ with the basepoint P_1, respectively, the consistency check is now easily finished.

We conclude with the basepoint P_2, where the Thomae formulae is as follows. For $n \equiv 1 \pmod 4$ and $\Delta = P_0^l P_3^{n-1-l}$, we define h_Δ as the expression

$$[(\lambda_0 - \lambda_1)(\lambda_2 - \lambda_3)]^{2n[(n^2+2n-3)/8-l(n-1-2l)]}[(\lambda_0 - \lambda_2)(\lambda_1 - \lambda_3)]^{2nl(n-1-2l)}$$

for $0 \le l \le (n-1)/2$, and

$$[(\lambda_0 - \lambda_1)(\lambda_2 - \lambda_3)]^{2n[(n^2+2n-3)/8-(n-l)(2l+1-n)]}[(\lambda_0 - \lambda_2)(\lambda_1 - \lambda_3)]^{2n(n-l)(2l+1-n)}$$

for $(n+1)/2 \le l \le n-1$. For the exceptional divisors $\Delta = P_0^{(n-1)/2}P_1P_3^{(n-3)/2}$ and $\Delta = P_1^{n-1}$, we define h_Δ to be

$$[(\lambda_0 - \lambda_1)(\lambda_2 - \lambda_3)]^{2n(n^2-6n+5)/8}[(\lambda_0 - \lambda_3)(\lambda_1 - \lambda_2)]^{2n(n-1)}.$$

If $n \equiv 3 \pmod 4$ and $\Delta = P_0^l P_1^{n-1-l}$, then we define h_Δ to be

$$[(\lambda_0 - \lambda_1)(\lambda_2 - \lambda_3)]^{2n[(n^2+2n+1)/8-l(n-1-2l)]}[(\lambda_0 - \lambda_2)(\lambda_1 - \lambda_3)]^{2nl(n-1-2l)}$$

for $0 \le l \le (n-1)/2$, and

$$[(\lambda_0 - \lambda_1)(\lambda_2 - \lambda_3)]^{2n[(n^2+2n+1)/8-(n-l)(2l+1-n)]}[(\lambda_0 - \lambda_2)(\lambda_1 - \lambda_3)]^{2n(n-l)(2l+1-n)}$$

for $(n+1)/2 \le l \le n-1$. If Δ is an exceptional divisor, $P_0^{(n-1)/2}P_1P_3^{(n-3)/2}$ or P_1^{n-1}, then let h_Δ be

$$[(\lambda_0 - \lambda_1)(\lambda_2 - \lambda_3)]^{2n(n^2-6n+9)/8}[(\lambda_0 - \lambda_3)(\lambda_1 - \lambda_2)]^{2n(n-1)}.$$

The Thomae formulae with the basepoint P_2 now claims that (for any odd n) the expression $\theta^{4n^2}[\Delta](0,\Pi)/h_\Delta$ with h_Δ thus defined is independent of the divisor Δ. Here we can obtain it by the symmetry argument, the symmetric point being P_1 (notice that interchange of P_0 with P_3 when doing so!). The other way to see it, is by repeating all the steps of the proof (either of Theorem 6.9 or of a construction similar to Sections B.2 and B.3, with the remark from the previous paragraph about the latter holding here as well), but with the basepoint taken as P_2. For the consistency check, Proposition 6.10 relates the characteristic $\varphi_{P_0}(P_1^l P_2^{n-1-l}) + K_{P_0}$ with $\varphi_{P_2}(P_0^{n-1-l/2}P_3^{l/2}) + K_{P_2}$ if $0 \le l \le n-1$ is even, and with $\varphi_{P_2}(P_0^{(n-2-l)/2}P_3^{(n+l)/2}) + K_{P_2}$ if $1 \le l \le n-2$ is odd; then substitute $(n-2-l)/2$ in the place of l in $l(n-1-2l)$, and $n-1-l/2$ in the place of l in $(n-l)(2l+1-n)$. The remaining divisors are the two exceptional ones, and since Proposition 6.10 relates $P_1^{n-3}P_2P_3$ and P_3^{n-1} with the basepoint P_0 with P_1^{n-1} and $P_0^{(n-1)/2}P_1P_3^{(n-3)/2}$ with the basepoint P_2 respectively, and the corresponding denominators are the same, the consistency check is done.

One can unite the Thomae formulae with the four branch points P_i, $0 \le i \le 3$ into one Thomae formulae written in a basepoint-independent form, but this form does not look any nicer than the results we have already written, so we shall not give it here. We conclude by a final remark about symmetry. In the case $n = 5$ we get also symmetry between the branch points of the pair of P_0 and P_3 and the branch points of the pair of P_1 and P_2. This comes from the fact that replacing w by the functions

$$\frac{w^2}{(z-\lambda_2)(z-\lambda_3)} \quad \text{or} \quad \frac{w^3}{(z-\lambda_1)(z-\lambda_2)(z-\lambda_3)^2},$$

gives functions whose 5th powers are

$$(z-\lambda_2)(z-\lambda_0)^2(z-\lambda_3)^3(z-\lambda_1)^4$$

or

$$(z-\lambda_1)(z-\lambda_3)^2(z-\lambda_0)^3(z-\lambda_2)^4,$$

respectively. The reader can draw the "graph of connections" between the divisors with the four branch points P_i, $0 \le i \le 3$, see that all the four graphs have the same shape, and confirm that this extra symmetry is consistent with all the other results. This can be considered as a special case of a much more general phenomenon, which is beyond the scope of this book.

References

[BR] Bershadski, M., Radul, A.: Conformal Field Theories with Additional Z_n Symmetry, *International Journal of Modern Physics* A**2-1**, 165–171 (1987).

[GDT] Gonzalez-Diez, G., Torres, D.: Z_n Curves Possessing No Thomae Formulae, preprint.

[EbF] Ebin, D., Farkas, H.M.: Thomae Formulae for Z_n Curves, *Journal D'Analyse* (to appear).

[EiF] Eisenmann, A., Farkas, H.M.: An Elementary Proof of Thomae's Formulae, *OJAC* **3** (2008).

[EG] Enolski, V., Grava, T.: Thomae Formulae for Singular Z_n Curves, *Letters in Math. Physics* **76**, 187–214 (2006).

[FK] Farkas, H.M., Kra, I.: *Riemann Surfaces* (second edition). Graduate Texts in Mathematics **71**, Springer–Verlag, 1992.

[F] Fay, J.D.: *Theta Functions on Riemann Surfaces*. Lecture Notes in Mathematics **352**, Springer, 1973.

[L] Lewittes, J.: Riemann Surfaces and the Theta Function, *Acta Mathematica* **111**, 37–61 (1964).

[M] Matsumoto, K.: Theta Constants Associated with the Cyclic Triple Coverings of the Complex Projective Line Branching at Six Points, *Pub. Res. Inst. Math. Sci.* **37**, 419–440, (2001).

[MT] Matsumoto, K., Terasoma, T.: Degenerations of Triple Coverings and Thomae's Formula. ArXiv:1001.4950v1 [Math. AG] (2010).

[N] Nakayashiki, A.: On the Thomae Formulae for Z_n Curves, *Pub Res Inst. Math Science* **33-6**, 987–1015 (1997).

[T1] Thomae, J.: Bestimmung von $d \log \theta(0,\dots,0)$ durch die Klassmoduln, *Crelle's Journal* **66**, 92–96 (1866).

[T2] Thomae, J.: Beitrag zur Bestimmung von $\theta(0,\dots,0)$ durch die Klassmoduln Algebraisher Funktionen, *Crelle's Journal* **71**, 201–222 (1870).

List of Symbols

∞_h	a point on a Riemann surface lying above ∞ by a given map
$\lvert Y \rvert, \lvert Z \rvert$, etc.	the cardinality of the finite set Y, Z, etc.
$\lfloor x \rfloor, \{x\}$	the integral value and fractional part of x
$Y \cup y, Y \setminus y$	shorthand for $Y \cup \{y\}$ and $Y \setminus \{y\}$ respectively
Y^k	the product of the kth powers of all the points in the (finite) set Y
$\prec, \preceq, \succ, \succeq$	relations based on the ordering (4.9)
α	a non-gap at a point on a Riemann surface
$\alpha, \alpha', \beta, \beta', \gamma, \gamma'$	vectors in $\mathbb{Z}^g$ appearing in characteristics of theta functions
α_i	a power in the general equation defining a Z_n curve, or a non-gap
$\Gamma, \Delta, \Lambda, \Xi, \Sigma, \Upsilon, \Psi$	divisors on a Riemann surface, usually a Z_n curve
$\Gamma_{P_i}, \Gamma_{Q_i}, \Gamma_S$	a divisor with the basepoint P_i, Q_i, S corresponding to Δ with the basepoint P_0
γ, γ_i	a gap at a point on a Riemann surface
$\delta, \delta', \varepsilon, \varepsilon', \mu, \mu', \rho, \rho'$	vectors in $\mathbb{R}^g$ appearing as characteristics of theta functions
δ_{ij}	the Kronecker delta symbol, 1 if $i = j$ and 0 otherwise
ζ	an element in $\mathbb{C}^g$ appearing as an argument of the theta function
θ	the theta function
$\theta\begin{bmatrix} \varepsilon \\ \varepsilon' \end{bmatrix}, \theta\begin{bmatrix} \delta \\ \delta' \end{bmatrix}$, etc.	theta function with characteristics
$\theta[\Delta], \theta[\Xi]$, etc.	theta function with characteristics $\varphi_{P_0}(\Delta) + K_{P_0}$, $\varphi_{P_0}(\Xi) + K_{P_0}$, etc.
λ_i, μ_i	the z-image of the branch point P_i or Q_i respectively
Π	an element in $\mathcal{H}_g$, usually the period matrix of a Riemann surface
$\varphi_{P_0}, \varphi_Q, \varphi_S$ etc.	the Abel–Jacobi map with the corresponding basepoint
$\Omega(\Delta)$	the space of meromorphic differentials on X whose divisors are multiples of Δ
$\Omega_k(\Delta)$	a certain subspace of $\Omega(\Delta)$ on a Z_n curve

ω	a differential on a Riemann surface
ω_k	the $\Omega_k(1)$-component of a differential ω
A, B, C_j, D_j	sets of branch points, with $C_{-1} = A \cup P_0$ and $D_{-1} = B$
$\widetilde{A}, \widetilde{B}, \widetilde{C}_j, \widetilde{D}_j,$	the sets A, B, C_j, D_j corresponding to $T_R(\Delta)$
$\widehat{A}, \widehat{B}, \widehat{C}_j, \widehat{D}_j,$	the sets A, B, C_j, D_j corresponding to $T_S(\Delta)$
a_i, b_i	elements of a canonical homology basis for a Riemann surface
$(C_j)_\Delta, (D_k)_\Xi$, etc.	the set C_j, D_k etc. corresponding to the divisor Δ, Ξ etc.
$\mathbb{C}$	the field of complex numbers
df, dz	the differential of the meromorphic function f or z
$\mathrm{div}(f), \mathrm{div}(\omega)$	the divisor of the function f or of the differential ω
E	the set $H \cup A = C_{-1} \cup D_0$
$\widetilde{E}, \widehat{E}$	same as E but corresponding to $T_R(\Delta)$ or $T_S(\Delta)$
$\mathbf{e}(z)$	the exponent $e^{2\pi i z}$
F	the set $C_0 \cup B$
$\widetilde{F}, \widehat{F}$	same as F but corresponding to $T_R(\Delta)$ or $T_S(\Delta)$
G	the set $C_{-1} \cup C_0$
$\widetilde{G}, \widehat{G}$	same as G but corresponding to $T_R(\Delta)$ or $T_S(\Delta)$, both equal G
g	the genus of a Riemann surface
g_Δ	the denominator corresponding to Δ in the PMT
H	the set $D_0 \cup P_0$
$\widetilde{H}, \widehat{H}$	same as H but corresponding to $T_R(\Delta)$ or $T_S(\Delta)$
$\mathscr{H}_g$	the Siegel upper halfplane of degree g
h_Δ	the denominator corresponding to Δ in the Thomae formulae
I	the set $J \setminus P_0 = D_{-1} \cup D_0$
$\widetilde{I}, \widehat{I}$	same as I but corresponding to $T_R(\Delta)$ or $T_S(\Delta)$, both equal I
$i(\Delta)$	the dimension of $\Omega(\Delta)$
J	the set $H \cup B$
$\widetilde{J}, \widehat{J}$	same as J but corresponding to $T_R(\Delta)$ or $T_S(\Delta)$, both equal J
K	the set $G \cup I$
$\widetilde{K}, \widehat{K}$	same as K but corresponding to $T_R(\Delta)$ or $T_S(\Delta)$, both equal K
K_{P_0}, K_Q, etc.	the vector of Riemann constants for the basepoint P_0, Q, etc.
$L(1/\Delta)$	the space of meromorphic functions on X whose poles are bounded by Δ
M	the basepoint change operator
M, M', T, T'	vectors in $\mathbb{Z}^g$ appearing as translations of arguments of theta functions
m	a parameter of a singular Z_n curve
$\mathbb{N}$	the set of natural numbers
N	an operator on the set of non-special divisors
n	the degree of the z function on a Z_n curve
$\mathrm{Ord}_P f, \mathrm{Ord}_P \omega$	the order of the function f, or the differential ω, at the point P
P	a point on a Riemann surface

$P\begin{bmatrix} \varepsilon \\ \varepsilon' \end{bmatrix}$ the polynomial corresponding to the characteristics $\begin{bmatrix} \varepsilon \\ \varepsilon' \end{bmatrix}$ in the Thomae formulae

P_i, Q_i branch points on a Z_n curve

p_Δ, q_Δ denominators corresponding to Δ on a singular Z_n curve

p_k a polynomial, usually used to define holomorphic differentials

$\mathbb{Q}$ the field of rational numbers

$\mathbb{R}$ the field of real numbers

r a parameter of a nonsingular Z_n curve

$r(1/\Delta)$ the dimension of $L(1/\Delta)$

T an operator on the set of non-special divisors on a nonsingular Z_n curve with $r = 1$

T_{P_i}, T_{Q_i}, etc. an operator on the set of non-special divisors

U a certain set branch points on a Z_n curve

V the switching operator N, T_R or T_S on a Z_n curve

$v_Q(\Gamma)$, $v_P(\Delta)$, etc. the order to which the point Q, P appears in the divisor Γ, Δ, etc.

W a certain operator on a nonsingular Z_n curve

w a defining meromorphic function on a Z_n curve

X a Riemann surface, usually a Z_n curve

Y_j, Z_j sets of branch points on a singular Z_n curve, either C_j or D_j with the same j

$\overline{Y}_j, \overline{Z}_j$ denotes D_j when Y_j or Z_j is C_j, and denotes C_j when Y_j or Z_j is D_j

$\mathbb{Z}$ the ring of integers

z a defining meromorphic function on a Z_n curve

Index